D0321889

SIXTH EDITION

ASTROPHYSICAL TECHNIQUES

SIXTH EDITION

ASTROPHYSICAL TECHNIQUES

C.R. KITCHIN

University of Hertfordshire, UK

CRC Press
Taylor & Francis Group
Boca Raton London New York

CRC Press is an imprint of the
Taylor & Francis Group, an **informa** business

A TAYLOR & FRANCIS BOOK

7160010

Taylor & Francis
Taylor & Francis Group
6000 Broken Sound Parkway NW, Suite 300
Boca Raton, FL 33487-2742

© 2014 by Taylor & Francis Group, LLC
Taylor & Francis is an Informa business

No claim to original U.S. Government works

Printed on acid-free paper
Version Date: 20131014

International Standard Book Number-13: 978-1-4665-1115-6 (Hardback)

Visit the Taylor & Francis Web site at
http://www.taylorandfrancis.com

and the CRC Press Web site at
http://www.crcpress.com

For Christine

Contents

Preface to the Sixth Edition

THE AIM OF THIS book is to provide a coherent state-of-the-art account of the instruments and techniques used in astronomy and astrophysics today. Whilst every effort has been made to make it as complete and up to date as possible, the author is only too well aware of the many omissions and skimpily treated subjects throughout the work. For some types of instrumentation it is possible to give full details of the instrument in its finally developed form. However, for the 'new astronomies' and even some aspects of established fields, development is occurring at a rapid pace and the details will change between the writing and publishing of this edition. For those areas of astronomy, therefore, a fairly general guide to the principles behind the techniques is given and this should enable the reader to follow the detailed designs in the scientific literature.

With this sixth edition, many new instruments and techniques are included for the first time and some topics have been eliminated on the grounds that they have not been used by either professional or amateur astronomers for many years. Other topics, while no longer employed by professional observers for current work, are included because archive material that is still in use was obtained using them and/or because amateur astronomers use the techniques. A few references to Internet sites have been included but not many because the sites change so frequently and search engines are now so good. However, this resource is usually the point of first call for most scientists when they have a query and much material, such as large sky surveys, is only available over the Internet. Furthermore, it is used for the operation of some remote telescopes and it forms the basis of virtual observatories discussed in the last chapter. The rapid development of techniques for the detection and study of exoplanets is reflected in several new sections dealing with those specialised techniques.

As in earlier editions, another aim has always been to try and reduce the trend towards fragmentation of astronomical studies and this is retained in this edition. The new techniques that are required for observing in exotic regions of the spectrum bring their own concepts and terminology with them. This can lead to the impression that the underlying processes are quite different, when in fact they are identical, but are merely being discussed from differing points of view. Thus, for example, the Airy disc and its rings and the polar diagram of a radio dish do not at first sight look related, but they are simply two different ways of presenting the same data. As far as possible, therefore, broad regions of the spectrum are dealt with as a single area, rather than as many smaller disciplines. The underlying unity of all of astronomical observation is also emphasised by the layout of the book; the pattern of detection → imaging → ancillary techniques has been adopted so that one stage of an observation is encountered together with the similar stages required for all other information carriers. This is not an absolutely hard and fast rule, however, and in some places it seemed appropriate to deal with topics out of sequence, either in order to prevent a multiplicity of very small sections or to keep the continuity of the argument going.

The treatment of the topics is at a level appropriate to a science-based undergraduate degree. As far as possible the mathematics and/or physics background that may be needed for a topic is developed or given within that section. In some places it was felt that some astronomy background would be needed as well, so that the significance of the technique under discussion could be properly realised. Although aimed at an undergraduate level, most of the mathematics should be understandable by anyone who has attended a competently taught mathematics course in their final years at school and some sections are non-mathematical. Thus, many amateur astronomers will find aspects of the book of use and interest. The fragmentation of astronomy, which has already been mentioned, means that there is a third group of people who may find the book useful and that is professional astronomers themselves. The treatment of the topics in general is at a sufficiently high level, yet in a convenient and accessible form, for those professionals seeking information on techniques in areas of astronomy with which they might not be totally familiar.

Last, I must pay a tribute to the many astronomers and other scientists whose work is summarised here. It is not possible to list them by name and to stud the text with detailed references would have ruined the intention of making the book readable. I would, however, like to take the opportunity afforded by this preface to express my deepest gratitude to them all.

Clear skies and good observing to you all!

C. R. Kitchin
November 2013

Standard Symbols

M OST OF THE SYMBOLS used in this book are defined when they are first encountered and that definition then applies throughout the rest of the section concerned. In a few cases the symbol may have a different meaning in another section – W, for example, is used as the symbol for the Wiener filter in Section 2.1 and for the linear spectral resolution in Section 4.1. Such duplication though is avoided as far as possible and the meaning in any remaining cases should be clear enough. Some symbols, however, have acquired such a commonality of use that they have become standard symbols amongst astronomers (and many other scientists). Some of these are listed below and the symbol will not then be separately defined when it is encountered in the text.

amu Atomic mass unit = 1.6605×10^{-27} kg

AU Astronomical Unit = 149,597,870,700 m (exact value by IAU definition – 1.4960×10^{11} m or even 1.5×10^{11} m suffices for most purposes)

c Velocity of light in a vacuum = 2.9979×10^8 m s^{-1}

e Charge on the electron = 1.6022×10^{-19} C

e^{-} Symbol for an electron

e^{+} Symbol for a positron

eV Electron volt = 1.6022×10^{-19} J

G Gravitational constant = 6.670×10^{-11} N m^2 kg^{-2}

h Planck's constant = 6.6262×10^{-34} J s

k Boltzmann's constant = 1.3806×10^{-23} J K^{-1}

ly Light year = 9.4607×10^{15} m (= 0.30660 pc = 6.3241×10^4 AU)

m$_{e}$ Mass of the electron = 9.1096×10^{-31} kg

n Symbol for a neutron

p^{+} Symbol for a proton

pc Parsec = 3.0857×10^{16} m (= 3.2616 ly = 2.0626×10^5 AU)

U,B,V Magnitudes through the standard UBV photometric system

$^{12}_{6}C$ Symbol for nuclei (normal carbon isotope given as the example). The superscript is the atomic mass (in amu) to the nearest whole number and the subscript is the atomic number

γ Symbol for a gamma ray

ε_o	Permittivity of free space = 8.85×10^{-12} F m^{-1}
λ	Symbol for wavelength
μ	Symbol for refractive index (also, n, is widely used)
μ_o	Permeability of a vacuum = $4\,\pi \times 10^{-7}$ H m^{-1}
ν	Symbol for frequency (also, f, is widely used)

CHAPTER 1

Detectors

1.1 OPTICAL DETECTION

1.1.1 Introduction

In this and the immediately succeeding sections, the emphasis is upon the detection of the radiation or other information carrier and upon the instruments and techniques used to facilitate and optimise that detection. There is inevitably some overlap with other sections and chapters and some material might arguably be more logically discussed in a different order from the one chosen. In particular in this section, telescopes are included as a necessary adjunct to the detectors themselves. The theory of the formation of an image of a point source by a telescope, which is all that is required for simple detection, takes us most of the way through the theory of imaging extended sources. Both theories are therefore discussed together even though the latter should perhaps be in Chapter 2. There are many other examples such as X-ray spectroscopy and polarimetry that appear in Section 1.3 instead of Sections 4.2 and 5.2. In general, the author has tried to follow the route *detection-imaging-ancillary techniques* throughout the book, but has dealt with items out of this order when it seemed more natural to do so.

The optical region is taken to include the mid-infrared (MIR), near-infrared (NIR), the visible and the long-wave ultraviolet regions and thus roughly covers the range from 100 μm to 10 nm (3 THz to 30 PHz*). The techniques and physical processes employed for investigations over this region bear at least a generic resemblance to each other and so may be conveniently discussed together.

1.1.2 Detector Types

In the optical region, detectors fall into two main groups: thermal and quantum (or photon) detectors. Both these types are incoherent – that is to say, only the amplitude of the electromagnetic wave is detected; the phase information is lost. Coherent detectors are common

---

* At longer wavelengths the frequency equivalent of the wavelength is listed (using $c = 300,000$ km/s for the conversion) because many astronomers researching in this region are accustomed to working with frequency. Usually this applies for wavelengths of 10 μm (30 THz) or longer, but the conversion is given at shorter wavelengths when (as here) it seems appropriate to do so. Similarly at short wavelengths the energy in eV is listed as well as the wavelength.

at long wavelengths (Section 1.2), where the signal is mixed with that from a local oscillator (heterodyne principle). Only recently have optical heterodyne techniques been developed in the laboratory and these have yet to be applied to astronomy.* We may therefore safely regard all optical detectors as incoherent in practice. With optical aperture synthesis (Section 2.5), some phase information may be obtained providing that three or more telescopes are used.

In quantum detectors, the individual photons of the optical signal interact directly with the electrons of the detector. Sometimes individual detections are monitored (photon counting), at other times the detections are integrated to give an analogue output like that of the thermal detectors. Examples of quantum detectors include the eye, photographic emulsion, photomultiplier, photodiode and CCDs.

Thermal detectors, in contrast, detect radiation through the increase in temperature that its absorption causes in the sensing element. They are generally less sensitive and slower in response than quantum detectors, but have a much broader spectral response. Examples include thermocouples, pyroelectric detectors and bolometers.

1.1.3 The Eye

This is undoubtedly the most fundamental of the detectors to a human astronomer although it is a very long way from being the simplest. It is now rarely used for primary detection, although there are still a few applications in which it performs comparably with or possibly even better than other detection systems. Examples of this could be very close double-star work and planetary observation, in which useful work can be undertaken even though the eye may have been superseded for a few specific observations by, say, interferometry and planetary space probes. Visual observing also finds a role in monitoring the behaviour of long-period variable stars and searching for new novae, supernovae and comets. More commonly it is necessary to find and/or guide on objects visually whilst they are being studied with other detectors. Plus there are, of course, the millions of people who gaze into the skies for pleasure – and that includes most professional astronomers. Thus, there is some importance in understanding how the eye and especially its idiosyncrasies and defects can influence these processes.

It is essential to realise that a simple consideration of the eye on basic optical principles will be misleading. The eye and brain act together in the visual process and this may give better or worse results than, say, an optically equivalent camera system, depending upon the circumstances. Thus, on the plus side, the image on the retina is inverted and suffers from chromatic aberration but the brain compensates for these effects. Likewise, unless we make a special effort, we are unaware of the lack of an image from the retina over the area where the optic nerve emerges (the blind spot). The brain also receives two slightly different

---

* At the time of writing, high-resolution heterodyne spectroscopy at frequencies up to a few terahertz has been demonstrated in the laboratory. During the lifetime of this sixth edition of *Astrophysical Techniques* the technique may well be developed to the point where it is used for observing real objects in the sky. Since a frequency of 3 THz corresponds to a wavelength of 100 μm, then on the definition of 'optical region' used here, we may hope to see the birth of coherent optical astronomical observing before the next edition appears.

   Within optical interferometers, the phase information in the incoming signals is used to obtain the instrument's results. But this is done by combining the radiation beams before they are picked up by the detectors. Even in optical interferometers, the detectors themselves remain incoherent devices.

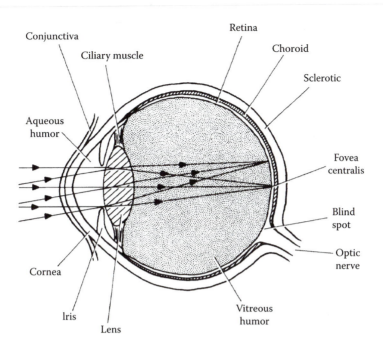

FIGURE 1.1   Optical paths in a horizontal cross section of the eye.

images from the two eyes, but instead of this causing us to see a blurred image we interpret the difference in terms of distance away from us.

Conversely, high-contrast objects or structures near the limit of resolution, such as planetary transits of the Sun and the Martian 'canals', may be interpreted falsely. A more widely experienced example is that the full moon appears to be larger when it is close to the horizon than when it is high in the sky. These various effects are partially physical and partially psychological. Thus, the difficulty in seeing a clear separation between the limb of the Sun and the silhouette of Mercury or Venus (the teardrop effect) during a transit probably arises through the cross connection of the eye's detector cells. The 'moon illusion' is almost certainly due to the effect of the brain's expectation about what it sees in the sky. Although the sky is often referred to as being hemispherical, the fact that clouds are closer to us when they are overhead than when near the horizon leads us to anticipate that the same will be true of all objects. Thus, although the full moon near the horizon has the same angular size as when it is higher in the sky,* we *think* that it is further away and thus *think* that it is larger. The Martian canals are an example of pareidolia whereby the brain interprets unfamiliar or random images in terms of things with which it is more familiar. Our ability to imagine faces in Rorschach ink blots, odd rock formations, inter-stellar nebulae (think of NGC 2392 – the Eskimo nebula) and so forth arises in this fashion.

The optical principles of the eye are shown in Figure 1.1. The receptors in the retina (Figure 1.2) are of two types: cones for colour detection (photopic vision) and rods for black

---

* In fact the angular size of the full moon near the horizon is slightly smaller than when it is higher in the sky. First, it is further away from the observer by about 6000 km (the Earth's radius) and second differential refraction in the Earth's atmosphere 'squashes' it slightly in the vertical direction.

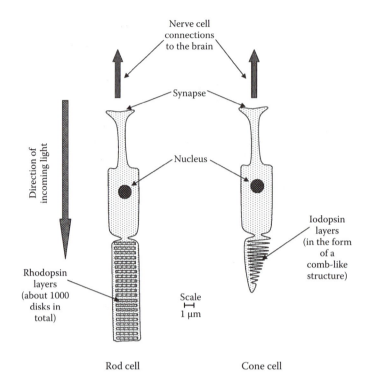

FIGURE 1.2   Retinal receptor cells.

and white reception at higher sensitivity (scotopic vision*). The light passes through the outer layers of nerve connections before arriving at these receptors. In the rods a pigment known as rhodopsin or, from its colour, visual purple, absorbs the radiation. It is a complex protein with a molecular weight of about 40,000 amu. It is arranged within the rods in layers about 20 nm thick and 500 nm wide and may comprise up to 35% of the dry weight of the cell. Under the influence of light a small fragment of the rhodopsin will split off. This fragment, or chromophore, is a vitamin A derivative called retinal or retinaldehyde and has a molecular weight of 286 amu. One of its double bonds changes from a *cis* to a *trans* configuration (Figure 1.4) within a picosecond of the absorption of the photon. The portion left behind is a colourless protein called opsin. The instant of visual excitation occurs at some stage during the splitting of the rhodopsin molecule, but its precise mechanism is not yet understood. The reaction causes a change in the permeability of the receptor cell's membrane to sodium ions, thus causing the electrical potential of the cell to become more negative (going from about –40 mV in the dark to –70 mV in the light). The change in potential then propagates through the nerve cells to the brain. The rhodopsin molecule is then slowly regenerated. The resulting response curve of the rod cells to light is shown in Figures 1.3 and 1.5.

---

* The intermediate situation is termed mesopic vision.

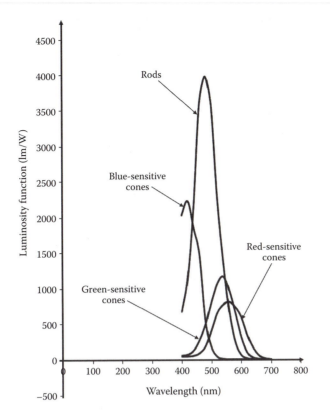

FIGURE 1.3   Intrinsic response curves of the retina's rod and cone cells. (Based upon data published by the Commission Internationale de l'Éclairage – CIE.)

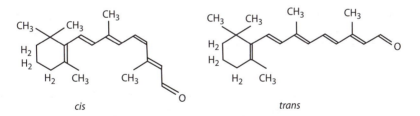

FIGURE 1.4   Transformation of retinaldehyde from *cis* to *trans* configuration by absorption of a photon near 500 nm.

The response of the cones is probably due to a similar mechanism to that of the rods. A pigment known as iodopsin is found in cones and this also contains the retinaldehyde group. Cone cells, however, are of three varieties with differing spectral sensitivities, peaking at 440 nm (blue), 535 nm (green) and 565 nm (often labelled 'red' although yellow/orange would be more accurate*). There are three slightly different iodopsins, called cyanolabe, chlorolabe and erythrolabe, respectively, which are responsible for the three

---

* The seven colours of the rainbow conventionally have central wavelengths of about 400 nm (violet), 450 nm (indigo), 470/480 nm (blue), 510 nm (green), 570 nm (yellow), 590 nm (orange) and 650 nm (red).

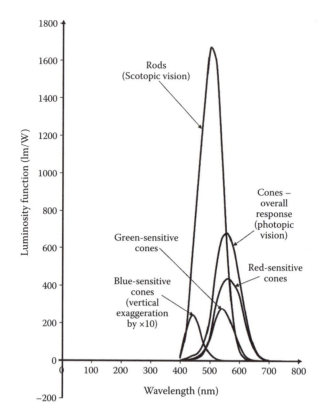

FIGURE 1.5 Overall response curves for the retina's rod and cone cells taking into account the absorption within the eye's lens and macula and the relative abundances of the three types of cone cell (assumed here to be in the ratio 5:30:65 for the blue-sensitive, green-sensitive and red-sensitive cells, respectively). The numbers of rod and cone cells are assumed to be equal. (Based upon data published by the CIE.) These curves show how the eye *actually* sees things, but can vary markedly over the retina, with age and between individuals.

differing responses. The three types of cone cell are not present in the retina in equal quantities and their relative proportions vary within the retina. Typically, the red-sensitive and green-sensitive cones each comprise between a third and three-fifths of all the cones in a particular area of the retina, whilst the blue-sensitive cones form only about a sixteenth to an eighth of the total. The intrinsic responses (Figure 1.3) of the cells also varies.

Whilst the commonly used criterion for the effectiveness of a detector – quantum efficiency – is not easy to assess for the eye, for comparison purposes the values of 3% (cones) and 10% (rods) at the peak of their response curves are probably about right.

In bright light much of the rhodopsin in the rods is split into opsin and retinaldehyde and their sensitivity is therefore much reduced. Vision is then provided primarily by the cones, although even at the peak, their sensitivity is only about 40% of the maximum for the rods. The three varieties of cones combine their effects to produce colour vision. At low light levels only the rods are triggered by the radiation and vision is then in black and white. The overall spectral sensitivities of the rods and cones differ (Figures 1.3 and 1.5), with that of the rods peaking at about 510 nm and the combined response of the cones peaking at 550 nm. This shift in sensitivity is called the Purkinje effect. It can be a problem for double-star observers since it may cause a hot and bright star to have its magnitude underestimated in comparison with a cooler and fainter star and vice versa.

Upon entering a dark observatory from a brightly illuminated room, the rhodopsin in the rods slowly re-forms over about half an hour; their sensitivity therefore improves concurrently. Thus, we have the well-known phenomenon of dark adaptation, whereby far more can be distinguished after a few minutes in the dark than can be detected initially. If sensitive vision is important for an observation, then for optimum results, bright lights should be avoided for half an hour before the observation is made.

As may be seen from Figures 1.3 and 1.5, the rods' response is effectively zero for wavelengths longer than about 600 nm. Lights radiating in this part of the spectrum will therefore not deplete the rods' rhodopsin, but may still be made bright enough to trigger photopic vision via the cones. Most observatories are therefore illuminated with red light in order to try and minimise any loss of dark adaptation. However, it should be noted that the lights must then be a very deep red indeed (a far deeper red than in the author's experience is actually used in most observatories) and have no emissions (leaks) at shorter wavelengths.

There is also a down side to dark adaptation. A small part of the change in the eye's response is due to the pupil expanding in size from about 2 mm diameter under high levels of illumination to about 7 mm in the dark. The change in pupil size typically takes about 4 seconds to occur (so most of dark adaptation results from the slower regeneration of the rhodopsin). However, the eye's lens suffers from spherical aberration which blurs the images produced by it and this effect worsens as the pupil's diameter increases. Thus, the images that we see when our eyes have become dark-adapted are fuzzier than normal experience would suggest and resolving fine details within the images is more difficult than we might expect.

Astronomical observation is generally due to rod vision. Usually with a dark-adapted eye, between one and ten photons are required to trigger an individual rod. However, several rods must be triggered in order to result in a pulse being sent to the brain. This arises because many rods are connected to a single nerve fibre. Most cones are also multiply connected, although a few, particularly those in the *fovea centralis*, have a one-to-one relationship with nerve fibres. The total number of rods is about 10^8 with about 6×10^6 cones and about 10^6 nerve fibres, so that upwards of a hundred receptors can be connected to a single nerve. In addition there are many cross connections between groups of receptors. The cones are concentrated towards the fovea centralis, where their separation may be as little as 2.3 μm and this is the region of most acute vision.

The rods form an increasing proportion of the light sensors of the eye towards the periphery of its field of view, peaking at about 30 rods for every cone at 8 to 10 mm from the fovea centralis. Towards the outer edge of the retina, the number of rods per square millimetre falls off so that their separation increases to some 14 μm. This variation in the relative proportions of types of rods and cones across the retina leads to the phenomenon called averted vision, whereby a faint object only becomes visible when *not* looked at directly. Its image is then falling onto a region of the retina richer in rods when the eye is averted from its direct line of sight. The combination of differing receptor sensitivities, change in pigment concentration and aperture adjustment by the iris mean that the eye is usable for illuminations* differing by a factor of around 10⁹ between the brightest and faintest.

The Rayleigh limit (Equation 1.34) of resolution of the eye is about 20 seconds of arc when the iris has its maximum diameter of 5 to 7 mm. But for two separate images to be distinguished, they must be separated on the retina by at least one unexcited receptor cell. So that even for images on the fovea centralis the actual resolution is between one or two minutes of arc. This is much better than elsewhere on the retina since the fovea centralis is populated almost exclusively by small tightly packed cones, many of which are singly connected to the nerve fibres. Its slightly depressed shape may also cause it to act as a diverging lens producing a slightly magnified image in that region. Away from the fovea centralis the multiple connection of the rods, which may be as high as a thousand receptors to a single nerve fibre, degrades the resolution far beyond this figure. Other aberrations and variations between individuals in their retinal structure means that the average resolution of the human eye lies between 5 and 10 minutes of arc for point sources. Linear sources such as an illuminated grating can be resolved down to 1 minute of arc fairly commonly. The effect of the granularity of the retina is minimised in images by rapid oscillations of the eye through a few tens of seconds of arc with a frequency of a few hertz, so that several receptors are involved in the detection when averaged over a short time.

With areas of high contrast, the brighter area is generally seen as too large, a phenomenon that is known as irradiation. We may understand this as arising from stimulated responses of unexcited receptors due to their cross connections with excited receptors. We may also understand the eye fatigue that occurs when staring fixedly at a source (for example when guiding on a star) as being due to depletion of the sensitive pigment in those few cells covered by the image. Averting the eye very slightly will then focus the image onto different cells, thereby reducing the problem. Alternatively, the eye can be rested momentarily by looking away from the eyepiece to allow the cells to recover.

The response of vision to changes in illumination is over the central part of the intensity range is approximately logarithmic (the Fechner law†). That is to say, if two sources, A and C, are observed to differ in brightness by a certain amount and a third source, B, appears to the eye as midway in brightness between them, then the energy from B will differ from that from A by the same *factor* as C differs from B. Thus, if we use L to denote the perceived luminosity and E to denote the actual energy of the sources then for

$$L_B = \frac{1}{2}\left(L_A + L_C\right) \tag{1.1}$$

---

* The upper limit is given by levels of illumination likely to cause damage to the eye. Safe viewing of the Sun requires its intensity to be reduced by a factor of about 1/30,000 (Section 5.3) and this is then about 0.5×10^9 times brighter than a magnitude 6 star or 2×10^9 times brighter than a magnitude 7.5 star.
† Sometimes called the Weber–Fechner law, although strictly the Weber law is a different formulation of the same physical phenomenon.

we have

$$\frac{E_A}{E_B} = \frac{E_B}{E_C} \quad \text{or} \quad \left[\log E_B = \frac{1}{2} (\log E_A + \log E_C) \right] \qquad (1.2)$$

This phenomenon is the reason for the nature of the magnitude scale used by astronomers to measure stellar brightnesses (Section 3.1), since that scale had its origins in the eye estimates of stellar luminosities by the ancient astronomers. The faintest stars visible to the dark-adapted naked eye from a good observing site on a good clear moonless night are of about magnitude six. This corresponds to the detection of about 3×10^{-15} W or about 8000 visible light photons per second. Special circumstances or especially sensitive vision may enable this limit to be improved upon by some one to one and a half stellar magnitudes (×2 to ×4). Conversely the normal ageing processes in the eye, such as a decreasing ability to dilate the eye's pupil and increasing numbers of 'floaters', etc., mean that the retina of a 60-year-old person receives only about 30% of the amount of light seen by a person half that age.* Eye diseases and problems such as cataracts may reduce this much further. Observers should expect a reduction in their ability to perceive very faint objects as time goes by.

1.1.4 Semiconductors

The photomultiplier, charge-coupled device (CCD) and several of the other detectors considered later derive their properties from the behaviour of semiconductors. Thus, some discussion of the relevant aspects of these materials is a necessary prerequisite to a full understanding of detectors of this type.

Conduction in a solid may be understood by considering the electron energy levels. For a single isolated atom they are unperturbed and are the normal atomic energy levels. As two such isolated atoms approach each other their interaction causes the levels to split (Figure 1.6). If further atoms approach, then the levels develop further splits, so that for N atoms in close proximity to each other, each original level is split into N sub-levels (Figure 1.7). Therefore, within a solid, each level becomes a pseudo-continuous band of permitted energies since the individual sub-levels overlap each other. The energy level diagram for a solid thus has the appearance shown in Figure 1.8. The innermost electrons remain bound to their nuclei, whilst the outermost electrons interact to bind the atoms together. They occupy an energy band called the valence band.

In order to conduct electricity through such a solid, the electrons must be able to move within the solid. From Figure 1.8 we may see that free movement could occur for electrons within the valence and higher bands. However, if the original atomic level that became the valence band upon the formation of the solid was fully occupied by electrons, then all the sub-levels within the valence band will still be fully occupied. If any given electron is to move under the influence of an electric potential, then its energy must increase. This it

---

* The eye is fully formed by the time a person is about age 13 and almost so from the age of about age 3 onwards. There are thus no equivalent age-related changes amongst young people.

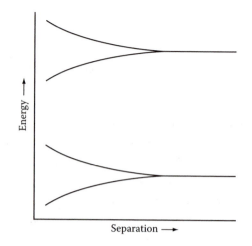

FIGURE 1.6 Schematic diagram of the splitting of two of the energy levels of an atom due to its proximity to another similar atom.

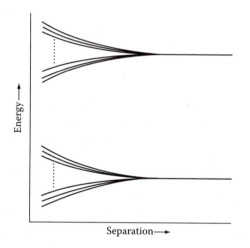

FIGURE 1.7 Schematic diagram of the splitting of two of the energy levels of an atom due to its proximity to many similar atoms.

cannot do since all the sub-levels are occupied and so there is no vacant level available for it at this higher energy. Thus, the electron cannot move after all. Under these conditions we have an electrical insulator.

If the material is to be a conductor, we can now see that there must be vacant sub-levels that the conduction electron can enter. There are two ways in which such empty sub-levels may become available. Either the valence band is unfilled, for example when the original atomic level had only a single electron in an s sub-shell, or one of the higher energy bands becomes sufficiently broadened to overlap the valence band. In the latter case at a temperature of absolute zero, all the sub-levels of both bands will be filled up to some energy that is called the Fermi level. Higher sub-levels will be unoccupied. As the temperature rises

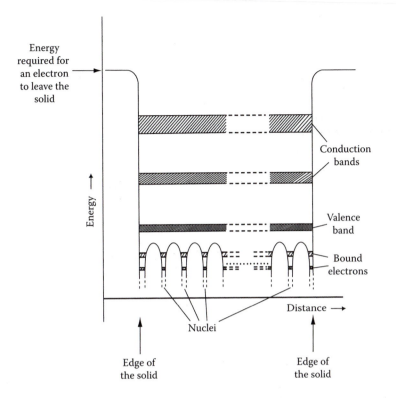

FIGURE 1.8   Schematic energy level diagram of a solid.

some electrons will be excited to some of these higher sub-levels, but will still leave room for conduction electrons.

A third type of behaviour occurs when the valence and higher bands do not actually overlap, but have only a small energy separation.* Thermal excitation may then be suffi-cient to push a few electrons into some of the higher bands. An electric potential can now cause the electrons in either the valence or the higher band to move. The material is known as a semiconductor since its conductivity is generally better than that of an insulator but considerably poorer than that of a true conductor. The higher energy bands are usually known as the conduction bands.

A pure substance will have equal numbers of electrons in its conduction bands and of spaces in its valence band. However, an imbalance can be induced by the presence of atoms of different elements from that forming the main semiconductor. If the valence band is full and one atom is replaced by another that has a larger number of valence electrons (a donor atom), then the excess electron(s) usually occupy new levels in the gap between the valence and conduction bands close to the bottom of the conduction band. From there they may more easily be excited into the conduction band by thermal motions or other energy sources. The semiconductor is then an n-type since any current within it will largely be carried by the (negative) electrons in the conduction band.

---

* The gaps in the widely used semiconductors silicon and germanium are 1.09 and 0.72 eV, respectively.

In the other case, when an atom is replaced with one that has fewer valence electrons (an acceptor atom), new, empty levels will be formed just above the top of the valence band. Electrons from the valence band can easily be excited into these new levels leaving energy gaps in the valence band. Physically this means that an electron bound to an atom within the crystal is now at a higher, but still bound, energy level. However, the excited electron no longer has an energy placing it within the valence band. Other electrons in the valence band can thus now take up the energy 'abandoned' by the excited electron. Physically this means that a valence electron is now able to move within the crystal. Without an externally applied voltage the electron movement will be random, but if there is an externally applied voltage then the electron will move from the negative potential side towards the positive potential side. The movement will comprise that of an electron from a neighbouring atom hopping over to the atom that has the gap in its electron structure. The moving atom will then be bound to its new atom but will have left a gap in the electron structure of its originating atom. An electron from the next atom over can now therefore hop into that gap. A third electron from the next atom over can then hop into this new gap – and so on. Since the hops occur rapidly, the appearance to an external observer would be that of the *absence* of an electron within the atoms' electron structures moving continuously from the positive side towards the negative side. The *absence* of a (negatively charged) electron, however, is equivalent to the *presence* of an equivalent positive charge. The process is thus usually visualised, not as comprising a series of electron hops, but as the continuous movement of a positively charged particle, called a hole, in the opposite direction. This type of semiconductor is called a p-type (from positive charge carrier) and its currents are thought of as mainly being transported by the movement of the positive holes in the valence band.

1.1.4.1 The Photoelectric Effect

The principle of the photoelectric effect is well known; the material absorbs a photon with a wavelength less than the limit for the material and an electron is then emitted from the surface of the material. The energy of that electron is a function of the energy of the photon, whilst the number of electrons depends upon the intensity of the illumination. In practice, the situation is somewhat more complex, particularly when what we are interested in is specifying the properties of a *good* photoemitter.

The main requirements are that the material should absorb the required radiation efficiently and that the mean free paths of the released electrons within the material should be greater than that of the photons. The relevance of these two requirements may be most simply understood from looking at the behaviour of metals in which neither condition is fulfilled. Since metals are conductors, there are many vacant sub-levels near their Fermi levels (see the earlier discussion). After absorption of a photon, an electron is moving rapidly and energetically within the metal. Collisions will occur with other electrons and since these electrons have other nearby sub-levels that they may occupy, they can absorb some of the energy from our photoelectron. Thus, the photoelectron is slowed by collisions until it may no longer have sufficient energy to escape from the metal even if it does reach the surface. For most metals the mean free path of a photon is about 10 nm and that of the released electron less than 1 nm, thus the number of electrons eventually emitted is very considerably reduced by collisional energy losses. Furthermore, the high reflectivity of metals results in only a small fraction of the suitable photons being absorbed and so the actual number of electrons emitted is only a very small proportion of those potentially available.

A good photoemitter must thus have low energy loss mechanisms for its released electrons whilst they are within its confines. The loss mechanism in metals (collision) can be eliminated by the use of semiconductors or insulators. Therefore, the photoelectron cannot lose significant amounts of energy to the valence electrons because there are no vacant levels for the latter to occupy, nor can it lose much energy to the conduction electrons because there are very few of these around. In insulators and semiconductors, the two important energy loss mechanisms are pair production and sound production. If the photoelectron is energetic enough, then it may collisionally excite other valence electrons into the conduction band thus producing *pairs* of electrons and holes. This process may be eliminated by requiring that E_1, the minimum energy to excite a valence electron into the conduction band of the material (Figure 1.9), is larger than E_2, the excess energy available to the photoelectron. Sound waves or phonons are the vibrations of the atoms in the material and can be produced by collisions between the photoelectrons and the atoms especially at discontinuities in the crystal lattice etc. Only 1% or so of the electron's energy will be lost at each collision because the atom is so much more massive than the electron. However, the mean free path between such collisions is only 1 or 2 nm so that this energy loss mechanism becomes very significant when the photoelectron originates deep in the material. The losses may be reduced by cooling the material since this then reduces the available number of quantised vibration levels in the crystal and also increases the electron's mean free path.

The minimum energy of a photon if it is to be able to produce photoemission is known as the work function and is the difference between the ionisation level and the top of the

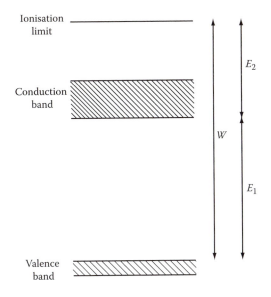

FIGURE 1.9   Schematic partial Grotrian diagram of a good photoemitter.

valence band (Figure 1.9). Its value may, however, be increased by some or all of the energy loss mechanisms mentioned above. In practical photoemitters pair production is particularly important at photon energies above the minimum and may reduce the expected flux considerably as the wavelength decreases. The work function is also strongly dependent upon the surface properties of the material; surface defects, oxidation products, impurities etc. can cause it to vary widely even amongst samples of the same substance. The work function may be reduced if the material is strongly p-type and is at an elevated temperature. Vacant levels in the valence band may then be populated by thermally excited electrons, bringing them nearer to the ionisation level and so reducing the energy required to let them escape. Since most practical photoemitters are strongly p-type, this is an important process and confers sensitivity at longer wavelengths than the nominal cut-off point. The long-wave sensitivity, however, is variable since the degree to which the substance is p-type is strongly dependent upon the presence of impurities and so is very sensitive to small changes in the composition of the material.

1.1.5 A Detector Index

After the natural detector formed by the eye, there are numerous types of artificial optical detectors. Before looking at some of them in more detail however, it is necessary to place them in some sort of logical framework or the reader is likely to become confused rather than enlightened by this section. We may idealise any detector as simply a device wherein some measurable property changes in response to the effects of electromagnetic radiation. We may then classify the types of detector according to the property that is changing and this is shown in Table 1.1.

Other processes may be sensitive to radiation but fail to form the basis of a useful detector. For example we can feel the (largely infrared) radiation from a fire directly on our

TABLE 1.1   Classification Scheme for Types of Detectors

Sensitive Parameter	Detector Names	Class
Voltage	Photovoltaic cells	Quantum
	Thermocouples	Thermal
	Pyroelectric detectors	Thermal
Resistance	Blocked impurity band device	Quantum
	Bolometer	Thermal
	Photoconductive cell	Quantum
	Phototransistor	Quantum
	Transition edge sensor	Thermal
Charge	Charge-coupled device	Quantum
Current	Superconducting tunnel junction	Quantum
Electron excitation	Photographic emulsion	Quantum
Electron emission	Photomultiplier	Quantum
	Television	Quantum
	Image intensifier	Quantum
Chemical composition	Eye	Quantum

skin but this process is not sensitive enough to form an infrared detector for any celestial objects other than the Sun. Conversely as yet unutilised properties of matter such as the initiation of stimulated radiation from excited atomic states as in the laser or maser may become the basis of detectors in the future.

1.1.6 Detector Parameters

Before resuming discussion of the detector types listed in Table 1.1, we need to establish the definitions of some of the criteria used to assess and compare detectors. The most important of these are listed in Table 1.2.

For purposes of comparison, D^* is generally the most useful parameter. For photomultipliers in the visible region it is around 10^{15} to 10^{16}. Figures for the eye and for photographic

TABLE 1.2   Criteria for Assessment and Comparison of Detectors

QE (quantum efficiency)	Ratio of the actual number of photons that are detected to the number of incident photons.
DQE (detective quantum)	Square of the ratio of the output signal/noise ratio to the input signal/noise efficiency ratio.
τ (time constant)	This has various precise definitions. Probably the most widespread is that τ is the time required for the detector output to approach to within $(1 - e^{-1})$ of its final value after a change in the illumination; i.e. the time required for about 63% of the final change to have occurred.
Dark noise	The output from the detector when it is un-illuminated. It is usually measured as a root-mean-square voltage or current.
NEP (noise equivalent detectable power)	The radiative flux as an input, that gives an output signal-to-noise ratio of power or minimum unity. It can be defined for monochromatic or black body radiation and is usually measured in watts.
D (detectivity)	Reciprocal of NEP. The signal-to-noise ratio for incident radiation of unit intensity.
D^* (normalised detectivity)	The detectivity normalised by multiplying by the square root of the detector area and by the electrical bandwidth. It is usually pronounced 'dee star'. $$D^* = \frac{(a\Delta f)^{1/2}}{NEP} \qquad (1.3)$$ The units, cm $Hz^{1/2}$ W^{-1}, are commonly used and it then represents the signal-to-noise ratio when 1 W of radiation is incident on a detector with an area of 1 cm^2 and the electrical bandwidth is 1 Hz.
R (responsivity)	Detector output for unit intensity input. Units are usually volts per watt or amps per watt. For the human eye it is called the luminosity function (Figures 1.3 and 1.5) and its unit is lumens per watt.
Dynamic range	Ratio of the saturation output to the dark signal. Sometimes only defined over the region of linear response.
Spectral response	The change in output signal as a function of changes in the wavelength of the input signal. Usually given as the range of wavelengths over which the detector is useful.
λ_m (peak wavelength)	The wavelength for which the detectivity is a maximum.
λ_c (cut-off wavelength)	There are various definitions. Amongst the commonest are wavelength(s) at which the detectivity falls to zero, wavelength(s) at which the detectivity falls to 1% of its peak value, wavelength(s) at which D^* has fallen to half its peak value.

emulsion are not directly obtainable, but values of 10^{12} to 10^{14} can perhaps be used for both to give an idea of their relative performances.

1.1.7 Cryostats

The noise level in many detectors can be reduced by cooling them to below ambient temperature. Indeed for some detectors, such as superconducting tunnel junctions (STJs) and transition edge sensors (TESs), cooling to very low temperatures is essential for their operation. Small CCDs produced for the amateur market are usually chilled by Peltier-effect coolers, but almost all other approaches require the use of liquid nitrogen or liquid helium as the coolant. Since these materials are both liquids, they must be contained in such a way that the liquid does not spill out as the telescope moves. The container is called a cryostat and whilst there are many different detailed designs, the basic requirements are much the same for all detectors. In addition to acting as a container for the coolant, the cryostat must ensure that the detector and sometimes pre-amplifiers, filters, optical components etc. are cooled whatever the position of the telescope, that the coolant is retained for as long as possible, that the evaporated coolant can escape, and that the detector and other cooled items do not ice up.

These requirements invariably mean that the container is a Dewar (vacuum flask) and that the detector is behind a window within a vacuum or dry atmosphere. Sometimes the window is heated to avoid ice forming on its surface. Most cryostats are designed like a bath and are essentially simply tanks containing the coolant with the detector attached to the outside of the tank or linked to it by a thermally conducting rod. Such devices can only be half filled with the coolant if none is to overflow as they are tilted by the telescope's motion. However, if the movement of the telescope is restricted, higher levels of filling may be possible. Hold times of a few days between refilling with coolant can currently be achieved. In cases where the detector is not tilted, such as when operating at Nasmyth or Coudé foci, or if the instrument is in a separate laboratory fed by fibre optics from the telescope, continuous flow cryostats can be used where the coolant is supplied from a large external reservoir and hold times of weeks, months or longer are then possible.

Closed cycle cryostats have the coolant contained within a completely sealed enclosure. The exhausted (warm) coolant is collected, recooled and reused. This type of cryostat design is mostly needed when (expensive) liquid or gaseous helium is the coolant. Open cycle cryostats allow the used coolant to escape from the system. It is then simply vented to the atmosphere or may, especially for helium, be collected and stored for future use in a separate operation.

The detector, its immediate electronics and sometimes filters or other nearby optical components are usually cooled to the lowest temperature. Other parts of the instrument such as heat shields, windows and more distant optical components may not need to be quite so cold. Hence, especially when helium is the main coolant, cheaper cryogens may be used for these less critical components. Thus, a Stirling engine used in reverse can routinely cool down to 70 K and in exceptional circumstances down to 40 K (a multi-stage variant on the Stirling system called a pulse tube refrigerator and utilising helium as its working fluid has recently achieved 1.7 K), liquid nitrogen is usually around 77 K, dry ice (solid carbon dioxide – not currently used much by astronomers – reaches 195 K and Peltier coolers can go down to around 220 K, depending upon the ambient temperature.

Bolometers, STJs and TESs require cooling to temperatures well below 1 K. Temperatures down to about 250 millikelvin (mK) can be reached using liquid $^3_2 He$. The $^3_2 He$ itself has to be cooled to below 2 K before it liquefies and this is achieved by using $^4_2 He$ under reduced pressure. Temperatures down to a few mK require a dilution refrigerator. This uses a mix of $^3_2 He$ and $^4_2 He$ at a temperature lower than 900 mK. The two isotopes partially separate out under gravity, but the lower $^4_2 He$ layer still contains some $^3_2 He$. The $^3_2 He$ is removed from the lower layer, distilling it off at 600 mK in a separate chamber. This forces some $^3_2 He$ from the upper layer to cross the boundary between the two layers to maintain the equilibrium concentration. However, crossing the boundary requires energy and this is drawn from the surroundings, thereby cooling them. The submillimetre common user bolometer array (SCUBA)-2, for example, (see bolometers below) uses a dilution refrigeration system in order to operate at 100 mK. An alternative route to millikelvin temperatures is the adiabatic demagnetisation refrigerator. The ions in a paramagnetic salt are first aligned by a strong magnetic field. The heat generated in this process is transferred to liquid helium via a thermally conducting rod. The rod is then moved from contact with the salt, the magnetic field is reduced and the salt cools adiabatically.

1.1.8 Charge Coupled Devices (CCDs)

Willard Boyle and George Smith invented CCDs in 1969 at the Bell telephone labs for use as a computer memory. The first application of CCDs within astronomy as optical detectors occurred in the late 1970s. Since then they have come to dominate completely the detection of optical radiation at professional observatories and are very widely used amongst amateur astronomers. Their popularity arises from their ability to integrate the detection over long intervals, their dynamic range ($>10^5$), linear response, direct digital output, high quantum efficiency, robustness, wide spectral range and from the ease with which arrays can be formed to give two-dimensional imaging. In fact, CCDs can only be formed as an array; a single unit is of little use by itself.

The basic detection mechanism is related to the photoelectric effect. Light incident on a semiconductor (usually silicon) produces electron-hole pairs, as we have already seen. These electrons are then trapped in potential wells produced by numerous small electrodes. There they accumulate until their total number is read out by charge coupling the detecting electrodes to a single read-out electrode.

An individual unit of a CCD is shown in Figure 1.10. The electrode is insulated from the semiconductor by a thin oxide layer. In other words, the device is related to the metal oxide-silicon (MOS) transistors. The electrode is held at a small positive voltage that is sufficient to drive the positive holes in the p-type* silicon away from its vicinity and to attract the electrons into a thin layer immediately beneath it. The electron-hole pairs produced in this depletion region by the incident radiation are thereby separated and the electrons accumulate in the storage region. Thus, an electron charge is formed whose magnitude is

---

* Boron is usually the doping element.

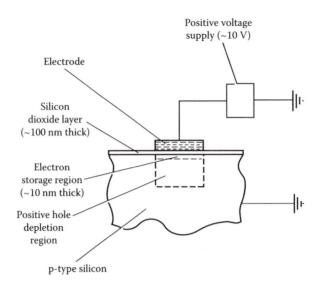

FIGURE 1.10 Basic unit of a CCD.

a function of the intensity of the illuminating radiation. In effect, the unit is a radiation-driven capacitor.

Now if several such electrodes are formed on a single silicon chip and zones of very high p-type doping insulate the depletion regions from each other, then each will develop a charge that is proportional to its illuminating intensity.* Thus, we have a spatially and electrically digitised reproduction of the original optical image (Figure 1.11).

All that remains is to retrieve this electron image in some usable form. This is accomplished by the charge coupling. Imagine an array of electrodes such as those we have already seen in Figure 1.11, but without their insulating separating layers. Then if one such electrode acquired a charge, it would diffuse across to the nearby electrodes. However, if the voltage of the electrodes on either side of the one containing the charge were reduced, then their hole depletion regions would disappear and the charge would once again be contained within two p-type insulating regions (Figure 1.12). This time, however, the insulating regions are not permanent but may be changed by varying the electrode voltage. Thus, the stored electric charge may be moved physically through the structure of the device by sequentially changing the voltages of the electrodes. Hence, in Figure 1.12, if the voltage on electrode C is changed to about +10 V, then a second hole depletion zone will form adjacent to the first. The stored charge will diffuse across between the two regions until it is shared equally. Now if the voltage on electrode B is gradually reduced to +2 V, its hole depletion zone will gradually disappear and the remaining electrons will transfer across to be stored under electrode C. Thus, by cycling the voltages of the electrodes as shown in Figure 1.13, the electron charge is moved from electrode B to electrode C.

With careful design the efficiency of this charge transfer (or coupling) may be made as high as 99.9999%. Furthermore, we may obviously continue to move the charge through

---

* Unless anti-blooming is used – see later in this chapter – the electron charge is linearly related to the optical intensity, at least until the number of electrons approaches the maximum number that the electrode can hold (known as the well capacity).

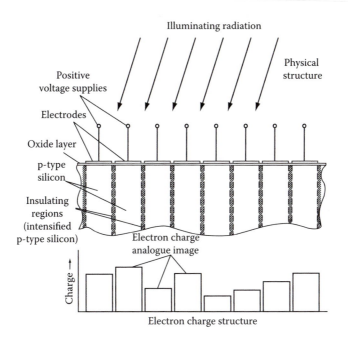

FIGURE 1.11 Array of CCD basic units.

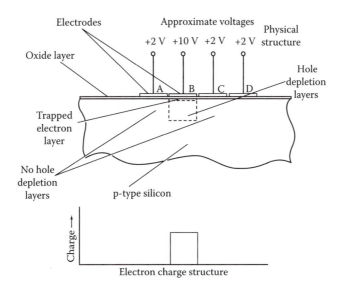

FIGURE 1.12 Active electron charge trapping in a CCD.

the structure to electrodes D, E, F etc. by continuing to cycle the voltages in an appropriate fashion. Eventually the charge may be brought to an output electrode from whence its value may be determined by discharging it through an integrating current metre or some similar device. In the scheme outlined here, the system requires three separate voltage cycles to the electrodes in order to move the charge and hence it is known as a three-phase CCD (variants on the basic CCD are discussed below and a quick-reference glossary of

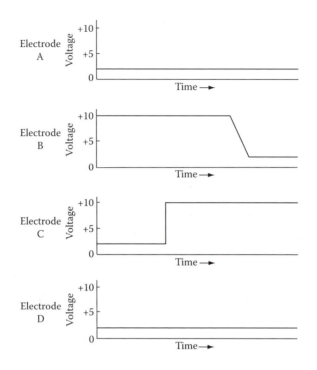

FIGURE 1.13   Voltage changes required to transfer charge from Electrode B to Electrode C in the array shown in Figure 1.12.

them may be found in Appendix D). Three separate circuits are formed, with each electrode connected to those three before and three after it (Figure 1.14). The voltage supplies – alpha, beta and gamma (Figure 1.14) – follow the cycles shown in Figure 1.15 in order to move charge packets from the left towards the right (Figure 1.16). Since only every third electrode holds a charge in this scheme the output follows the pattern shown schematically at the bottom of Figure 1.15. The order of appearance of an output pulse at the output electrode is directly related to the position of its originating electrode in the array. Thus, the

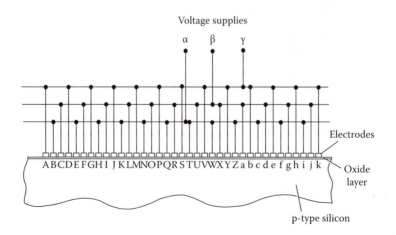

FIGURE 1.14   Connection diagram for a three-phase CCD.

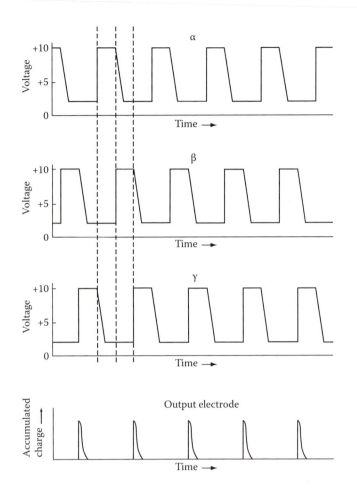

FIGURE 1.15   Voltage and output cycles for a three-phase CCD.

original spatial charge pattern and hence the original optical image may easily be inferred from the time-varying output.

The complete three-phase CCD is a combination of the detecting and charge transfer systems. Each pixel* has three electrodes and is isolated from pixels in adjacent columns by insulating barriers (Figure 1.17). During an exposure electrodes B are at their full voltage and the electrons from the whole area of the pixel accumulate beneath them. Electrodes A and C meanwhile are at a reduced voltage and so act to isolate each pixel from its neighbours along the column. When the exposure is completed, the voltages in the three electrode groups are cycled as shown in Figure 1.15 until the first set of charges reaches the end of the column. At the end of the column a second set of electrodes running orthogonally to the columns (Figure 1.17) receives the charges into the middle electrode for each column. That electrode is at the full voltage and its neighbours are at reduced voltages, so that each charge package retains its identity. The voltages in the read-out row of electrodes are then cycled to move the charges to the output electrode where they appear as a series of pulses.

---

* Pixel is a commonly used term for an individual detecting unit within an array detector of any type. It is derived from 'picture element'.

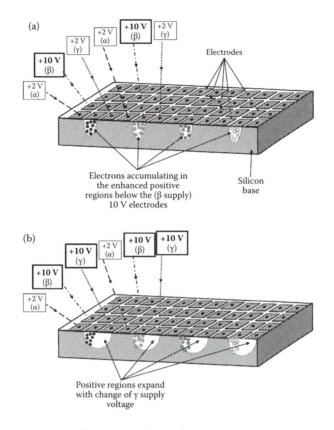

FIGURE 1.16 The movements of several packets of electrons through the physical structure of a three-phase CCD array. For clarity only six electrical connections are shown, but all the electrodes are connected to one of the three power supplies (α, β or γ), also the full CCD would comprise upwards of a million such electrodes. (a) Power supply β is at +10 V and power supplies α and γ at +2 V. Every third electrode thus has an enhanced positive region (hole depletion zone) below it into which the photoelectrons accumulate. (b) Power supplies β and γ are at +10 V and power supply α is at +2 V. The enhanced positive regions have expanded to cover the original electrodes and their right-hand neighbours. (c) Power supplies β and γ are at +10 V and power supply α is at +2 V. The electrons move physically through the silicon substrate until they are shared between each pair of +10-V electrodes. (d) Power supply γ is at +10 V and power supplies α and β at +2 V. The enhanced positive regions under the original electrodes disappear and the electrons continue to move through the silicon substrate until they are all accumulated below the right-hand neighbours of the original electrodes. Every third electrode thus again has an enhanced positive region below it containing the photoelectrons. We are back to the situation shown in Figure 1.16a except that all the electron packages have been moved one electrode to the right of their previous positions. Further, similar voltage cycles will continue to move the electron packages towards the right (movement to the left would simply require a slightly different phasing of the power supply changes).

When the first row of charges has been read out, the voltages on the column electrodes are cycled to bring the second row of charges to the read-out electrodes and so on until the whole image has been retrieved.

In the early days a small number of basic CCD units were simply strung together to form a linear array. However, whilst these have uses (such as in bar code readers) they can only

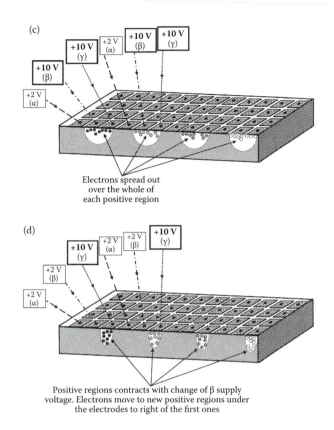

(c)

Electrons spread out
over the whole of
each positive region

(d)

Positive regions contracts with change of β supply
voltage. Electrons move to new positive regions under
the electrodes to right of the first ones

FIGURE 1.16 **(Continued)** The movements of several packets of electrons through the physical structure of a three-phase CCD array. For clarity only six electrical connections are shown, but all the electrodes are connected to one of the three power supplies (α, β or γ), also the full CCD would comprise upwards of a million such electrodes. (a) Power supply β is at +10 V and power supplies α and γ at +2 V. Every third electrode thus has an enhanced positive region (hole depletion zone) below it into which the photoelectrons accumulate. (b) Power supplies β and γ are at +10 V and power supply α is at +2 V. The enhanced positive regions have expanded to cover the original electrodes and their right-hand neighbours. (c) Power supplies β and γ are at +10 V and power supply α is at +2 V. The electrons move physically through the silicon substrate until they are shared between each pair of +10-V electrodes. (d) Power supply γ is at +10 V and power supplies α and β at +2 V. The enhanced positive regions under the original electrodes disappear and the electrons continue to move through the silicon substrate until they are all accumulated below the right-hand neighbours of the original electrodes. Every third electrode thus again has an enhanced positive region below it containing the photoelectrons. We are back to the situation shown in Figure 1.16a except that all the electron packages have been moved one electrode to the right of their previous positions. Further, similar voltage cycles will continue to move the electron packages towards the right (movement to the left would simply require a slightly different phasing of the power supply changes).

produce two-dimensional images if scanned orthogonally to their long axes. However, two-dimensional arrays can easily be made by stacking linear arrays side by side. The operating principle is unchanged and the correlation of the output with position within the image is only slightly more complex than before.

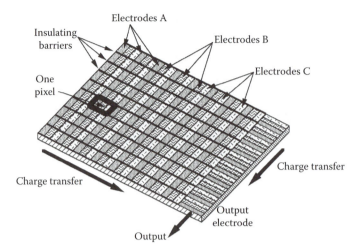

FIGURE 1.17   Schematic structure of a three-phase CCD.

The largest single CCD arrays currently produced are 9216 × 9216 pixels. In these arrays each pixel is 8.75 μm square, giving a physical size for the whole device of about 81 mm square. Thus, the dark energy detector Javalambre-PAU camera (JPCam) uses fourteen 9k × 9k* CCDs giving nearly 1.2 giga-pixels in total. However, most astronomical CCD arrays are 2k × 4k or 4k × 4k. For an adaptive optics telescope operating at 0.2″ resolution, a 2k × 4k CCD covers only 400″ × 800″ of the sky if the resolution is to be preserved. Many applications require larger areas of the sky than this to be imaged, so several such CCD arrays must then be formed into a mosaic. However, the normal construction of a CCD with electrical connections on all four edges means that there will then be a gap of up to 10 mm between each device, resulting in large dead-space areas in the eventual image. To minimise the dead space, three-edge-buttable CCDs are used. These have all the connections brought to one of the short edges, reducing the gaps between adjacent devices to about 0.2 mm on the other three edges. It is thus possible to form mosaics with two rows and as many columns as desired with a minimum of dead space. The largest such mosaic in operation at the time of writing is used in the camera on the first of four Panoramic Survey Telescope & Rapid Response System (Pan-STARRS) telescopes. The camera has a square mosaic of sixty 4800 × 4800 pixel orthogonal transfer CCDs giving a total of 1.38 giga-pixels and a field of view of 3° × 3°. The European Space Agency's (ESA) Gaia[†] spacecraft, due for launch in October 2013, will have a hundred and six CCD arrays each 1966 × 4500 pixels making a total of 1.41 giga-pixels (Figure 1.18). However, only 62 of the arrays will be for direct imaging, resulting in a 0.7° × 0.7° field of view for the instrument. Both of these will lose any records they may hold around 2020 when first light is scheduled for the Large Synoptic Survey Telescope (LSST). This 8.4 m telescope will use a 3.17-giga-pixel camera to

---

[*] The k in this context does not denote the number 1000 but the number 1024 (= 2^{10}). These arrays are thus actually 9216 × 9216 pixels in size.

[†] The name originated as an acronym for Global Astrometric Interferometer for Astronomy. It is not now planned to use interferometry for the project but the name has been retained.

FIGURE 1.18   Part of the Gaia CCD mosaic during its assembly. (Courtesy of Astrium.)

image an area of the sky 3.5° × 3.5°. The CCD mosaic will contain one hundred and eighty-nine 4k × 4k arrays with 10-μm pixels giving it nominal resolution of 0.2″, although the atmosphere will limit this to 0.7″ most of the time.*

With the larger format CCD arrays, the read-out process can take some time (typically several hundred milliseconds†). In order to improve observing efficiency, some devices therefore have a storage area between the imaging area and the read-out electrodes. This is simply a second CCD that is not exposed to radiation or half of the CCD is covered by a mask (frame transfer CCD). The image is transferred to the storage area in less than 0.1 ms and whilst it is being read out from there, the next exposure can commence on the detecting part of the CCD. Even without a separate storage area, reading the top half of the pixels in a column upwards and the other half downwards will halve read-out times. Column parallel CCDs (CPCCD) have independent outputs for each column of pixels thus allowing read-out times as short as 50 μs. More rapid read-out of a CCD can also be achieved by binning. In this process the electron charges from two or more pixels are added together before they are read out. There is, of course, a consequent reduction in the spatial resolution of the CCD and so the procedure will rarely if ever be encountered in astronomical applications. If binning is needed for other reasons, such as noise reduction, it is easily undertaken as a part of subsequent image processing (Section 2.9).

A two-phase CCD requires only a single clock, but needs double electrodes to provide directionality to the charge transfer (Figure 1.19). The second electrode, buried in the oxide layer, provides a deeper well than that under the surface electrode and so the charge accumulates under the former. When voltages cycle between 2 V and 10 V (Figure 1.20), the stored charge is attracted over to the nearer of the two neighbouring surface electrodes and

---

* Although multi-conjugate adaptive optics has recently enabled fields of view of up to 4′ × 4′ to be imaged at near diffraction-limited resolution, it seems unlikely that this will have improved to 3.5° × 3.5° by 2020.

† The delay, of course, is insignificant when an exposure has a duration of 10 seconds or more. However, for some applications, such as adaptive optics, images have to be obtained, processed and the correcting optics adjusted on a time scale of a millisecond or so. It is then vital to have a detector with the shortest possible read-out time.

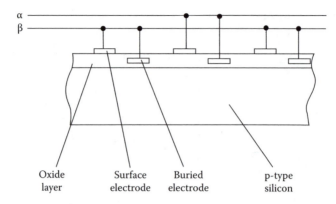

FIGURE 1.19   Physical structure of a two-phase CCD.

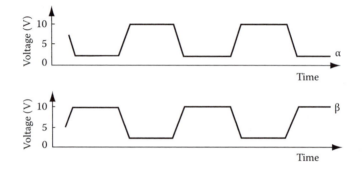

FIGURE 1.20   Voltage cycles for a two-phase CCD.

then accumulates again under the buried electrode. Thus, cycling the electrode voltages, which may be done from a single clock, causes the charge packets to move through the structure of the CCD, just as for the three-phase device.

A virtual phase CCD requires just one set of electrodes. Additional wells with a fixed potential are produced by p and n implants directly into the silicon substrate. The active electrode can then be at a higher or lower potential as required to move the charge through the device. The active electrodes in a virtual phase CCD are physically separate from each other, leaving parts of the substrate directly exposed to the incoming radiation. This enhances their sensitivity, especially at short wavelengths.

Non-tracking instruments such as the Carlsberg Meridian Telescope (Figure 5.4), liquid mirror, Hobby-Eberly and Southern African Large Telescope (SALT) telescopes can follow the motion of objects in the sky by transferring the charges in their CCD detectors at the same speed as the image drifts across their focal planes (time delayed integration [TDI]). To facilitate this, orthogonal transfer CCDs (OTCCD) are used. These can transfer the charge in up to eight directions (up/down, left/right and at 45° between these directions). OTCCDs can also be used for active image motion compensation arising from scintillation, wind shake etc. Other telescopes use OTCCDs for the nod and shuffle technique. This permits the almost simultaneous observation of the object and the sky background through the same light path by a combination of moving the telescope slightly (the nod) and the charge packets within the CCD (the shuffle). The recently commissioned 1-degree imager on the WIYN* 3.5-metre telescope uses four thousand and ninety-six 512 × 512 pixel OTCCDs to cover its 1 degree-square field of view. Sixty-four of the individual arrays at a time are grouped into 8 × 8 mosaics called an array packages and 64 array packages in an 8 × 8 mosaic then form the complete detector, comprising 1.07 giga-pixels in total.

For faint objects a combination of photomultiplier/image intensifier (Sections 2.1 and 2.3) and CCD, known as an electron bombarded CCD (EBCCD), electron multiplying CCD (EMCCD)[†] or an intensified CCD (ICCD)[‡] may be used. This places a negatively charged photocathode before the CCD. The photoelectron from the photocathode is accelerated by the voltage difference between the photocathode and the CCD and hits the CCD at high energy, producing many electron-hole pairs in the CCD for each incident photon. This might appear to give the device a quantum efficiency of over 100% but it is in effect merely another type of amplifier; the signal-to-noise ratio remains that of the basic device (or worse) and so no additional information is obtained.

Confusingly, low light level CCDs (LLLCCD or L3CCD) are also known as EMCCDs because they have a high on-chip amplification and so their basic output is many electrons for a single photon input. Unlike the EMCCDs mentioned above, where the detection of the photon occurs at the photocathode, the photon detection in an L3CCD is within the CCD as usual and the amplification occurs after the detection. In a similar manner to the frame-transfer CCD, the output from the detector is moved to a storage area called the extended output register. The extended output register is operated at up to 40 V and this voltage is high enough for electrons entering it to gain sufficient energy that they may collide with silicon atoms and occasionally liberate a second electron-hole pair – therefore effectively giving an amplification of the original detection by a factor of two (cf. avalanche photodiodes below). The voltage is adjusted so that the probability of such an electron-hole-producing collision is only 1% or 2%. The average amplification is thus reduced to about ×1.01 or ×1.02. The extended output register, however, has up to 600 stages, each of which has this average intrinsic amplification. The output from the extended output register is thus hundreds or thousands of times the number of electrons produced by the incoming photons directly in the CCD ($1.01^{600} \approx 400$ whilst $1.02^{600} \approx 150,000$). The output from the extended output register then goes through the remaining output stages required for a 'normal' CCD. The read noise (see below) of an L3CCD is thus the same as that for any other variety of CCD, but when superimposed upon a basic output which is hundreds or thousands of times larger than that of the normal CCD, it becomes almost negligible. The main noise source in an L3CCD is the variation in the amplification factor arising from the stochastic nature of the amplification process. The detections of several otherwise identical single photons will thus be a number of pulses of varying amplitudes. As with the photomultiplier tube (below) this will not be a problem if those individual pulses can be individually counted. However, if the number of photons increases to the point where

* The name WIYN comes from the operating institutes: University of Wisconsin–Madison, Indiana University, Yale University and the National Optical Astronomy Observatory.
† This term is also used for a quite different type of CCD, as discussed in the following paragraph.
‡ This term may also be used to designate a CCD being fed by a normal image intensifier.

the pulses merge into each other, then this pulse strength variation will reduce the signal-to-noise ratio by about a factor of 1.4 ($= \sqrt{2}$) – effectively halving the device's quantum efficiency. L3CCDs are thus best suited to counting individual photons and so must either be used for observing very faint sources or must be operated at very high frame rates.* The high amplification of L3CCDs also means that other noise sources may become important. In particular, electrons produced by induction from the clock-driven power supplies can be amplified and appear as bright pixels anywhere within the image. This clock-induced noise is intrinsic to all CCDs but only becomes obtrusive in L3CCDs. L3CCD arrays are currently produced up to about 1k × 1k in size.

Interline transfer CCDs have an opaque column adjacent to each detecting column. The charge can be rapidly transferred into the opaque columns and read out from there more slowly whilst the next exposure is obtained using the detecting columns. This enables rapid exposures to be made, but half the detector is dead space. They are mainly used for digital video cameras and rarely find astronomical applications although the Zurich Imaging Polarimeter (ZIMPOL) (see Section 5.2) does use one.

To an astronomer working 50 years ago and struggling to hypersensitise photographic emulsions using noxious chemicals, dry ice and acetone mixtures or potentially dangerous hydrogen-air combinations, a detector with all the advantages of a CCD would have seemed to be beyond his or her wildest dreams. Yet despite their superiority, CCDs are not quite the ultimate perfect detector. Their outputs are affected by the general sources of noise that bedevil all forms of measurement but they also have some problems that are unique to themselves and those we consider here.

The electrodes on or near the surface of a CCD can reflect some of the incident light, thereby reducing the quantum efficiency and changing the spectral response. To overcome this problem several approaches have been adopted. First, transparent polysilicon electrodes replace the metallic electrodes used in early devices. Second, the CCD may be illuminated from the back so that the radiation does not have to pass through the electrode structure at all. This, however, requires that the silicon forming the CCD be very thin so that the electrons produced by the incident radiation are collected efficiently. Nonetheless, even with thicknesses of only 10 to 20 μm, some additional cross talk (see below) may occur. More importantly, the process of thinning the CCD chips (by etching away the material with an acid) is risky and many devices may be damaged during the operation. Successfully thinned CCDs are therefore expensive in order to cover the cost of the failures. They are also very fragile and can become warped, but they do have the advantage of being less affected by cosmic rays than thicker CCDs. With a suitable anti-reflection coating, a back-illuminated CCD can now reach a quantum efficiency of 90% in the red and near-infrared. Other methods of reducing reflection losses include using extremely thin electrodes or an open electrode structure (as in a virtual phase CCD) that leaves some of the silicon exposed directly to the radiation. At longer wavelengths, the thinned CCD may become

---

* CCDs used as the detectors within instruments on board spacecraft may be subject to intense ionising radiation from the Earth's Van Allen belts, solar flares etc. One way of radiation-hardening such detectors is to use very fast read-out rates so that detections of individual photons do not have time to be affected by the radiation.

semi-transparent. Not only does this reduce the efficiency because not all the photons are absorbed, but interference fringes may occur between the two faces of the chip.* These can become very obtrusive and have to be removed as a part of the data reduction process.

The CCD as so far described suffers from loss of charge during transfer because of imperfections at the interface between the substrate and the insulating oxide layer. This has the most negative effect on the faintest images since the few electrons that have been accumulated are physically close to the interface. Artificially illuminating the CCD with a low-level uniform background during an exposure will ensure that even the faintest parts of the image have the surface states filled. This offset or 'fat zero' can then be removed later in signal processing. Alternatively, a positively charged layer of n-type† silicon between the substrate and the insulating layer may be added to force the charge packets away from the interface, producing a buried-channel CCD. A variant on the buried-channel CCD, termed a peristaltic CCD (from a rather fanciful analogy between the way electrons move in the device and the way food is moved down the oesophagus) intensifies the transfer speed of the electrons by using additional implanted electrodes. This allows the devices to operate at up to frequencies of 100 MHz and perhaps in the future, after further development, at frequencies of 1 GHz or more.

Charge transfer efficiencies for CCDs now approach 99.9999%, leading to typical read-out noise levels of 1 to 2 electrons per pixel. Using a non-destructive read-out mechanism and repeating and averaging many read-out sequences can reduce the read-out noise further. To save time, only the lowest intensity parts of the image are repeatedly read out, with the high-intensity parts being skipped. The devices have therefore become known as skipper CCDs.

At some point and usually within the CCD integrated circuit itself, the electron charge packages are converted to digital voltage signals. The analogue-to-digital converter(s) (ADCs) used for this do not normally respond to single electrons but only groups of them. Typically, 10 electrons might be needed to give a unit output from the ADC. The group of electrons required for a unit output is termed the analogue-to-digital unit (ADU) and the number of electrons is termed the gain.‡ The gain (and the ADU) is usually determined by matching the maximum value that can be output by the ADC to the well capacity of a single pixel. Thus, a 16-bit ADC has an output range from 0 to 65,535. If such an ADC is used for a CCD with a well capacity of 500,000 electrons, then the gain must be around 7 or 8 (500,000/65,536 = 7.6). The rounding (up or down) of the number of electrons to the number of ADUs is one component of the read noise from the CCD. The others are the errors in the conversion of electron numbers to ADUs and noise from the ADCs' electronics and other electronic components, such as amplifiers, incorporated into the CCD chip. Typically, in a science-grade CCD, the read noise corresponds to a few (≤10) electrons. The read noise is effectively the dark signal of the CCD. The dynamic range (Table 1.2) of a CCD with a well capacity of 500,000 electrons, but a read noise of 10 electrons, is thus ×50,000 (= 500,000/10). The use of a 16-bit ADC for such a CCD, as cited previously, is thus

---

* The effect is sometimes called 'etaloning' since similar internal reflections are utilised productively in etalons (Section 4.1).
† Phosphorous is usually the doping element.
‡ An odd terminology since the number of ADUs is *lower* than the number of photons originally detected by the factor given by the 'gain' (except in L3CCDs and EBCCDs).

not quite as contrary as it might seem. The read noise would need to reduce to 2 electrons before 18-bit conversion would be justified. With L3CCDs the read noise is similar to that of other CCDs, but the amplification that occurs before read-out means that the read noise is superimposed upon a much larger signal. In proportion to the number of originally detected photons, the read noise in L3CCDs can thus be down to 0.01 electrons per pixel.

The spectral sensitivity of CCDs ranges from 400 nm to 1100 nm, with a peak near 750 nm where the quantum efficiency can approach 90%. The quantum efficiency drops off further into the infrared as the silicon becomes more transparent. Short-wave sensitivity can be conferred by coating the device with an appropriate phosphor to convert the radiation to longer wavelengths.

Personal digital video and still cameras automatically obtain their images in colour. To do this, small filters are placed over each CCD element. In the commonly used Bayer arrangement, for each set of four pixels, two have green-transmission filters, one a blue transmission filter and one a red transmission filter (allowing very roughly for the eye's intrinsic spectral response – Figure 1.5). This means that the spatial resolution of the image is degraded – a 10-megapixel colour camera only has the resolution of a 2.5 mega-pixel monochromatic camera. More importantly for rigorous scientific analysis of the images, the three colours do not come from exactly the same locality within the original object. If the properties of the original object are varying significantly on a size scale equivalent to an individual pixel, then the colour image will give false results. When images at two or more wavelengths are needed for astronomical purposes (including the beautiful colour representations of nebulae and galaxies etc. ornamenting many astronomy books) it is usual to obtain individual images at each wavelength through appropriate filters and then combine them into a colour (or false-colour) final image. The full spatial resolution of the detector is thus retained and, if wanted, a truer visual representation of the appearance of the object may be obtained by better relative weightings of the different images than that given by the crude Bayer system.[*]

The CCD regains sensitivity at very short wavelengths because X-rays are able to penetrate the device's surface electrode structure. For both X-ray and infrared detection (and possibly also for the direct detection of low-energy dark matter particles – see Section 1.7) a deep-depletion CCD[†] may provide higher sensitivity. These devices have a thick silicon substrate which has a high resistivity and uses a bias voltage. The hole depletion zones are deeper than normal so that there is more opportunity for penetrating photons to interact with the material and so produce electron-hole pairs. As an example, the Soft X-ray Imager camera (due for launch on the Japanese Astro-H spacecraft in 2014) will use four 640 × 640 pixel deep depletion CCDs to observe an area of the sky 38′ square to a resolution of 1.74″ in the soft X-ray region (energy < 10 keV,[‡] wavelength > 0.1 nm).

---

[*] Of course, if the object is varying on a *time* scale faster than the interval between the separate images, then the standard astronomical approach will also give false results. When an object is changing rapidly on both the spatial and time scales, such as with a solar flare, then special techniques have to be devised.

[†] Also known as pn-CCDs or p-channel CCDs.

[‡] The electron volt (eV) equals 1.6×10^{-19} J and is a convenient unit for use in this spectral region (and also when discussing cosmic rays in Section 1.4). Since the values of the corresponding frequencies are 10^{17} Hz and above, they are not commonly encountered in the literature.

For integration times longer than a fraction of a second,* it is usually necessary to cool the device in order to reduce its dark signal. Using liquid nitrogen for this purpose and operating at around 170 K, integration times of minutes to hours are easily attained. Small commercial CCDs produced for the amateur astronomy market (which often also find applications at professional observatories for guiding etc.) usually use Peltier coolers to get down to about 50° below the ambient temperature. Subtracting a dark frame from the image can further reduce the dark signal. The dark frame is in all respects (exposure length, temperature etc.) identical to the main image, except that the camera shutter remains closed whilst it is obtained. It therefore just comprises the noise elements present within the detector and its electronics. Because the dark noise itself is noisy, it may be better to use the average of several dark frames obtained under identical conditions.

For typical noise levels and pixel capacities, astronomical CCDs have dynamic ranges of 50,000 to 500,000 which results in a usable magnitude range in accurate brightness determination of up to 14.5^m. This therefore compares very favourably with the dynamic range of less than 1000 (brightness range of less than 7.5^m) available from a typical photographic image.

A major problem with CCDs used as astronomical detectors is the noise introduced by cosmic rays. A single cosmic ray particle passing through one of the pixels of the detector can cause a large number of ionisations. The resulting electrons accumulate in the storage region along with those produced by the photons. It is usually possible for the observer to recognise such events in the final image because of the intense spike that is produced. Replacing the signal from the affected pixel by the average of the eight surrounding pixels improves the *appearance* of the image, but does *not* retrieve the original information. This correction often has to be done by hand and is a time-consuming process. When two or more images of the same area are available, automatic removal of the cosmic ray spikes is possible with reasonable success rates.

Another serious defect of CCDs is the variation in background noise between pixels. This takes two forms. There may be a large-scale variation of 10%–20% over the whole sensitive area and there may be individual pixels with permanent high background levels (hot spots). The first problem has been much reduced by improved production techniques and may largely be eliminated by flat fielding during subsequent signal processing, if the effect can be determined by observing a uniform source. Commonly used sources for the flat field include the twilit sky and a white screen inside the telescope dome illuminated by a single distant light source. The flat field image is divided into the main image after dark frame subtraction from both images to reduce the large-scale sensitivity variations. The effect of a single hot spot may also be reduced in signal processing by replacing its value with the mean of the four or eight surrounding pixels. However, because the hot spots are additionally often poor transferors of charge, all preceding pixels are then affected as their charge packets pass through the hot spot or bad pixel introducing a spurious line into the image. There is little that can be done to correct this last problem, other than to buy a new CCD. Even the 'good' pixels do not have 100% charge transfer efficiency, so that images containing very bright stars show a tail to the star caused by the electrons lost to other pixels as the star's image is read out.

---

* For astronomical purposes even CCDs used for short exposures will be cooled.

Yet another problem is that of cross talk or blooming. This occurs when an electron strays from its intended pixel to one nearby. It affects rear-illuminated CCDs because the electrons are produced at some distance from the electrodes and is the reason why such CCDs have to be thinned. It can also occur for any CCD when the accumulating charge approaches the maximum capacity of the pixel. Then the mutual repulsion of the negatively charged electrons may force some over into adjacent pixels. The well capacity of each pixel depends upon its physical size and is around half a million electrons for 25 micron-sized pixels. Some CCDs have extra electrodes to enable excess charges to be bled away before they spread into nearby pixels. Such electrodes are known as drains and their effect is often adjustable by means of an anti-blooming circuit. Anti-blooming should not be used if you intend making photometric measurements on the final image, but otherwise it can be very effective in improving the appearance of images containing both bright and faint objects.

The size of the pixels in a CCD can be too large to allow a telescope to operate at its limiting resolution.* In such a situation, the images can be dithered to provide improved resolution. Dithering consists simply of obtaining multiple images of the same field of view, with a shift of the position of the image on the CCD between each exposure by a fraction of the size of a pixel. The images are then combined to give a single image with sub-pixel resolution. A further improvement is given by drizzling (also known as variable pixel linear reconstruction) in which the pixel size is first shrunk, leaving gaps between the pixels. The images are then rotated before mapping onto a finer scale output grid.

One of the major advantages of a CCD over a photographic emulsion is the improvement in the sensitivity. However, widely different estimates of the degree of that improvement may be found in the literature, based upon different ways of assessing performance. At one extreme, there is the ability of a CCD to provide a recognisable image at very short exposures because of its low noise. Based upon this measure, the CCD is perhaps 1000 times faster than photographic emulsion. At the other extreme, one may use the time taken to reach the mid-point of the dynamic range. Because of the much greater dynamic range of the CCD, this measure suggests CCDs are about 5 to 10 times faster than photography. The most sensible measure of the increase in speed, however, is based upon the time required to reach the same signal-to-noise ratio in the images (i.e. to obtain the same information). This latter measure suggests that in their most sensitive wavelength ranges (around 750 nm for CCDs and around 450 nm for photographic emulsion), CCDs are 20 to 50 times faster than photographic emulsions.

1.1.9 Avalanche Photodiodes

The avalanche photodiode is finding increasing use as an optical detector within astronomy. It is a variant on the basic photodiode (which can also be an optical detector but which is not now employed by astronomers as a primary detector). However, to understand avalanche photodiodes we need to know something about the basic photodiode's operating principles and so we look at those first.

---

* The well capacity of a CCD pixel varies with its size. Therefore, for a given noise level, a larger dynamic range requires physically larger pixels. Most CCDs have pixel sizes in the range 10 to 50 μm.

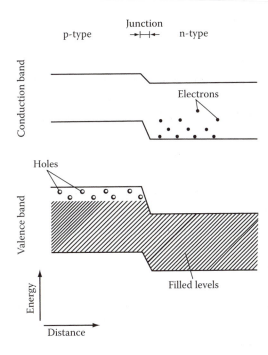

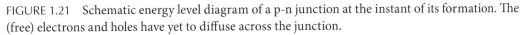

FIGURE 1.21 Schematic energy level diagram of a p-n junction at the instant of its formation. The (free) electrons and holes have yet to diffuse across the junction.

1.1.9.1 Photodiodes

Photodiodes are also known as photovoltaic cells, photoconductors and barrier junction detectors. They rely upon the properties of a p-n junction* in a semiconductor. The energy level diagram of such a junction is shown in Figure 1.21. Electrons in the conduction band of the n-type material are at a higher energy than the holes in the valence band of the p-type material. Electrons therefore diffuse across the junction and produce a potential difference across it. Equilibrium occurs when the potential difference is sufficient to halt the electron flow. The two Fermi levels are then coincident, the potential across the junction is equal to their original difference and there is a depletion zone containing neither type of charge carrier across the junction. The n-type material is positive, the p-type negative and we have a simple p-n diode.

Now, if light of sufficiently short wavelength falls onto such a junction then it can generate electron-hole pairs in both the p- and the n-type materials. The electrons in the conduction band in the p region will be attracted towards the n region by the intrinsic potential difference across the junction and they will be quite free to flow in that direction. The holes in the valence band of the p-type material will be opposed by the potential across the junction and so will not move. In the n-type region the electrons will be similarly trapped whilst the holes will be pushed across the junction. Thus, a current is generated by the illuminating radiation and this may simply be monitored and used as a measure of the light intensity. For use as a radiation detector the p-n junction often has a region of undoped (or

---

* That is, the union between a p-doped semiconductor and an n-doped semiconductor, usually with the same bulk semiconductor on both sides of the join.

intrinsic) material between the p and n regions in order to increase the size of the detecting area. These devices are then known as p-i-n junctions. Their operating principles do not differ from those of the simple p-n junction.

The response of a p-n junction to radiation is shown in Figure 1.22. It can be operated under three different regimes. The simplest, that is labelled B in Figure 1.22, has the junction short-circuited through a low-impedance metre. The current through the circuit is the measure of the illumination. In regime C the junction is connected to a high impedance so that the current is very close to zero and it is the change in voltage that is the measure of the illumination.

Finally, in regime A, the junction is back (or reverse) biased – the p-type material is made more negative and the n-type material made more positive. When an incident photon produces an electron-hole pair, electrons in the conduction band of the p-type material are strongly attracted over to the n-type material. Similarly, holes in the valence band of the n-type material are strongly attracted towards the p-type material. The movement of holes in the valence band of the p-type material and of electrons in the conduction band of the n-type material is prevented by the biasing. In regime A, the voltage across a load resistor in series with the junction measures the radiation intensity. In this mode the device is known as a photoconductor.

The construction of a typical photovoltaic cell is shown in Figure 1.23. The material in most widespread use for the p and n semiconductors is silicon that has been doped with appropriate impurities. Solar power cells are of this type. The silicon-based cells have a peak sensitivity near 900 nm and cut-off wavelengths near 400 and 1100 nm. Their quantum efficiency can be up to 50% near their peak sensitivity and D* can be up to 10^{12}.

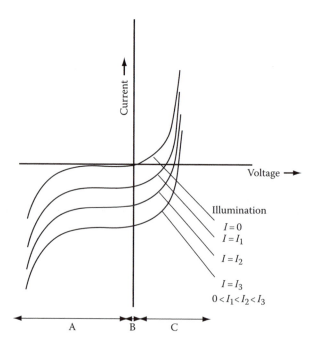

FIGURE 1.22   Schematic V/I curves for a p-n junction under different levels of illumination.

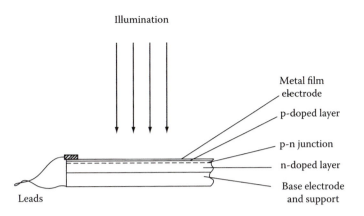

FIGURE 1.23   Cross section through a p-n junction photovoltaic detector.

Indium arsenide, indium selenide, indium antimonide, gallium arsenide (GaAs) and indium gallium arsenide can also all be fabricated into photodiodes. They are particularly useful in the infrared where germanium doped with gallium out-performs bolometers for wavelengths up to 100 μm.

1.1.9.2 Avalanche Photodiode

If a p-n junction is reverse-biased to more than half its breakdown voltage then an avalanche photodiode* (APD) results. The original electron-hole pair produced by the absorption of a photon will be accelerated by the applied field sufficiently to cause further pair production through inelastic collisions. These secondary electrons and holes can in their turn produce further ionisations and so on (cf. L3CCDs above and Geiger and proportional counters – see Sections 1.3 and 1.4). Eventually, an avalanche of carriers is created, leading to an intrinsic gain in the signal by a factor of 100 or more. The typical voltage used is 100 to 200 V and the response of the device is linear.

The basic APD has a physical structure which has the radiation entering through a thin layer of p-doped semiconductor and then being absorbed in a thicker intrinsic semiconductor layer. The n-doped semiconductor forms the bottom layer of the device. The electrons accelerate towards the n-doped layer producing the avalanche along the way. The semiconductor material that is used varies with the wavelength of the radiation to be detected. Silicon is used for the 200 nm to 1.1 μm region, gallium nitride operates down to around 250 nm in the ultraviolet (UV) region, indium-gallium-arsenide is used for the region from 1.0 to 2.6 μm with high gains, germanium for 800 nm to 1.7 μm and mercury-cadmium-telluride-alloy (HgCdTe) based APDs can reach out to 14 μm. Quantum efficiencies approach 50% in the optimum regions of the spectrum for each material. Response times to the detection of an incoming photon are generally in the region of tens to hundreds of picoseconds. A major drawback though for APDs is that the gain depends very sensitively upon the bias. The power supply typically has to be stable to a factor of 10 better than the desired accuracy of the output. The devices also have to be cooled and maintained

---

* Sometimes called a solid photomultiplier (see later in the chapter).

at a stable temperature in order to reduce noise. Arrays of APDs are coming into production but are still small at the time of writing.

Once an incident photon has produced an electron-hole pair, both are accelerated by the back biasing voltage. However, materials are generally chosen for construction of APDs that make collisions by the electrons more likely to produce more electrons than the collisions by the holes are likely to produce more holes.* The ratio of the likelihood of hole production to that of electron production within the avalanche is called the k-factor. For some materials such as HgCdTe the value of k can reach zero (i.e. the avalanche contains only electrons). For silicon, $k \approx 0.02$ whilst for germanium, $k \approx 0.9$. APDs with very low or zero values of k are termed electron-APDs or e-APDs and they can have very high gains at low bias voltages. E-APDs have yet to find astronomical applications, but at the time of writing their possible use is being investigated at several observatories. Hybrid APDs are just APD versions that have amplifiers and other electronic circuits integrated onto their chips.

APDs can also be used for direct X-ray detection although they have not yet been so used for studying astronomical sources.† The physical structure of these devices differs somewhat from that of the basic APD and allows much higher bias voltages to be employed. The difference lies in the presence of a large drift region between the layer in which the electron-hole pairs are produced and the layer in which the avalanche occurs. The highest bias voltages are to be found in the bevelled-edge APDs. The bevelled edge acts to reduce the electrical field along the edges thus preventing unwanted electrical break-downs. Bias voltages of up to 2000 V and gains of up to ×10,000 are possible. Reach-through APDs have the avalanche layer at the back of the structure, are operated at up to 500 V, and offer gains of up to ×200. Reverse or buried junction APDs have the avalanche layer immediately behind the X-ray entry window, are operated at less than 500 V, and offer gains that are generally less than ×200.

Once started, the avalanche is quenched (because otherwise the current would continue to flow) by connecting a resistor of several hundred kilo-ohms in series with the diode. Then as the current starts to flow the voltage across the diode is reduced and the breakdown ended. However, quenching the avalanche using a series resistor is a relatively slow process and leads to long dead times before the APD recovers and is ready to detect another photon. Most modern APDs therefore employ active quenching which reduces the dead time to a few tens of nanoseconds. In active quenching there is a circuit that detects the onset of the avalanche and which rapidly reduces and then restores the bias voltage.

1.1.9.3 Single Photon Avalanche Photodiodes

Single photon avalanche photodiodes (SPADs – also known as Geiger-mode APDs) are versions of the basic APD in which the bias voltage is well above the breakdown voltage of the semiconductor in use (typically > 30,000 V/mm). Like the Geiger counter (Section 1.3), the electron avalanche saturates whatever the number of photons may be that may have

---

* If this seems counter-intuitive then remember that the motions of the holes in one direction are actually due to bound electrons jumping from one atom within the crystal to another in the opposite direction. Their motions and interactions may thus be expected to differ in some ways from those of the free electrons.

† APDs are to form part of the active shield (Section 1.3) for the soft γ-ray detector to be launched on board the Astro-H spacecraft in 2014. They will not detect the γ-rays directly, however, but pick up the visible light flashes produced in the bismuth germanate (BGO) scintillator that surrounds the main Compton detector.

entered the device to initiate the avalanche. The gain is thus in the region of $\times 10^8$ to 10^{12} or more, but the response of the detector is non-linear. This though is unimportant since the devices are used to detect single photons and so only operate in circumstances (low light levels and/or high frame rates) where the average of the arrival time between individual photons is at least 10 times longer than the dead time of the device. Since electron-hole pairs produced by any means, including thermal motions, will result in avalanches, these form a major noise source in SPADs.

Although not yet widely used within astronomy, APDs have been used for some years used in the University of Hertfordshire's extra-solar planet polarimeter (Section 5.2) and in the wavefront sensor for the European Southern Observatory's (ESO) Multi-Application Curvature Adaptive Optics (MACAO) adaptive optics systems for the Very Large Telescope (VLT) instruments. SPADs have been used to observe the Crab nebula pulsar using Asagio Quantum Eye (AquEYE) on the 1.8-metre telescope at the Asagio observatory with a timing accuracy of 50 ps. They are also being investigated for use within instruments being considered for the 30-metre telescope (see below).

1.1.10 Photography

Photography will be dealt with as a part of imaging in Section 2.2. Here it is sufficient to point out that the basic mechanism involved in the formation of the latent image is electron-hole pair production. The electrons excited into the conduction band are trapped at impurities, whilst the holes are removed chemically. Unsensitised emulsion is blue-sensitive as a consequence of the minimum energy required to excite the valence electrons in silver bromide.

1.1.11 Photomultipliers

Electron multiplier phototubes (or photomultipliers [PMTs] as they are more commonly but less accurately known) were at one time the workhorses of optical photometry (see Chapter 3). For this purpose they have largely been superseded by CCDs. However, they continue to be used for UV measurements in the 10-nm to 300-nm region where CCDs are insensitive. One of their primary astronomy-related uses now is within cosmic ray detectors such as HESS and the Telescope Array Project (see Section 1.4) and neutrino detectors like Super Kamiokande (see Section 1.5) where, since PMTs can be made physically large, they can cover extensive areas relatively cheaply. A brief discussion of their principles of operation and properties is included here to cover these applications (fuller coverage of the devices may be found from sources in Appendix E).

Photomultipliers detect photons through the photoelectric effect. The basic components and construction of the device are shown in Figure 1.24. The photoemitter is coated onto the cathode and this is at a negative potential of some 1000 V. Once a photoelectron has escaped from the photoemitter, it is accelerated by an electric potential until it strikes a second electron emitter. The primary electron's energy then goes into pair production and secondary electrons are emitted from the substance in a manner analogous to photo-electron emission. Since typically 1 eV of energy is required for pair production and the primary's energy can reach 100 eV or more by the time of impact, several secondary electron emissions result from a single primary electron.

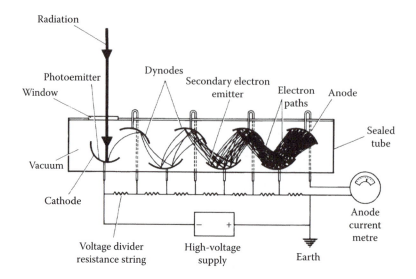

FIGURE 1.24   Schematic arrangement for a photomultiplier.

The secondary emitter is coated onto dynodes that are successively more positive than the cathode by 100 V or so for each stage. The various electrodes are shaped and positioned so that the electrons are channelled towards the correct next electrode in the sequence after each interaction. The final signal pulse may contain 10^6 electrons for each incoming photon and after arriving at the anode it may be further amplified and detected in any of the usual ways. This large intrinsic amplification of the photomultiplier is one of the major advantages of the device.

The microchannel plate (MCP) (see Section 1.3 and Figure 1.91) is closely related to the photomultiplier tube in terms of it operating principles and can also be used at optical wavelengths. With an array of anodes to collect the clouds of electrons emerging from the plate, it provides an imaging detector with a high degree of intrinsic amplification. Such devices are often referred to as multi-anode microchannel array detectors (MAMAs).

Noise in the signal from a photomultiplier arises from many sources. The amplification of each pulse can vary by as much as a factor of 10 through the sensitivity variations and changes in the number of secondary electrons lost between each stage. The final registration of the signal can be by analogue means and then the pulse strength variation is an important noise source. Alternatively, individual pulses may be counted and then it is less significant. Indeed, when using pulse counting, even the effects of other noise sources can be reduced by using a discriminator to eliminate the very large and very small pulses which do not originate in primary photon interactions. Unfortunately, however, pulse counting is limited in its usefulness to faint sources, otherwise the individual pulses start to overlap. Electrons can be emitted from either the primary or secondary electron emitters through processes other than the required ones and these electrons then contribute to the noise of the system. The most important of these processes are thermionic emission and radioactivity. Cosmic rays and other sources of energetic particles and gamma rays contribute to the noise in several ways. Their direct interactions with the cathode or dynodes early in the chain can expel electrons that are then amplified in the normal manner. Alternatively, electrons or ions may be produced from the residual gas or from other components of the structure of the photomultiplier. The most important interaction, however, generally is Ĉerenkov radiation produced in the window or directly in the cathode. Such Ĉerenkov pulses can be up to a hundred times more intense than a normal pulse. They can thus be easily discounted when using the photomultiplier in a pulse counting mode, but make a significant contribution to the signal when simply its overall strength is measured.

1.1.12 Superconducting Tunnel Junction Devices (STJs)

A possible replacement for the CCD in a few years, at least at well-equipped major observatories, is the STJ detector. The STJ can operate from the UV to long-wave infrared and also in the X-ray region, can detect individual photons, has a very rapid response and, perhaps most importantly, provides an intrinsic spectral resolution (Section 4.1) of around 500 or 1000 in the visible. Its operating principle is based upon a Josephson junction. This has two superconducting layers separated by a very thin insulating layer. Electrons are able to tunnel across the junction because they have a wave-like behaviour as well as a particle-like behaviour and so a current may flow across the junction despite the presence of the insulating layer. Within the superconductor, the lowest energy state for the electrons occurs when they link together to form Cooper pairs. The current flowing across the junction due to paired electrons can be suppressed by a magnetic field.

The STJ detector therefore comprises a Josephson junction based upon tantalum, hafnium, niobium, aluminium etc. placed within a magnetic field to suppress the current and having an electric field applied across it. It is cooled to about a tenth of the critical temperature of the superconductor – normally less than 1 K. A photon absorbed in the superconductor may split one of the Cooper pairs. This requires an energy of a millielectron-volt or so compared with about an electron volt for pair production in a CCD. Potentially therefore the STJ can detect photons with wavelengths up to a millimetre. Shorter wavelength photons will split many Cooper pairs, with the number split being dependent upon the energy of the photon, hence the device's intrinsic spectral resolution. The free electrons are able to tunnel across the junction under the influence of the electric field and produce a detectable burst of current.

STJ detectors and the arrays made from them are still very much under development at the time of writing, but have recently been tried on the William Herschel telescope (WHT). The ESA/European Space Research and Technology Centre's (ESA/ESTEC) S Cam 3 used a 10 × 12 array of 35-μm square STJs and although somewhat damaged made successful observations whilst on the WHT and also with the 1-metre Optical Ground Telescope. Later versions of S-Cam are under development at the time of writing using slightly different operating systems and with the aims of increasing the array sizes and of using simpler (closed-cycle helium) cryostats.

STJs are also used as heterodyne receivers in the 100 GHz to 1 THz frequency range (see Section 1.2) and in some dark matter detectors (see Section 1.7). They will also detect X-rays directly up to energies of several kiloelectron volts and arrays of up to 48 × 48 basic units have been considered as possibilities for some future X-ray spacecraft observatories. Outside astronomy they find applications within quantum computing, as voltage standards and as magnetometers.

1.1.13 Microwave Kinetic Inductance Detectors or Kinetic Inductance Detectors

A second device currently showing great future potential as a wide band detector and also based upon photon interactions within a superconductor is the microwave kinetic inductance detector (MKID) or kinetic inductance detector (KID). These detectors potentially operate from the sub-millimetre region to X-rays. The use of the word 'microwave' in one version of the name for the devices refers to the way in which they operate, not to the e-m radiation region that they detect.

A superconducting circuit has a non-zero impedance (resistance to an alternating current) which arises from the energy that is required to accelerate the electrons (in the form of Cooper pairs) in one direction during the 0° to 90° phase part of the AC cycle, to decelerate them to zero velocity during the 90° to 180° part of the cycle, accelerate them in the opposite direction during the 180° to 270° part and then to decelerate them again to zero velocity during the 270° to 360° (0°) part. This kinetic inductance increases as the density of the Cooper pairs of electrons decreases. A photon interacting within the superconductor will disrupt one or more Cooper pairs thus decreasing their density and so increasing the kinetic inductance. By combining the device with a capacitor, a resonant circuit is formed whose frequency lies in the microwave region (10^8 to 10^{11} Hz or so – hence the 'M' in MKIDs). The change in kinetic inductance lowers the resonant frequency and it is this change that is registered externally and provides the detection signal.

Like STJs, MKIDs can detect photons into the far infrared (FIR) and even microwave regions because the energy required to split the Cooper pairs of electrons is so small. The number of Cooper pairs split by an incoming photon will also be proportional to that photon's energy, giving the devices an intrinsic spectral resolution. Their quantum efficiencies are currently up to around 75%. Another major advantage of MKIDs is their relatively easy fabrication into large arrays. This arises because several hundred or more of the basic units can be constructed so that each has a slightly different resonant frequency. They can then all be read using a single output and fed to a single amplifier. Aluminium is commonly used as the superconductor material together with an absorber appropriate to the spectral region of interest (tin, for example, at long wavelengths). Also like STJs, MKIDs have to be operated at temperatures below 1 K so that their use is likely to be restricted to major observatories where suitable cryogenic facilities can be provided.

Recently, an array camera for optical to near-infrared spectrophotometry (ARCONS), a 32 × 32 pixel MKID array, was successfully tried out on the 5-metre Mount Palomar telescope. ARCONS covered the 400-nm to 1.1-μm region using MKIDs fabricated from titanium nitride. The intrinsic spectral resolution was around 10% and the arrival times of photons were recorded to better than 2 microseconds. MKIDs have also been used for astronomical observations at longer wave detections and potentially they may be developed for use into the UV region and even for X-ray detection.

1.1.14 Future Possibilities

As a general rule we may expect those detectors and techniques currently in use to be developed and improved (better quantum efficiency, wider spectral coverage, faster, bigger, or smaller if that is more desirable, cheaper, larger dynamic range, lower noise etc.) on time scales of a few years or so – and (without underplaying the hard and brilliant work that is needed for such developments) we may regard this process as being routine, or at least as normal. Equally and clearly, detectors and techniques that rely upon the invention or discovery of some radically new technical process or scientific principle cannot currently be speculated about. We are left therefore to look at the in-between possibilities.

Given that we already have detectors such as STJs and MKIDs that are capable of detecting individual photons with good quantum efficiencies over a very wide spectral range, with reasonable spectral and time resolutions, and perhaps reaching array sizes within a few years time comparable with those of present-day CCDs, our optical detectors are getting pretty close to the ideal detector conceivable for astronomy (if only they operated at room temperature and were a lot cheaper!). What, therefore, apart from (relatively) minor improvements in these various present-day attributes of optical detectors, could be demanded of an ideal optical astronomical detector? The answer is intrinsic sensitivities to

- The phase of the incoming radiation

- Its state of polarisation

- Radiation from 1000-km radio wavelengths to GeV γ-rays

The first two of these desirable attributes of these may not be too far distant – we have seen earlier that heterodyne detection (and hence information of the phase of the radiation) at frequencies of a few terahertz may become a reality within a decade or so. Given the normal way in which technologies, once discovered, then have their boundaries pushed to further and further limits, we may expect heterodyne detection to become possible at shorter and shorter wavelengths as time goes by. Visible light or UV heterodyne detectors, however, are probably still several decades away. Many crystals do absorb or reflect light that is polarised in one fashion better (or worse) than light polarised in a different fashion (see Section 1.3). Incorporating such materials into CCDs, STJs, TESs (see below) etc. could endow them with an intrinsic sensitivity to polarisation and perhaps this might be possible in years rather than decades.

A bit more speculatively, carbon nanotubes might be developable to act as antennas and waveguides for visible light (they are at least of roughly the right size). Then all the techniques, including the direct detection of phase and polarisation which are currently the tools of radio astronomers, would become available to optical astronomy.

Extreme wideband sensitivity, however, is probably a long way off if, indeed, it is ever possible. The reason for this pessimism is that although electromagnetic radiation has the same basic nature whether we consider the highest energy γ-rays or the longest wavelength radio waves, the ways in which that radiation interacts with matter *does* vary from one part of the spectrum to another. Thus (to a first approximation),

- Radio waves interact with matter by direct induction of electric currents in conductors

- FIR and millimetre waves interact with the vibrations and rotations of molecules

- NIR, visible and UV light interacts with the electrons in the outer levels of atoms and molecules

- Far UV light and soft-X-rays interact with the inner electrons of atoms

- Hard X-rays and γ-rays interact with the particles within atoms' nuclei

- The highest energy γ-rays interact with themselves (producing particle/anti-particle pairs of various subatomic particles)

It is thus most unlikely that high-quality* detectors capable of picking up (say) hard X-rays and NIR photons simultaneously and via the same physical process will be found, simply because the two varieties of e-m radiation interact with the material forming the detector in different ways.

1.1.15 Infrared Detectors

Many of the detectors just considered have some infrared sensitivity, especially out to 1 μm. However, at longer wavelengths, other types of detectors are needed, although if the STJ or MKID fulfils their promise, one or the other may replace some of these devices in the future.

In astronomy the infrared region is conventionally divided into three sections[†]; NIR, 0.7 to 5 μm (4.3 PHz–600 THz), mid-infrared (MIR), 5 to 30 μm (600–100 THz) and FIR, 30 to 1000 μm (100 THz–300 GHz). At the long-wavelength end of the FIR region, there is over-lap with the sub-millimetre region or as it is now quite frequently labelled, the terahertz region (radiation of frequency 1 THz has a wavelength of 300 μm) and so there is some duplication with the techniques considered in Section 1.2.

All infrared detectors need to be cooled,[‡] and the longer the operating wavelength, the lower the required temperature. Thus, in the NIR, liquid nitrogen (77 K) generally suffices, in the MIR, liquid helium (4 K) is needed, whilst in the FIR, temperatures down to 100 mK are used. Currently, there are two main types of infrared detector; the photoconductor

---

* Thermal detectors, such as thermocouples, can detect over very wide ranges of the spectrum provided only that they can absorb and so be heated by the e-m radiation that is involved. They are of low sensitivity but have in the past been used to cross-calibrate more sensitive detectors operating in different parts of the spectrum.

   Future developments in the superconducting detectors that operate by detecting photons via the disruption of Cooper pairs of electrons (STJs, MKIDs, SNSPDs and QCDs) may give the lie to this statement. However, it still seems likely that different absorbers would be needed for the devices when operating in different spectral regions even though the detect-ing mechanism may be the same in each case. So effectively, different detectors will still be used.

[†] Much narrower subdivisions are used within photometry (Section 3.1), especially for the NIR. There are thus seven defined photometric bands (I, Z, J, H, K, L and M) within the 780-nm to 4.75-μm region for the JCG photometric sys-tem alone. Also, there are many other conventions in use – for example in military and surveillance applications the definitions, short-wavelength IR (SWIR: 1–3 μm), middle-wavelength IR (MWIR: 3–5 μm), long wavelength IR (LWIR: 8–14 μm) and very long wavelength IR (VLWIR:14–30 μm) are likely to be encountered.

[‡] NASA's Spitzer infrared space observatory was launched in August 2003 and for almost 6 years its instruments were cooled to below 3 K using liquid helium, thus enabling observations to be made over the wavelength range from 3 to 180 μm. The helium coolant was exhausted in May 2009. The spacecraft then entered the 'warm' phase of its mission, observing just at 3.6 and 4.5 μm. This terminology however is misleading – the instruments are operating at 31 K – so 'warm' in this context is still decidedly chilly.

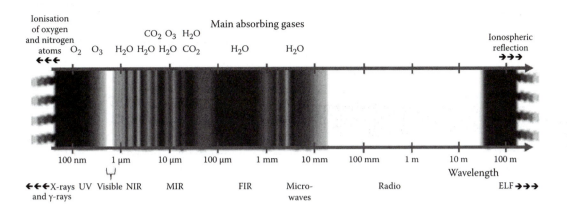

FIGURE 1.25   The transparent and partially transparent spectral regions (windows) of the Earth's atmosphere. The main, but not the only causes, of the opaque regions are listed at the top of the diagram. The spectrum goes off to both the left (X-rays and γ-rays) and right (extremely low frequencies [ELFs]), theoretically to infinity, but there are no further windows.

for the NIR and MIR and somewhat into the FIR and the bolometer for the FIR. As just discussed, various varieties of superconductor-based detectors seem likely to add a third strand to this list in the future.

The Earth's atmosphere is opaque over much of the infrared region, although there are narrow wavelength ranges (windows – Figure 1.25) where it becomes transparent to a greater or lesser degree. The windows can be enhanced by observing from high altitude, dry sites or by flying telescopes on balloons or aircraft. Nonetheless, the sky background can still be sufficiently high that images have to be read out several hundred times a second so that they do not saturate. Much of the observing therefore has to be done from spacecraft. Conventional reflecting optics can be used for the telescope and the instruments, though the longer wavelength means that lower surface accuracies are adequate. Refractive optics, including achromatic lenses, can be used in the NIR, using materials such as barium, lithium and strontium fluoride, zinc sulphate or selenide and infrared-transmitting glasses.

1.1.15.1 Photoconductive Cells

Photoconductive cells exhibit a change in conductivity with the intensity of their illumination. The mechanism for that change is the absorption of the radiation by the electrons in the valence band of a semiconductor and their consequent elevation to the conduction band. The conductivity therefore increases with increasing illumination and is monitored by a small bias current. There is a cut-off point determined by the minimum energy required to excite a valence electron over the band gap. A very wide variety of materials may be used, with widely differing sensitivities, cut-off wavelengths, operating temperatures etc. The semiconductor may be intrinsic, such as silicon, germanium, lead sulphide, indium antimonide or HgCdTe. For the NIR and the short wavelength end of the MIR, the HgCdTe alloy approaches being the ideal material for detectors. It is a mix of cadmium and mercury tellurides. Its sensitivity can be tailored to requirements over the range from 1 to 30 µm (300 to 10 THz) by adjusting the relative proportion of cadmium in the mix and it has a quantum efficiency, with anti-reflection coatings, of up to 90%. By removing the substrate after its manufacture the detector can be made simultaneously sensitive to visible and infrared radiation.

The band gaps in intrinsic semiconductors, however, tend to be large, restricting their use to the NIR. Doping of an intrinsic semiconductor produces an extrinsic semiconductor with the electrons or gaps from the doping atom occupying isolated levels within the band gap. These levels can be just above the top of the valence band, or close to the bottom of the conduction band so that much less energy is needed to excite electrons to or from them. Extrinsic semiconductors can therefore be made that are sensitive across most of the infrared. Doping is normally carried out during the melt stage of the formation of the material; however, this can lead to variable concentrations of the dopant and so to variable responses for the detectors. For germanium doped with gallium (Ge(Ga)), the most widely used detector material at wavelengths longer than 50 µm, extremely uniform doping has been achieved by exposing pure germanium to a flux of thermal neutrons in a nuclear reactor. Some of the germanium nuclei absorb a neutron and become radioactive. The $^{70}_{32}Ge$ nucleus transmutes to $^{71}_{32}Ge$ which in turn decays to $^{71}_{31}Ga$ via β decay. Arsenic, an n-type dopant, is also produced from $^{74}_{32}Ge$ during the process; however, only at 20% of the rate of production of the gallium. The process is known as neutron transmutation doping (NTD). The response of Ge(Ga) detectors may additionally be changed by applying pressure or by stressing the material along one of its crystal axes. The pressure is applied by a spring (Figure 1.28) and can change the detectivity range of the material (which is normally from ~40 to ~115 µm: 7.5 to 2.6 THz) to ~80 to ~240 µm (3.8 to 1.3 THz).

Ge(Ga) along with silicon doped with arsenic (Si(As)) or antimony (Si(Sb)) is also one of the materials used in the relatively recently developed blocked impurity band (BIB*) detectors. These use a thin layer of very heavily doped semiconductor to absorb the radiation. Such heavy doping would normally lead to high dark currents but a layer of undoped semiconductor blocks these. Ge(Ga) BIB detectors are sensitive out to about 180 µm (1.7 THz) and are twice as sensitive as normal stressed Ge(Ga) photoconductors around the 140 µm (2.1 THz) region. Further extension of the cut-off wavelength limit of these devices to 300 µm (1 THz) may soon be achievable using GaAs(Te), Ge(Sb) and/or Ge(Ga).

Photoconductive detectors that do not require the electrons to be excited all the way to the conduction band have been made using alternating layers of GaAs and indium gallium arsenide phosphide (InGaAsP) or aluminium gallium arsenide (AlGaAs), with each layer being only 10 or so atoms thick. The detectors are known as quantum well infrared photodetectors (QWIPs). The lower energy required to excite the electron gives the devices a wavelength sensitivity ranging from 1 to 12 µm. The sensitivity region is quite narrow and can be tuned by changing the proportions of the elements. Recently, National Aeronautics and Space Administration (NASA) has produced a broadband 1k × 1k QWIP with sensitivity from 8 to 12 µm by combining over a hundred different layers ranging

---

* Also known as Impurity Band Conduction (IBC) detectors.

TABLE 1.3   Materials Used for Infrared Photoconductors

Material	Cut-off Wavelength/Frequency or Wavelength/Frequency Range	
	μm	THz
Silicon (Si)	1.11	270
Germanium (Ge)	1.8	167
Gold-doped germanium (Ge(Au))	1–9	300–33
Mercury-cadmium-telluride (HgCdTe)	1–30	300–10
Gallium arsenide QWIPS (GaAs + InGaAsP or AlGaAs)	1–12	300–25
Lead sulphide (PbS)	3.5	86
Mercury-doped germanium (Ge(Hg))	4	75
Indium antimonide (InSb)	6.5	46
Copper-doped germanium (Ge(Cu))	6–30	50–10
Gallium-doped silicon (Si(Ga))	17	18
Arsenic-doped silicon BIB (Si(As))	23	13
Gallium-doped germanium (Ge(Ga))	~40–~115	~7.5–~2.6
Gallium-doped germanium stressed (Ge(Ga))	~80–~240	~3.8–~1.3
Boron-doped germanium (Ge(B))	120	2.5
Gallium-doped germanium BIB (Ge(Ga))	~180	~1.7
Antimony-doped germanium (Ge(Sb))	130	2.3

from 10 to 700 atoms thick. Quantum dot infrared photodetectors (QDIPs) have recently been developed wherein the well is replaced by a dot (i.e. a region that is confined in all spatial directions). It remains to be seen if QDIPs have any advantages for astronomy over QWIPs.

The superlattice has a similar structure to that of QWIPs with alternating layers of semiconductor materials each a few nanometres thick. In the type II superlattice the conduction and valence bands do not overlap so that the electrons and holes are trapped. These structures may be developed into infrared detectors suitable for astronomical applications in the future since they can have quantum efficiencies of up to 30%.

Details of some of the materials used for infrared photoconductors are listed in Table 1.3. Those in current widespread use are italicized.

1.1.15.2 Bolometers

A bolometer is simply a device that changes its electrical resistivity in response to heating by illuminating radiation. At its simplest, two strips of the material are used as arms of a Wheatstone bridge. When one is heated by the radiation its resistance changes and so the balance of the bridge alters. Two strips of the material are used to counteract the effect of slower environmental temperature changes, since they will both vary in the same manner under that influence.

Cooled semiconductor bolometers were once used as astronomical detectors throughout most of the infrared region. Photoconductive cells have now replaced them for NIR and MIR work, but they are still used for the FIR (~100 μm to a few mm, 3 THz to 100 GHz or so). Germanium doped with gallium (a p-type dopant) is widely used for the bolometer material with a metal-coated dielectric as the absorber. The bolometer is cooled to around 100 mK when in operation to reduce the thermal noise. Germanium doped with beryllium, silicon and silicon nitride are other possible bolometer materials.

The current state of the art uses an absorber that is a mesh of metallised silicon nitride like a spider's web, with a much smaller bolometer bonded to its centre. This minimises the noise produced by cosmic rays, but since the mesh size is much smaller than the operating wavelength, it still absorbs all the radiation. Arrays of bolometers up to 30 × 30 pixels in size can now be produced.

1.1.15.3 Other Types of Detectors

A recent development that promises to result in much larger arrays is the TES. These detectors are thin films of a superconductor, such as tungsten, held at their transition temperature from the superconducting to the normally conducting state. There is thus a very strong dependence of the resistivity upon temperature in this region. The absorption of a photon increases the temperature slightly and so increases the resistance. The resistance is monitored through a bias voltage. Their quantum efficiency can reach 95%. MKIDs, also based upon thin superconducting films, are already in use as astronomical infrared detectors and promise to find much wider application as their array sizes increase.

Two more superconducting detectors (in addition to STJs and MKIDs) whose operating principle is based upon the disruption of Cooper pairs are the superconducting nanowire single-photon detector (SNSPD) and the quantum capacitance detector (QCD). These devices have yet to be applied to astronomy, but may find such application in the future.

The basic element of the SNSPD is a tightly concertinaed niobium nitride wire whose dimensions are typically 10 nm × 100 nm × 100 μm. The wire is cooled to well below its critical temperature and has a dc bias current running through it that is just below the critical current. Absorption of a photon breaks some of the Cooper pairs of electrons, reducing the critical current to below the bias current. The resulting transition to a non-superconducting state is used to shunt the bias current to an amplifier and the ensuing voltage pulse indicates the absorption of the photon. Its speed of operation is higher than that of TES detectors and it has most of the other advantages of superconducting detectors.

In the QCD the electrons from Cooper pairs that have been disrupted by an absorbed photon tunnel through to a microwave resonator. They change the capacitance of the resonator and so also its resonant frequency. The read-out of the detection is similar to that in the MKID. The device's advantage over the MKID may be that of improved sensitivity.

For detection from the visual out to 30 μm (10 THz) or so, a solid-state photomultiplier can be used. This is closely related to the avalanche photodiode. It uses a layer of Si(As) on a layer of undoped silicon. A potential difference is applied across the device. An incident photon produces an electron-hole pair in the doped silicon layer. The electron drifts under the potential difference towards the undoped layer. As it approaches the latter, the increasing potential difference accelerates it until it can ionise further atoms. An electrode collects the resulting avalanche of electrons on the far side of the undoped silicon layer. A gain of 10^4 or so is possible with this device. It has found little application in astronomy to date.

A surprising recent development is the revival of the Golay cell as an FIR detector. The Golay cell comprises a chamber containing a gas. One wall of the chamber is a thin flexible membrane with a reflective coating. Radiation entering the chamber heats the gas, thus raising its pressure. The increased pressure causes the membrane to balloon outwards by a small amount. The membrane's movement is then detected by reflecting a beam of light off it. Golay cells were used as astronomical IR detectors in the early days but were quickly replaced by superior devices. Their modern application arises because they do not need cooling so they are being investigated as possible detectors of terahertz continuum radiation from solar flares for instruments on board balloons or spacecraft.

Platinum silicide acting as a Schottky diode can operate out to 5.6 μm. It is easy to fabricate into large arrays, but is of low sensitivity compared with photoconductors. Its main application is for terrestrial surveillance cameras.

Large* (2k × 2k pixel) arrays can now be produced for some of the NIR detectors and 1k × 1k pixel arrays for some MIR detectors, although at the long-wave end of the MIR, arrays are 256 × 256 at maximum. In the FIR, array sizes are limited to a maximum of around 100 × 100 pixels at the short-wave end and to a few tens of pixels (not always in rectangular arrangements) in total at the long-wave end. Unlike CCDs, infrared arrays are read out pixel by pixel. Although this complicates the connecting circuits, there are advantages; the pixels are read out non-destructively and so can be read out several times and the results averaged to reduce the noise, there is no cross talk (blooming) between the pixels and one bad pixel does not affect any of the others. The sensitive material is usually bonded to a silicon substrate that contains the read-out electronics.

A recent technological development may, however, lead to cheaper and larger arrays. Typically, a 2k × 2k IR array currently costs up to $500,000. One of the major constraints on building larger sizes is the mismatch between the properties of the silicon substrate and that of the IR detector materials – particularly the inter-atomic spacings in the crystals and their thermal expansion coefficients. Molecular beam epitaxy,† already a technique widely used in the semiconductor industry, is starting to be used to apply HgCdTe to a silicon wafer substrate. Since silicon wafers can come in sizes up to 300 mm, arrays up to 14k × 14k could potentially be produced within a few years and at much lower costs than the smaller arrays available today.

In the NIR, large format arrays have led to the abandonment of the once common practice of alternately observing the source and the background (chopping), but at longer wavelengths the sky is sufficiently bright that chopping is still needed. Chopping may be via a rotating 'windmill' whose blades reflect the background (say) onto the detector, whilst the gaps between the blades allow the radiation from the source to fall onto the

---

* For two or more decades the sizes of infrared arrays have been doubling every 18 to 20 months. This is similar to Moore's law growth in the capabilities of computers (named after Gordon Moore). However, the sizes of individual CCD arrays have not followed such a law in recent years and have perhaps reached a naturally useful size limit at around 4k × 4k (increasingly large mosaics, of course, *do* carry on with the Moore's law trend). Since some infrared arrays are now approaching that size, their growth may also soon slow down or come to a halt.

† A process whereby a layer of a different material is deposited onto a substrate within a vacuum chamber at a rate slow enough that the new layer adopts the crystalline structure of the substrate.

detector directly. Alternatively, some telescopes designed specifically for infrared work have secondary mirrors that can be oscillated to achieve this switching.

It might seem that the larger the array, the better. However, with limited resources (i.e. money!) this may not always be the case. The relative noise within the combined output of n detectors is reduced by a factor of $n^{1/2}$ compared with the relative noise from a single detector, but if the noise level from a single detector can be reduced, the improvement in the performance is directly proportional to the degree of that noise reduction. Thus, it may be better to spend the available money on a few good detectors rather than on lots of poorer detectors. Of course, lots of good detectors are even better. In the early 1990s, J. N. Bahcall attempted to quantify this concept with the astronomical capability of a system. This he defined as

$$\text{Astronomical capability} = \frac{\text{Lifetime} \times \text{Efficiency} \times \text{Number of Pixels}}{(\text{Sensitivity})^2} \qquad (1.4)$$

If used with some caution, this is a useful parameter for deciding between the different options that may be available during the early planning stages of a new instrument. For example the potential astronomical capability of the Mid-Infrared Instrument (MIRI) on the James Webb Space Telescope (JWST) is four to five magnitudes ($\times 10^4$ to $\times 10^5$) better than the capability of the Spitzer spacecraft's instruments.

1.1.15.4 Applications

There are many recent, current and possible future examples of the use of infrared detectors within astronomy. Some selected examples are briefly discussed below.

The JWST (currently due for launch in 2018) will carry two main instruments for NIR observations (and one for the MIR – see below). The NIR camera (NIRCAM) will use four 2k × 2k HgCdTe arrays for the range 600 nm to 2.3 μm and one 2k × 2k HgCdTe array for the 2.4- to 5.0-μm region. Whilst the NIR spectrograph ([NIRSPEC] – see Section 4.2) will use two 2k × 2k HgCdTe arrays for the 1- to 5.0-μm region (and will be able to extend this down to 600 nm at lower spectral resolution).

The Visible and Infrared Survey Telescope for Astronomy (VISTA), which had first light in 2009 (Figure 1.26), uses sixteen 2k × 2k HgCdTe arrays that are sensitive to the range 800 nm to 2.5 μm and has a field of view 1.65° across. The individual arrays are separated by large dead spaces so that six exposures with slight shifts between each are needed to give complete coverage of one area of the sky. VISTA's science programme until 2016 is to survey the whole of the southern sky in the NIR with selected areas being observed more intensively.

An instrument for ESO's VLT, the K-Band Multi-Object Spectrograph (KMOS) has recently achieved first light. It is an NIR multi-object, integral field spectrograph (see Section 4.1) covering the 1- to 2.5-μm region. The instrument uses three separate spectrographs, each of which employs a single 2k × 2k HgCdTe detector.

The first generation instrument, FLITECAM, for the 2.5-metre airborne Stratospheric Observatory for Infrared Astronomy (SOFIA) telescope uses a 1k × 1k indium antimonide

FIGURE 1.26 **(See color insert.)** ESO's 4.1-metre VISTA telescope. The large metal cylinder mounted at and projecting beyond the top ring of the telescope is the infrared camera containing the cryostat, baffles, filters and, of course, the 67.1-mega-pixel detector array. (Reproduced by kind permission of ESO.)

array to observe over the 1- to 5-μm region and this is currently being commissioned. The Wide-field Infrared Survey Explorer (WISE) spacecraft operated from December 2009 to February 2011 (when its hydrogen coolant ran out) observing at 3.4 and 4.6 μm using four 1k × 1k HgCdTe arrays. The Indian Chandrayaan-1 mission to the Moon that was launched in 2008 carried a low-resolution spectrograph that used a 480 × 640 substrate-removed HgCdTe array covering the region from 700 nm to 3.0 μm in order to study lunar mineralogy and to look for water (which it found). The Spitzer spacecraft, launched in August 2003 and now in its 'warm' phase, is still operating with an NIR camera using 256 × 256 indium antimonide arrays to observe at 3.6 and 4.5 μm.

QWIPs have been used for terrestrial and atmospheric remote sensing and for commercial and military purposes, but whilst a 256 × 256 QWIP array has been used on the 5-metre Hale telescope and others in the Search for Extraterrestrial Intelligence (SETI), they have found few other astronomical applications so far, perhaps because of their low quantum efficiencies (around 10%).

In the MIR, the JWST will carry MIRI. The MIRI imager will cover the 5- to 28-μm (60–11 THz) region using a 1k × 1k Si(As) BIB array. The MIRI spectrograph will cover the same spectral region with low and high spectral resolution options using two 1k × 1k Si(As) BIB arrays. The VLT Mid-Infrared Imager and Spectrometer (VISIR) which is used on ESO's VLT for imaging and spectroscopy between 8 to 13 μm (38–23 THz) and 16.5 to 24.5 μm (18–12 THz) has two 256 × 256 BIB detectors, whilst SOFIA's Faint Object Infrared Camera for the SOFIA Telescope (FORCAST) MIR detector uses 256 × 256 Si(As) and Si(Sb) BIB arrays to observe over the 5- to 40-μm (60–7.5 THz) range.

In the FIR, ESA's Herschel Space Observatory was launched in May 2009 to occupy the second Lagrangian point (L2) of the Earth–Sun system some 1.5 million km away from the Earth. It carried a 3.5-metre infrared telescope to observe in the FIR (60 to 670 μm, 5 THz–450 GHz) and was cooled to below 2 K using liquid helium. The coolant finally evaporated in April 2013. Three instruments were carried on board the spacecraft: the Photodetector Array Camera and Spectrometer (PACS), the Spectral and Photometric Imaging Receiver (SPIRE) and the Heterodyne Instrument for the Far Infrared (HIFI) (see Section 1.2).

PACS comprised two sub-instruments. The first was an imaging photometer operating over three bands from 60 to 210 μm (5–1.4 THz) and using 32 × 64 pixel and 16 × 32 pixel monolithic silicon bolometer arrays. The second was an integral field spectrograph (see Section 4.2) covering the 51- to 220-μm (5.9–1.4 THz) region with two 16 × 25 pixel Ge(Ga) photodetector arrays. An image slicer (see Section 4.2) was used to re-format the 5 × 5 pixel field of view to 1 × 25 pixels in order to match the arrays' shapes. SPIRE also comprised two sub-instruments. The first of these was again an imaging photometer. It observed bands centred at 250, 350 and 500 μm (1.2 THz, 860 and 600 GHz) using 9 × 15/16, 7 × 12/13 and 5 × 8/9 pixel hexagonally packed NTD spiderweb germanium bolometer arrays with feed horns for each pixel. The second instrument was a Fourier transform spectrometer (see Section 4.1) covering the range from 194 to 671 μm (1.5 THz to 450 GHz) and with 19-pixel and 37-pixel hexagonally shaped arrays of NTD spiderweb germanium bolometers.

The Large APEX Bolometer Camera (LABOCA) (Figure 1.27) on the 12-metre Atacama Pathfinder Experiment (APEX) telescope situated at the Atacama Large Millimeter/sub-millimeter Array (ALMA) (see Section 2.5) site on the Chajnantor plateau in Chile comprises 295 NTD germanium bolometers mounted on silicon-nitride membranes to observe around 870 μm (340 GHz).

SCUBA on the 15-metre James Clerk Maxwell Telescope (JCMT) was decommissioned in 2004. It operated over the 350–450 μm (860–670 GHz) and 750–850 μm (400–350 GHz) bands using NTD germanium bolometer arrays with 91 and 37 pixels, respectively. The individual detectors operated at a temperature of 100 mK and were fed by polished horns whose size matched the telescope's resolution. The current array, SCUBA-2, uses eight 32 × 40 pixel molybdenum/copper (Mo/Cu) TES arrays to observe at 450 μm (670 GHz) and 850 μm (350 GHz). Feed horns are not used, allowing closer spacing of the detectors.

FIGURE 1.27 The bolometer array for the LABOCA instrument on the APEX telescope. Left: the backs of the bolometers and the wiring assembly. Right: the feed horn array on the other side. (Reproduced by kind permission of Giorgio Siringo.)

POLAR-1 will be a 1.6-metre microwave telescope which is planned to start operating at the South Pole in 2013. It will follow on from the five smaller telescopes constituting the Keck array in studying the polarisation of the cosmic microwave background (CMB) radiation at a wavelength of 2 mm (150 GHz) using sixteen 288 pixel TES arrays.

One of the possible instruments for the planned Space Infrared Telescope for Cosmology and Astrophysics (SPICA) (with a 3.2-metre telescope and a possible launch date in 2017) is the Background-Limited Infrared-Submillimeter Spectrograph (BLISS). This is to cover the 35- to 430-μm (8.6 THz–700 GHz) band using 4200 TES detectors.

The Spitzer spacecraft, before running out of coolant, used 256 × 256 Si(As) BIB detectors for 5.8 and 8.0 μm, a 32 × 32 array of unstressed Ge(Ga) detectors to observe at 70 μm (4.3 THz) and a 2 × 20 array of stressed Ge(Ga) detectors to observe at 160 μm (1.9 THz; Figure 1.28).

A 24 × 24 array of MKIDs forms the heart of the Multiband Submillimeter Inductance Camera (MUSIC) used on the 10.4-metre Leighton telescope of the Caltech Submillimeter Observatory to observe simultaneously at 850 μm, 1.03 mm, 1.36 mm and 2 mm (350, 290, 220 and 150 GHz). APEX MKID (A-MKID) is an MKID-based camera under development for APEX. It is planned to observe at 350 and 860 μm (860 and 350 GHz) with arrays totalling 20,000 pixels and a possible start date in 2013. A prototype with 88 pixels has already been tried out on the telescope, achieving a sensitivity comparable with LABOCA. The Néel IRAM KID Arrays (NIKA) is similarly under development and has been tested on the 30-metre Institute de Radioastronomie Millimétrique (IRAM) telescope observing at 1.25 and 2 mm (240 and 150 GHz) with 16 × 16 and 12 × 12 pixel MKID arrays, respectively.

FIGURE 1.28 The 160-μm stressed Ge(Ga) detector array for Spitzer's Multiband Imaging Photometer (MIPS). Leaf springs apply 500 N to each of the 1-mm³ detectors in order to extend their response out to the required wavelength. (Reproduced by kind permission of G. Rieke.)

Bolometers may make a comeback at shorter wavelengths in the future as detectors for signals from spacecraft. Many spacecraft, particularly planetary probes, generate large quantities of data, which the current radio receivers and transmitters can send back to Earth only very slowly. Faster communications requires shorter wavelengths to be used and transmission speeds thousands of times of that possible in the radio region could be attained if the 1.55-μm infrared radiation commonly used to transmit broadband signals along optical fibres on the Earth was to be utilised. A recent development has a superconductor bolometer using a coil of extremely thin wire placed within a mirrored cavity (photon trap) that bounces unabsorbed photons back to the coil of wire to increase their chances of being absorbed. Just such a nano-wire detector, produced by the Massachusetts Institute of Technology (MIT), has now achieved a 57% detection efficiency, perhaps enabling lasers with power requirements low enough for them to be used on spacecraft to be employed for communications in the future.

1.1.16 UV Detectors

The UV spectral region is conventionally defined as extending from 10 to 400 nm (124 to 3 eV). The normal human eye can thus see slightly into the UV region since its short-wave limit is about 380 nm (Figure 1.5). Persons who have been born without a lens in the eye (aphakia) or who have lost it (for example through surgery for the treatment of cataracts) reportedly can see down to about 350 nm. The new 'colour' appears as white tinged with blue or violet.

There are a number of subdivisions of the UV which may be encountered. Whilst usage does vary, the definitions listed in Table 1.4 are widely employed.

Ground-based telescopes can observe down to a wavelength of about 300 to 340 nm depending upon altitude, state of the ozone layer and other atmospheric conditions (i.e. the UV-A and UV-B regions as they are popularly known). Wavelengths shorter than this can only be detected from high-flying aircraft or balloons or by using rockets and spacecraft.

As with detectors for infrared radiation, some of the detectors that we have already reviewed are intrinsically sensitive to short-wave radiation, although modification from their standard optical or infrared forms may be required. For example, photomultipliers will require suitable photoemitters and windows that are transparent to the required

TABLE 1.4   Regions Within the UV

Name	Abbreviation	Wavelength Range (nm)	Energy Range (eV)
Ultraviolet-A	UV-A	315–400	3.94–3.10
Near ultraviolet	NUV	300–400	4.14–3.10
Ultraviolet-B	UV-B	280–315	4.43–3.94
Middle ultraviolet	MUV	200–300	6.20–4.14
Ultraviolet-C	UV-C	100–280	12.40–4.43
Far ultraviolet	FUV	122–200	10.16–6.20
Vacuum ultraviolet	VUV	10–200	123.98–6.20
Extreme ultraviolet	EUV or XUV	10–121	123.98–10.25

wavelengths. Lithium fluoride and sapphire are common materials for such windows. Thinned, rear-illuminated CCDs have a moderate intrinsic sensitivity into the long-wave UV region. EBCCDs with an appropriate UV photocathode can also be used. At shorter wavelengths microchannel plates (see Section 1.3) take over. Detectors sensitive to the visible region as well as to the UV need filters to exclude the usually much more intense longer wavelengths. Unfortunately, the filters also absorb some of the UV radiation and can bring the overall quantum efficiency down to a few per cent. The term 'solar blind' is used for detectors or detector/filter combinations that are *only* UV-sensitive.

Another common method of short-wave detection is to use a standard detector for the visual region and to add a fluorescent or short-glow phosphorescent material to convert the radiation to longer wavelengths. Sodium salycylate and tetraphenyl butadiene are the most popular such substances since their emission spectra are well matched to standard photocathodes. Sensitivity down to 60 nm can be achieved in this manner. Additional conversions to longer wavelengths can be added for matching to CCD and other solid-state detectors whose sensitivities peak in the red and infrared. Ruby ($Al_2 O_3$) is suitable for this and its emission peaks near 700 nm. In this way, the sensitivity of CCDs can be extended down to about 200 nm.

Recently, a variant on the CCD, much used in cell-phone and web cameras, has found application as an astronomical UV detector. It is called the complementary metal-oxide-semiconductor – active pixel sensor (CMOS-APS). It has the same detection mechanism as the CCD, but the pixels are read out individually. CMOS-APS detectors are less power-hungry than CCDs, do not require mechanical shutters during the read-out process and are easier to harden against radiation. The dead spaces on CMOS-APS's though are larger than for CCDs because of the areas occupied by the read-out electronics. In other respects, including sensitising to the UV via fluorescent coatings, CMOS-APSs are similar to CCDs.

MCPs are discussed more fully in Section 1.3. Here, we just note that as well as being X-ray detectors, they can also be used as UV detectors for wavelengths shorter than about 200 nm (6.2 eV).

Many of the superconducting detectors discussed earlier (TESs, STJs, MKIDs, SNSPSs and QCDs) potentially have the ability to detect into the UV and X-ray regions and some of these are currently starting to be so used in the laboratory.

1.1.16.1 Applications

During the HST's 2009 servicing mission, the Cosmic Origins Spectrograph (COS) was installed to undertake spectroscopy over the 115- to 320-nm region (10.8 to 3.9 eV). It uses two windowless $1k \times 16\,k$ MCPs as the detectors for its far UV channel (115 to 205 nm, 10.8 to 6.0 eV) and a single $1k \times 1k$ caesium telluride MCP for its near UV (170 to 320 nm, 7.3 to 3.9 eV) channel. Also installed in 2009, the Wide Field Camera (WFC3) has two channels and covers the region from 200 nm to 1.7 μm. The UV and visible channel (200 nm to 1.0 μm) has two thinned and UV-optimised $2k \times 4k$ CCDs as its detectors.

ESA's microsatellite, Project for Onboard Autonomy (PROBA-2) launched in November 2009, carries an EUV imager to observe the lower solar corona at 17.4 nm (71 eV). Named Sun Watcher with Active Pixels and Image Processing (SWAP), the instrument uses a radiation-hardened CMOS-APS $1k \times 1k$ pixel array as its detector.

PROBA-2 was followed 3 months later by the launch of NASA's Solar Dynamics Observatory (SDO) carrying two UV instruments – the Atmospheric Imaging Assembly (AIA) and the Extreme Ultraviolet Variability Experiment (EVE).

AIA comprises four separate telescopes (Figure 1.29) and observes the Sun at nine UV wavelengths ranging from 9.4 to 170 nm (132 to 7.3 eV) as well as in white light. The telescopes are of Ritchey-Chrétien design with 0.2-metre diameter, normal incidence, primary mirrors coated with multiple alternating layers of silicon and molybdenum. The mirror surfaces are accurate to 0.3 nm (about 200 times better than a typical visual telescope). The primary mirrors are coated in two halves with the reflecting layers in each sector optimised for a different spectral region – effectively giving the instrument eight telescopes. There are filters at the front of each telescope to reject the visible and infrared radiation which are also in two halves. Five of these sectors have thin metallic aluminium filters, two have zirconium filters and one has a magnesium fluoride filter. Each telescope has a 4k × 4k pixel CCD array as its detector.

EVE observes the Sun from 0.1 to 105 nm (1240 to 11.8 eV). The UV component of this range is measured by the two channels of the Multiple EUV Grating Spectrograph (MEGS) instrument. The two channels both use 1k × 2k split-frame transfer CCDs as their detectors.

ESA/NASA's planned Solar Orbiter (SolO) spacecraft (scheduled for launch in 2017), will carry the Spectral Imaging of the Coronal Environment (SPICE) EUV spectrograph which will use two MCP intensified 1k × 1k APS arrays to cover the 70.2- to 79.2-nm (17.7 to 15.7 eV) and 97.2- to 105-nm (12.8 to 11.8 eV) spectral regions.

India's ASTROSAT is scheduled for launch in 2013 and will carry (amongst other instruments) two 0.375-metre Ritchey-Chrétien telescopes for UV and visual observations. The Ultraviolet Imaging Telescope (UVIT) instrument will have three channels covering the range from 130 to 530 nm and all three of these will have 512 × 512 pixel MCP intensified CMOS arrays as their detectors.

FIGURE 1.29 The AIA instrument for the SDA spacecraft during its assembly. The image shows the four 0.2-metre telescopes with their protective front covers in place. The CCD array detectors are at the far ends of the telescopes. (Reproduced by kind permission of NASA.)

For ground-based observers, ESO's X-shooter spectrograph, which started work in 2009, covers the 300-nm to 2.48-μm nm spectral region using three channels. The UVB channel (300 to 560 nm) employs a 2048 × 4102 pixel CCD as its detector.

1.1.17 Noise, Uncertainties, Errors, Precision and Accuracy

Noise is also often called the measurement uncertainty or error. The latter term implies that the investigator is somehow to blame for the problem. However, unless the investigator has made some sort of mistake (e.g. running a cooled detector at the wrong temperature) then the noise is usually intrinsic to the source and/or to the transmission path and/or to the measuring/detecting instrument. It may additionally be introduced or worsened during data reduction and analysis, but in that case it *is* an error and the reduction/analysis process should be improved.

In the absence of noise any detector would be capable of detecting any source, however, faint. Noise, however, is never absent and generally provides the major limitation on detector performance. A minimum signal-to-noise ratio of unity is required for reliable detection. However, most research work requires signal-to-noise ratios of at least 10 and preferably 100 or 1000.

Noise sources in photomultipliers and CCDs have already been mentioned and the noise for an unilluminated detector (dark signal) is a part of the definitions of DQE, NEP, D* and dynamic range (see Table 1.2). Now we must look at the nature of detector noise in more detail.

We may usefully separate noise sources into four main types: *intrinsic noise* (noise originating in the detector), *signal noise* (noise arising from the character of the incoming signal, particularly its quantum nature), *external noise* such as spurious signals from cosmic rays etc. and *processing noise*, arising from amplifiers etc. used to convert the signal from the detector into a usable form. We may generally assume processing noise to be negligible in any good detection system. Likewise, external noise sources should be reduced as far as possible by external measures. Thus, an infrared detector should be in a cooled enclosure and be allied to a cooled telescope to reduce thermal emission from its surroundings. In the NIR, recent developments in fibre optics hold out the hope of almost completely suppressing the sky background noise. Atmospheric emission from the hydroxide (OH) molecule forms about 98% of background noise in the NIR and between the lines, the sky is very clear. The lines, however, are numerous and closely spaced so that existing filters cannot separate the clear regions from the emission regions, but by manufacturing a fibre whose refractive index varies rapidly along its length in a sinusoidal manner a Bragg grating (see Section 1.3) can be formed. Such a grating produces a narrow absorption band at a wavelength dependent upon the spacing of the refractive index variations. The absorption band can thus be tuned to centre on one of the OH emission lines' wavelengths. By making the refractive index variations aperiodic many absorption bands can be produced, each aligned with an OH emission line. In this way the Australian Astronomical Observatory (AAO) has recently manufactured OH-suppressing infrared fibres covering 36 OH emission lines over the 1.5- to 1.57-μm region. If the potential of this technology can be extended, then NIR observing could become as easy as that in the visible. Similarly, a photomultiplier used in a photon-counting mode should employ a discriminator to eliminate the large pulses arising from Ĉerenkov radiation from cosmic rays and so on. Thus, we are left with intrinsic and signal noise to consider further.

1.1.17.1 Intrinsic Noise

Intrinsic noise in photomultipliers has already been discussed and arises from sources such as variation in photoemission efficiency over the photocathode, thermal electron emission from the photocathode and dynode chain etc. Noise in photographic emulsion, such as chemical fogging, is discussed later (see Section 2.2), whilst irradiation etc. in the eye was covered earlier.

In solid-state devices, intrinsic noise comes from four sources, described below.

Thermal noise, also known as Johnson or Nyquist noise (see Section 1.2), arises in any resistive material. It is due to the thermal motion of the charge carriers. These motions give rise to a current, whose mean value is zero, but which may have non-zero instantaneous values. The resulting fluctuating voltage is given by Equation 1.84.

Shot noise (also called quantum noise or the quantum limit) arises from the quantum nature of the signal (usually the signal is either a flow of electrons or of photons). It occurs, for example, in junction devices and is due to variation in the diffusion rates in the neutral zone of the junction because of random thermal motions. The general form of the shot noise current is

$$i = (2eI\Delta f + 4eI_0\Delta f) \tag{1.5}$$

where e is the charge on the electron, Δf is the measurement frequency bandwidth, I is the diode current and I_0 is the reverse bias or leakage current. When the detector is reverse-biased, this equation simplifies to

$$i = (2eI_0\Delta f) \tag{1.6}$$

g-r noise (generation-recombination) is caused by fluctuations in the rate of generation and recombination of thermal charge carriers, which in turn leads to fluctuations in the device's resistivity. *g-r noise* has a flat spectrum up to the inverse of the mean carrier lifetime and then decreases roughly with the square of the frequency.

Flicker noise, or 1/f noise, occurs when the signal is modulated in time, either because of its intrinsic variations or because it is being 'chopped' (i.e. the source and background or comparison standard are alternately observed). The mechanism of flicker noise is unclear but its amplitude follows an f^n power spectrum where f is the chopping frequency and n lies typically between 0.75 and 2.0. This noise source may obviously be minimised by increasing f. Furthermore, operating techniques such as phase-sensitive detection (see Section 3.2) are commonplace especially for infrared work, when the external noise may be many orders of magnitude larger than the desired signal. Modulation of the signal is therefore desirable or necessary on many occasions. However, the improvement of the detection by

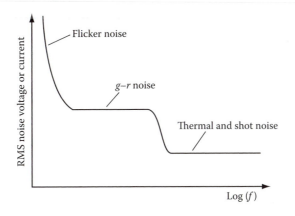

FIGURE 1.30 Relative contributions of various noise sources (schematic).

faster chopping is usually limited by the response time of the detector. Thus, an optimum chopping frequency normally needs to be found for each type of detector, which is about 100 Hz for bolometers and 1000 Hz or more for most photoconductive and photovoltaic cells.

The relative contributions of these noise sources are shown in Figure 1.30.

1.1.17.2 Signal Noise

Noise can be present in the signal for a variety of reasons. One obvious example is background noise. The source under observation will generally be superimposed upon a signal from the sky due to scattered terrestrial light sources, scattered starlight, diffuse galactic emission, zodiacal light, microwave background radiation etc. The usual practice is to reduce the importance of this noise by measuring the background and subtracting it from the main signal. Often the source and its background are observed in quick succession (chopping, see *flicker noise* and Section 3.2). Alternatively, there may only be measurements of the background at the beginning and end of an observing run. In either case, some noise will remain, due to fluctuations of the background signal about its mean level. This noise source also reduces as the resolution of the telescope improves. If the resolution is 1″, then a point source has to have an energy equal to that coming from a square second of arc of the background sky in order to have a signal-to-noise ratio of unity. But if the resolution were to be 0.1″, then the same point source would have a signal-to-noise ratio of 100 since it is only 'competing' with 0.01 square seconds of arc of the background. Since the light grasp of a telescope increases as the diameter squared, for *diffraction-limited* telescopes, the signal-to-noise ratio for point sources thus improves as D^4.

Noise also arises from the quantum nature of light. At low signal levels photons arrive at the detector sporadically. A Poisson distribution gives the probability of arrival and this has a standard deviation of $\sqrt{n}$ (where n is the mean number of photons per unit time). Thus, the signal will fluctuate about its mean value. In order to reduce the fluctuations to less than $x\%$, the signal must be integrated for $10^4/(nx^2)$ times the unit time. At high photon densities, photons tend to cluster more than a Poisson distribution would suggest because they are subject to Bose-Einstein statistics. This latter noise source may dominate at radio wavelengths (see Section 1.2), but is not normally of importance over the optical region.

1.1.17.3 Digitisation

Signals are digitised in two ways, signal strength and time. The effect of the first is obvious; there is an uncertainty (i.e. noise) in the measurement corresponding to plus or minus half the measurement resolution. The effect of sampling a time varying signal is more complex. The well-known sampling theorem (see Section 2.1) states that the highest frequency in a signal that can be determined is half the measurement frequency. Thus, if a signal is bandwidth-limited to some frequency, f, then it may be completely determined by sampling at $2f$ or higher frequencies. In fact sampling at higher than twice the limiting (or Nyquist) frequency is a waste of effort; no further information will be gained. However, in a non-bandwidth-limited signal, or a signal containing components above the Nyquist frequency, errors or noise will be introduced into the measurements through the effect of those higher frequencies. In this latter effect, or aliasing (see Section 1.2) as it is known, the beat frequencies between the sampling rate and the higher frequency components of the signal appear as spurious low-frequency components of the signal.

1.1.17.4 Errors and Uncertainties in Data Reduction, Analysis and Presentation

One source of error (not noise) arises if the equipment is in some way faulty, so that the results obtained from it are inaccurate. Clearly, equipment that is known to be faulty will not normally be used unless there is some overriding reason to do so. Thus, instruments on board a spacecraft may degrade over time through the effects of radiation but continue to be used because there is no alternative. In such circumstances it may be possible to compensate for the fault, perhaps by calibrating the instrument's later results against data obtained earlier, or comparing the faulty data with data from another information channel. An example of another type of common fault is a prism-based spectrograph. The dispersion of such a spectrograph is non-linear (see Section 4.1), but the non-linearity is anticipated in the design of the instrument through the provision of the comparison spectrum, which enables the correct wavelengths of lines in the spectrum still to be determined.

Faults of this type are called systematic errors. Where the error is well known and understood (as with the prism spectrograph), its effects may be corrected during data reduction. The systematic error does not then form a part of the uncertainty in a measurement. However, the correction of the systematic error itself contains uncertainties and these will contribute to the uncertainty in the actual measurement. Of more importance are systematic errors that are not well understood, or maybe even not known to exist. These errors will then contribute to the uncertainty in the measurement. The possibility of the presence of this latter type of error is the main reason why scientists seek to confirm measurements by a second approach that is independent of the manner in which the initial data were obtained.

The presence of unknown systematic errors in an instrument is one of many reasons why its measurements may be inaccurate even though they may also be precise. The difference between accuracy and precision is illustrated in Figure 1.31. An inaccurate but precise (or consistent) piece of apparatus always produces the same wrong result with little scatter in the measurements because the problem always affects the measurements in the same way. An accurate but imprecise instrument simply has large uncertainties in its results arising from random processes.

It is also possible for errors to arise through a faulty theoretical understanding of the situation. Ptolemy's geocentric model of the solar system for example, with its epicycles,

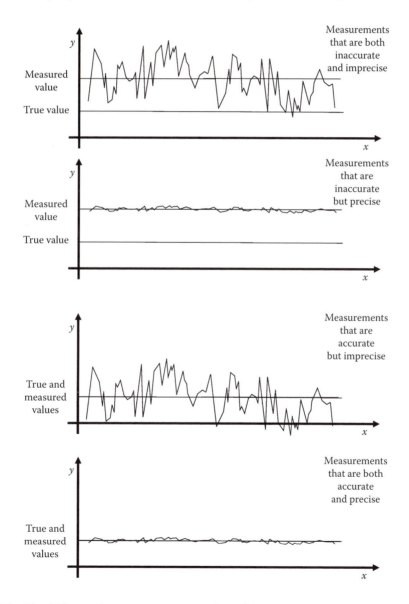

FIGURE 1.31   The difference between accuracy and precision.

deferents, eccentrics and equants, was the most successful scientific theory ever devised in that it continued to be accepted as correct and to make useful predictions of planetary positions for around one and a half millennia. Yet we now know that physically the idea was completely incorrect. This book, however, is (mostly) concerned with practical astronomy and so further consideration of theoretical misunderstandings will be left for the reader to pursue elsewhere.

In presenting results – whether they are simply raw data or the outcome from a complex and sophisticated piece of analysis and synthesis – the effects of noise/uncertainty/error on those results should always be included. Sometimes, especially in unique or unexpected situations such as the arrival of neutrinos from the 1987 LMC supernova, it may be difficult to quantify some or all the sources of uncertainty, but they should always be estimated or discussed. More usually, the uncertainties will be reasonably well understood and the measurement is then given in the form

$$x \pm \Delta x \tag{1.7}$$

where x is the result and Δx is its uncertainty.

When repeated measurements of the same unchanging phenomenon can be obtained, then their grouping often approximates towards a Gaussian (or normal) distribution. The Gaussian takes the form of a bell-shaped curve (Figure 1.32), which when normalised to have unit total area under the curve is given mathematically by

$$P(x) = -\frac{1}{\sigma\sqrt{2\pi}} e^{(x-\mu)^2/2\sigma^2} \tag{1.8}$$

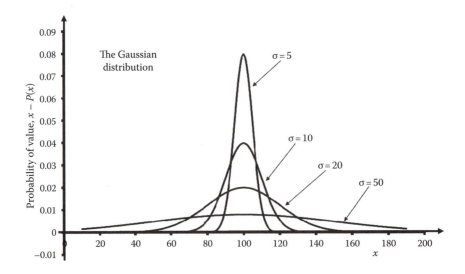

FIGURE 1.32   Gaussian distributions for a value of μ of 100 and for values of σ of 5, 10, 20 and 50.

TABLE 1.5   Areas Under the Gaussian Curve

x Lying between Values of	Relative Area under the Curve (%) between the Values of x	Relative Area under the Curve (%) outside the Values of x
$\mu \pm 0.5\sigma$	38.293	61.707
$\mu \pm 0.67449\sigma$	50.000	50.000
$\mu \pm \sigma$	68.269	31.731
$\mu \pm 1.5\sigma$	86.639	13.361
$\mu \pm 2\sigma$	95.450	4.5500
$\mu \pm 2.5\sigma$	98.758	1.2419
$\mu \pm 3\sigma$	99.730	2.6998×10^{-1}
$\mu \pm 4\sigma$	99.994	6.3343×10^{-3}
$\mu \pm 5\sigma$	100.000	5.7330×10^{-5}
$\mu \pm 6\sigma$	100.000	1.9731×10^{-7}
$\pm \infty$	100 (exact)	0 (exact)

where $P(x)$ is the probability of value, x, μ is the value of x at the peak of the distribution and σ is a constant discussed below.

The constant, σ, is called the standard deviation of the curve. From the mathematical definition of the Gaussian distribution, σ's value can be seen to equal the half-width of the curve between the points where the height of the curve is $e^{-0.5}$ ($\approx 60.7\%$) of its peak value. The areas under the curve between points equidistant from the curve's peak ($x = \mu$) on each side of that peak are listed in Table 1.5.

In a real scientific investigation, the number of independent measurements of a quantity will be limited and their actual distribution will only approximate towards the shape of a Gaussian. The standard deviation in these circumstances is given by[*]

$$\sigma = \sqrt{\dfrac{\sum\limits_{i=1}^{n}(\bar{x} - x_i)^2}{n-1}} \tag{1.9}$$

where n is the number of measurements, $\bar{x}$ is the mean value of the measurements and x_i is the ith measurement.

The interpretation of the standard deviation is also now approximate, but is likely to be based upon the values given in Table 1.5. The area under the curve between a given range of values of x gives the probability that x will be found between those values. Thus, we will usually find that about 68%[†] of the individual measurements will lie between $\bar{x} \pm \sigma$, about

---

[*] An approximation to this equation that is useful when the number of measurements is large replaces the $n - 1$ with 'n'. Many calculators and computers offer both versions. If working with small numbers of measurements then always ensure that you select the correct equation.

[†] The reader should always bear in mind that statistics is a house built on shifting sands. Thus, not only is the interpretation of the values in Table 1.5 approximate, but the value of σ obtained from Equation 1.9 has its own uncertainties. These are typically ±10% when $n \approx 100$, ±1% when $n \approx 10,000$ and ±0.1% when $n \approx 1,000,000$. Rarely, therefore, can the value of σ when based upon experiment measurements, justifiably be given to more than one significant Figure.

95% between $\bar{x} \pm 2\sigma$ and about 99% between $\bar{x} \pm 2.5\sigma$. Only about one measurement in two million might be expected to lie outside the range, $\bar{x} \pm 5\sigma$. These Figures form the basis by which results of experimental data are usually presented in refereed journals. The value of Δx in Equation 1.8 is usually taken as 2.5σ.* If this is not the case, then the definition of the uncertainty will need to be specified (and it is not bad practice always to specify the definition being used).

When a number of measurements have already been made and an additional one is obtained, then the significance of that new result may be examined by using the mean and standard deviation obtained from the previous set of results. If, for example, a variable star is being observed and a single new measurement of its brightness differs from the previously obtained mean value, then the question will arise, has the star really changed in brightness or is the difference just the result of the measurement uncertainties? If the new measurement differs from the mean of the old measurements by more than ±2σ, then the difference is conventionally termed to be significant, and if by more than ±2.5σ, then the difference is termed to be highly significant. The chances that the star has *not* changed in brightness are then around 5% and 1%, respectively. A result at the 5σ level (a chance that the star has *not* changed of around 0.00006%) is likely to be as close to being certain as any practising scientist can hope to achieve during his or her lifetime!

If a number of measurements of a phenomenon have been obtained, then it is usually the mean value of those measurements that is used in subsequent analysis or published as the result. The uncertainty in the mean value of a set of individual measurements is called the standard error of the mean and commonly symbolised as, *S*, with

$$S = \frac{\sigma}{\sqrt{n}} \tag{1.10}$$

and the result should be given in the form

$$\bar{x} \pm S \tag{1.11}$$

Equation 1.10 is the basis of the common assertion that the uncertainty in a result improves as the inverse-square-root of the number of measurements. However, this rule only applies when then uncertainties are random and approximate to a Gaussian distribution. There are other circumstances where the rule will be misleading; in particular, systematic errors are unlikely to improve much, if at all, with repeated measurements.

The practising researcher will need to understand far more about handling uncertainties in measurements than it is possible to include in this brief introduction. For example even

---

* Another measure of uncertainty is called the probable error and is given by 0.67449σ. It gives equal chances of a result lying inside or outside the range μ ± 0.67449σ. Its use may be encountered but it is not a preferred way of presenting results.

in the early years of an undergraduate science degree, the student will need to know how uncertainties propagate through formulae, how to compare two average values (Student's *t*-test), how to find the best (linear) formula to fit a set of measurements containing uncertainties (linear regression), whether the variations in two sets of data correlate with each other (correlation coefficient) and so on. However, that is left to be found in other sources (see Appendix E and the author's book *Telescopes and Techniques*).

1.1.18 Telescopes

The *apologia* for the inclusion of telescopes in the chapter entitled 'Detectors' has already been given, but it is perhaps just worth repeating that for astronomers the use of a detector in isolation is almost unknown. Some device to restrict the angular acceptance zone and possibly to increase the radiation flux as well, is almost invariably a necessity. In the optical region this generally implies the use of a telescope and so they are discussed here as a required ancillary to detectors.

1.1.18.1 Telescopes from the Beginning

The telescope may justly be regarded as the symbol of the astronomer, for notwithstanding the heroic work done with the naked eye in the past and the more recent advances in 'exotic' regions of the spectrum, our picture of the universe is still to a large extent based upon optical observations made through telescopes.

The invention of the telescope is nowadays usually attributed to a German spectacle maker, Hans Lipperhey (or Lippershey, 1570?–1619). Lipperhey settled during the late sixteenth century in Middelburg, the capital of Zeeland in the southwest of the Netherlands. Sometime during the summer of 1608 Lipperhey applied to the Zeeland government for a patent for "a certain device by means of which all things at a very great distance can be seen as if they were nearby." The Zeeland authorities sent Lipperhey to the States General (the national government) at The Hague to plead his case and provided him with a letter of support dated 25th September 1608 (Gregorian calendar). That letter is the first unequivocal record of a genuine and usable telescope. The application was discussed at the States General on 2nd October but was eventually rejected partly because the device was too simple to be kept secret and partly because a Dutchman, Jacob Metius (1571?–1628) of Alkmaar, applied for a patent on a telescope design just a few weeks after Lipperhey's request. Later it would emerge that another Middelburg spectacle maker, Zacharias Janssen (c. 1580–c. 1638) also had a design for or even an actual telescope at about the same time as Lipperhey's patent application.

Lipperhey's telescopes have not survived although he was commissioned to produce several, including binocular versions, by the Stadholder, Prince Maurits of Orange and by others. However, the telescopes were almost certainly of the design that we now call, rather unfairly, the Galilean refractor. This design uses a converging lens as the objective and a diverging lens as the eyepiece and produces an upright image. Lipperhey's telescopes had magnifications of around three or four times. Lipperhey's design and probably also those of Metius and Janssen were almost certainly the result of a serendipitous combination of actual lenses – indeed one account, though probably apocryphal, has the invention occurring whilst Lipperhey's children played with some of his spare lenses. The following year, however, Galileo Galilei (1564–1642), then living in Padua in Italy, heard of what Lipperhey's invention could do but not the details of its construction. His studies of optics enabled him to design a telescope from theoretical principles and then to manufacture the required lenses. He was soon constructing telescopes of the Galilean refractor design that magnified up to 30 times and these enabled him go on to make his epic astronomical observational discoveries such as lunar craters and the four major satellites of Jupiter.

How is it that within a period of just a few months in 1608/1609 the telescope was effectively invented four times? Essentially by the start of the seventeenth century the telescope was a device whose time had come and indeed it had possibly already been invented several more times during the preceding decades. If Lipperhey, Metius, Janssen and Galileo had not produced their instruments when they did, then someone else would have done so soon afterwards. The accidental discovery of the telescope requires a supply of lenses and/or curved mirrors with reasonable optical quality and a variety of focal lengths. The design of a telescope from theoretical principles requires a good understanding of the laws of optics. Both of these requirements had been achieved by the middle of the fifteenth century.

The earliest known example of a lens, made from natural quartz (rock crystal) and dated at around 640 BC, was found during excavations at Nineveh. Whilst the earliest written mention seems to be that of burning glasses (very short focal length lenses used to concentrate sunlight sufficiently to start fires) in Aristophanes' 424 BC play *The Clouds*, the Romans also knew of burning glasses. Furthermore, numerous examples made from quartz, possibly turned on pole lathes and dating from around the tenth century AD onwards, have been excavated from Viking graves in Gotland, Sweden.

Convex lenses for use as magnifiers came into use in the thirteenth century and had been incorporated into frames to make the first reading spectacles by around 1350. Spectacles using concave lenses to correct short sight followed about a century later. On the theoretical side, the Arabian mathematician Ibn Sahl (c. 940–1000) first used the law of refraction, now known as Snell's law, to design lenses in the late tenth century and wrote about the use of burning lenses and mirrors. Whilst Ibn al-Haitham (965–1039) explained the formation of images within the eye soon afterwards, Galileo's own work on optics added significantly to the understanding of the principles of optical devices.

Thus, by 1450 or soon afterwards the telescope was potentially capable of being invented; that it took another century and a half to be invented is the surprising aspect of the story, not that four people invented it quasi-simultaneously. The long delay may be due to suitable lenses not being available – the Galilean refractor for example requires a long focal length converging lens for the objective and a short focal length diverging lens for the eyepiece – or potential inventors may not have known of the earlier Arabian writings. However, it seems more likely that the delay is illusory and that the telescope was indeed invented several times during the centuries before 1608, but the resulting devices did not become widely known and the inventions were soon lost. Thus, Bishop Robert Grosseteste (c. 1175–1253) writing in *De Iride* (The Rainbow) says

"This part of optics, when well understood, shows us how we may make things a very long distance off appear as if placed very close and large near things appear very small and how we may make small things placed at a distance appear any size we want, so that it may be possible for us to read the smallest letters at incredible distances, or to count sand, or seed, or any sort of minute objects."

At least one reasonable interpretation of this text is that it is describing a telescope, used both normally and in reverse (looking through the objective) and perhaps also a compound microscope, the invention of which is usually attributed to Zacharias Janssen sometime between 1590 and 1609. In 1266 Roger Bacon (c. 1214–1294) published his *Opus Majus* (Great Work) and writes

"The wonders of refracted vision are still greater; for it is easily shown … that very large objects can be made to appear small and the reverse and very distant objects will seem very close at hand and conversely. For we can so shape transparent bodies and arrange them in such a way with respect to our sight and objects of vision, that the rays will be refracted and bent in any direction that we desire and under any angle we wish we shall see the object near or at a distance."

The *Opus Majus*, however, was a compendium of Bacon's knowledge rather than a report of his own work. Since Bacon certainly knew of Grosseteste's work, the similarity to Grosseteste's writing suggests that Bacon is reporting on that rather than describing a new invention of his own. A genuine independent invention of the telescope may be attributable to Leonard Digges (1520–1559). His son Thomas (1546–1595) writes in 1571 in *Pantometrica* that his father

"… sundry times hath by proportional glasses duly situate in convenient angles, not only discovered things far off, read letters, numbered pieces of money with the very coin and subscription thereof, cast by some of his friends of purpose upon downs in open fields, but also seven miles off declared what hath been done at that instant in private places."

One plausible interpretation of this (if it is not a gross exaggeration by Thomas) is as the description of some type of telescope, perhaps using a concave mirror as the eyepiece. Yet another possible inventor is Giambattista della Porta (c. 1535–1615). He is credited with the invention of the camera obscura and claimed also to have invented the telescope, but he died before he could give any account of his invention and his claim is now regarded with some scepticism.

Using the modern practice of attributing the credit for a discovery to the first person to publish a clear account of it within the public domain, Lipperhey is correctly identified as the inventor of the telescope, but only by the skin of his teeth and perhaps he was not genuinely the first person to make and use a telescope.

The Galilean design of telescope is difficult to manufacture with high magnifications. Since magnification in a simple telescope is just the focal length of the objective divided by the focal length of the eyepiece (Equation 1.71), high magnification requires a long focal length for the objective and/or a short focal length for the eyepiece. A long focal length for the objective makes the whole telescope long and unwieldy and the field of view (Equation 1.73) very small so that it becomes difficult to find and then to follow objects as they move across the sky. The negative lens required for the eyepiece must have very deep curves for its surfaces if it is to have a short focal length and these were hard to create with the techniques available four centuries ago.

A telescope design that was easier to construct was needed and just 3 years after Lipperhey's invention, Johannes Kepler (1571–1630) provided it, publishing in his 1611 book *Dioptrice* a design that used two converging lenses. The Keplerian telescope, or as it now more commonly known, the astronomical refractor, uses a long focal length converging lens for the objective and a short focal length converging lens for the eyepiece. The eyepiece is placed after the focal point of the objective, resulting in an inverted image. The inverted image is a major drawback if the telescope is to be used for everyday purposes and had such an instrument been accidentally put together, the way that Lipperhey's design may have been, then it would probably have been discarded as being of no use. However, it matters little whether we see the stars and planets in the same way that they appear in the sky to the unaided eye, or inverted. Kepler's design, therefore, being far easier to produce with high magnifications, eventually replaced the Galilean telescope for astronomical observing. The first to employ an astronomical refractor to observe the sky was probably Christoph Scheiner (1573–1650) sometime around 1613, although its use does not seem to have spread widely until after the publication of Scheiner's *Rosa Ursina* in 1630. The addition of a third lens, the relay or erector lens, to an astronomical refractor enables it to produce an upright image, resulting in the terrestrial telescope design, which for non-astronomical uses also quickly replaced the Galilean design. The Galilean design survives today only in the form of opera glasses, where high magnifications are not needed and the shorter length of the instrument compared with that of a comparable terrestrial telescope makes it more convenient to use.

Once invented, the astronomical refractor followed a pattern of development that has been repeated several times since – opticians successively manufacture bigger and better telescopes following the design until some flaw in or difficulty with the design makes further improvement impractical or too expensive. Additional progress then has to await a new design or technique and when that appears, the same pattern of development recurs. The optical telescope has passed through four of these major phases of development, each of which has caused a quantum jump in astronomical knowledge. We are now at the start of the fifth phase when very large diffraction-limited telescopes can be formed from smaller mirrors or segments with active control of their alignment etc. We may hope for a similar rapid development of our knowledge as these instruments come on stream in the next few decades.

The development of the astronomical refractor was limited by the optical aberrations of the simple lenses then in use, especially chromatic aberration (Figure 1.42). The effects of the aberrations could be reduced somewhat by minimising the curvatures of the surfaces of the lenses, but this meant long focal lengths, even for the eyepiece lens. Useful levels of magnification (×100 or so) thus required very long focal lengths for the objective. By 1656 Christiaan Huygens (1629–1695) had produced a telescope 7 metres long and Johannes Hevelius (1611–1687) had a 43 metre long telescope by the early 1670s. Such lengthy instruments were known as aerial telescopes and dispensed with the telescope tube. The lenses were simply mounted on the ends of a long pole, one end of which was attached to a tower and which could be moved up and down whilst the observer moved the other end (with the eyepiece) around the tower to acquire and follow whatever was being observed. For the very longest aerial telescopes even the pole was dispensed with and replaced by a cord linking the objective and eyepiece. The observer then had to keep the string in tension to align the two lenses whilst simultaneously performing his (at that time there were few if any female observers) ballet around the tower to find and track the object. As may be imagined observations with such instruments were exceedingly difficult and the slightest breath of wind made them impossible. Nonetheless Giovanni Domenico Cassini (1625–1712) made several significant discoveries using aerial telescopes, including in 1675 that of 'his' division (or gap) in Saturn's rings.

Most observers, however, did not have Cassini's skill or patience so it was fortunate that as the astronomical refractor in the form of the aerial telescope was reaching the practical limit of the first phase of its development, several new designs appeared. These new designs all employed a mirror for the telescope objective. Thus, at one stroke the effects of chromatic aberration were eliminated because mirrors reflect all wavelengths equally (although lenses continued to be used for the eyepieces, so chromatic aberration did not disappear entirely). The mirror forming the objective reflects the light back towards the direction from which it came, so the eyepiece cannot be simply be placed behind the objective's focal point because the observer's head would then block the incoming light. Mirror-based (or reflecting) telescope designs thus have to use a second mirror to reflect the light from the objective (primary) mirror to an accessible point. The secondary mirror is much smaller than the primary and so only obstructs a small and acceptable proportion of the light gathered by the primary.

The French polymath, Father Marin Mersenne* (1588–1648), came within a hair's-breadth of inventing reflecting telescopes in 1637 with the publication of his *L' Harmonie Universelle*. In that book he described two optical systems which we would now call beam compressors (used in many modern instruments). The two systems were basically Gregorian and Cassegrain telescopes (see below) with the secondary mirrors adjusted so that the light beam exiting through the hole in the primary mirror was a parallel beam of light, not a focussed one. In effect the designs are reflecting telescopes using the secondary mirrors as the eyepieces (cf. Leonard Digges' possible instrument, above). Had Mersenne constructed instruments to either of his designs and looked through them (without needing to use a lens-based eyepiece), he would have been able to see distant objects appearing to be much closer, just as with any other telescope. However, there is no evidence that Mersenne ever

---

* Best known for his work on acoustics and music and for the Mersenne prime numbers.

attempted to construct real instruments to his designs. It is also possible that the Milanese mathematician Bonaventura Cavalieri (1598–1647) may have been describing Cassegrain and Newtonian type reflecting telescope designs in his 1632 book *Lo Specchio Ustorio* which was concerned with the theory of mirrors of parabolic, elliptical and hyperbolic shapes, but again no practical instruments resulted from his work.

Chronologically, therefore, the first true mirror-based (or reflecting) telescope design was devised by the Scottish mathematician, James Gregory (1638–1675) and published in his *Optica Promota* in 1663. His design used a concave paraboloidal primary mirror (the paraboloidal shape eliminates spherical aberration, Figure 1.32) and a concave ellipsoidal mirror placed beyond the primary's focal point as the secondary mirror. The secondary reflected the light through a hole in the centre of the primary mirror to the eyepiece placed behind that mirror. Gregory had a prototype instrument constructed but it never seems to have worked satisfactorily. The failure of the telescope is perhaps not surprising since there was at that time no way of ensuring that the two mirrors were of the correct shape, and another century would pass before paraboloidal mirrors could be made reliably.

The first person thus to design *and* build (or have built) a functioning reflecting telescope was Sir Isaac Newton (1642–1727) who showed a working instrument to the Royal Society in 1671. It had a 2-inch (50 mm) primary mirror, was some 6 inches (150 mm) in length and magnified by about 25 times (Figure 1.33). It was more powerful, as Newton boasted, than a refracting telescope 10 or 12 times larger (large at this date meaning the length). Newton had actually first constructed a reflecting telescope in 1668 and he avoided the necessity for a second curved mirror and a hole in the primary mirror required for the Gregorian telescope by making the secondary mirror flat and angled at 45° to the incoming

FIGURE 1.33 A full-size replica of Newton's first telescope. It was to be found facing the visitors' gallery of the 2.5-metre Isaac Newton telescope during the 1970s when that instrument was still in the United Kingdom. At the bottom of the case (and not visible to the visitors) can be seen the instruction 'When main telescope fails – Break glass'. (Copyright 1977 C. R. Kitchin.)

light. The light reflected from the primary mirror was thus reflected to the side of the telescope by the secondary and thence into the eyepiece (Figure 1.59). The simplicity of the design of the Newtonian telescope made it very popular and it is still widely used today for small to medium-sized instruments, especially by amateur astronomers.

In 1672, yet a third design appeared. It is very similar to the Gregorian telescope except that the secondary mirror is convex and hyperboloidal and positioned before the primary mirror's focus (Figure 1.53). The design is known as the Cassegrain telescope after its inventor although very little is known about him beyond his surname. His nationality was French and his first name is variously suggested to be Laurent, Guillaume, Jacques or Nicolas and his profession either a Catholic priest and teacher or a sculptor and metal founder. He may have been born in 1625 or 1629 and died in 1693 or 1712 – or quite possibly none of those years.

The optician, James Short (1710–1768), who also produced Gregorian and Newtonian telescopes, probably made the first working telescopes to Cassegrain's design during the mid-1700s. A surviving example of Short's work is a 9-inch (225 mm) diameter reflector manufactured around 1767/1768 that has several secondary mirrors enabling it to be used in all three modes at magnifications up to ×400. It is a great pity that so little is known of Cassegrain since his design, far more so than Newton's, dominates modern observational astronomy. The Cassegrain design and its variants and extensions such as the Ritchey-Chretién, the Coudé (Figure 1.58) and the Nasmyth are used for every major and many smaller telescopes. Even the very popular small Schmidt-Cassegrain and Maksutov (Figure 1.64) telescopes, mostly sold for amateur use, are just a Cassegrain design with the addition of a thin correcting lens. Large telescopes (over 2 metres or so in diameter) are often used at prime focus (i.e. without a secondary mirror) but in almost all cases can also be converted to a Cassegrain mode of operation. The popularity of the Cassegrain design arises primarily from the way in which the secondary mirror expands the cone of light from the primary mirror, enabling the effective focal length (and so the magnification or image scale) to be several times the instrument's actual physical size. This size reduction reduces the construction costs of both the telescope and its dome by as much as a factor of 10 or more.

Apart from the difficulty of manufacturing the required ellipsoidal, paraboloidal or hyperboloidal surfaces for the mirrors, the early reflecting telescope suffered from another problem. The mirrors had to be made from speculum metal, an alloy with around three parts of copper to one part of tin. Even when freshly polished, speculum metal has a reflectivity of only around 60%. Thus, after the two reflections required in the Cassegrain, Newtonian and Gregorian designs, only a little better than a third of the light entering the telescope will be delivered to the eye. Furthermore, speculum metal tarnishes and requires regular re-polishing, but that process could and very likely did change the shape of the surface of the mirror so that the quality of the images deteriorated each time unless care was taken to ensure that the correct surface shapes were maintained.

Despite these problems, the early reflecting telescopes underwent the same 'telescope race' experienced by the early refractors, rapidly becoming larger, longer and with higher and higher magnifications. By 1789, Sir William Herschel (1738–1822) had produced a 1.2-metre diameter reflector with a primary mirror weighing nearly 1000 kg. Herschel's telescopes did not use any of the designs mentioned so far, but to avoid the loss of light when the secondary mirror was used, he tilted the primary so that he could look directly at the image from the top of the telescope without his head getting in the way. Unfortunately, unless an off-axis-paraboloid shape can be produced for the primary mirror (as happens today with many TV satellite dish receivers and the components of segmented mirrors; see Figure 1.53), the quality of the image seen in this manner is severely degraded by aberrations. Herschel, however, accepted this drawback for the sake of the improved brightness of the images. Reflecting telescopes using speculum metal mirrors reached their pinnacle with the construction in 1845 of the Leviathan of Parsonstown. William Parsons (Lord Rosse, 1800–1867) had the telescope constructed within the grounds of his castle in central Ireland with a primary mirror 1.8 metres in diameter and weighing around 4000 kg. Supported between substantial parallel walls of masonry, the telescope made one major discovery – that some of the nebulous objects seen in the sky were spiral in shape – the first hint of the existence of galaxies outside the Milky Way.

At about the same time as speculum-metal reflecting telescopes were being developed, a lens that was far less affected by chromatic aberration was invented by Chester Moor Hall (1703–1771). In 1729, Hall suggested that combining two lenses made from different types of glass would reduce the colour problems arising from simple lenses. The achromatic lens (Figure 1.43) was first manufactured commercially by John Dollond (1706–1761) in 1758 and reduced the chromatic aberration compared with that of a simple lens by about a factor of 20. Dollond and his successors* went on to construct many refractors using achromatic lenses for their objectives and other opticians soon followed. The achromatic refractor could be of a reasonable length and was quickly adopted by the general public as well as by astronomers – as it still is. The achromatic refractor followed the development pattern of other types of telescopes culminating in 1897 with the 1-metre Yerkes telescope. Refractors larger than this were never built successfully† because lenses can only be supported at their edges and their weight leads to distortions, ruining their optical quality (mirrors can be supported on their backs as well as around their edges and so their distortions can be kept within bounds).

By the end of the nineteenth century both achromatic refractors and speculum-metal mirror reflectors had thus reached the limits of practical development. Fortunately, a technological breakthrough came to the rescue. Glass mirrors with a reflective backing of a tin and mercury amalgam had been produced for domestic use in Venice from the early sixteenth century onwards. The chemical process required to deposit a thin layer of silver onto glass was discovered by Justus von Liebig (1803–1873) in 1835 and is known within chemistry as the silver mirror test. In 1857, Léon Foucault started manufacturing telescope mirrors from glass and used Liebig's process to put a reflective coating of silver onto their front surfaces. The silver coating was a far better reflector than speculum metal, but like the latter it quickly tarnished. However, the great advantage of silvered mirrors was that the silver could be simply removed chemically without changing the shape of the surface of

---

* A firm of opticians incorporating 'Dollond' in their name survived in the United Kingdom until the early part of the twenty-first century.

† A refractor with a lens 1.25 metres in diameter was exhibited at the Paris Universal Exhibition of 1900, but it was never used.

FIGURE 1.34 (a) The 8.2-metre monolithic primary mirror of Yepun, one of the four main instruments comprising ESO's VLT. (Reproduced by kind permission of ESO/G. Huedepohl.) (b) The 10-metre mirror of the Keck II telescope showing its component hexagonal segments. (Reproduced by kind permission of L. Hatch. © 2007 Laurie Hatch. See: www.lauriehatch.com for many more of her beautiful astro-images.)

the mirror, as re-polishing a speculum metal mirror was liable to do. Furthermore, a fresh, untarnished coating of silver could be placed onto the glass with just an hour or so's work – and the mirror would be as good as new. Thus, the era of metal-on-glass reflecting telescopes began. Today aluminium is normally used as the reflecting metal and it is deposited onto a low-expansion ceramic or quartz substrate by evaporation inside a vacuum chamber, but the principle is unchanged. Casting large blanks of material to make such mirrors is not easy but throughout the twentieth century opticians entered the telescope race yet again with a will. The practical limit of development has probably been reached for monolithic mirrors with ESO's VLT and its four 8.2-metre diameter primary mirrors (Figure 1.34), the two Gemini 8.1-metre telescopes and the Large Binocular Telescope (LBT's) two 8.4-metre mirrors.

Writing during the first years of the twenty-first century, we are at the start of the fifth tele-scope race. Just when monolithic mirrors have reached their limits, technology generally and especially information technology have developed to the point where larger mirrors can be built by putting many smaller mirrors together. Multi-mirror telescopes like the Keck instru-ments with their 10-metre mirrors each made up from thirty-six 1.8-metre hexagonal smaller mirrors (Figure 1.34) and the 91 component mirrors of the 11 × 10-metre Hobby-Eberly tele-scope represent the starting points for the next great leap forward. Since these telescopes and related developments such as real-time atmospheric compensation are covered later in this chapter, their further discussion will be left until then except to ask where will the race end this time? With plans for 30-metre and larger multi-mirror instruments already in the initial funding stages and with more speculative suggestions for instruments up to 100 metres in diameter, we can note that the largest fully-steerable *radio* dishes are only just over 100 metres in size. Their sizes are limited by gravitational and wind stresses and of course by costs. Such considerations would apply even more to 100-metre optical telescopes, so they may be the end of the line for multi-mirror instruments. Anything even more powerful will have to await fur-ther, as yet unknown, developments in designs and/or technology, although radio astronomy's Square Kilometre Array (SKA) (see Section 1.2) may hint at one possible way forward.

1.1.18.2 Optical Theory

Before looking further at the designs of telescopes and their properties it is necessary to cover some of the optical principles upon which they are based. It is assumed that the reader is familiar with the basics of the optics of lenses and mirrors; such terminology as focal ratio, magnification, optical axis etc. with laws such as Snell's law, the law of reflection and so on, and with equations such as the lens and mirror formulae and the lens-maker's formula. If it is required, however, any basic optics book or reasonably comprehensive physics book will suffice to provide this background.

We start by considering the resolution of a lens (the same considerations apply to mir-rors, but the light paths etc. are more easily pictured for lenses). That this is limited at all is due to the wave nature of light. As light passes through any opening it is diffracted and the wavefronts spread out in a shape given by the envelope of the Huygens' secondary wavelets (Figure 1.35). Huygens' secondary wavelets radiate spherically outwards from all points on a wavefront with the velocity of light in the medium concerned. Three examples are shown in Figure 1.35. Imaging the wavefront after passage through a slit-shaped aperture produces an image whose structure is shown in Figure 1.36.

The fringes shown in Figure 1.36 are due to interference between waves originating from different parts of the aperture. The paths taken by such waves to arrive at a given point will differ and so the distances that they have travelled will also differ. The waves will be out of step with each other to a greater or lesser extent depending upon the magnitude of this path difference. When the path difference is a half wavelength or an integer number of wavelengths plus a half wavelength, then the waves will be 180° out of phase with each other and so will cancel out. When the path difference is a whole number of wavelengths, the waves will be in step and will reinforce each other. Other path differences will cause intermediate degrees of cancellation or reinforcement.

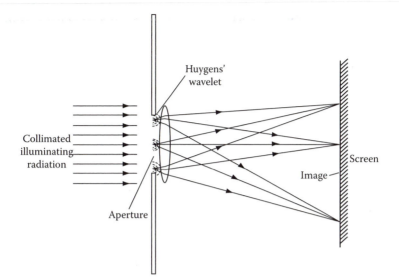

FIGURE 1.35   Fraunhofer diffraction at an aperture.

FIGURE 1.36   Image of a narrow slit.

The central maximum arises from the reinforcement of many waves. Whilst the first minimum occurs when the path difference for waves originating at opposite edges of the aperture is one wavelength, then for every point in one half of the aperture there is a point in the other half such that their path difference is half a wavelength and all the waves cancel out completely. The intensity at a point within the image of a narrow slit may be obtained from the result that the diffraction pattern of an aperture is the power spectrum of the Fourier transform of its shape and is given by

$$I_\theta = I_0 \frac{\sin^2(\pi d \sin\theta/\lambda)}{(\pi d \sin\theta/\lambda)^2} \qquad (1.12)$$

where θ is the angle to the normal from the slit, d is the slit width and I_0 and I_θ are the intensities within the image on the normal and at an angle θ to the normal from the slit, respectively. With the image focused onto a screen at distance F from the lens, the image structure is as shown in Figure 1.37, where d is assumed to be large when compared with the wavelength λ. For a rectangular aperture with dimensions $d \times l$, the image intensity is similarly

$$I(\theta,\phi)=I_0\frac{\sin^2(\pi d\sin\theta/\lambda)}{(\pi d\sin\theta/\lambda)^2}\frac{\sin^2(\pi l\sin\phi/\lambda)}{(\pi l\sin\phi/\lambda)^2} \tag{1.13}$$

where ϕ is the angle to the normal from the slit measured in the plane containing the side of length l. To obtain the image structure for a circular aperture, which is the case normally of interest in astronomy, we must integrate over the surface of the aperture. The image is then circular with concentric light and dark fringes. The central maximum is known as Airy's disc after George Biddell Airy, the Astronomer Royal who first succeeded in completing the integration. The major difference from the previous case occurs in the position of the fringes.

Consider a circular aperture of radius r, illuminated by a beam of light normal to its plane (Figure 1.38) and consider also the light which is diffracted at an angle θ to the normal to the aperture from a point P, whose cylindrical coordinates with respect to the centre of the aperture C and the line AB which is drawn through C parallel to the plane containing the incident and diffracted rays, are (ϕ, ρ). The path difference, Δ, between diffracted rays from P and A is then

$$\Delta = (r - \rho \cos \phi) \sin \theta \tag{1.14}$$

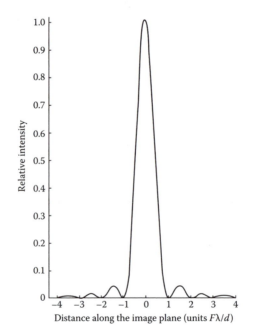

FIGURE 1.37 Cross section through the image of a narrow slit.

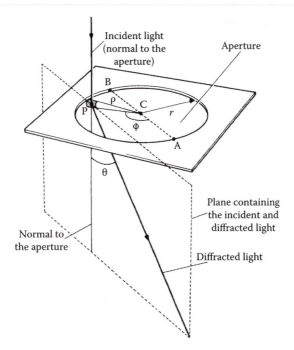

FIGURE 1.38 Diffraction by a circular aperture.

and the phase difference is

$$\frac{2\pi\Delta}{\lambda} = \frac{2\pi}{\lambda}(r - \rho\cos\phi)\sin\theta \tag{1.15}$$

The elemental area at P is just

$$dA = \rho\,d\phi\,d\rho \tag{1.16}$$

So the contribution to the electric vector of the radiation in the image plane by the elemental area around P for an angle θ to the normal is proportional to

$$\sin\left(\omega t + (2\pi/\lambda)(r - \rho\cos\phi)\sin\theta\right)\rho\,d\phi\,d\rho \tag{1.17}$$

where $\omega/2\pi$ is the frequency of the radiation and the net effect is obtained by integrating over the aperture

$$\int_0^{2\pi}\int_0^r \sin\left[\omega t + \left(\frac{2\pi r\sin\theta}{\lambda}\right) - \left(\frac{2\pi\rho\cos\phi\sin\theta}{\lambda}\right)\right]\rho\,d\rho\,d\phi$$

$$= \sin\left(\omega t + \frac{2\pi r\sin\theta}{\lambda}\right)\int_0^{2\pi}\int_0^r \rho\cos\left(\frac{2\pi\rho\cos\phi\sin\theta}{\lambda}\right)d\rho\,d\phi \tag{1.18}$$

$$- \cos\left(\omega t + \frac{2\pi r\sin\theta}{\lambda}\right)\int_0^{2\pi}\int_0^r \rho\sin\left(\frac{2\pi\rho\cos\phi\sin\theta}{\lambda}\right)d\rho\,d\phi$$

The second integral on the right-hand side is zero, which we may see by substituting

$$s = \frac{2\pi\rho\cos\phi\sin\theta}{\lambda} \tag{1.19}$$

so that

$$\cos\left(\omega t + \frac{2\pi r\sin\theta}{\lambda}\right) \int_0^{2\pi}\int_0^r \rho\sin\left(\frac{2\pi\rho\cos\phi\sin\theta}{\lambda}\right) d\rho\, d\phi$$

$$= \cos\left(\omega t + \frac{2\pi r\sin\theta}{\lambda}\right)\int_0^r \rho \int_{s=2\pi\rho\sin\theta/\lambda}^{s=2\pi\rho\sin\theta/\lambda} \frac{-\sin s}{\left[(4\pi^2\rho^2\sin^2\theta/\lambda^2)-s^2\right]^{1/2}} ds\, d\phi \tag{1.20}$$

$$= 0 \tag{1.21}$$

since the upper and lower limits of the integration with respect to s are identical. Thus, we have the image intensity $I(\theta)$ in the direction θ to the normal

$$I(\theta) \propto \left[\sin\left(\omega t + \frac{2\pi r\sin\theta}{\lambda}\right)\int_0^{2\pi}\int_0^r \rho\cos\left(\frac{2\pi\rho\cos\phi\sin\theta}{\lambda}\right) d\rho\, d\phi\right]^2 \tag{1.22}$$

$$\propto \left(r^2\int_0^{2\pi} \frac{\sin(2m\cos\phi)}{2m\cos\phi} d\phi - \frac{1}{2}r^2\int_0^{2\pi} \frac{\sin^2(m\cos\phi)}{(m\cos\phi)^2} d\phi\right)^2 \tag{1.23}$$

where

$$m = \frac{\pi r\sin\theta}{\lambda} \tag{1.24}$$

Now

$$\frac{\sin(2m\cos\phi)}{2m\cos\phi} = 1 - \frac{(2m\cos\phi)^2}{3!} + \frac{(2m\cos\phi)^4}{5!} - \cdots \tag{1.25}$$

and

$$\frac{\sin^2(m\cos\phi)}{(m\cos\phi)^2} = 1 - \frac{2^3(m\cos\phi)^2}{4!} + \frac{2^5(m\cos\phi)^4}{6!} - \cdots \tag{1.26}$$

so that

$$I(\theta) \propto \left(r^2 \sum_{0}^{\infty} (-1)^n \int_{0}^{2\pi} \frac{(2m\cos\phi)^{2n}}{(2n+1)!} d\phi - \frac{1}{2} r^2 \sum_{0}^{\infty} (-1)^n \int_{0}^{2\pi} \frac{2^{2n+1}(m\cos\phi)^{2n}}{(2n+2)!} d\phi \right)^2 \qquad (1.27)$$

Now

$$\int_{0}^{2\pi} \cos^{2n} \phi \, d\phi = \frac{(2n)!}{2(n!)^2} \pi \qquad (1.28)$$

and so

$$I(\theta) \propto \pi^2 r^4 \left[\sum_{0}^{\infty} (-1)^n \frac{1}{n+1} \left(\frac{m^n}{n!} \right)^2 \right]^2 \qquad (1.29)$$

$$\propto \frac{\pi^2 r^4}{m^2} \left(J_1(2m) \right)^2 \qquad (1.30)$$

where $J_1(2m)$ is the Bessel function of the first kind of order unity. Thus, $I(\theta)$ is a maximum or zero accordingly as $J_1(2m)/m$ reaches an extremum or is zero. Its variation is shown in Figure 1.39. The first few zeros occur for values of m of

$$m = 1.916, \ 3.508, \ 5.087,\ldots \qquad (1.31)$$

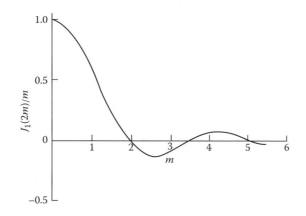

FIGURE 1.39   Variation of $J_1(2m)/m$ with m.

and so in the image of a point source produced by a circular aperture (Figure 1.41), the dark fringes occur at values of θ given by

$$\sin\theta = \frac{1.916\lambda}{\pi r}, \frac{3.508\lambda}{\pi r}, \frac{5.087\lambda}{\pi r}, \ ....$$ (1.32)

or for small θ

$$\theta \approx \frac{1.220\lambda}{d}, \frac{2.233\lambda}{d}, \frac{3.238\lambda}{d}, \ ....$$ (1.33)

where d is now the diameter of the aperture. The image structure along a central cross section is therefore similar to that of Figure 1.37, but with the minima expanded outwards to the points given in Equation 1.33 and the fringes, of course, are circular.

Many of the recently constructed large telescopes have primary mirrors made up of numerous hexagonal segments (Figure 1.34) or of several circular monolithic mirrors (e.g. LBT: two 8.4-metre mirrors, GMT: seven 8.4-metre mirrors). The matrix of hexagonal segments may approximate to the shape of a circle, an oval or a hexagon, but will still have a sawtooth edge. The diffraction patterns from point sources in these cases will differ slightly from those for simple circular apertures, but not by very much. There will also be diffraction effects arising from the edges of the hexagonal segments themselves and for all reflecting telescopes from the secondary and tertiary mirrors and their supports and/or from equipment such as cameras, photometers or spectroscopes that may obstruct parts of the light beam.

Most of these diffraction effects will show up as minor additional features superimposed upon the basic Airy diffraction pattern (in terms of the Fourier transform, they are the high frequency components of the pattern). The spikes that often appear on images of bright stars are probably the best known example of these effects and the spikes arise from diffraction at the edges of the supports for the secondary mirror (Figure 1.40). In further discussions in this book and without much loss of generality, the diffraction pattern in these cases will be taken to be the appropriate Airy disk for a simple circular aperture with the same collecting area as the telescope concerned.

If we now consider two distant point sources separated by an angle α, then two circular diffraction patterns will be produced and will be superimposed. There will not be any interference effects between the two images since the sources are mutually incoherent and so their intensities will simply add together. The combined image will have an appearance akin to that shown in Figure 1.41. When the centre of the Airy disc of one image is superimposed upon the first minimum of the other image (and vice versa), then we have Rayleigh's criterion for the resolution of a lens (or mirror). This is the normally accepted measure of the theoretical resolution of a lens. It is given in radians by (from Equation 1.33)

$$\alpha = \frac{1.220\lambda}{d}$$ (1.34)

FIGURE 1.40 Diffraction spikes on images of bright stars arising from the secondary mirror supports. Detail from an image of the globular cluster NGC 2257 obtained with the Wide Field Imager instrument on the 2.2-metre MPG/ESO telescope at La Silla. (Reproduced by kind permission of ESO.)

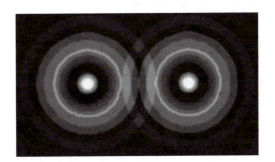

FIGURE 1.41 Image of two distant point sources through a circular aperture.

It is a convenient measure, but it is quite arbitrary. For sources of equal brightness the image will appear non-circular for separations of about one third of α, whilst for sources differing appreciably in brightness, the separation may need to be an order of magnitude larger than α for them to be resolved. For visual observing, the effective wavelength is 510 nm for faint images so that the resolution of a telescope used visually is given by

$$R = \frac{0.128}{d} \tag{1.35}$$

where d is the objective's diameter in metres and R is the angular resolution in seconds of arc. An empirical expression for the resolution of a telescope of 0.116″/d, known as Dawes' limit, is often used by amateur astronomers since the slightly better value it gives for the resolution takes some account of the abilities of a skilled observer. To achieve either resolution in practice, the magnification must be sufficient for the angular separation of the images in the eyepiece to exceed the resolution of the eye. Taking this to be an angle, β, we have the minimum magnification required to realise the Rayleigh limit of a telescope, M_m

$$M_m = \frac{\beta d}{1.220\lambda} \qquad (1.36)$$

so that for β equal to 3 minutes of arc, which is about its average value

$$M_m = 1300d \qquad (1.37)$$

where d is again measured in metres.

Of course most astronomical telescopes are actually limited in their resolution by the atmosphere. A 1-metre telescope might reach its Rayleigh resolution on one night a year from a good observing site. On an average night, scintillation will spread stellar images to about 2 seconds of arc so that only telescopes smaller than about 0.07 metres can regularly attain their diffraction limit. Since telescopes are rarely used visually for serious work, such high magnifications as are implied by Equation 1.37 are hardly ever encountered today. However, some of William Herschel's eyepieces still exist and these, if used on his 1.2-metre telescope, would have given magnifications of up to 8000 times.

The theoretical considerations of resolution that we have just seen are only applicable if the lens or mirror is of sufficient optical quality that the image is not already degraded beyond this limit. There are many effects that will blur the image and these are known as aberrations. With one exception they can all affect the images produced by either lenses or mirrors. The universal or monochromatic aberrations are known as the Seidel aberrations after Ludwig von Seidel who first analysed them. The exception is chromatic aberration (Figure 1.42) and the related second-order effects of transverse chromatic aberration and secondary colour and these affect only lenses.

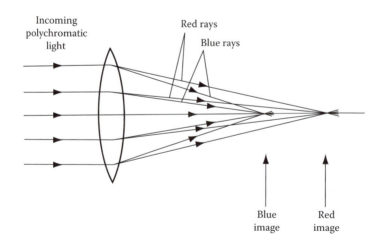

FIGURE 1.42 Chromatic aberration.

Chromatic aberration arises through the change in the refractive index of glass or other optical material with the wavelength of the illuminating radiation. Some typical values of the refractive index of some commonly used optical glasses are tabulated in Table 1.6.

The degree to which the refractive index varies with wavelength is called the dispersion and is measured by the constringence, ν

$$\nu = \frac{\mu_{589} - 1}{\mu_{486} - \mu_{656}} \tag{1.38}$$

where μ_{λ} is the refractive index at wavelength λ. The three wavelengths that are chosen for the definition of ν are those of strong Fraunhofer lines: 486 nm – the F line ($H\beta$); 589 nm – the D lines (Na); 656 nm – the C line ($H\alpha$). Thus, for the glasses listed earlier, the constringence varies from 57 for the crown glass to 33 for the dense flint (note that the *higher* the value of the constringence, the *less* that rays of different wavelengths diverge from each other).

The effect of the dispersion upon an image is to string it out into a series of different-coloured images along the optical axis (Figure 1.42). Looking at this sequence of images with an eyepiece, then at a particular point along the optical axis, the observed image will consist of a sharp image in the light of one wavelength surrounded by blurred images of varying sizes in the light of all the remaining wavelengths. To the eye, the best image occurs when yellow light is focused since the retina is less sensitive to the red and blue light. The image size at this point is called the circle of least confusion. The spread of colours along the optical axis is called the longitudinal chromatic aberration, whilst that along the image plane containing the circle of least confusion is called the transverse chromatic aberration.

Two lenses of different glasses may be combined to reduce the effect of chromatic aberration. Commonly in astronomical refractors, a biconvex crown glass lens is allied to a plano-concave flint glass lens to produce an achromatic doublet. In the infrared, achromats can be formed using barium, lithium and strontium fluorides, zinc sulphate or selenide and infrared-transmitting glasses. In the sub-millimetre region (i.e. wavelengths of several hundred microns) crystal quartz and germanium can be used. The lenses are either cemented together or separated by only a small distance (Figure 1.43). Despite its name there is still some chromatic aberration remaining in this design of lens since it can only bring two wavelengths to a common focus. If the radii of the curved

TABLE 1.6　Typical Values of the Refractive Index of Commonly Used Optical Glasses

Glass Type	Refractive Index at the Specified Wavelengths				
	361 nm	486 nm	589 nm	656 nm	768 nm
Crown	1.539	1.523	1.517	1.514	1.511
High dispersion crown	1.546	1.527	1.520	1.517	1.514
Light flint	1.614	1.585	1.575	1.571	1.567
Dense flint	1.705	1.664	1.650	1.644	1.638

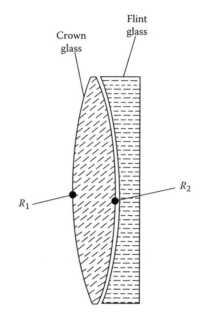

FIGURE 1.43   An achromatic doublet.

surfaces are all equal, then the condition for two given wavelengths, λ_1 and λ_2, to have coincident images is

$$2\Delta\mu_C = \Delta\mu_F \qquad (1.39)$$

where $\Delta\mu_C$ and $\Delta\mu_F$ are the differences between the refractive indices at λ_1 and λ_2 for the crown glass and the flint glass, respectively. More flexibility in design can be attained if the two surfaces of the converging lens have differing radii. The condition for achromatism is then

$$\frac{|R_1| + |R_2|}{|R_1|}\Delta\mu_C = \Delta\mu_F \qquad (1.40)$$

where R_2 is the radius of the surface of the crown glass lens that is in contact with the flint lens (note that the radius of the flint lens surface is almost invariably R_2 as well, in order to facilitate alignment and cementing) and R_1 is the radius of the other surface of the crown glass lens. By a careful selection of λ_1 and λ_2 an achromatic doublet can be constructed to give tolerable images. For example, by achromatising at 486 and 656 nm, the longitudinal chromatic aberration is reduced when compared with a simple lens of the same focal length by a factor of about 30. Nevertheless since chromatic aberration varies as the square of the diameter of the objective and inversely with its focal length, achromatic refractors larger than about 0.25 metres still have obtrusively coloured images. More seriously, if filters are used then the focal position will vary with the filter. Similarly, the scale on photographic plates will alter with the wavelengths of their sensitive regions. Further lenses may be added to produce apochromats that have three corrected wavelengths and super-apochromats with four corrected wavelengths. But such designs are impossibly expensive for telescope objectives of any size, although eyepieces and camera lenses may have eight or ten components and achieve very high levels of correction.

A common and severe aberration of both lenses and mirrors is spherical aberration. In this effect, annuli of the lens or mirror that are of different radii have different focal lengths. This is illustrated in Figure 1.44 for a spherical mirror. For rays parallel to the optical axis it can be eliminated completely by deepening the sphere to a paraboloidal surface for the mirror. It cannot be eliminated from a simple lens without using aspheric surfaces, but for a given focal length it may be minimised. The shape of a simple lens is measured by the shape factor, q

$$q = \frac{R_2 + R_1}{R_2 - R_1} \tag{1.41}$$

where R_1 is the radius of the first surface of the lens and R_2 is the radius of the second surface of the lens. The spherical aberration of a thin lens then varies with q as shown in Figure 1.45, with a minimum at $q = +0.6$. The lens is then biconvex with the radius of the surface nearer to the image three times the radius of the other surface. Judicious choice of surface radii in an achromatic doublet can lead to some correction of spherical aberration whilst still retaining the colour correction. Spherical aberration in lenses may also be reduced by using high refractive index glass since the curvatures required for the lenses' surfaces are lessened, but this is likely to increase chromatic aberration. Spherical aberration increases as the cube of the aperture.

The deepening of a spherical mirror to a paraboloidal one in order to correct for spherical aberration unfortunately introduces a new aberration called coma. This also afflicts

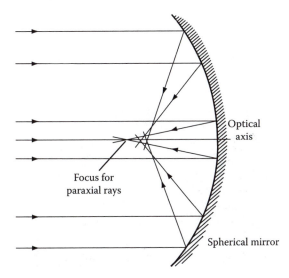

FIGURE 1.44   Spherical aberration.

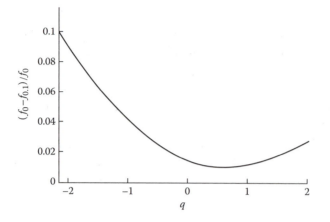

FIGURE 1.45 Spherical aberration in thin lenses. f_x is the focal length for rays parallel to the optical axis, and at a distance x times the paraxial focal length away from it.

mirrors of other shapes and lenses. It causes the images for objects away from the optical axis to consist of a series of circles that correspond to the various annular zones of the lens or mirror and which are progressively shifted towards or away from the optical axis (Figure 1.46). The severity of coma is proportional to the square of the aperture. It is zero in a system that obeys Abbe's sine condition

$$\frac{\sin\theta}{\sin\phi} = \frac{\theta_p}{\phi_p} = \text{constant} \tag{1.42}$$

where the angles are defined in Figure 1.47. A doublet lens can be simultaneously corrected for chromatic and spherical aberrations and for coma within acceptable limits if the two lenses can be separated. Such a system is called an aplanatic lens. A parabolic mirror can be corrected for coma by adding thin correcting lenses before or after the mirror (discussed in more detail under the heading 'Telescope Designs'). The severity of the coma at a given angular distance from the optical axis is inversely proportional to the square of the focal ratio. Hence, using as large a focal ratio as possible can also reduce its effect. In Newtonian

FIGURE 1.46 Shape of the image of a point source due to coma.

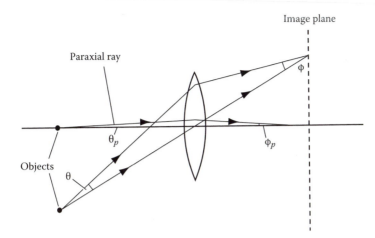

FIGURE 1.47　Parameters for Abbe's sine condition.

reflectors a focal ratio of f8 or larger gives acceptable coma for most purposes. At f3, coma will limit the useful field of view to about 1 minute of arc so that prime focus imaging almost always requires the use of a correcting lens to give reasonable fields of view.

Astigmatism (Figure 1.48) is an effect whereby the focal length differs for rays in the plane containing an off-axis object and the optical axis (the tangential plane), in comparison with rays in the plane at right angles to this (the sagittal plane). It decreases more

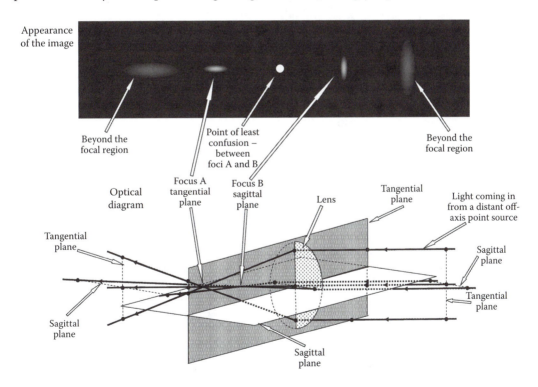

FIGURE 1.48　Shape of the image of a point source due to astigmatism, at different points along the optical axis.

slowly with focal ratio than coma so that it may become the dominant effect for large focal ratios. There is a point between the tangential and sagittal foci, known as the point of least confusion, where the image is at its sharpest. It is possible to correct astigmatism, but only at the expense of introducing yet another aberration, field curvature. This is simply that the surface containing the sharply focused images is no longer a flat plane but is curved. A system in which a flat image plane is retained and astigmatism is corrected for at least two radii is termed an anastigmatic system.

The final aberration is distortion and this is a variation in the magnification over the image plane. An optical system will be free of distortion only if the condition

$$\frac{\tan\theta}{\tan\phi} = \text{constant} \tag{1.43}$$

holds for all values of theta (see Figure 1.49). Failure of this condition to hold results in pincushion or barrel distortion (Figure 1.50) accordingly as the magnification increases or

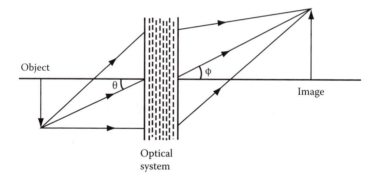

FIGURE 1.49   Terminology for distortion.

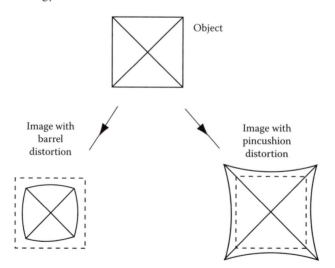

FIGURE 1.50   Distortion.

decreases with distance from the optical axis. A simple lens is very little affected by distortion and it can frequently be reduced in more complex systems by the judicious placing of stops within the system.

There are higher-order aberrations than the ones just discussed which the reader may encounter, especially if venturing to study advanced optical design. These have names such as Trefoil, Quadrafoil, Pentafoil... etc. Trefoil has effects upon images that are symmetrical about a three-fold axis (quadrafoil – four-fold axis, pentafoil – five-fold axis and so on). However, interested readers will have to seek explanations of these from more specialised texts than this one (although the effects of trefoil aberration at least will be familiar to any amateur astronomer possessing a telescope with a mirror that is held in place by three supporting bolts and who has over-tightened those bolts).

A fault of optical instruments as a whole is vignetting. This is not an attribute of a lens or mirror and so is not included amongst the aberrations. It arises from the uneven illumination of the image plane, usually due to obstruction of the light path by parts of the instrument. Normally it can be avoided by careful design, but it may become important if stops are used in a system to reduce other aberrations.

This long catalogue of faults of optical systems may well have led the reader to wonder if the Rayleigh limit can ever be reached in practice. However, optical designers have a wide variety of variables to play with; refractive indices, dispersion, focal length, mirror surfaces in the form of various conic sections, spacings of the elements, number of elements and so on, so that it is usually possible to produce a system which will give an adequate image for a particular purpose. Other criteria such as cost, weight, production difficulties etc. may well prevent the realisation of the system in practice even though it is theoretically possible. Multipurpose systems can usually only be designed to give lower quality results than single-purpose systems. Thus, the practical telescope designs that are discussed later in this section are optimised for objects at infinity and if they are used for nearby objects their image quality will deteriorate.

The method of designing an optical system is still largely an empirical one. The older approach requires an analytical expression for each of the aberrations. For example, the third-order approximation for spherical aberration of a lens is

$$\frac{1}{f_x} - \frac{1}{f_p} = \frac{x^2}{8f^3\mu(\mu-1)}\left[\frac{\mu+2}{\mu-1}q^2 + 4(\mu+1)\left(\frac{2f}{v}-1\right)q \right.$$
$$\left. +(3\mu+2)(\mu-1)\left(\frac{2f}{v}-1\right)^2 + \frac{\mu^2}{\mu-1}\right] \tag{1.44}$$

where f_x is the focal distance for rays passing through the lens at a distance x from the optical axis, f_p is the focal distance for paraxial rays from the object (paraxial rays are rays which are always close to the optical axis and which are inclined to it by only small angles), f is the focal length for paraxial rays which are initially parallel to the optical axis and v is

the object distance. For precise work it may be necessary to involve fifth-order approximations of the aberrations and so the calculations rapidly become very unwieldy.

The alternative and now far more widely used method is via ray tracing. The concepts
involved are much simpler since the method consists simply of accurately following the
path of a selected ray from the object through the system and finding its arrival point on
the image plane. Only the basic formulae are required; Snell's law for lenses

$$\sin i = \frac{\mu_1}{\mu_2} \sin r \tag{1.45}$$

and the law of reflection for mirrors

$$i = r \tag{1.46}$$

where i is the angle of incidence, r is the angle of refraction or reflection as appropriate and
μ_1 and μ_2 are the refractive indices of the materials on either side of the interface.

The calculation of i and r for a general ray requires knowledge of the ray's position and
direction in space and the specification of this is rather more cumbersome. Consider first
of all a ray passing through a point P, which is within an optical system (Figure 1.51). We
may completely describe the ray by the spatial coordinates of the point P, together with the
angles that the ray makes with the coordinate system axes (Figure 1.52). We may without
any loss of generality set the length of the ray, l, equal to unity and therefore write

$$\gamma = \cos \theta \tag{1.47}$$

$$\delta = \cos \phi \tag{1.48}$$

$$\varepsilon = \cos \psi \tag{1.49}$$

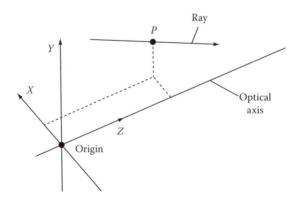

FIGURE 1.51   Ray tracing coordinate system.

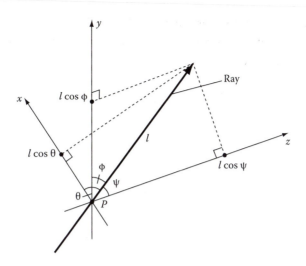

FIGURE 1.52   Ray tracing angular coordinate system.

The angular direction of the ray is thus specified by the vector

$$\mathbf{v} = (\gamma, \delta, \varepsilon) \tag{1.50}$$

The quantities γ, δ and ε are commonly referred to as the direction cosines of the ray. If we now consider the point P as a part of an optical surface, then we require the angle of the ray with the normal to that surface in order to obtain i and thereafter, r. Now we may similarly specify the normal at P by its direction cosines, forming the vector, $\mathbf{v}'$,

$$\mathbf{v}' = (\gamma', \delta', \varepsilon') \tag{1.51}$$

and we then have

$$\cos i = \frac{\mathbf{v}.\mathbf{v}'}{\|\mathbf{v}\|\|\mathbf{v}'\|} \tag{1.52}$$

or

$$i = \cos^{-1}(\gamma\gamma', + \delta\delta', + \varepsilon\varepsilon') \tag{1.53}$$

The value of r can now be obtained from Equations 1.45 or 1.46. We may again specify its track by direction cosines (the vector $\mathbf{v}''$)

$$\mathbf{v}'' = (\gamma'', \delta'', \varepsilon'') \tag{1.54}$$

and we may find the values of the components as follows. Three simultaneous equations can be obtained; the first from the angle between the incident ray and the refracted or reflected ray, the second from the angle between the normal and the reflected or refracted ray and the third by requiring the incident ray, the normal and the refracted or reflected ray to be coplanar and these are

$$\gamma\gamma'' + \delta\delta'' + \varepsilon\varepsilon'' = \cos 2r \quad \text{(reflection)} \tag{1.55}$$

$$= \cos (i - r) \quad \text{(refraction)} \tag{1.56}$$

$$\gamma'\gamma'' + \delta'\delta'' + \varepsilon'\varepsilon'' = \cos r \tag{1.57}$$

$$(\varepsilon\delta' - \varepsilon'\delta)\gamma'' + (\gamma\varepsilon' - \gamma'\varepsilon)\delta'' + (\delta\gamma' - \delta'\gamma)\varepsilon'' = \cos r \tag{1.58}$$

After a considerable amount of manipulation one obtains

$$\varepsilon'' = \frac{\left\{\left[(\varepsilon\delta' - \varepsilon'\delta)\delta' - (\gamma\varepsilon' - \gamma'\varepsilon)\gamma'\right](\gamma'\cos\alpha - \gamma\cos r) - (\gamma'\delta - \gamma\delta')(\varepsilon\delta' - \varepsilon'\delta)\cos r\right\}}{\left\{\left[(\varepsilon\delta' - \varepsilon'\delta)\delta' - (\gamma\varepsilon' - \gamma'\varepsilon)\gamma'\right](\gamma'\varepsilon - \gamma\varepsilon') - \left[(\varepsilon\delta' - \varepsilon'\delta)\varepsilon' - (\delta\gamma' - \delta'\gamma)\gamma'\right](\gamma'\delta - \gamma\delta')\right\}} \tag{1.59}$$

$$\delta'' = \frac{\gamma'\cos\alpha - \gamma\cos r - (\gamma'\varepsilon - \gamma\varepsilon')\varepsilon''}{(\gamma'\delta - \gamma\delta')} \tag{1.60}$$

$$\gamma'' = \frac{\cos\alpha - \delta\delta'' - \varepsilon\varepsilon''}{\gamma} \tag{1.61}$$

where α is equal to $2r$ for reflection and $(i - r)$ for refraction.

The direction cosines of the normal to the surface are easy to obtain when its centre of curvature is on the optical axis

$$\gamma' = \frac{x_1}{R} \tag{1.62}$$

$$\delta' = \frac{y_1}{R} \tag{1.63}$$

$$\varepsilon' = \frac{(z_1 - z_R)}{R} \tag{1.64}$$

where (x_1, y_1, z_1) is the position of P and $(0, 0, z_R)$ is the position of the centre of curvature. If the next surface is at a distance s from the surface under consideration, then the ray will arrive on it at a point (x_2, y_2, z_2) still with the direction cosines γ'', δ'' and ε'', where

$$x_2 = \frac{\gamma'' s}{\varepsilon''} + x_1 \tag{1.65}$$

$$y_2 = \frac{\delta'' s}{\varepsilon''} + y_1 \tag{1.66}$$

$$z_2 = s + z_1 \tag{1.67}$$

We may now repeat the calculation again for this surface and so on. Ray tracing has the advantage that all the aberrations are automatically included by it, but it has the disadvantage that many rays have to be followed in order to build up the structure of the image of a point source at any given place on the image plane, and many images have to be calculated in order to assess the overall performance of the system. Ray tracing, however, is eminently suitable for programming onto a computer, whilst the analytical approach is not, since it requires frequent value judgements.

The approach of a designer, however, to an optical design problem is similar whichever method is used to assess the system. The initial prototype is set up purely on the basis of what the designer's experience suggests may fulfil the major specifications such as cost, size, weight, resolution etc. A general rule of thumb is that the number of optical surfaces needed will be at least as many as the number of aberrations to be corrected. The performance is then assessed either analytically or by ray tracing. In most cases it will not be good enough, so a slight alteration is made with the intention of improving the performance and it is reassessed. This process continues until the original prototype has been optimised for its purpose. If the optimum solution is within the specifications then there is no further problem. If it is outside the specifications, even after optimisation, then the whole procedure is repeated starting from a different prototype. The performances of some of the optical systems favoured by astronomers are considered in the next subsection.

Even after a design has been perfected, there remains the not inconsiderable task of physically producing the optical components to within the accuracies specified by the design. The manufacturing steps for both lenses and mirrors are broadly similar, although the details may vary. The surface is roughly shaped by moulding or by diamond milling. It is then matched to another surface formed in the same material whose shape is its inverse, called the tool. The two surfaces are ground together with coarse carborundum or other grinding powder between them until the required surface begins to approach its specifications. The pits left behind by this coarse grinding stage are removed by a second grinding stage in which finer powder is used. The pits left by this stage are then removed in turn by a third stage using still finer powder and so on. As many as eight or ten such stages may be necessary. When the grinding pits are reduced to a micron or so in size, the surface may be polished. This process employs a softer powder such as iron oxide or cerium oxide that is embedded in a soft matrix such as pitch. Once the surface has been polished it can be tested for the accuracy of its fit to its specifications. Since, in general, it will not be within the specifications after the initial polishing, a final stage, which is termed figuring, is necessary. This is simply additional polishing to adjust the surface's shape until it is correct. The magnitude of the changes involved during this stage is only about a micron or two, so that if the alteration that is needed is larger than this, it may be necessary to return to a previous stage in the grinding to obtain a better approximation. There are a number of tests that can determine the shape of the mirrors surface to a precision of ±50 nm or better, such as the Foucault, Ronchi, Hartmann and Null tests. The details of these tests are beyond the scope of this book but may be found in books on optics and telescope making.

Small mirrors can be produced in this way by hand and many an amateur astronomer has acquired his or her telescope at very low cost by making their own optics. Larger mirrors require machines that move the tool (now smaller than the mirror) in an epicyclic fashion. The motion of the tool is similar to that of the planets under the Ptolemaic model of the solar system and so such machines have become known as planetary polishers. The epicyclic motion can be produced by a mechanical arrangement, but commercial production of large mirrors now relies on computer-controlled planetary polishers. The deepening of the surface to a parabolic, hyperbolic or other shape is accomplished by preferential polishing and sometimes by the use of a flexible tool whose shape can be adjusted. The mirror segments for large instruments like the 10-metre Keck telescopes are clearly small, off-axis parts of the total hyperbolic shape. These and also the complex shapes required for Schmidt telescope corrector lenses (Section 1.1.19.2), have been produced using stressed polishing. The blank is deformed by carefully designed forces from a warping harness and then polished to a spherical shape (or flat for Schmidt corrector plates). When the deforming forces are released, the blank springs into the required non-spherical asymmetric shape. For the Keck and Hobby-Eberly mirror segments the final polishing is undertaken using ion beams. The ion beam is an accelerated stream of ions, such as argon, that wears away the surface an atom at a time. Since it applies almost no force to the mirror, the process can be used to correct defects left by normal polishing techniques such as deformations at the edge and print-through of the honeycomb back on lightweight mirrors.

Recently, the requirements for non-axi-symmetric mirrors for segmented mirror telescopes (Section 1.1.19.2) and for glancing incidence X-ray telescopes (see Section 1.3) have led to the development of numerically controlled diamond milling machines which can produce the required shaped and polished surface directly to an accuracy of 10 nm or better.

The defects in an image that are due to surface imperfections on a mirror will not exceed the Rayleigh limit if those imperfections are less than about one-eighth of the wavelength of the radiation for which the mirror is intended. Thus, we have the commonly quoted $\lambda/8$ requirement for the maximum allowable deviation of a surface from its specifications. Sometimes the resulting deviations of the wavefront are specified rather than those of the optical surface. This is twice the value for the surface (i.e. a limit of $\lambda/4$). The restriction on lens surfaces is about twice as large as those of mirror surfaces since the ray deflection is distributed over the two lens faces. However, as we have seen, the Rayleigh limit is an arbitrary one and for some purposes the fit must be several times better than this limit. This is particularly important when viewing extended objects such as planets or for high-contrast images and improvements in the signal-to-noise ratio can continue to be obtained by figuring surfaces to $\lambda/20$ or better. The surfaces of the JWST segments deviate from their correct form, for example, by generally less than 20 nm. To put this in perspective: if the Earth's surface were as smooth as the JWST's mirror, then the largest hills or valleys would be only 40 mm high (or deep).

The surface must normally receive its reflecting coating after its production. The vast majority of astronomical mirror surfaces have a thin layer of aluminium evaporated onto them by heating aluminium wires suspended over the mirror inside a vacuum chamber. The initial reflectivity of an aluminium coating in the visual region is around 90%. This, however, can fall to 75% or less within a few months as the coating ages. Mirrors therefore have to be re-aluminised at regular intervals. The intervals between re-aluminising can be lengthened by gently cleaning the mirror every month or so. The currently favoured methods of cleaning are rinsing with de-ionised water and/or drifting carbon dioxide snow across the mirror surface. Mirrors coated with a suitably protected silver layer can achieve 99.5% reflectivity in the visible and are becoming more and more required since some modern telescope designs can have four or five reflections. With ancillary instrumentation the total number of reflections can then reach 10 or more. A silver coating also lowers the infrared emission from the mirror itself from around 4% at a wavelength of 10 μm for an aluminium coating to around 2%. For this reason the 8.1-metre Gemini telescopes use protected silver coatings that are applied to the surfaces of the mirrors by evaporation inside a vacuum chamber. Other materials, such as silicon carbide, are occasionally used as the reflective layer, especially for UV work, since the reflectivity of aluminium falls off below 300 nm. For infrared telescopes, such as the JWST (Figure 1.53), the mirror coating can be gold.

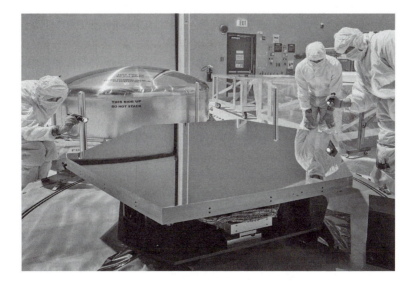

FIGURE 1.53 **(See color insert.)** One of the 18 mirror segments for the JWST. The segment is 1.32 metres in diameter (between the flat edges) and weighs around 20 kg. The gold reflective coating can be clearly seen. (Reproduced by kind permission of NASA/Chris Gunn.)

Lens surfaces also normally receive a coating after their manufacture, but in this case the purpose is to reduce reflection. Uncoated lenses reflect about 5% of the incident light from each surface so that a system containing, say, five uncoated lenses could lose 40% of its available light through this process. To reduce the reflection losses a thin layer of material covers the lens, for which

$$\mu' = \sqrt{\mu} \qquad (1.68)$$

where μ' is the refractive index of the coating and μ is the refractive index of the lens material. The thickness, t, should be

$$t = \frac{\lambda}{4\mu'} \qquad (1.69)$$

This gives almost total elimination of reflection at the selected wavelength, but some will still remain at other wavelengths. Lithium fluoride and silicon dioxide are commonly used materials for the coatings. Recently, it has become possible to produce anti-reflection coatings that are effective simultaneously over a number of different wavelength regions through the use of several interfering layers.

Mirrors only require the glass as a support for their reflecting film. Thus, there is no requirement for it to be transparent; on the other hand it is essential for its thermal expansion coefficient to be low. Many of today's large telescope mirrors are therefore made from materials other than glass. The coefficient of thermal expansion of ordinary glass is about 9×10^{-6} K^{-1}, that of Pyrex™ about 3×10^{-6} K^{-1} and for fused quartz it is about 4×10^{-7} K^{-1}.

Amongst the largest telescopes, the only ordinary glass mirror is the one for the 2.5-metre Hooker telescope[*] on Mount Wilson in California. After its construction, borosilicate glasses[†] such as Pyrex became the favoured material until the last few decades, when quartz or artificial materials with a similar low coefficient of expansion such as CerVit and Zerodur etc. but which are easier to manufacture have been used. These low-expansion materials have a ceramic glass structure that is partially crystallised. The crystals are around 50 nm across and contract as the temperature rises, whilst the surrounding matrix of amorphous material expands. An appropriate ratio of crystals to matrix provides very small expansion (10^{-7} K^{-1}) over a wide range of temperatures. Ultra-low expansion fused silica (ULE) has an even smaller coefficient of expansion (3×10^{-8} K^{-1}).

Borosilicate glasses, however, have returned to favour in the last 10 years with monolithic mirrors up to 8.4 metres in diameter being produced for the LBT, LSST, SASIR[‡] and

---

[*] The Hooker telescope was completed in 1917 and the first borosilicate glass, Duran, was invented in 1893. Pyrex was produced from about 1915 onwards. The telescope is still in use today.

[†] These glasses are typically composed of around 70% silica with the remaining 30% made up by boron, sodium and potassium oxides in varying proportions.

[‡] Synoptic All-Sky Infrared Imaging Survey. The survey will use the 6.5-metre San Pedro Mártir telescope that is currently under construction for the San Pedro Mártir Observatory in Baja California, Mexico.

the Magellan I and II* instruments amongst others. The Giant Magellan Telescope (GMT), due for completion around 2020, will also use seven 8.4-metre monolithic borosilicate glass mirrors on a single mounting. The reasons for the material's current popularity include easy workability at (relatively) low temperatures thus facilitating the construction of honey-combed mirror blanks by the spin-casting process (see below), relatively low cost and resistance to the chemicals used to remove the reflective coating when it has to be replaced.

A radically different alternative to that of forming mirrors from low expansion materials is to use a material with a very high thermal conductivity, such as silicon carbide, graphite epoxy, steel, beryllium or aluminium. The mirror is then always at a uniform temperature and its surface has no stress distortion. ESA's infrared Herschel Space Observatory launched in 2009 for example carries a 3.5-metre silicon carbide mirror, which, at the time of its manufacture, was the largest silicon carbide structure in the world. Most metallic mirrors, however, have a relatively coarse crystalline surface that cannot be polished adequately. They must therefore be coated with a thin layer of nickel before polishing and there is always some risk of this being worn through if the polishing process extends for too long.

Provided that they are mounted properly, small solid mirrors can be made sufficiently thick that once ground and polished, their shape is maintained simply by their mechanical rigidity. However, the thickness required for such rigidity scales as the cube of the size so that the weight of a solid rigid mirror scales as D^5. Increasing solid mirror blanks to larger sizes thus rapidly becomes expensive and impractical. Most mirrors larger than about 0.5 to 1 metre therefore have to have their weight reduced in some way. There are two main approaches to reducing the mirror weight; thin mirrors and honeycomb mirrors. The thin mirrors are also subdivided into monolithic and segmented mirrors. In both cases, active support of the mirror is required in order to retain its correct optical shape. The 8.2-metre mirrors of ESO's VLT (Figure 1.34) are an example of thin monolithic mirrors; they are 178 mm thick Zerodur, with a weight of 23 tonnes each, but need 150 actuators to maintain their shape. The 10-metre Keck telescopes use thin, segmented Zerodur mirrors (Figure 1.53). There are 36 individual hexagonal segments in each main mirror, with each segment 1.8 metres across and about 70 mm thick, giving a total mirror weight of just 14.4 tonnes (compared with 14.8 tonnes for the 5-metre Hale telescope mirror).

Honeycomb mirrors are essentially thick solid blanks that have had a lot of the material behind the reflecting surface removed, leaving only thin struts to support that surface. The struts often surround hexagonal cells of removed material giving the appearance of honeycomb, but other cell shapes such as square, circular or triangular can be used. The mould for a honeycomb mirror blank has shaped cores protruding from its base that produce the empty cells, with the channels between the cores filling with the molten material to produce the supporting ribs. Once the blank has solidified and been annealed it may be ground and polished as though it were solid. However, if the original surface of the blank is flat, then considerable amounts of material will remain near the edge after grinding,

---

* The Magellan I and II telescopes are twin 6.5-metre instruments at the Las Campanas Observatory which had first light early this millennium. They should not be confused with the GMT which is currently under construction and which will also be based at Las Campanas.

adding to the final weight of the mirror. Roger Angel of the University of Arizona has therefore pioneered the technique of spin-casting honeycomb mirror blanks. The whole furnace containing the mould and molten material is rotated so that the surface forms a parabola whose shape is close to that which is finally required. The surface shape is preserved as the furnace cools and the material solidifies. Honeycomb mirrors up to nearly eight and a half metres diameter may be produced by this method. Howsoever it may have been produced, thinning the ribs and the underside of the surface may further reduce the weight of the blank. The thinning is accomplished by milling and/or etching with hydrogen fluoride. In this way the 1-metre diameter Zerodur secondary mirrors produced for the 8.1-metre Gemini telescopes have supporting ribs just 3 mm wide and weigh less than 50 kg.

Whatever approach is taken to reducing the mirror's weight, it will be aided if the material used for the blank has an intrinsic high stiffness (resistance to bending). Beryllium has already been used to produce the 1.1-metre secondary mirrors for the VLT, whose weights are only 42 kg. Silicon carbide and graphite epoxy are other high stiffness materials that may be expected to be used more in the future.

A quite different approach to producing mirrors that, perhaps surprisingly gives low weights, is to use a rotating bath of mercury. Isaac Newton was the first to realise that the surface of a steadily rotating liquid would take up a paraboloidal shape under the combined influences of gravity and centrifugal acceleration. If that liquid reflects light, like mercury, gallium, gallium-indium alloy or an oil suffused with reflecting particles, then it can act as a telescope's primary mirror. Of course the mirror has to remain accurately horizontal, so that it always points towards the zenith, but with suitable, perhaps active, correcting secondary optics the detector can be moved around the image plane to give a field of view currently tens of minutes of arc wide and conceivably in the future up to 8° across. Moving the electrons across the CCD detector at the same speed as the image movement enables short time exposures to be obtained (TDI). The Earth's rotation enables such a zenith telescope to sample up to 7% of the sky for less than 7% of the cost of an equivalent fully steerable telescope. The bath containing the mercury is in fact a lightweight structure whose surface is parabolic to within 0.1 mm. Only a thin layer of mercury is thus needed to produce the accurate mirror surface and so the mirror overall is a lightweight one despite the high density of mercury. The bath is rotated smoothly around an accurately vertical axis at a constant angular velocity using a large air bearing. Mercury is toxic, so that suitable precautions have to be taken to protect the operators. Also, its reflectivity is less than 80% in the visible, but since all the other optics in the instrument can be conventional mirrors or lenses, this penalty is not too serious.

The Large Zenith Telescope (LZT), in British Columbia, saw first light in 2003 using a 6-metre diameter liquid mirror. There are also proposals for an 8-metre liquid mirror telescope on the same site and a rather ambitious concept, the Large Aperture Mirror Array (LAMA), which might comprise sixty-six 6.15-metre liquid mirror telescopes gathering the same amount of light as a single 50-metre dish. An even more ambitious concept places a 20- to 100-metre liquid mirror telescope on the Moon. Since mercury evaporates in a vacuum, the proposal suggests using liquid salts with a thin surface layer of silver to act as the reflector. Currently under construction, the International Liquid Mirror Telescope (ILMT) is to be a 4-metre instrument located at Devasthal on the Himalayan foothills (Northern India) imaging a 0.5-degree square of the sky using a 4k × 4k TDI CCD array. In space a liquid mirror of potentially almost any size could be formed from a ferromagnetic liquid confined by electromagnetic forces.

An approach to producing mirrors whose shape can be rapidly adjusted is to use a thin membrane with a reflecting coating. The membrane forms one side of a pressure chamber and its shape can be altered by changing the gas pressure inside the chamber. A λ/2 surface quality or better can currently be achieved over mirror diameters up to 5 mm and such mirrors have found application in optical aperture synthesis systems (see Section 2.5).

1.1.19 Telescope Designs

1.1.19.1 Background

Most serious work with telescopes uses equipment placed directly at the focus of the telescope. But for visual work such as finding and guiding on objects, an eyepiece is necessary. Often it matters little whether the image produced by the eyepiece is of a high quality. Ideally, however, the eyepiece should not degrade the image noticeably more than the main optical system. There are an extremely large number of eyepiece designs, whose individual properties can vary widely. For example, one of the earliest eyepiece designs of reasonable quality is the Kellner. This combines an achromat and a simple lens and typically has a field of view of 40° to 50°. The Plössl uses two achromats and has a slightly wider field of view. More recently, the Erfle design employs six or seven components and gives fields of view of 60° to 70°, whilst the current state-of-the-art is represented by designs such as the Nagler with eight or more components and fields of view* up to 85°. Details of these and other designs may generally be found in books on general astronomy or on optics or from the manufacturers. For small telescopes used visually, a single low-magnification, wide-angle eyepiece may be worth purchasing for the magnificent views of large objects like the Orion Nebula that it will provide. There is little point in the higher power eyepieces being wide-angle ones (which are very expensive), since these will normally be used to look at angularly small objects. The use of a Barlow lens provides an adjustable magnification for any eyepiece. The Barlow lens is an achromatic negative lens placed just before the telescope's focus. It increases the telescope's effective focal length and so the magnification.

For our purposes, only four aspects of eyepieces are of concern: light losses, eye relief, exit pupil and angular field of view. Light loss occurs through vignetting when parts of the light beam fail to be intercepted by the eyepiece optics, or are obstructed by a part of the structure of the eyepiece and also through reflection, scattering and absorption by the optical components. The first of these can generally be avoided by careful eyepiece design and selection, whilst the latter effects can be minimised by anti-reflection coatings and by keeping the eyepieces clean.

---

* The individual healthy human eye has a field of view that is about 150° across in the horizontal plane and about 130° in the vertical plane. The periphery of this region though only serves to detect something, especially if it is moving and acts mainly as a defence mechanism, thus alerting the observer to the need to move his/her head and/or eyes to look more directly at that object. The field of sharp vision is only about 5° across.

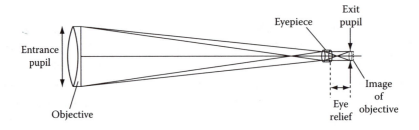

FIGURE 1.54   Exit pupil and eye relief.

The exit pupil is the image of the objective produced by the eyepiece (Figure 1.54). All the rays from the object pass through the exit pupil, so that it must be smaller than the pupil of the human eye if all of the light gathered by the objective is to be utilised. Its diameter, E, is given by

$$E = \frac{F_e D}{F_o} \tag{1.70}$$

where D is the objective's diameter, F_e is the focal length of the eyepiece and F_o is the focal length of the objective. Since magnification is given by

$$M = \frac{F_o}{F_e} \tag{1.71}$$

and the diameter of the pupil of the dark-adapted eye is 6 or 7 mm, we must therefore have

$$M \geq \sim 170 \, D \tag{1.72}$$

where D is in metres, if the whole of the light from the telescope is to pass into the eye. In practice, for small telescopes being used visually, the maximum useable magnification is around ×2000 D (×300 for a 0.15-metre telescope etc.).

The eye relief is the distance from the final lens of the eyepiece to the exit pupil. It should be about 6 to 10 mm for comfortable viewing. If you wear spectacles, however, then the eye relief may need to be up to 20 mm.

The angular field of view is defined by the acceptance angle of the eyepiece, θ'. Usually this is about 40°–60°, but it may be up to 90° for wide-angle eyepieces. The angular diameter of the area of sky which is visible when the eye is positioned at the exit pupil and which is known as the angular field of view, θ, is then just

$$\theta = \frac{\theta'}{M} \tag{1.73}$$

The brightness of an image viewed through a telescope is generally expected to be greater than when it is viewed directly. However, this is not the case for *extended* objects. The naked eye brightness of a source is proportional to the eye's pupil diameter squared, whilst that of the image in a telescope is proportional to the objective diameter squared. If the eye looks at that image, then its angular size is increased by the telescope's magnification. Hence, the increased brightness of the image is spread over a greater area. Thus, we have

$$R = \frac{\text{brightness through a telescope}}{\text{brightness to the naked eye}} = \frac{D^2}{M^2 P^2} \tag{1.74}$$

where D is the objective's diameter, P is the diameter of the pupil of the eye and M is the magnification. But from Equations 1.70 and 1.71 we have

$$R = 1 \tag{1.75}$$

when the exit pupil diameter is equal to the diameter of the eye's pupil and

$$R < 1 \tag{1.76}$$

when it is smaller than the eye's pupil.

If the magnification is less than ×170 D, then the exit pupil is larger than the pupil of the eye and some of the light gathered by the telescope will be lost. Since the telescope will generally have other light losses due to scattering, imperfect reflection and absorption, the surface brightness of an extended source is always fainter when viewed through a telescope than when viewed directly with the naked eye. This result is in fact a simple consequence of the second law of thermodynamics; for if it were not the case, then one could have a net energy flow from a cooler to a hotter object. The apparent increase in image brightness when using a telescope arises partly from the increased angular size of the object so that the image on the retina must fall to some extent onto the regions containing more rods, even when looked at directly and partly from the increased contrast resulting from the exclusion of extraneous light by the optical system. Even with the naked eye, looking through a long cardboard tube enables faint extended sources such as M31 to be seen more easily.

The analysis that we have just seen does not apply to images that are physically smaller than the detecting element. For this situation, the image brightness is proportional to D^2. Again, however, there is an upper limit to the increase in brightness that is imposed when the energy density at the image is equal to that at the source. This limit is never approached in practice since it would require 4π steradians for the angular field of view. Thus, stars may be seen through a telescope which are fainter than those visible to the naked eye by a factor called the light grasp. The light grasp is simply given by D^2/P^2. Taking $+6^m$ as the magnitude of the faintest star visible to the naked eye (see Section 3.1); the faintest star that may be seen through a telescope has a magnitude, m_l, which is the limiting magnitude for that telescope

$$m_l = 17 + 5 \log_{10} D \tag{1.77}$$

where D is in metres. If the stellar image is magnified to the point where it spreads over more than one detecting element, then we must return to the analysis for the extended sources. For an average eye, this upper limit to the magnification is given by

$$M \approx 850\, D \tag{1.78}$$

where D is again in metres.

1.1.19.2 Designs

Probably the commonest format for large telescopes is the Cassegrain system, although most large telescopes can usually be used in several alternative different modes by interchanging their secondary mirrors. The Cassegrain system is based upon a paraboloidal primary mirror and a convex hyperboloidal secondary mirror (Figure 1.55). The nearer focus of the conic section which forms the surface of the secondary is coincident with the focus of the primary and the Cassegrain focus is then at the more distant focus of the secondary mirror's surface. The major advantage of the Cassegrain system lies in its telephoto characteristic; the secondary mirror serves to expand the beam from the primary mirror so that the effective focal length of the whole system is several times that of the primary mirror. A compact and hence rigid and relatively cheap mounting can thus be used to hold the optical components whilst retaining the advantages of long focal length and large image scale. The Cassegrain design is afflicted with coma and spherical aberration to about the same degree as an equivalent Newtonian telescope, or indeed to just a single parabolic

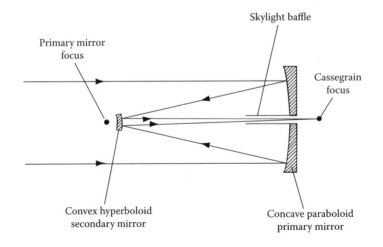

FIGURE 1.55 Cassegrain telescope optical system.

mirror with a focal length equal to the effective focal length of the Cassegrain. The beam expanding effect of the secondary mirror means that Cassegrain telescopes normally work at focal ratios between 12 and 30, although their primary mirror may be f3 or f4 (or even f1 as in the 4-metre VISTA telescope). Thus, the images remain tolerable over a field of view that may be several tenths of a degree across (Figure 1.56). Astigmatism and field curvature are stronger than in an equivalent Newtonian system, however.

Focusing of the final image in a Cassegrain system is customarily accomplished by moving the secondary mirror along the optical axis. The amplification due to the secondary means that it only has to be moved a short distance away from its optimum position in order to move the focal plane considerably. The movement away from the optimum position, however, introduces severe spherical aberration. For the 0.25-metre f4/f16 system whose images are shown in Figure 1.56, the secondary mirror can only be moved by 6 mm either side of the optimum position before even the on-axis images without diffraction broadening become comparable in size with the Airy disc. Critical work with Cassegrain telescopes should therefore always be undertaken with the secondary mirror at or very near to its optimum position.

A great improvement to the quality of the images may be obtained if the Cassegrain design is altered slightly to the Ritchey-Chrétien system. The optical arrangement is identical with that shown in Figure 1.53, except that the primary mirror is deepened to a hyperboloid and a stronger hyperboloid is used for the secondary. With such a design, both coma and spherical aberration can be corrected and we have an aplanatic system. The improvement in the images can be seen by comparing Figure 1.57 with Figure 1.56. It should be noted, however, that the improvement is in fact considerably more spectacular since we have a 0.5-metre Ritchey-Chrétien and a 0.25-metre Cassegrain with the same effective

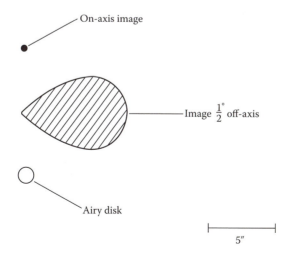

FIGURE 1.56   Images in a 0.25-metre f4/f16 Cassegrain telescope. The images were obtained by ray tracing. Since this does not allow for the effects of diffraction, the on-axis image in practice will be the same as the Airy disc and the 0.5° off-axis image will be blurred even further.

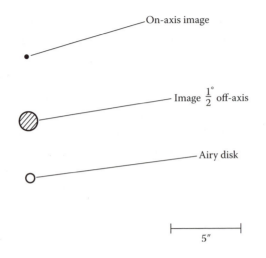

On-axis image

Image $\frac{1}{2}^{\circ}$ off-axis

Airy disk

5"

FIGURE 1.57   Ray tracing images in a 0.5-metre f3/f8 Ritchey-Chrétien telescope.

focal lengths. A 0.5-metre Cassegrain telescope would have its off-axis image twice the size of that shown in Figure 1.56 and its Airy disc half the size shown there.

Another variant on the Cassegrain system is the Dall-Kirkham telescope that has a concave ellipsoidal main mirror and spherical convex secondary mirror. The design is popular with amateur telescope makers since the curves on the mirrors are easier to produce. However, it suffers badly from aberrations and its field of sharp focus is only about a third that of an equivalent conventional Cassegrain design.

Alternatively, a Cassegrain or Ritchey-Chrétien system can be improved by the addition of correctors just before the focus. The correctors are low or zero-power lenses whose aberrations oppose those of the main system. There are numerous successful designs for correctors although many of them require aspheric surfaces and/or the use of exotic materials such as fused quartz. Images can be reduced to less than the size of the seeing disc over fields of view of up to 1 degree, or sometimes over even larger angles. The 3.9-metre Anglo-Australian Telescope (AAT) for example uses correcting lenses to provide an unvignetted 2° field of view (two-degree field [2dF]), over which 400 optical fibres can be positioned to feed a spectroscope. The 4-metre VISTA telescope uses a modified Ritchey-Chrétien design that leaves some significant aberrations in images produced by the mirrors alone. However, by using three correcting lenses a fully corrected image over 2° across in the visible and 1.6° across in the infrared is produced. Reflective corrective optics are also possible. The corrective optics may be combined with a focal reducer to enable the increased field of view to be covered by the detector array. A focal reducer is the inverse of a Barlow lens and is a positive lens, usually an apochromatic triplet, placed just before the focal point of the telescope that decreases the effective focal length and so gives a smaller image scale.

Another telescope design that is again very closely related to the Cassegrain is termed the Coudé system. It is in effect a very long focal length Cassegrain or Ritchey-Chrétien whose light beam is folded and guided by additional flat mirrors to give a focus whose position is fixed in space irrespective of the telescope position. One way of accomplishing this

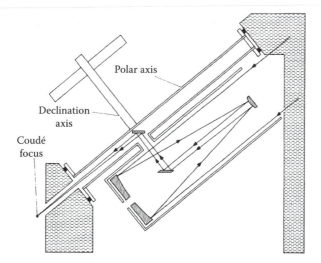

FIGURE 1.58   Coudé system for a modified English mounting.

(there are many others) is shown in Figure 1.58. After reflection from the secondary, the light is reflected down the hollow declination axis by a diagonal flat mirror and then down the hollow polar axis by a second diagonal. The light beam then always emerges from the end of the polar axis whatever part of the sky the telescope may be inspecting. Designs with similar properties can be devised for most other types of mountings although additional flat mirrors may be needed in some cases. With altazimuth (alt-az) mountings the light beam can be directed along the altitude axis to one of the two Nasmyth foci on the side of the mounting. These foci still rotate horizontally as the telescope changes its azimuth, but this poses far fewer problems than the changing altitude and attitude of a conventional Cassegrain focus. On large modern telescopes, platforms of considerable size are often constructed at the Nasmyth foci allowing large ancillary instruments to be used. The fixed or semi-fixed foci of the Coudé and Nasmyth systems are a very great advantage when bulky items of equipment, such as high dispersion spectrographs, are to be used, since these can be permanently mounted in a nearby separate laboratory and the light brought to them rather than have to have the equipment mounted on the telescope. The design also has several disadvantages: the field of view rotates as the telescope tracks an object across the sky and is very tiny due to the large effective focal ratio (f25 to f40) which is generally required to bring the focus through the axes, and finally the additional reflections will cause loss of light.

The simplest of all designs for a telescope is a mirror used at its prime focus. That is, the primary mirror is used directly to produce the images and the detector is placed at the top end of the telescope. The largest telescopes have a platform or cage which replaces the secondary mirror and which is large enough for the observer to ride in whilst he or she operates and guides the telescope from the prime focus position. With smaller instruments, too much light would be blocked, so they must be guided using a separate guide telescope or use a separate detector monitoring another star (the guide star) within the main telescope's field of view. The image quality at the prime focus is usually poor even a few tens of seconds

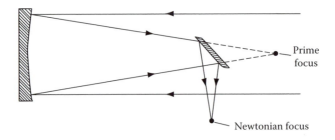

FIGURE 1.59   Newtonian telescope optical system.

of arc away from the optical axis, because the primary mirror's focal ratio may be as short as f3 or less in order to keep the length of the instrument to a minimum. Thus, correcting lenses are invariably essential to give acceptable images and reasonably large fields of view. These are similar to those used for correcting Cassegrain telescopes and are placed immediately before the prime focus.

A system that is almost identical to the use of the telescope at prime focus and which was the first design to be made into a working reflecting telescope is one that is due to Newton and hence is called the Newtonian telescope. A secondary mirror is used which is a flat diagonal placed just before the prime focus. This reflects the light beam to the side of the telescope from where access to it is relatively easy (Figure 1.59). The simplicity and cheapness of the design make it very popular as a small telescope for the amateur market, but it is rarely encountered in telescopes larger than about 1 metre. There are several reasons for its lack of popularity for large telescope designs; the main ones being that it has no advantage over the prime focus position for large telescopes since the equipment used to detect the image blocks out no more light than the secondary mirror; the secondary mirror introduces additional light losses and the position of the equipment high on the side of the telescope tube causes difficulties of access and counterbalancing. The images in a Newtonian system and at prime focus are very similar and are of poor quality away from the optical axis as shown in Figure 1.60.

As has already been noted, if n aberrations are to be corrected, then the optical design will generally need to involve n optical surfaces. Thus, three-mirror telescope designs are increasingly being considered for future telescope projects. The (many) individual design variants permitted by three optical surfaces are usually merged together as Korsch* telescopes or as three-mirror anastigmats (TMAs), although specific designs such as the Paul, Paul-Baker and Willstrop (Figure 1.61) telescopes are still named for their originators. These designs have close to zero spherical aberration, coma and astigmatism. The JWST is already in production and will use a Korsch-type design for its telescope with a 6.5-metre segmented elliptical primary mirror, a hyperbolic secondary and an elliptical tertiary mirror. Other future applications of the designs will include the LSST, the European Extremely Large Telescope (E-ELT) and ESA's Euclid space-based survey mission.

---

* After Dietrich Korsch, who developed a general set of solutions for three-mirror telescopes in 1972.

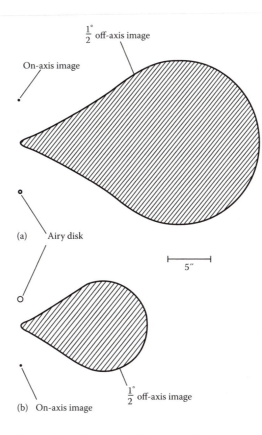

FIGURE 1.60   Images in Newtonian telescopes. (a) 1-metre f4 telescope; (b) 0.5-metre f8 telescope.

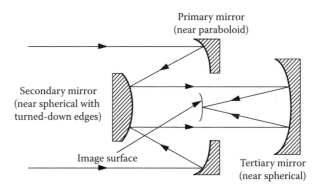

FIGURE 1.61   The Willstrop three-mirror optical system.

A very great variety of other catoptric (reflecting objective) telescope designs abound, but most have few or no advantages over the groups of designs which are discussed above. Some find specialist applications; for example the Gregorian design is similar to the Cassegrain except that the secondary mirror is a concave ellipsoid and is placed after the prime focus. Thus the two 8.4-metre instruments of the LBT on Mount Graham include Gregorian secondary mirrors for infrared work. The Pfund-type telescope uses a siderostat

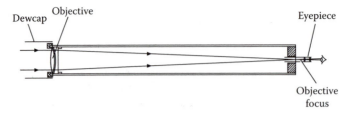

FIGURE 1.62   The astronomical refractor.

to feed a fixed paraboloidal main mirror that reflects light through a hole in the flat mirror of the siderostat to a focus behind it. The ISI interferometer (see Section 2.5) used Pfund-type telescopes but other examples are uncommon. A few such specialised designs may also be built as small telescopes for amateur use because of minor advantages in their production processes, or because of the challenge to the telescope maker's skills, or for space-saving reasons as in the folded Schiefspieglers and the Loveday design that utilises a double reflection from the primary mirror, but most such designs will be encountered very rarely.

In the radio and microwave regions (see Section 1.2) off-axis Gregorian designs are used (for example the Green Bank Radio telescope, Figure 1.80) since the secondary mirror is kept clear of the incoming beam of light. For the same reason, variations on the off-axis Gregorian design, known as Dragonian* telescopes and using a concave paraboloidal primary mirror and a concave hyperboloidal secondary mirror, are also to be found.

Of the dioptric (refracting objective) telescopes, only the basic refractor using an achromatic doublet, or very occasionally a triplet, as its objective is in any general use (Figure 1.62). Apart from the large refractors that were built towards the end of the nineteenth century, most refractors are now found as small instruments for the amateur market or as the guide telescopes of larger instruments. The enclosed tube and relatively firmly mounted optics of a refractor means that they need little adjustment once aligned with the main telescope.

The one remaining class of optical telescopes is the catadioptric group, of which the Schmidt camera is probably the best known. A catadioptric system uses both lenses and mirrors in its primary light gathering section. Very high degrees of correction of the aberrations can be achieved because of the wide range of variable parameters that become available to the designer in such systems. The Schmidt camera uses a spherical primary mirror so that coma is eliminated by not introducing it into the system in the first place! The resulting spherical aberration is eliminated by a thin correcting lens at the mirror's radius of curvature (Figure 1.63). The only major remaining aberration is field curvature and the effect of this is eliminated through the use of additional correcting lenses (field flatteners). The correcting lens can introduce small amounts of coma and chromatic aberration, but is usually so thin that these aberrations are negligible. Diffraction-limited performance over fields of view of several degrees with focal ratios as fast as f1.5 or f2 is possible.

The need to use a lens in the Schmidt design limits the sizes possible for them. Larger sizes can be achieved using a correcting mirror. The largest such instrument is the 4-metre

---

* After the design's originators Corrado Dragone, Yoshihiko Mizuguchi and Masataka Akagawa.

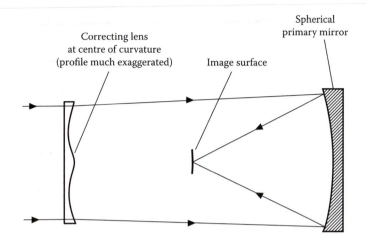

FIGURE 1.63   The Schmidt camera optical system.

Large Sky Area Multi-Object Fibre Spectroscopic Telescope (LAMOST) at the Xinglong Station of the Chinese National Astronomical Observatory. It operates using a horizontally fixed primary mirror fed by a coelostat (see below). The coelostat mirror is actively controlled and also acts as the corrector. With up to 4000 optical fibres, the instrument is one of the most rapid spectroscopic survey tools in the world and it started observations in May 2007. Other large Schmidt cameras include the 1.35-metre Tautenburg Schmidt and the Oschin Schmidt at Mount Palomar and the U.K. Schmidt camera at the AAO, both of which have entrance apertures 1.2 metres across.

The Schmidt design suffers from having a tube length at least twice its focal length. Also, the Schmidt camera cannot be used visually since its focus is inaccessible. There are several modifications of its design, however, that produce an external focus whilst retaining most of the beneficial properties of the Schmidt. One of the best of these is the Maksutov design (Figure 1.64) that originally also had an inaccessible focus, but which is now a kind of Schmidt-Cassegrain hybrid. All the optical surfaces are spherical and spherical aberration, astigmatism and coma are almost eliminated, whilst the chromatic aberration is almost negligible. A similar system is the Schmidt-Cassegrain telescope itself. This uses a thin correcting lens like the Schmidt camera and has a separate secondary mirror. The focus is accessible at the rear of the primary mirror as in the Maksutov system. Schmidt-Cassegrain telescopes are now available commercially in sizes up to 0.4 metre diameter. They are produced in large numbers by several firms at very reasonable costs for the amateur and education markets. They are also finding increasing use in specific applications, such as site testing for professional astronomy.

Although it is not a telescope, there is one further system that deserves mention here and that is the coelostat. This comprises two flat mirrors that are driven so that a beam of light from any part of the sky is directed towards a fixed direction. It is used particularly in conjunction with solar telescopes whose extremely long focal lengths make them impossible to move. One mirror of the coelostat is mounted on a polar axis and driven at half sidereal rate. The second mirror is mounted and driven to reflect the light into the fixed

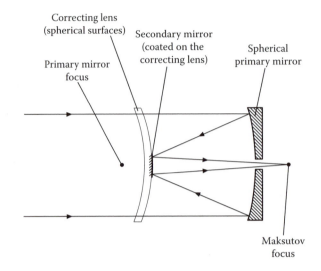

FIGURE 1.64   The Maksutov optical system.

telescope. Various arrangements and adjustments of the mirrors enable differing declinations to be observed. A related system is the siderostat which uses just a single flat mirror to feed a fixed telescope. The flat mirror has to be driven at variable speeds in altitude and azimuth to track an object across the sky, but even a small computer can undertake the calculations needed for this purpose and control the driving motors.

No major further direct developments in the optical design of large telescopes seem likely to lead to improved resolution, since such instruments are already limited by the atmosphere. Better resolution therefore requires that the telescope be lifted above the Earth's atmosphere, or that the distortions that the atmosphere introduces be overcome by a more subtle approach. Telescopes mounted on rockets, satellites and balloons are discussed later in this section, whilst the four ways of improving the resolution of Earth-based telescopes such as: interferometry, speckle interferometry, occultations and real-time compensation are discussed in Sections 2.5, 2.6, 2.7 and later in this section, respectively. A fairly common technique for improving resolution, deconvolution, is discussed in Section 2.1. This requires that the precise nature of the degradations of the image be known so that they can be removed. Since the atmosphere's effects are changing in a fairly random manner on a time scale of milliseconds, it is not an appropriate technique for application here, although it should be noted that one approach to speckle interferometry is in fact related to this method and deconvolution may be needed on the compensated image.

A technique termed apodisation (literally 'removal of the feet') (see also Sections 2.5 and 4.1) gives an apparent increase in resolution. It is actually, however, a method of enabling the telescope to achieve its normal resolution and sometimes even reduces that resolution slightly. For optical telescopes, apodisation is usually accomplished through the use of masks over the entrance aperture of the telescope. These have the effect of changing the diffraction pattern (see discussion about the Airy disk and also Section 2.1). The nominal resolution given by the size of the central disk may then worsen, but the outer fringes disappear or become of a more convenient shape for the purpose in mind. A couple of examples will illustrate the process. When searching for a faint object next to a bright one, such as Sirius B next to Sirius A, or for an exoplanet next to a star, a square aperture with fuzzy edges or a four-armed star aperture will produce fringes in the form of a cross and suppress the normal circular fringes. If the aperture is rotated so that the faint object lies in one of the regions between the arms of the cross, then it will have a much higher signal-to-noise ratio than if superimposed upon a bright fringe from a circular aperture and so be more easily detected. Second, for observing low-contrast features on planets, a mask that is a neutral density filter varying in opacity in a Gaussian fashion from clear at the centre to nearly opaque at the edge of the objective may be used. This has the effect of doubling the size of the Airy disk, but also of eliminating the outer fringes completely. The resulting image will be significantly clearer since the low-contrast features on extended objects are normally swamped by the outer fringes of the point spread function (PSF) (see Section 2.1). A professionally made neutral density filter of this type would be extremely expensive, but a good substitute may be made using a number of masks formed out of a mesh. Each mask is an annulus with an outer diameter equal to that of the objective and an inner diameter that varies appropriately in size. When they are all superimposed on the objective and provided that the meshes are *not* aligned, then they will obstruct an increasingly large proportion of the incoming light towards the edge of the objective. The overall effect of the mesh masks will thus be similar to the Gaussian neutral density filter.

Now although resolution may not be improved in any straightforward manner (see discussion on real-time atmospheric compensation), the position is less hopeless for the betterment of the other main function of a telescope, which is light gathering. The upper limit to the size of an individual mirror is probably being approached with the 8-metre telescopes presently in existence. A diameter of 10 to 15 metres for a monolithic metal-on-glass mirror would almost certainly be about the ultimate possible with present-day techniques and their foreseeable developments.* Greater light-gathering power may therefore be acquired only by the use of several mirrors, by the use of a single mirror formed from separate segments (see discussions elsewhere in this section), by aperture synthesis (see Section 2.5) or perhaps in the future by using liquid mirrors.

The accuracy and stability of surfaces, alignments and so forth required for both the multi-mirror and the segmented mirror systems are a function of the operating wavelength. For infrared and short microwave work, far less technically demanding limits are therefore possible and their telescopes are correspondingly easier to construct. Specialist infrared telescopes do, however, have other requirements that add to their complexity. First, it is often necessary to chop between the source and the background in order to subtract the latter. Some purpose-built infrared telescopes such as the 3.8-metre UKIRT oscillate the secondary mirror through a small angle to achieve this (a mass equal to that of the chopping mirror may need to be moved in the opposite direction to avoid vibrating the telescope). The UKIRT

---

* An additional practical constraint on mirror size is that imposed by the need to transport the mirror from the manufacturer to the telescope's observing site. Even with 8-metre mirrors, roads may need widening, obstructions levelling, bends have to have their curvature reduced etc. and this can add very significantly to the cost of the project.

FIGURE 1.65   The 2.5-metre SOFIA telescope seen through the open observing hatch in the side of its Boeing 747 aircraft in flight. (Reproduced by kind permission of NASA/Carla Thomas.)

secondary mirror can also be moved to correct for flexure of the telescope tube and mounting and for buffeting by the wind. The increasing size of infrared detector arrays, however, (Section 1.1.15.3) has now reduced the necessity for chopping in this manner.

For NIR observations, the telescope must be sited above as much of the atmospheric water vapour as possible, limiting the sites to places like Mauna Kea on Hawaii, the Chilean altiplano, or the Antarctic plateau. For MIR and some FIR work it may also be necessary to cool the telescope, or parts of it, to reduce its thermal emission and such telescopes have to be lifted above most of the atmosphere by balloon or aircraft or flown on a spacecraft.

SOFIA (Figure 1.65) is a 2.5-metre telescope flown on board a Boeing 747 aircraft which flies at altitudes up to 13,700 m and which recently started observing at infrared wavelengths. Its instruments can cover wavelengths from 300 nm to 1.6 mm.

The surface accuracy requirements may also be relaxed if the system is required as a light bucket rather than for imaging. Limits of ten times the operating wavelength may still give acceptable results when a photometer or spectroscope is used on the telescope, providing that close double stars of similar magnitudes and extended sources are avoided. The Observatoire de Haute Provence's CARLINA telescope concept, for example, envisages utilising a natural bowl-shaped valley on the Plateau de Calern* and lining it with up to a hundred smallish spherical mirrors to synthesise an unfilled spherical mirror up to 100 metres in diameter. The detectors and correcting optics would be suspended from a helium balloon at the focal point of the mirror. If operated with adaptive optics (see below)

---

* Compare this with the Arecibo radio telescope, Section 1.2.

and used as an interferometer such an instrument might achieve milliarc second (mas*) resolutions.

The atmosphere is completely opaque to radiation shorter than about 320 nm. Telescopes designed for UV observations therefore have to be launched on rockets or carried onboard spacecraft. Except in the EUV (~6 to ~90 nm), where glancing optics are needed (see Section 1.3), conventional telescope designs are used. Aluminium suffices as the mirror coating down to 100 nm. At shorter wavelengths, silicon carbide may need to be used.

1.1.20 Telescopes in Space

The most direct way to improve the resolution of a telescope that is limited by atmospheric degradation is to lift it above the atmosphere, or at least above the lower, denser parts of the atmosphere. The three main methods of accomplishing this are to place the telescope onto an aircraft, a balloon-borne platform or an artificial satellite. The Kuiper Airborne Observatory (KAO) (a 0.9-metre telescope on board a C141 aircraft that operated until 1995) and SOFIA are examples of the first, but aircraft are limited to maximum altitudes of about 15 km.

Balloon-borne telescopes, of which the 0.9-metre Stratoscope II is the best known but far from the only example, have the advantages of relative cheapness and that their images may be recorded by fairly conventional means. Their disadvantages include low maximum altitude (40 km) and short flight duration (a few days). There have also been problems with crash landings of the instruments upon their return to the Earth's surface, with ensuing extensive damage or even their complete destruction. For example, the Balloon-borne Large Aperture Submillimeter Telescope (BLAST), a 2-metre diameter sub-millimetre telescope, was so badly damaged after its third flight that it had to be completely re-built, and High-Energy Replicated Optics (HERO), an X-ray telescope, was a total write-off after a crash in 2010. BLAST was recently launched on its fifth and final mission and Sunrise, a 1-metre Gregorian solar telescope, successfully observed the Sun in 2009. Balloon-borne telescopes are often launched from near the North or South Poles where the wind patterns tend to cause them to move in circular paths around the pole. This both facilitates recovery and enables long-duration observations of single objects to be undertaken. Thus, the E and B Experiment (EBEX) telescope was recently launched from Antarctica. It comprises a 1.4-metre Dragone telescope with 1320 TES detectors observing the CMB between 150 and 450 GHz (2 mm and 670 μm).

Many small UV, visual and infrared telescopes have already been launched or are due for launch soon on satellites and mention has already been made of several of them: HST, Herschel, JWST, Spitzer and Gaia. Other examples include Kepler, launched in 2009 with a 0.95-metre telescope to search for exoplanets, Hipparcos (1989–1993), an astrometric

---

* The units of milliarc seconds (mas) and microarc seconds (μas) are widely used in astronomy. 1″ = 1000 mas and 1 mas = 1000 μas. To put these units in perspective, 1 μas is about the size of this letter 'O' when at the distance of the Moon (i.e. given a telescope with microarc second resolution you could easily read the headlines in a newspaper carried by an astronaut walking on the Moon). One milliarc second is about the size of an astronaut on the Moon seen from the Earth. The units of microradians (μrad) may also be encountered, and 1 μrad ≈ 0.206″ ≈ 206 mas and is about the size of the smallest crater that can be seen on the Moon from the Earth without correcting for scintillation (400 metres).

spacecraft carrying a 0.29-metre telescope and WISE (2009–2011) with a 0.4-metre telescope. Of course, in addition to these examples, most planetary exploration spacecraft such as Cassini, Messenger, New Horizons and Mars Reconnaissance Orbiter carry small optical telescopes. In the future it is possible that ambitious plans for the direct imaging of exoplanets using numerous spacecraft to form interferometers may be realised, but at the time of writing all such proposals remain simply concept studies.

1.1.21 Mountings

The functions of a telescope mounting are simple – to hold the optical components in their correct mutual alignment and to direct the optical axis towards the object to be observed. The problems in successfully accomplishing this to within the accuracy and stability required by astronomers are so great, however, that the cost of the mounting can be the major item in funding a telescope. We may consider the functions of a mounting under three separate aspects; first, supporting the optical components, second, preserving their correct spatial relationships and thirdly acquiring and holding the object of interest in the field of view.

Mounting the optical components is largely a question of supporting them in a manner that ensures their surfaces maintain their correct shapes. Lenses may only be supported at their edges and it is the impossibility of doing this adequately for large lenses that limits their practicable sizes to a little over a metre. There is little difficulty with smaller lenses and mirrors since most types of mount may grip them firmly enough to hold them in place without at the same time straining them. However, large mirrors require very careful mounting. They are usually held on a number of mounting points to distribute their weight and these mounting points, especially those around the edges, may need to be active so that their support changes to compensate for the different directions of the mirror's weight as the telescope moves. The active support can be arranged by systems of pivoted weights, especially on older telescopes or more normally nowadays by computer control of the supports. As discussed earlier some recently built telescopes and many of those planned for the future have active supports for the primary mirror that deliberately stress the mirror so that it retains its correct shape whatever the orientation of the telescope or the temperature of the mirror. Segmented mirror telescopes additionally need the supports to maintain the segments in their correct mutual alignments. On the 10-metre Keck telescopes for example, the active supports adjust the mirror segment's positions twice a second to maintain their places correctly to within ±4 nm.

The optical components are held in their relative positions by the telescope tube. With most large telescopes, the tube is in fact an open structure, but the name is still retained. For small instruments the tube may be made sufficiently rigid that its flexure is negligible. But this becomes impossible as the size increases. The solution then is to design the flexure so that it is identical for both the primary and secondary mirrors. The optical components then remain aligned on the optical axis, but are no longer symmetrically arranged within the mounting. The commonest structure in use that allows this equal degree of flexure and also maintains the parallelism of the optical components is the Serrurier truss (Figure 1.66). However, this design is not without its own problems. The lower trusses may need

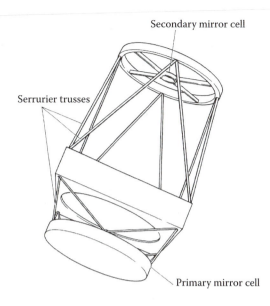

Secondary mirror cell

Serrurier trusses

Primary mirror cell

FIGURE 1.66   Telescope tube construction based on Serrurier trusses.

to be excessively thin or the upper ones excessively thick in order to provide the required amount of flexure. Even so, many of the large reflecting telescopes built in the latter half of the twentieth century have used Serrurier truss designs for their tubes. More recently, computer-aided design has enabled other truss systems to be developed. This, however, is mostly for economic reasons; the design objectives of such supports remain unchanged.

A widely used means of mounting the telescope tube so that it may be pointed at an object and then moved to follow the object's motion across the sky is the equatorial mounting. This is a two-axis mounting, with one axis, the polar axis, aligned parallel with the Earth's rotational axis and the other, the declination axis, perpendicular to the polar axis. The design has the enormous advantage that only a single constant velocity motor is required to rotate the mounting around the polar axis in order to track an object. It is also very convenient in that angular read-outs on the two axes give the hour angle or right ascension and declination directly.

A large variety of different arrangements for the basic equatorial mounting exist, but these will not be reviewed here. Books on telescopes and most general astronomy books list their details by the legion and should be checked if the reader requires further information.

The alt-az mounting which has motions in altitude and azimuth is the other main two-axis mounting system. Structurally it is a much simpler form than the equatorial and has therefore long been adopted for most large radio telescope dishes. Its drawbacks are that the field of view rotates with the telescope motion* and that it needs driving continuously

---

* Large alt-az-mounted telescopes use field de-rotators, often in the form of counter-rotating Dove prisms, to overcome this problem. Some manufacturers of small telescopes also offer field de-rotators as accessories, but they add considerably to the instrument's cost. A cheaper solution to obtaining sharp time-exposure images using small alt-az telescopes is to note that there are periods when the field rotation ceases for any object whose declination is nearer to the equator than the observer's latitude. Details of how to find such opportunities for an observer at a particular latitude and for an object at a particular declination are given, for example, in the author's *Telescopes and Techniques*, 3rd Edition, 2013.

in both axes and with variable speeds in order to track an object. For the last two decades most new large optical telescopes have also used alt-az mountings and the reduction in price and increase in capacity of small computers means that smaller telescopes even down to a few centimetres in size frequently have such a mounting as an option.

Telescopes up to 1 metre in diameter constructed for or by amateur astronomers are often mounted on a Dobsonian alt-azimuth mounting. This may be simply and cheaply constructed from sheets of plywood or similar material. The telescope is supported by two circles, which rest in two semi-circular cut-outs in a three-sided cradle shaped like an inverted Greek letter pi (Π). These allow the telescope to move in altitude. The cradle then pivots at its centre to give motion in azimuth. Telescopes on such mountings are usually undriven and are moved by hand to find and follow objects in the sky. Small telescopes on any alt-azimuth design of mounting can be given a tracking motion for a short interval of time by placing them on a platform that is itself equatorially driven. There are several designs for such equatorial platforms using sloping planes or inclined bearings and details may be found in modern books aimed at the amateur observer.

Large telescopes, whether on equatorial or alt-azimuth mountings, almost universally use hydrostatic bearings for the main moving parts. These bearings use a thin layer of pressurised oil between the moving and static parts and the oil is continuously circulated by a small pump. The two solid surfaces are thus never in direct contact and movement round the bearing is very smooth and of low friction.

Several telescopes have fixed positions. These include liquid-mirror telescopes that always point near the zenith and the 11-metre Hobby-Eberly and SALT telescopes. Tracking is then accomplished by moving the detector and any correcting optics to follow the motion of the image, or for short exposures, by moving the charges along the pixels of a CCD detector (TDI). Zenith-pointing telescopes are generally limited to a few degrees either side of the zenith, but the Hobby-Eberly telescope can point to objects with declinations ranging from –11° to +71°. It accomplishes this because it points to a fixed zenith angle of 35°, but can be rotated in azimuth between observations. There are also a few instruments mounted on altitude-altitude mountings. These use a universal joint (such as a ball and socket) and tip the instrument in two orthogonal directions to enable it to point anywhere in the sky.

With any type of mounting, acquisition and tracking accuracies of a second of arc or better are required, but these may not be achieved, especially for the smaller Earth-based telescopes. Thus, the observer may have to search for his or her object of interest over an area tens to hundreds of seconds of arc across after initial acquisition and then guide, either himself or herself, or with an automatic system, to ensure that the telescope tracks the object sufficiently closely for his or her purposes. With balloon-borne and space telescopes such direct intervention is difficult or impossible. However, the reduction of external disturbances and possibly the loss of weight mean that the initial pointing accuracy is higher. For space telescopes tracking is easy since the telescope will simply remain pointing in the required direction once it is fixed, apart from minor perturbations such as light pressure, the solar wind, gravitational anomalies and internal disturbances from the spacecraft. Automatic control systems can therefore usually be relied upon to operate the telescope.

Earlier spacecraft such as the International Ultraviolet Explorer (IUE, 1978–1996), however, had facilities for transmitting a picture of the field of view after the initial acquisition to the observer on the ground, followed by corrections to the position in order to set onto the desired object.

With terrestrial telescopes the guiding may be undertaken via a second telescope that is attached to the main telescope and aligned with it, or a small fraction of the light from the main telescope may be diverted for the purpose. The latter method may be accomplished by beam splitters, dichroic mirrors, or by intercepting a fraction of the main beam with a small mirror, depending upon the technique that is being used to study the main image. The whole telescope can then be moved using its slow-motion controls to keep the image fixed, or additional optical components can be incorporated into the light path whose movement counteracts the image drift, or the detector can be moved around the image plane to follow the image motion or the charges can be moved within a CCD chip (TDI). Since far less mass needs to be moved with the latter methods, their response times can be much faster than that of the first method. The natural development of the second method leads to the active surface control which is discussed later in this section and which can go some way towards eliminating the effects of scintillation.

If guiding is attempted by the observer then there is little further to add, except to advise a plentiful supply of black coffee in order to avoid going to sleep through the boredom of the operation. Automatic guiding has two major problem areas. The first is sensing the image movement and generating the error signal, whilst the second is the design of the control system to compensate for this movement. Detecting the image movement has been attempted in two main ways – CCDs and quadrant detectors.

With a CCD guide, the guide image is read out at frequent intervals and any movement of the object being tracked is detected via software that also generates the correction signals. The guide CCD is adjacent to the main detector (which may or may not be a CCD itself) and thus it guides on objects other than those being imaged. This is termed off-set guiding and often has advantages over guiding directly on the object of interest. The off-set guide star can be brighter than the object of interest and if the latter is an extended object, then guiding on it directly will be difficult. However, for comets, asteroids and other moving objects, the object itself must be followed. In some small CCD systems aimed at the amateur market, it is the main image that is read out at frequent intervals and the resulting multiple images are then shifted into mutual alignment before being added together. This enables poor tracking to be corrected provided that the image shift is small during any single exposure.

A quadrant detector is one whose detecting area is divided into four independent sectors, the signal from each of which can be separately accessed. If the image is centred on the intersection of the sectors, then their signals will all be equal. If the image moves, then at least two of the signals will become unbalanced and the required error signal can then be generated. A variant of this system uses a pyramidal prism. The image is normally placed on the vertex and so is divided into four segments each of which can then be separately detected. When guiding on stars by any of these automatic methods it may be advantageous to increase the image size slightly by operating slightly out of focus; scintillation jitter then becomes less important.

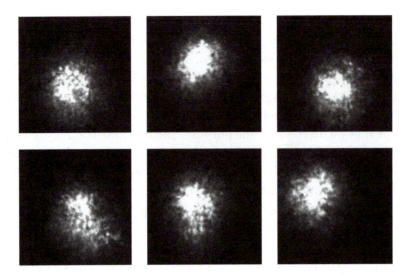

FIGURE 1.67 Scintillation and seeing. Images of a star obtained in July 2000 on the William Herschel telescope. The changing brightness (scintillation) and the image movement (seeing) are both apparent. The smallest of the individual speckles forming the images (see Section 2.6) are diffraction-limited images of the star from the 4.2-metre instrument. (Reproduced by kind permission of Nick Wooder. The full animation can be seen at http://apod.nasa.gov/apod/ap000725.html.)

The problem of the design of the control system for the telescope to compensate for the image movement should be handed to a competent control engineer. If it is to work adequately, then proper damping, feedback, rates of motion and so on, must be calculated so that the correct response to an error signal occurs without hunting or excessive delays.

1.1.22 Real-Time Atmospheric Compensation

The Earth's atmosphere causes astronomers many problems – it absorbs light from the objects that you want to look at, it scatters light into the telescope from objects that you do not want to look at, and its clouds, rain and snow are distinctly inconvenient at times. Here though we are concerned with two other effects that arise from turbulence and inhomogeneity within the atmosphere – scintillation (or twinkling) and seeing. Scintillation is the rapid change in the brightness of a stellar image* as more or less light from the incoming beam is scattered out of the beam by irregularities within the atmosphere (Figure 1.67). Seeing is the slight movement of the stellar image away from its true position arising from refraction at boundaries between different layers within the atmosphere.† Scintillation and seeing both change on time scales ranging from a hundredth to several seconds. Images obtained with exposures longer than around 0.01 s will therefore be blurred into the 'seeing

---

* Small fragments of the images of extended objects, like planets, also undergo scintillation. However, since the overall image is made up from many such fragments, some of which will be brightening whilst others will be getting fainter, the scintillation is much less apparent. The widely believed maxim that you can distinguish a star from a planet with the unaided eye because the latter does not twinkle arises from this phenomenon – but even planets will twinkle when the atmospheric turbulence is high enough.

† In the widely used Kolmogorov model of the atmosphere, the refractive index variations arise within thin layers of the atmosphere, each of which has a different intensity for the turbulence.

disk' whose diameter can be from 0.5 to 5 or more seconds of arc, irrespective of the actual resolution of the telescope's optics. Therefore, for about the last two decades increasingly successful attempts have been made to reduce or eliminate the effects of scintillation and seeing so that large telescopes can perform to their theoretical capabilities. One procedure for doing this is known as real-time atmospheric compensation (or sometimes as adaptive optics).* Alternative approaches such as interferometry, speckle interferometry and occultations are covered in Sections 2.5, 2.6, and 2.7, respectively.

The resolution of ground-based telescopes of more than a fraction of a metre in diameter is limited by the turbulence in the atmosphere. The maximum diameter of a telescope before it becomes seriously affected by atmospheric turbulence is given by Fried's coherence length, r_o,

$$r_o \approx 0.114 \left(\frac{\lambda \cos z}{550} \right)^{0.6} m \qquad (1.79)$$

where λ is the operating wavelength in nanometers and z is the zenith angle. Fried's coherence length, r_o, is the distance over which the phase difference is one radian. Thus, for visual work, telescopes of more than about 11.5 cm (4.5 inch) diameter will always have their images degraded by atmospheric turbulence.

In an effort to reduce the effects of this turbulence, many large telescopes are sited at high altitudes or placed on board high-flying aircraft or balloons. The ultimate, though very expensive solution of course, is to orbit the telescope beyond the Earth's atmosphere out in space. Recently, significant reductions in atmospheric blurring have been achieved through lucky imaging. In this process numerous short exposure images are obtained of the object (cf. speckle interferometry, Section 2.6). Only the sharpest are then combined to produce the final image. In this way the final image can approach the diffraction limit for 2- or 3-metre telescopes at visible wavelengths. However, up to 99% of the images may need to be discarded, making the process expensive in telescope time.

An alternative approach to obtaining diffraction-limited performance for large telescopes, especially in the NIR and MIR, which is relatively inexpensive and widely applicable, is to correct the distortions in the incoming light beam produced by the atmosphere. This atmospheric compensation is achieved through the use of adaptive optics. In such systems, one or more of the optical components can be changed rapidly and in such a manner that the undesired distortions in the light beam are reduced or eliminated. Although a relatively recent development in its application to large telescopes, such as the VLT, Gemini, Keck, William Herschel and Subaru, adaptive optics is actually a very ancient technique familiar to us all. That is because the eye operates via an adaptive optic system in order to keep objects in focus, with the lens being stretched or compressed by the ciliary muscle (Figure 1.1).

The efficiency of an adaptive optics system is measured by the Strehl ratio. This quantity is the ratio of the intensity at the centre of the corrected image to that at the centre of a

---

* Related procedures applicable in the microwave and radio regions are discussed in Section 1.2.

perfect diffraction-limited image of the same source. The normalised Strehl ratio is also used. This is the Strehl ratio of the corrected image divided by that for the uncorrected image. A Strehl ratio of close to 100% (perfection) has been realised at 10-μm wavelengths by the 6.5-metre MMT using a deformable secondary mirror, whilst recently the LBT has achieved 95% at 5 μm and 60% to 90% at 1.8 μm.

Adaptive optics is not a complete substitute for spacecraft-based telescopes, however, because in the visual and NIR, the correction only extends over a very small area (the isoplanatic patch, see below). Thus, for example, if applied to producing improved visual images of Jupiter, only a small percentage of the planet's surface would be seen sharply. Additionally, of course, ground-based telescopes are still limited in their wavelength coverage by atmospheric absorption.

There is often some confusion in the literature between adaptive optics and active optics. However, the most widespread usage of the two terms is that an adaptive optics system is a fast closed-loop system and an active optics system a more slowly operating open- or closed-loop system. The division is made at a response time of a few seconds. Thus, the tracking of a star by the telescope drive system can be considered as an active optics system, that is open-loop if no guiding is used and closed-loop if guiding is used. Large thin mirror optical telescopes and radio telescopes may suffer distortion due to buffeting by the wind at a frequency of a tenth of a hertz or so, they may also distort under gravitational loadings or thermal stresses and have residual errors in their surfaces from the manufacturing process. Correction of all of these effects would also be classified under active optics. There is additionally the term active support that refers to the mountings used for the optical components in either an adaptive or an active optics system.

An atmospheric compensation system contains three main components – a sampling system, a wave front sensor and a correcting system. We will look at each of these in turn.

1.1.22.1 Sampling System

The sampling system provides the sensor with the distorted wavefront or an accurate simulacrum thereof. For astronomical adaptive optics systems, a beam splitter is commonly used. This is just a partially reflecting mirror that typically diverts about 10% of the radiation to the sensor, whilst allowing the other 90% to continue on to form the image. A dichroic mirror can also be used which allows all the light at the desired wavelength to pass into the image whilst diverting light of a different wavelength to the sensor. However, atmospheric effects change with wavelength and so this latter approach may not give an accurate reproduction of the distortions unless the operating and sampling wavelengths are close together.

Since astronomers go to great lengths and expense to gather photons as efficiently as possible, the loss of even 10% to the sensor is to be regretted. Many adaptive optics systems therefore use a guide star rather than the object of interest to determine the wavefront distortions. This becomes essential when the object of interest is a large extended object, since most sensors need to operate on point or near-point images. The guide star must be very close in the sky to the object of interest, or its wavefront will have undergone different atmospheric distortion (Figure 1.68). For solar work small sunspots or pores can be used

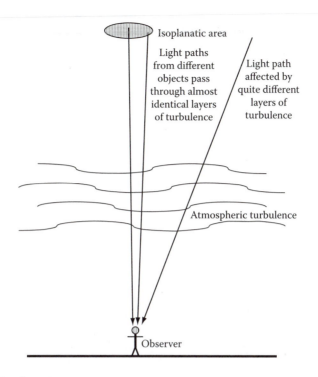

Isoplanatic area

Light paths
from different
objects pass
through almost
identical layers
of turbulence

Light path
affected by
quite different
layers of
turbulence

Atmospheric turbulence

Observer

FIGURE 1.68   The isoplanatic area.

as the guide object. The region of the sky over which images have been similarly affected by the atmosphere is called the isoplanatic area or patch. It is defined by the distance over which the Strehl ratio improvement due to the adaptive optics halves. In the visible it is about 15″ across. The size of the isoplanatic patch scales as $\lambda^{1.2}$, so it is larger in the infrared, reaching 80″ at 2.2 μm (K band, see Section 3.1). With several guide stars (Multi-Conjugate Adaptive Optics [MCAO] and Multi-Object Adaptive Optics [MOAO]) (see below), corrected areas of 1′ to 2′ are currently possible and plans for the EAGLE instrument on the E-ELT envisage corrected areas in the NIR up to 7′ across.

The small size of the isoplanatic area means that few objects have suitable guide stars. Less than 1% of the sky can be covered using real stars as guides, even in the infrared. Recently therefore, artificial guide stars have been produced. This is accomplished using a powerful laser* pointed skywards. The laser is tuned to one of the sodium D line frequencies and excites the free sodium atoms in the atmosphere at a height of about 90 km. The glowing atoms appear as a star-like patch that can be placed as near in the sky to the object of interest as required (Figure 1.69).

Guide stars at lower heights and at other wavelengths can be produced through back-scattering by air molecules of a laser beam. Since these latter guide stars are produced via

---

* Continuous wave lasers with powers of up to 50 W may be used. These radiate sufficient energy to cause skin burns and retinal damage. Care therefore has to be taken to ensure that the beam does not intercept an aircraft. The VLT for example uses a pair of cameras and an automatic detection system to close a shutter over the laser beam should an aircraft approach. The lasers are also powerful enough to damage the optics on board some spacecraft. Similar precautions thus need to be taken in this respect, although these are somewhat easier to accomplish since the positions of potentially vulnerable satellites in the sky can be predicted well ahead of time.

Rayleigh scattering, they are sometimes called Rayleigh stars. They can be of any wavelength, but a laser operating in the green is often chosen. The laser light can be sent out through the main telescope or more usually using an auxiliary telescope mounted on the main telescope.

Laser-produced guide stars have two problems that limit their usefulness. First, for larger telescopes, the relatively low height of the guide star means that the light path from it to the telescope differs significantly from that for the object being observed (the cone problem) (see Figure 1.70). At 1 μm, the gain in Strehl ratio is halved through this effect and in the visible it results in almost no improvement at all to the images. Second, the outgoing laser beam is affected by atmospheric turbulence and therefore the guide star moves with respect to the object. This limits the correction of the overall inclination (usually known as the tip-tilt) of the wavefront resulting in a blurred image on longer exposures. A real star can, however, be used to determine the tip-tilt of the wavefront separately. The wavefront sensor simply needs to detect the motion of the star's image and since the whole telescope

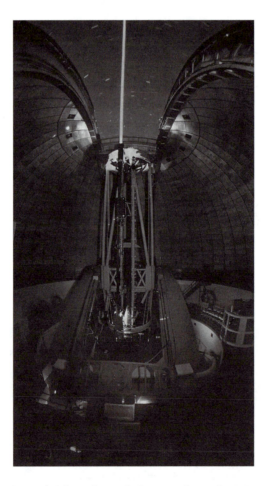

FIGURE 1.69 **(See color insert.)** A laser beam being sent from the 3.05-metre Shane telescope at the Lick Observatory to produce an artificial guide star for its adaptive optics system. The exposure is five minutes. (Reproduced by kind permission of L. Hatch. © 2002 Laurie Hatch. See: www.lauriehatch.com for many more of her beautiful astro-images.)

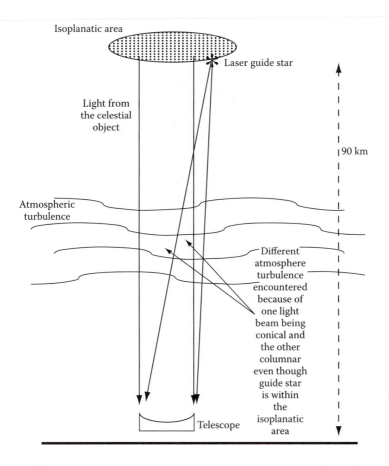

FIGURE 1.70 Light paths from a celestial object and a laser guide star to the telescope.

aperture can be utilised for this purpose, very faint stars can be observed. Most objects have a suitable star sufficiently nearby to act as a tip-tilt reference.

An adaptive optics system using two or more (up to a dozen) guide stars, MCAO, eliminates the cone effect and produces an isoplanatic patch up to 120″ across. The guide stars can be artificial and/or real stars and are separated by small angles and detected by separate wavefront sensors. This enables the atmospheric turbulence to be modelled as a function of altitude. Two or three subsidiary deformable mirrors then correct the wavefront distortion. A recent variation on MCAO is MOAO. This latter system also uses several guide stars and may be able to correct over areas up to 7 or more arc minutes across. However, the correction is only undertaken for a few selected smaller areas within the overall field of view – the selected smaller areas, of course, being chosen to coincide with the positions of the objects of interest to the observer.

1.1.22.2 Wavefront Sensing
The wavefront sensor detects the residual and changing distortions in the wavefront provided by the sampler after reflection from the correcting mirror. The Hartmann (also known as the Hartmann-Shack or the Shack-Hartmann) sensor is in widespread use in

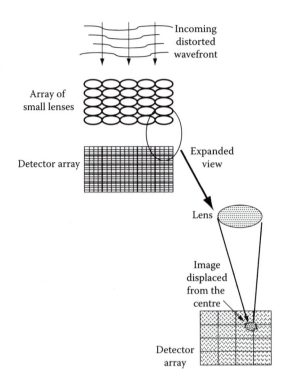

FIGURE 1.71   The Hartmann sensor.

astronomical adaptive optics systems. This uses a two-dimensional array of small lenses (Figure 1.71). Each lens produces an image that is sensed by an array detector. In the absence of wavefront distortion, each image will be centred on each detector. Distortion will displace the images from the centres of the detectors and the degree of displacement and its direction is used to generate the error signal. An alternative sensor is based upon the shearing interferometer (Figure 1.72). This is a standard interferometer but with the mirrors marginally turned so that the two beams are slightly displaced with respect to each other when they are recombined. The deformations of the fringes in the overlap region then provide the slopes of the distortions in the incoming wavefront. The shearing inter-ferometer was widely used initially for adaptive optics systems, but has now largely been replaced by the Hartmann sensor. A new sensor has been developed recently, known as the curvature sensor, which detects the wavefront distortions by comparing the illumination variations across slightly defocused images just inside and outside the focal point. A vibrat-ing mirror is used to change the focus at kilohertz frequencies. In the future for solar work and perhaps for other extended sources, a Mach-Zehnder interferometer* may be usable to estimate the wavefront distortions from the interference patterns between the central and outer parts of the wavefront.

---

* Essentially this is a double Michelson type interferometer that is used to determine the phase shift introduced into one of two light beams originating from a coherent source and arising from a sample placed within that light beam.

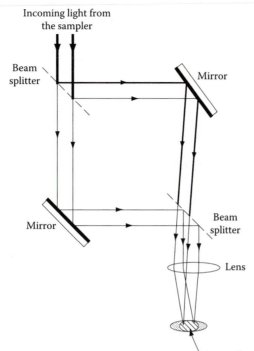

FIGURE 1.72　The shearing interferometer.

1.1.22.3 Wavefront Correction

In most astronomical adaptive optics systems, the correction of the wavefront is achieved by distorting a subsidiary mirror. Since the atmosphere changes on a time scale of 10 ms or so, the sampling, sensing and correction have to occur in a millisecond or less. In the simplest systems only the overall tip and tilt of the wave front introduced by the atmosphere is corrected. That is accomplished by suitably inclining a plane or segmented mirror placed in the light beam from the telescope in the opposite direction (Figure 1.73). Tip-tilt correction systems for small telescopes are now available commercially at about 20% of the cost of a 0.2-metre Schmidt-Cassegrain telescope and improve image sharpness by about a factor of two. An equivalent procedure, since the overall tilt of the wavefront causes the image to move, is shift and add. Multiple short exposure images are shifted until their brightest points are aligned and then added together. Even this simple correction, however, can result in a considerable improvement of the images.

More sophisticated approaches provide better corrections; either just of the relative displacements within the distorted wavefront, or of both displacement and fine scale tilt. In some systems the overall tilt of the wavefront is corrected by a separate system using a flat mirror whose angle can be changed. Displacement correction would typically use a thin mirror capable of being distorted by up to 100 piezo-electric or other actuators placed underneath it. The error signal from the sensor is used to distort the mirror in the opposite manner to the distortions in the incoming wavefront. The reflected wavefront is therefore

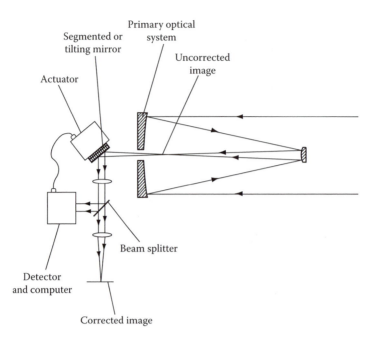

FIGURE 1.73   Schematic optical system for real-time atmospheric compensation.

almost flat. The non-correction of the fine scale tilt, however, does leave some small imperfections in the reflected wavefront. Nonetheless, currently operating systems using this approach can achieve diffraction-limited performance in the NIR for telescopes of 3 or 4 metres diameter (i.e. about 0.2″ at 2-μm wavelength).

Most large telescopes now have an adaptive optics image correction system available and many have MCAO systems available or under construction at the time of writing. The 6.5-metre MMT uses an adaptive secondary mirror 0.64 metres across for example. It is a 2-mm thick Zerodur plate attached to a thick base plate by 336 actuators. At MIR wavelengths a Strehl ratio of 0.98 can be achieved, compared with about 0.6 if the adaptive optics system is not used. The 10-metre Keck telescopes started operating with a single laser guide star in 2004 and now can achieve near diffraction-limited resolution in the NIR. An MCAO system – New Generation Adaptive Optics (NGAO) – is currently being planned for these telescopes that will provide diffraction-limited images from 800 nm to 2.4 μm over a 120″ field of view. The Gemini Multi-Conjugate Adaptive Optics System (GeMS) recently commissioned on the Gemini-South telescope provides a diffraction-limited field of view over 60″ across in the NIR using five laser guide stars. Whilst the 8.2-metre Subaru telescope uses a deformable mirror with 188 actuators in combination with a laser guide star and a second generation stellar coronagraph (High Contrast Instrument for the Subaru Next generation Adaptive Optics [HiCIAO]) (see Section 5.3) in an attempt to image extra-solar planets directly. Raven is due to be commissioned on Subaru as a demonstrator in 2013 and is a MOAO system with a planned field of view of 120″ using one laser guide star and three natural guide stars. For the Thirty Metre Telescope (TMT) it is planned to commission the Narrow Field Infrared Adaptive Optics System (NFIRAOS) to provide diffraction limited images over a 30″ area.

A recent proposal to improve the correcting mirrors is for a liquid mirror based upon reflective particles floating on a thin layer of oil. The oil contains nanometre-sized magnetic grains (a ferro-fluid) and its surface can be shaped by the use of small electromagnets, but this has yet to be applied to astronomical image correction. Potentially such mirrors could offer strokes (the distance moved up or down) measured in millimetres instead of the few tens of micrometres offered by current flexible mirrors. At visual wavelengths, reductions of the uncorrected image size by about a factor of 10 are currently being reached in the laboratory. Further improvements can sometimes be achieved by applying blind or myopic deconvolution (see Section 2.1) to the corrected images after they have been obtained.

Correction of fine scale tilt within the distorted wavefront as well as displacement is now being investigated in the laboratory. It requires an array of small mirrors rather than a single 'bendy' mirror. Each mirror in the array is mounted on four actuators so that it can be tilted in any direction as well as being moved linearly. At the time of writing it is not clear whether such systems will be applied to many large astronomical telescopes, because at visual wavelengths, few of them have optical surfaces good enough to take advantage of the improvements such a system would bring. Plans for future 50- and 100-metre telescopes, however, include adaptive secondary or tertiary mirrors up to 8 metres in diameter, requiring up to 500,000 actuators.

1.1.23 Future Developments

There are several telescopes significantly larger than 10 metres either now in operation or under construction. The VLT is currently starting to act as an aperture synthesis system (see Section 2.5) with a sensitivity equal to that of a 16-metre telescope and an unfilled aperture diameter of 100 metres. For some years the Keck telescopes operated as an aperture synthesis system with a sensitivity equal to that of a 14-metre telescope and an unfilled aperture diameter of 85 metres but their interferometer mode (not the telescopes) was mothballed in 2012. The Giant Magellan Telescope (GMT) has started construction with a planned completion date of 2020. It will be sited on Cerro Las Campanas in Chile and will comprise seven 8.4-metre monolithic mirrors in a close-packed array 25.4 metres across.

There are also two detailed proposals underway for even larger telescopes. The European Extremely Large Telescope is to have a 39-metre diameter primary mirror formed from seven hundred and ninety-eight 1.4-metre hexagonal segments (Figure 1.74) and has a planned completion date of early in the 2020s. The United States and Canada are pursuing the TMT project that is expected to have a primary mirror formed from 492 hexagonal segments and perhaps to start operations around 2020 (Figure 1.74). Less certain of completion is the already mentioned LAMA with a possible 42-metre equivalent aperture and based upon liquid mirrors.

With MCAO correction of the images (which might require 100,000 active elements), a 40-metre class telescope would have a resolution of 1 mas in the visible region. The limiting

FIGURE 1.74 (a) An artist's concept of the TMT in its dome. (Reproduced by kind permission of the TMT project.) (b) A mock-up showing the scale of the mirror for the E-ELT. (Reproduced by kind permission of the ESO. Faces have been obscured in line with the ESO's image use policy.)

magnitude for such a telescope might be 35^m in the visible, not just because of its increased light grasp, but because a 1-mas stellar image only has to equal the brightness of a 1-mas^2 area of the background sky in order to be detectable. This would enable the telescope to observe Jupiter-sized exoplanets directly out to a distance of about 100 pc.

As a general guide to further into the future we may look to the past. Figure 1.75 shows the way in which the collecting area of optical telescopes has increased with time.

Simple extrapolation of the trends shown there (dotted line) suggests that an optical telescope with a diameter of 100 m (or the equivalent total area) might be working around the year 2150. A quadratic extrapolation (solid line) suggest that target might be reached by 2080. However, there appear to be no fundamental technical differences between

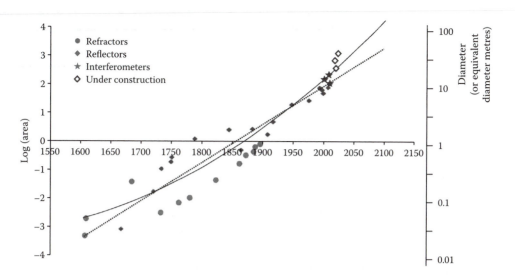

FIGURE 1.75   Optical telescope collecting area as a function of time. The logarithm to base 10 of the area in square metres is plotted over the last four centuries for a selection of telescopes that were, in their time, the largest examples of their type. Linear (dotted line) and quadratic (solid line) curves have been fitted to the data. For multi-mirror telescopes (like the GMT) and for interferometers (like the VLT interferometer [VLTI]), the diameter is that for the equivalent single circular aperture with the same total area.

constructing 10-metre segmented mirrors like those for the Keck telescopes and 30-metre or even 100-metre segmented mirrors. It is just a case of doing the same thing more times and finding the increased funding required. As discussed earlier, the fact that the largest fully steerable radio telescopes are about 100 metres in diameter, suggests that this sort of size is likely to be the upper limit for individual optical telescopes as well. Of course interferometers and aperture synthesis systems (see Section 2.5) conceivably could be made thousands of kilometres across like the radio very long baseline interferometers.

Traditionally the cost of a telescope is expected to rise as $D^{2.6}$ or thereabouts. Based upon the costs of the large telescopes built in the 1960s and 1970s, this would suggest a price tag in today's (2013) money for a 100-metre optical telescope of $200 billion to $300 billion. In 1992, however, the Keck I 10-metre telescope was built for $100 million – a quarter of the cost suggested by the $D^{2.6}$ formula. The estimated costs for the TMT and the E-ELT are currently around $1.3 billion and $1.2 billion, respectively. However, using the $D^{2.6}$ formula on the cost today ($165 million) of the Keck I telescope suggests that the more realistic estimates of $3 billion and $6 billion for these two instruments. Taking today's (probably optimistic) costings, however, puts the price tag on the 100-metre telescope at around $10 billion to $15 billion. Since the JWST is now expected to come in at some $10 billion, the price of the 100-metre telescope is not necessarily prohibitive. In addition to the capital costs however, the running costs of major telescopes need to be taken into account. These are usually put at around 5% per year of their capital costs – operating a 100-metre telescope would thus be likely to cost in excess of $500 million per year.

The multi-mirror concept may be extended or made easier by the development of fibre optics. Fused silica fibres can now be produced whose light losses and focal ratio degradation are acceptable for lengths of tens of metres. Thus, mirrors on independent mounts with a total area of hundreds of square metres which all feed a common focus will become technically feasible within a decade. Recently, the two Keck telescopes were linked by two 300-metre fluoride glass cables thus potentially providing an alternative to the Keck interferometer (see Section 2.5). A quite different future use of fibre optics may be the suppression of atmospheric emission lines in the NIR mentioned earlier.

At the other end of the scale, there is still a use for small telescopes – some of them being very small indeed. Super Wide Angle Search for Planets (SuperWASP) for example uses four 0.11-metre wide-angle lenses coupled to CCD cameras to search for exoplanets via transits. Likewise the Trans-Atlantic Exoplanet Survey (TrES) comprises three 0.1-metre instruments and Stellar Astrophysics and Research on Exoplanets (STARE) sited on Tenerife, Sleuth on Mount Palomar, Vulcan at the Lick Observatory and Planet Search Survey Telescope (PSST) at the Lowell observatory, Arizona, monitor thousands of stars at a time with small telescopes. A fifth exoplanet finder takes the record for being the baby amongst all research instruments. The Kilodegree Extremely Little Telescope (KELT) uses a 0.05-metre camera lens – the same size as the telescope that Newton showed to the Royal Society in 1761.

A technique that is in occasional use today but which may well gain in popularity is daytime observation of stars. Suitable filters and large focal ratios (that diminish the sky background brightness, but not that of the star) (see Section 1.1.19.1 and the discussion relating to Equation 1.74) enable useful observations to be made of stars as faint as the seventh magnitude. The technique is particularly applicable in the infrared where the scattered solar radiation is lower. If diffraction-limited 100-metre telescopes are ever built, then with 10-nm bandwidths they would be able to observe stars (or other sub-mas sources, but not extended objects) down to a visible magnitude of 23^m during the day.

Lenses seem unlikely to make a comeback as primary light-gathering components. However, they are still used extensively in eyepieces and in ancillary equipment. Three major developments seem likely to affect these applications, principally by making lens systems simpler and/or cheaper. The first is the use of plastic to form the lens. High-quality lenses then can be cheaply produced in quantity by moulding. The second advance is related to the first and it is the use of aspherical surfaces. Such surfaces have been used when they were essential in the past, but their production was expensive and time-consuming. The investment in an aspherical mould, however, would be small if that mould were then to produce thousands of lenses. Hence, cheap aspherical lenses could become available. The final possibility is for lenses whose refractive index varies across their diameter or with depth. Several techniques exist for introducing such variability to the refractive index of a material such as the diffusion of silver into glass and the growing of crystals from a melt whose composition varies with time. Lenses made from liquid crystal are now being made in small sizes. The refractive index in this material depends upon an externally applied electric field and can thus be changed rapidly. Such lenses may find application to wavefront correction in adaptive optics systems in due course.

At THz frequencies (tens to hundreds of microns wavelengths), lenses have recently been constructed from metal meshes embedded within a dielectric such as polypropylene. By using several meshes of variable grid sizes the refractive index can be varied radially resulting in a flat lens (sometimes called a Wood's lens*). The Luneberg lens (see Section 1.2) is a radio-frequency version of a Wood's lens in the shape of a sphere and with the refractive index varying radially from its centre to its surface. Yet another design for a flat lens relies upon variable phase shifts to the radiation across its surface. It is potentially usable from the NIR to the sub-millimetre region and could be largely aberration-free. At present, only available in the laboratory, it comprises a layer of gold a few nanometres thick deposited onto a silicon substrate. The gold layer is etched into closely spaced V-shaped ridges which act as antennas, receiving and then re-emitting the signal and so introducing a brief delay to it. The delays are tuned across the device's surface so that the radiation is brought to a focus, just as with a normal lens. Highly corrected and relatively cheap lens systems may thus become available within a few years through the use of some or all of these techniques and possibly for use in many regions of the spectrum.

Space telescopes seem likely to follow the developments of terrestrially based telescopes towards increasing sizes, although if terrestrial diffraction-limited telescopes with diameters of several tens of metres or more become available, the space telescopes will only have advantages for spectral regions outside the atmospheric windows. Other developments may be along the lines of the Laser Interferometer Space Antenna (LISA) (see Section 1.6) with two or more separately orbiting telescopes forming very high resolution interferometers.

A quite different approach that may yield great dividends in the future is the application of photonics to astronomical instrumentation. Photonics is the science of using e-m radiation to perform some of the functions traditionally associated with electronics. Most astronomical applications that can be classed under this heading so far are related to fibre optics, but beam combiners for infrared interferometers (see Section 2.5) have recently been fabricated by laser writing onto chalcogenide glass (infrared-transmitting glass containing sulphur, selenium, tellurium etc.).

1.1.24 Observing Domes, Enclosures and Sites

Any permanently mounted optical telescope requires protection from the elements. Traditionally this has taken the form of a hemispherical dome with an aperture through which the telescope can observe. The dome can be rotated to enable the telescope to observe any part of the sky and the aperture may be closed to protect the telescope during the day and during inclement weather. Many recently built large telescopes have used cylindrical or other shapes for the moving parts of the enclosure for economic reasons; however, such structures are still clearly related to the conventional hemisphere. The dome, however, is an expensive item and it can amount to a third of the cost of the entire observatory, including the telescope. Domes and other enclosures also cause problems through heating up during the day so inducing convection currents at night and through the generation of eddies

---

* Robert Wood produced cylindrical lenses with variable refractive indices in 1905 via a gelatine dipping technique.

as the wind blows across the aperture. Future very large telescopes may therefore operate without any enclosure at all, just having a movable shelter to protect them when not in use. This will expose the telescopes to wind buffeting, but active optics can now compensate for that at any wind speed for which it is safe to operate the telescope.

The selection of a site for a major telescope is at least as important for the future usefulness of that telescope as the quality of its optics. The main aim of site selection is to minimise the effects of the Earth's atmosphere, of which the most important are usually scattering, absorption and scintillation.

Scattering by dust and molecules causes the sky background to have a certain intrinsic brightness and it is this that imposes a limit upon the faintest detectable object through the telescope. The main source of the scattered light is artificial light and most especially street lighting. Thus, a first requirement of a site is that it be as far as possible from built-up areas. If an existing site is deteriorating due to encroaching suburbs etc. then some improvement may be possible for some types of observation by the use of a light pollution rejection (LPR) filter which absorbs in the regions of the most intense sodium and mercury emission lines. Scattering can be worsened by the presence of industrial areas upwind of the site or by proximity to deserts, both of which inject dust into the atmosphere.

Absorption is due mostly to the molecular absorption bands of the gases forming the atmosphere. The two well-known windows in the spectrum, wherein radiation passes through the atmosphere relatively unabsorbed, extend from about 360 nm to 100 μm and from 10 mm to 100 m (Figure 1.25). But even in these regions there is some absorption, so that visible light is decreased in its intensity by 10% to 20% for vertical incidence. The infrared region is badly affected by water vapour, OH and other molecules to the extent that portions of it are completely obscured. Thus, the second requirement of a site is that it be as high an altitude as possible to reduce the air paths to a minimum and that the water content be as low as possible. We have already seen how balloons and spacecraft are used as a logical extension of this requirement and a few high-flying aircraft are also used for this purpose. The possibility of using a fibre-optic Bragg grating to reduce the sky background in the NIR by a factor of up to 50 has been mentioned earlier.

Scintillation has been discussed above in relation to real-time atmospheric compensation. It is the primary cause of the low resolution of large telescopes, since the image of a point source is rarely less than half a second of arc across due to this blurring. Thus, the third requirement for a good site is a steady atmosphere. The ground-layer effects of the structures and landscape in the telescope's vicinity may worsen scintillation. A rough texture to the ground around the dome, such as low-growing bushes, seems to reduce scintillation when compared with that found for smooth (e.g. paved) and very rough (e.g. tree-covered) surfaces. Great care also needs to be taken to match the dome, as telescope and mirror temperatures to the ambient temperature or the resulting convection currents can worsen the scintillation by an order of magnitude or more. Cooling the telescope during the day to the predicted nighttime temperature is helpful, provided that the weather forecast is correct. Forced air circulation by fans, and thermally insulating and siting any ancillary equipment that generates heat as far from the telescope as possible, or away from the dome completely, can also reduce convection currents.

These requirements for an observing site restrict the choice very considerably and lead to a clustering of telescopes on the comparatively few optimum choices. Most are now found at high altitudes and on oceanic islands, or with the prevailing wind from the ocean. Built-up areas tend to be small in such places and the water vapour is usually trapped near sea level by an inversion layer. The long passage over the ocean by the winds tends to minimise dust and industrial pollution. The Antarctic plateau with its cold, dry and stable atmospheric conditions is also a surprisingly good observing site. The Antarctic plateau is also optimum for radio studies of the aurora etc. with, for example, the European Incoherent Scatter (EISCAT) fixed and movable dishes operating at wavelengths of a few hundreds of millimetres.

The recent trend in the reduction of the real cost of data transmission lines, whether these are via optic fibres, cables or satellites, is likely to lead to the greatly increased use of remote control of telescopes. In a decade or two, most major observatories are likely to have only a few permanent staff physically present at the telescope, wherever that might be in the world, and the astronomer would only need to travel to a relatively nearby control centre for his or her observing shifts. Indeed it seems quite likely that in some cases the astronomer will be able to sit in his/her normal office and operate a multi-metre telescope thousands of kilometres away via the Internet in the near future.

There are also already a few observatories with completely robotic telescopes, in which all the operations are computer-controlled, with no staff on site at all during their use. Many of these robotic instruments are currently used for long-term photometric monitoring programs or for teaching purposes. However, their extension to more exacting observations such as spectroscopy is now starting to occur. The Liverpool telescope on La Palma and the two Faulkes telescopes in Hawaii and Australia are 2-metre robotic telescopes that can be linked to respond to gamma-ray burst (GRB) alerts in under 5 minutes. Five per cent of the observing time on the Liverpool telescope and most of the time on the Faulkes telescopes is devoted to educational purposes with students able to control the instruments via the Internet. Other examples of robotic telescopes include the Rapid Eye Mount (REM) a 0.6-metre instrument designed for GRB observing and sited at La Silla, Panchromatic Robotic Optical Monitoring and Polarimetry Telescopes (PROMPT) that uses six 0.4-metre telescopes covering the violet to NIR regions and designed for GRB and supernova observations as well as for education, and Peters Automated Infrared Imaging Telescope (PAIRITEL) located on Mount Hopkins that utilises the 1.3-metre telescope constructed for the 2-Micron All Sky Survey (2MASS) and is also for GRB observing.

EXERCISES

1.1 Calculate the effective theoretical resolutions of the normal and the dark-adapted eye, taking their diameters to be 2 and 6 mm, respectively.

1.2 Calculate the focal lengths of the lenses of a cemented achromatic doublet, whose overall focal length is 4 m. Assume the contact surfaces are of equal radius, the second surface of the diverging lens is flat and that the components are made from crown and dense flint glasses. Correct it for wavelengths of 486 and 589 nm.

1.3 By ray tracing, calculate the physical separation in the focal plane of two rays that are incident onto a parabolic mirror of focal length 4 m. The rays are parallel prior to reflection and at an angle of 1° to the optical axis. The first ray intersects the mirror at its centre, whilst the second intersects it 0.5-metres from the optical axis and in the plane containing the rays and the optical axis (Figure 1.76).

1.4 Calculate the maximum usable eyepiece focal length for the 5-metre f3.3/f16 Mount Palomar telescope at its Cassegrain focus and hence its lowest magnification when used visually.

1.5 Taking Newton's reflecting telescope to have a diameter of 0.05 metres and the dark-adapted eye to respond to an effective wavelength of 510 nm, what would be the maximum zenith distance that Newton could have observed objects in the sky without being limited by atmospheric turbulence?

1.6 If the E-ELT, using adaptive optics, were to achieve its diffraction limited resolution at a wavelength of 1.5 μm would it be able to resolve the exoplanet α Cen B b from its host star α Cen B? (Ignore problems of contrast and whether the planet would be bright enough to be detected.)

Data – exoplanet's orbital radius – 0.004 AU, distance of the α Cen B system from the Earth – 1.4 pc.

1.7 Calculate the mean value, standard deviation and standard error of the mean of the following set of estimates of the distance to a supernova remnant that have been obtained by a variety of different methods. Express the final result correctly.

Data – distance estimates (pc):- 730, 890, 1100, 639, 670, 745, 900, 760

1.8 Calculate Fried's coherence length for visual observations (i.e. λ = 500 nm) for altitudes of 20°, 40°, 60°, 80° and 90°. What is the minimum altitude that a 100-mm telescope could be used at its diffraction limit (all other things permitting)?

FIGURE 1.76 Optical arrangement for Exercise 1.3.

1.2 RADIO AND MICROWAVE DETECTION

1.2.1 Introduction

Radio astronomy is the oldest of the 'new' astronomies since it is now approaching a century since Karl Jansky first observed radio waves from the Milky Way galaxy. It has passed beyond the developmental stage wherein some of the other new astronomies – neutrino astronomy, gravity wave astronomy etc. – still remain. It thus has quite well established instruments and techniques that are not likely to change much.

The reason why radio astronomy developed relatively early is that radio radiation penetrates to ground level (Figure 1.25). For wavelengths from about 10 mm to 10 m, the atmosphere is almost completely transparent. The absorption becomes almost total at about 0.5 mm wavelength and between 0.5 and 10 mm there are a number of absorption bands that are mainly due to oxygen and water vapour, with more or less transparent windows between the bands. The scale height for water vapour in the atmosphere is about 2000 m, so that observing from high altitudes reduces the short-wave absorption considerably. Radiation with wavelengths longer than about 50 m again fails to penetrate to ground level, but this time the cause is reflection by the ionosphere.

Thus, this section is concerned with the detection of radiation with wavelengths longer than 0.1 mm. That is frequencies less than 3×10^{12} Hz, or photon energies less than 2×10^{-21} J (0.01 eV). The detection of radiation in the 0.1 to a few millimetres wavelength region using bolometers is covered under infrared detectors in Section 1.1. Also, this section is primarily concerned with individual radio telescopes, whilst two or more individual telescopes acting together as interferometers and aperture synthesis systems are covered in Section 2.5. However, there is inevitably some overlap and aspects of telescope arrays such as ALMA and SKA are included in this section where it seems appropriate. The detection of the radio and microwave emission from cosmic ray showers in the Earth's atmosphere is covered in Section 1.4. Whilst the detection of neutrinos, via their induced radio emission, is covered in Section 1.5.

The unit of intensity that is commonly used at radio wavelengths as a measure of the intensity of point sources is the jansky (Jy)

$$1 \text{ Jy} = 10^{-26} \text{ Wm}^{-2} \text{ Hz}^{-1} \tag{1.80}$$

and detectable radio sources vary from about 10^{-6} to 10^6 Jy. Most radio sources of interest to astronomers generate their radio flux as thermal radiation, when the Rayleigh-Jeans law gives their spectrum

$$F_\nu = \frac{2\pi k}{c^2} T \nu^2 \tag{1.81}$$

or as synchrotron radiation from energetic electrons spiralling around magnetic fields, when the spectrum is of the form

$$F_\nu \propto \nu^{-\alpha} \tag{1.82}$$

where F_ν is the flux per unit frequency interval at frequency ν. α is called the spectral index of the source and is related to the energy distribution of the electrons. For many sources $0.2 \leq \alpha \leq 1.2$.

For extended sources the unit of Jy sr^{-1} is sometimes used, or more frequently, the brightness temperature. The latter is defined as the temperature of a black body that would emit the same intensity of radiation as the observed object at the selected frequency.

1.2.2 Detectors and Receivers

The detection of radio signals is a two-stage process in which the sensor produces an electrical signal that then has to be processed until it is in a directly usable form. Coherent detectors, which preserve the phase information of the signal, are available for use over the whole of the radio spectrum, in contrast to the optical and infrared detectors discussed in Section 1.1 that respond only to the total power of the signal. The detection of sub-millimetre and millimetre radiation (FIR) via the use of bolometers is covered in Section 1.1. In the megahertz radio region, the sensor is normally a dipole placed directly at the focus of the telescope, such as the half-wave dipole shown in Figure 1.77, although this can only be optimised for one wavelength and has very restricted bandwidths. The two halves of such a dipole are each a quarter of a wavelength long. Connection to the remainder of the system is by coaxial cable.

In the gigahertz and higher frequency regions, a horn antenna is normally used to collect the radiation, usually with waveguides for the connection to the rest of the system, though plastic, quartz, metal grid and other lenses may be used at very high frequencies. Recent developments in the design of horn antennas have enabled them to have much wider

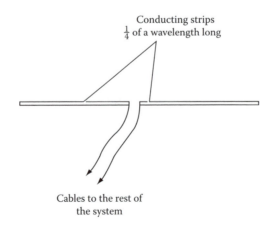

Conducting strips
$\frac{1}{4}$ of a wavelength long

Cables to the rest of
the system

FIGURE 1.77   A half-wave dipole.

bandwidths than earlier designs. These developments include using a corrugated internal surface for the horn and reducing its diameter in a series of steps or a smooth surface with step changes in the cone angle, giving a bandwidth that covers a factor of two in terms of frequency. Bandwidths with factors of nearly eight in frequency are possible with dielectric loaded horns. These are smooth-walled horns filled with an appropriate dielectric except for a small gap between the filling and the walls of the horn through which the wave propagates. At high frequencies the feed horns may need cooling to cryogenic temperatures.

The sensor at the higher frequencies is nowadays normally a superconductor-insulator-superconductor (SIS) device. In an SIS detector, an electron in one superconducting film absorbs a photon, giving the electron enough energy to tunnel through the insulating barrier into the other superconducting film (cf. STJs, Section 1.1). This process, known as photon-assisted-tunnelling, produces one electron for every absorbed photon. Modern devices are based upon two niobium layers separated by an insulating region of aluminium oxide around 1 nm thick and the whole cooled to 4 K or less. Such SIS devices can operate up to about 700 GHz (400 µm). SIS devices can be used up to 1.2 THz (250 µm) using niobium titanium nitride and this may be extended in the future to 2 THz (150 µm). Even higher frequencies require the use of Schottky diodes, or more recently TES devices (Section 1.1) based upon niobium and generally known amongst radio astronomers as hot electron bolometers (HEBs).

The 12-metre APEX telescope, for example, started observing with its CO N+ Deuterium Observation Receiver (CONDOR)* in 2005. This detector uses a niobium-titanium nitride HEB cooled to 4K to convert the THz radiation to GHz frequencies and thereby observe over the 1.25 to 1.52 THz (240 to 200 µm) region. In addition to SCUBA-2 (Section 1.1) the 15-metre JCMT uses a 4 × 4 array of SIS detectors, known as Heterodyne Array Receiver Programme (HARP), to obtain spectra and images over the 325- to 375-GHz (920 to 800 µm) region. Recently, in the laboratory, an HEB-based heterodyne receiver using a laser as the local oscillator has successfully operated at nearly 5 THz (60 µm). MKIDs and STJs (see Section 1.1) have the potential to become detectors of microwave and even radio radiation. At the time of writing, MKIDs are being investigated as possible detectors for radiation of a few hundreds of megahertz (sub-metre wavelengths).

The signal from the sensor is carried to the receiver whose purpose is to convert the high-frequency electrical currents into a convenient form. The behaviour of the receiver is governed by five parameters: sensitivity, amplification, bandwidth, receiver noise level and integration time.

The sensitivity and the other parameters are very closely linked, for the minimum detectable brightness, B_{min}, is given by

$$B_{min} = \frac{2k\nu^2 K T_s}{c^2 \sqrt{t \Delta \nu}}$$
(1.83)

---

* The CONDOR was a visitor instrument from the Max-Planck-Institut für Radioastronomie and it now forms the low-frequency part of the German Receiver for Astronomy at Terahertz Frequencies (GREAT) on board SOFIA.

where T_s is the noise temperature of the system, t is the integration time, $\Delta\nu$ is the frequency bandwidth and K is a constant close to unity that is a function of the type of receiver. The bandwidth is usually measured between output frequencies whose signal strength is half the maximum when the input signal power is constant with frequency. The amplification and integration time are self-explanatory, so that only the receiver noise level remains to be explained. This noise originates as thermal noise within the electrical components of the receiver and may also be called Johnson or Nyquist noise (see also Section 1.1). The noise is random in nature and is related to the temperature of the component. For a resistor, the root mean square (RMS) voltage of the noise per unit frequency interval, $\bar{V}$, is given by

$$\bar{V} = 2\sqrt{kTR} \tag{1.84}$$

where R is the resistance and T is the temperature. The noise of the system is then characterised by the temperature T_s that would produce the same noise level for the impedance of the system. It is given by

$$T_s = T_1 + \frac{T_2}{G_1} + \frac{T_3}{G_1 G_2} + \dots \frac{T_n}{G_1 G_2 \cdots G_{n-1}} \tag{1.85}$$

where T_n is the noise temperature of the nth component of the system and G_n is the gain (or amplification) of the nth component of the system. It is usually necessary to cool the initial stages of the receiver with liquid helium in order to reduce T_s to an acceptable level. Other noise sources that may be significant include shot noise resulting from random electron emission, g-r noise due to a similar effect in semiconductors (see Section 1.1), noise from the parabolic reflector or other collector that may be used to concentrate the signal onto the antenna, radiation from the atmosphere and last but by no means least, spill-over from radio taxis, microwave ovens and other artificial sources.

Many types of receiver exist; the simplest is a development of the heterodyne system employed in the ubiquitous transistor radio. The basic layout of a heterodyne receiver is shown in block form in Figure 1.78. The pre-amplifier operates at the signal frequency and will typically have a gain of 10 to 1000. It is often mounted close to the feed and cooled to near absolute zero to minimise its contribution to the noise of the system. The most widely used amplifiers today are based upon cooled gallium arsenide and indium phosphide high electron mobility transistors (HEMTs), also known as heterostructure field effect transistors (HFETs), in which the current-carrying electrons are physically separated from the donor atoms. The current-carrying electrons are restricted to a thin (10 nm) layer of undoped material producing a fast, low-noise device. Above 40 GHz (7.5 mm) the mixer must precede the pre-amplifier in order to decrease the frequency before it can be amplified; a second lower frequency local oscillator is then employed to reduce the frequency even further.

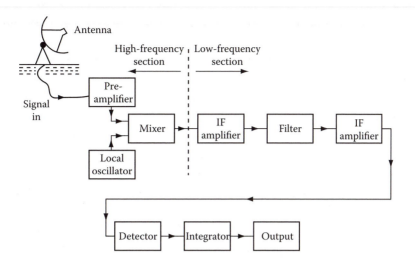

FIGURE 1.78 Block diagram of a basic heterodyne receiver.

The local oscillator produces a signal that is close to but different from the main signal in its frequency. Thus, when the mixer combines the main signal and the local oscillator signal, the beat frequency between them (intermediate frequency [IF]) is at a much lower frequency than that of the original signal. The relationship is given by

$$\nu_{\text{SIGNAL}} = \nu_{\text{LO}} \pm \nu_{\text{IF}} \tag{1.86}$$

where ν_{SIGNAL} is the frequency of the original signal (i.e. the operating frequency of the radio telescope), ν_{LO} is the local oscillator frequency and ν_{IF} is the IF. Normally, at lower frequencies, only one of the two possible signal frequencies given by Equation 1.86 will be picked up by the feed antenna or passed by the pre-amplifier. At high frequencies, both components may contribute to the output.

The power of the IF emerging from the mixer is directly proportional to the power of the original signal. The IF amplifiers and filter determine the pre-detector bandwidth of the signal and further amplify it by a factor of 10^6 to 10^9. The detector is normally a square-law device; that is to say, the output voltage from the detector is proportional to the square of the input voltage. Thus, the output *voltage* from the detector is proportional to the input *power*. In the final stages of the receiver, the signal from the detector is integrated, usually for a few seconds, to reduce the noise level. Then it is fed to an output device, usually an analogue-to-digital input to a computer for further processing. Further advances in speed and noise reduction come from combining the sensor and much of the electronics onto monolithic microwave integrated circuits (MMICs). With MMICs and other coherent devices there is a noise source, termed the quantum limit (see also shot noise, Section 1.1) that arises from fluctuations in the number of photons (quanta) being collected. The equivalent temperature is given by

$$T_{\text{Quantum limit}} = \frac{h\nu}{k} \approx 5 \times 10^{-11} \nu \, K \tag{1.87}$$

Thus, at gigahertz frequencies, the quantum limit temperature is generally too low to be significant, but at terahertz frequencies it rises to 50 to 50,000 K and becomes the dominant noise source. MMICs currently perform at about five to ten times the quantum limit for frequencies less than 150 GHz (2 mm) and it is hoped to reduce this to two or three times the quantum limit in the near future. MMICs are cheap and are used for example in the SETI to construct low-noise amplifiers, filters and local oscillators.

The basic heterodyne receiver has a high system temperature and its gain is unstable. The temperature may be lowered by applying an equal and opposite voltage in the later stages of the receiver and the stability of the gain may be greatly improved by switching rapidly from the antenna to a calibration noise source and back again, with a phase-sensitive detector (see Section 3.1) to correlate the changes. Such a system is then sometimes called a Dicke radiometer. The radiometer works optimally if the calibration noise source level is the same as that of the signal, and so it may be further improved by continuously adjusting the noise source to maintain the balance and it is then termed a null-balancing Dicke radiometer. Since the signal is only being detected half the time the system is less efficient than the basic receiver, but using two alternately switched receivers will restore its efficiency. The value of T_s for receivers varies from 10 K at metre wavelengths to 10,000 K at millimetre wavelengths. The noise sources must therefore have a comparable range, and at long wavelengths are usually diodes, whilst at the shorter wavelengths a gas discharge tube inside the waveguide and inclined to it by an angle of about 10° is used.

Heterodyne detectors have now been developed to operate at sub-millimetre wavelengths and one such is HIFI that is carried on board the Herschel spacecraft (Figure 1.79). HIFI is a spectrometer that observes over seven wavebands in the region from 480 GHz to 1.91 THz (625 to 160 μm).

Receivers are generally sky background limited just like terrestrial optical telescopes. The Earth's atmosphere radiates at 100 K and higher temperatures below a wavelength of about 3 mm. Only between 30 and 100 mm does its temperature fall as low as 2 K. Then,

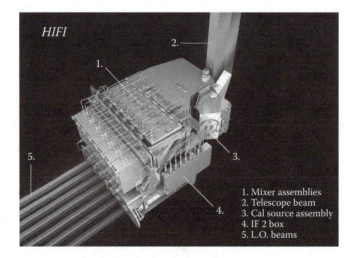

FIGURE 1.79   An artist's impression of HIFI. (Reproduced by kind permission of ESA.)

at longer wavelengths, the galactic emission becomes important, rising to temperatures of 10^5 K at wavelengths of 30 m.

Spectrographs at radio frequencies can be obtained in several different ways. In the past, the local oscillator has been tuned, producing a frequency-sweeping receiver, or the receiver been a multichannel device so that it registered several different discrete frequencies simultaneously. For pulsar observations such filter banks may have several hundred channels over a bandwidth of a few megahertz. Today, most radio spectroscopy is carried out by auto-correlation, even at the highest frequencies. Successive delays are fed into the signal that is then cross-correlated with the original signal in a computer. The spectrum is obtained from the Fourier transform of the result. The polarisation of the original signal is determined by separately detecting the orthogonal components and cross-correlating the electrical signals later within the receiver.

Alternatively, the radio signal may be converted into a different type of wave and the variations of this secondary wave studied instead. This is the basis of the acousto-optical radio spectrometer (AOS). The radio signal is converted into an ultrasonic wave whose intensity varies with that of the radio signal and whose frequency is also a function of that of the radio signal. In the first such instruments, water was used for the medium in which the ultrasound propagated and the wave was generated by a piezo-electric crystal driven either directly from the radio signal or by a frequency-reduced version of the signal. More recently, materials such as fused silica, lithium niobate and lead molybdate have replaced water as the acoustic medium in order to improve the available spectral range. A laser illuminates the cell containing the acoustic medium and a part of the light beam is diffracted by the sound wave. The angle of diffraction depends upon the sound wave's frequency, whilst the intensity of the diffracted light depends on the sound wave's intensity. Thus, the output from the device is a fan beam of light, the position within which ultimately depends upon the observed radio frequency and whose intensity at that position ultimately depends upon the radio intensity. The fan beam may then simply be detected by a linear array of optical detectors or by scanning and the spectrum inferred from the result. The AOS initially found application to the observation of solar radio bursts (see Section 5.3) at metre wavelengths, but is now employed more in the sub-millimetre and infrared regions, since auto-correlation techniques have replaced it at the longer wavelengths and are even starting to do so in the millimetre region.

A major problem at all frequencies in radio astronomy is interference from artificial noise sources. In theory, certain regions of the spectrum are reserved partially or exclusively for use by radio astronomers. An up-to-date listing of the reserved frequencies may be found on the Committee on Radio Astronomy Frequencies' (CRAF) web site at http://www.craf.eu/. But leakage from devices such as microwave ovens, incorrectly tuned receivers and illegal transmissions often overlap into these bands. The use of highly directional aerials reduces the problem to some extent. But it is likely that radio astronomers will have to follow their optical colleagues to remote parts of the globe or place their aerials in space if their work is to continue in the future. Even the latter means of escape may be threatened by solar power satellites with their potentially enormous microwave transmission intensities.

1.2.3 Radio Telescopes

The antenna and receiver, whilst they are the main active portions of a radio detecting system, are far less physically impressive than the large structures that serve to gather and concentrate the radiation and to shield the antenna from unwanted sources (Figure 1.80). Before going on, however, to the consideration of these large structures that form most people's ideas of what comprises a radio telescope, we must look in a little more detail at the physical background of antennae.

The theoretical optics of light and radio radiation are identical, but different traditions within the two disciplines have led to differences in the mathematical and physical formulations of their behaviours. Thus, the image in an optical telescope is discussed in terms of its diffraction structure (Figure 1.37 for example), whilst that of a radio telescope is discussed in terms of its polar diagram. However, these are just two different approaches to the presentation of the same information. The polar diagram is a plot, in polar coordinates, of the sensitivity or the voltage output of the telescope, with the angle of the source from the optical axis. (Note that the polar diagrams discussed herein are all far-field patterns, i.e. the response for a distant source, near-field patterns for the aerials may differ from those shown here.) The polar diagram may be physically realised by sweeping the telescope past a point source, or by using the telescope as a transmitter and measuring the signal strength around it.

The simplest antenna, the half-wave dipole (Figure 1.77), accepts radiation from most directions and its polar diagram is shown in Figure 1.81. This is only a cross-section through the beam pattern; the full three-dimensional polar diagram may be obtained by rotating the pattern shown in Figure 1.81 about the dipole's long axis and thus has the appearance

FIGURE 1.80 (**See color insert.**) The world's largest fully steerable radio telescope, the 100 m × 110 m Robert C. Byrd Green Bank Telescope. (Reproduced by kind permission of NRAO/AUI.)

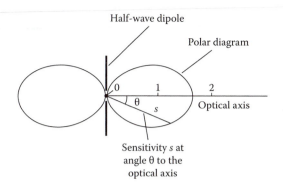

FIGURE 1.81   Polar diagram of a half-wave dipole.

of a toroid (ring doughnut) that is filled-in to the centre. The polar diagram and hence the performance of the antenna may be described by four parameters: the beam width at half-power points (BWHP), the beam width at first nulls (BWFN), the gain and the effective area. The first nulls are the positions either side of the optical axis where the sensitivity of the antenna first decreases to zero and the BWFN is just the angle between them. Thus, the value of the BWFN for the half-wave dipole is 180°. The first nulls are the direct equivalent of the first fringe minima in the diffraction pattern of an optical image and for a dish aerial type of radio telescope, their position is given by Equation 1.33, thus,

$$BWFN = 2 \times \frac{1.22\lambda}{D} \tag{1.88}$$

The Rayleigh criterion of *optical* resolution may thus be similarly applied to *radio* telescopes; two point sources are resolvable when one is on the optical axis and the other is in the direction of a first null. The half-power points may be best understood by regarding the radio telescope as a transmitter; they are then the directions in which the broadcast power has fallen to one half of its peak value. The BWHP is just the angular separation of these points. For a receiver they are the points at which the output voltage has fallen by a factor of the square root of 2 and hence the output power has fallen by half. The maximum gain or directivity of the antenna is also best understood in terms of a transmitter. It is the ratio of the peak value of the output power to the average power. In a receiver it is a measure of the output from the system compared with that from a comparable (and hypothetical) isotropic receiver. The effective area of an antenna is the ratio of its output power to the strength of the incoming flux of the radiation that is correctly polarised to be detected by the antenna, i.e.

$$A_e = \frac{P_v}{F_v} \tag{1.89}$$

where A_e is the effective area, P_ν is the power output by the antenna at frequency ν, and F_ν is the correctly polarised flux from the source at the antenna at frequency ν. The effective area and the maximum gain, g, are related by

$$g = \frac{4\pi}{c^2} \nu^2 A_e \qquad (1.90)$$

For the half-wave dipole, the maximum gain is about 1.6 and so there is very little advantage over an isotropic receiver.

The performance of a simple dipole may be improved by combining the outputs from several dipoles that are arranged in an array. In a collinear array, the dipoles are lined up along their axes and spaced at intervals of half a wavelength (Figure 1.82). The arrangement is equivalent to a diffraction grating and so the sensitivity at an angle θ to the long axis of the array, $s(\theta)$, is given by

$$s(\theta) = s_o \left(\frac{\sin(n\pi \sin\theta)}{\sin(\pi \sin\theta)} \right) \qquad (1.91)$$

where n is the number of half-wave dipoles and s_o is the maximum sensitivity (cf. Equation 4.8). Figure 1.83 shows the polar diagrams for 1-, 2- and 4-dipole arrays, their three-dimensional structure can be obtained by rotating these diagrams around a vertical axis

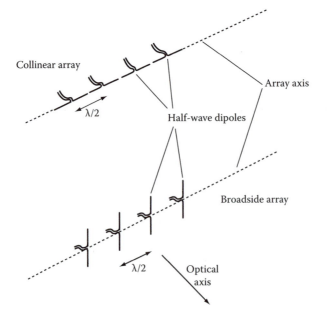

Collinear array

Array axis

$\lambda/2$

Half-wave dipoles

Broadside array

$\lambda/2$

Optical axis

FIGURE 1.82 Dipole arrays.

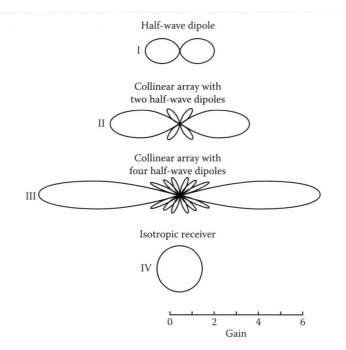

FIGURE 1.83   Polar diagrams for collinear arrays.

so that they become lenticular toroids. The resolution along the axis of the array, measured to the first null is given by

$$\alpha = \sin^{-1}\left(\frac{1}{n}\right) \tag{1.92}$$

The structure of the polar diagrams in Figure 1.83 shows a new development. Apart from the main lobe whose gain and resolution increase with n as might be expected, a number of smaller side lobes have appeared. Thus, the array has a sensitivity to sources that are at high angles of inclination to the optical axis. These side lobes correspond precisely to the fringes surrounding the Airy disc of an optical image (Figure 1.37). Although the resolution of an array is improved over that of a simple dipole along its optical axis, it will still accept radiation from any point perpendicular to the array axis. The use of a broadside array in which the dipoles are perpendicular to the array axis and spaced at half-wavelength intervals (Figure 1.82) can limit this 360° acceptance angle somewhat. For a 4-dipole broadside array, the polar diagram in the plane containing the optical and array axes is given by polar diagram number II in Figure 1.83, whilst in the plane containing the optical axis and the long axis of an individual dipole, the shape of the polar diagram is that of a single dipole (number I in Figure 1.83), but with a maximum gain to match that in the other plane. The three-dimensional shape of the polar diagram of a broadside array

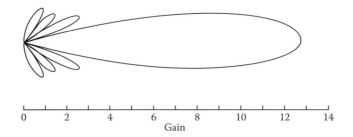

FIGURE 1.84   Polar diagram for a four-element collinear array with a mesh reflector.

thus resembles a pair of squashed balloons placed end to end. The resolution of a broadside array is given by

$$\alpha = \sin^{-1}\left(\frac{2}{n}\right) \tag{1.93}$$

along the array axis and is that of a single dipole (i.e. 90°) perpendicular to this. Combinations of broadside and collinear arrays can be used to limit the beam width further if necessary.

   With the arrays as shown, there is still a two-fold ambiguity in the direction of a source that has been detected, however, narrow the main lobe may have been made, due to the forward and backward components of the main lobe. The backward component may easily be eliminated, however, by placing a reflector behind the dipole. This is simply a conducting rod about 5% longer than the dipole and unconnected electrically with it. It is placed parallel to the dipole and about one eighth of a wavelength behind it. For an array, the reflector may be a similarly placed electrically conducting screen. The polar diagram of a four-element collinear array with a reflector is shown in Figure 1.84 to the same scale as the diagrams in Figure 1.83. It will be apparent that not only has the reflector screened out the backward lobe, but it has also doubled the gain of the main lobe. Such a reflector is termed a parasitic element since it is not a part of the electrical circuit of the antenna. Similar parasitic elements may be added in front of the dipole to act as directors. These are about 5% shorter than the dipole. The precise lengths and spacings for the parasitic elements can only be found empirically, since the theory of the whole system is not completely understood. With a reflector and several directors we obtain the parasitic or Yagi antenna, familiar from its appearance on so many rooftops as a television aerial. A gain of up to 12 is possible with such an arrangement and the polar diagram is shown in Figure 1.85. The Rayleigh resolution is 45° and it has a bandwidth of about 3% of its operating frequency. The main use of parasitic antennae in radio astronomy is as the receiving element (sometimes called the feed) of a larger reflector such as a parabolic dish.

   The use of a single dipole or even several dipoles in an array is the radio astronomy equivalent of a naked-eye observation. The Netherlands-centred Low Frequency Array

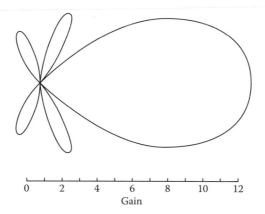

0   2   4   6   8   10   12
Gain

FIGURE 1.85   Polar diagram of a parasitic aerial.

(LOFAR) array, which has recently been making its first observations over the 10- to 250-MHz (30 to 1.2 m) band, uses simple dipole antennas. It has 42 out of a possible 48 of its stations operating with some 5000 individual antennas and a maximum baseline of 1500 km. The Long Wavelength Array (LWA) and the low frequency part of the SKA (see below) will also use simple dipole receivers.

However, just as in optical astronomy, the observations may be greatly facilitated by the use of some means of concentrating the signal from over a wide area onto the antenna. The most familiar of these devices are the large parabolic dishes that are the popular conception of a radio telescope. These are directly equivalent to an optical reflecting telescope. They are usually used at the prime focus or at the Cassegrain focus or recently as off-axis Gregorians (see Section 1.1).

The gain may be found roughly by substituting the dish area for the effective area in Equation 1.90. The Rayleigh resolution is given by Equation 1.34. The size of the dishes is so large because of the length of the wavelengths being observed; for example, to obtain a resolution of 1° at a wavelength of 0.1 m requires a dish 7 metres across, which is larger than most optical reflectors for a resolution over 10^4 times poorer. The requirement on surface accuracy is the same as that for an optical telescope – deviations from the paraboloid to be less than $\lambda/8$ if the Rayleigh resolution is not to be degraded. Now, however, the longer wavelength helps since it means that the surface of a telescope working at 0.1 m, say, can deviate from perfection by over 10 mm without seriously affecting the performance.* In practice, a limit of $\lambda/20$ is often used, for as we have seen, the Rayleigh limit does not represent the ultimate limit of resolution.

These less stringent physical constraints on the surface of a radio telescope ease the construction problems greatly; more importantly however, it also means that the surface need not be solid, because a wire mesh with spacings less than $\lambda/20$ will function equally well as

---

* Interferometers working in the terahertz region, such as ALMA, are affected by atmospheric scintillation and just like their optical counterparts need real-time atmospheric compensation to reach their diffraction limited resolutions (see Section 2.5).

a reflector. The weight and wind resistance of the reflector are thus reduced by very large factors. At the shorter radio wavelengths a solid reflecting surface may be more convenient, though and at very short wavelengths (<1 mm) active surface control to retain the accuracy is used along somewhat similar lines to the methods discussed in Section 1.1 for optical telescopes, the shape of the mirror being monitored holographically. The dishes are usually of very small focal ratio, f0.5 is not uncommon and the reason for this is so that the dish acts as a screen against unwanted radiation as well as concentrating the desired radiation.

The feed antennae for radio telescope dishes may be made of such a size that they intercept only the centre lobe of the telescopes' responses (i.e. the Airy disc in optical terms). The effects of the side lobes are then reduced or eliminated. This technique is known as tapering the antenna and is the same as the optical technique of apodisation (see Section 4.1). The tapering function may take several forms and may be used to reduce background noise as well as eliminating the side lobes. Tapering reduces the efficiency and resolution of the dish, but this is usually more than compensated for by the improvement in the shape of the response function.

Fully steerable dishes up to 100 metres across have been built such as the Green Bank (Figure 1.80) and Effelsberg telescopes, whilst the Arecibo telescope in Puerto Rico* is a fixed dish 300 metres across. This latter instrument acts as a transit telescope and has some limited ability to track sources and look at a range of declinations by moving the feed antenna around the image plane. Such fixed telescopes may have spherical surfaces in order to extend this facility and use a secondary reflector to correct the resulting spherical aberration.

In the microwave region, the largest dishes are currently the 50-metre Large Millimetre Telescope which started operations in 2006 and is located near Mexico City observing from 75 to 350 GHz (4 to 850 mm), the 45-metre telescope at Nobeyama in Japan and the 30-metre IRAM instrument on Pico Veleta in Spain.

With a single feed, the radio telescope is a point-source detector only. Images have to be built up by scanning (see Section 2.4) or by interferometry (see Section 2.5). Scanning used to be accomplished by pointing the telescope at successive points along the raster pattern and observing each point for a set interval. This practice, however, suffers from fluctuations in the signal due to the atmosphere and the electronics. Current practice is therefore to scan continuously or on-the-fly. The atmospheric or electronic variations are then on a longer time scale than the changes resulting from moving over a radio source and can be separated out.

True imaging can be achieved through the use of cluster or array feeds (radio cameras). These are simply multiple individual feeds arranged in a suitable array at the telescope's focus. Each feed is then the equivalent of a pixel in a CCD or other type of detector. The number of elements in such cluster feeds currently remains small compared with their optical equivalents. For example, the 14-metre telescope at the Five College Radio Astronomy Observatory (FCRAO) in Massachusetts uses the Second Quabbin Optical Imaging Array (SEQUOIA), a 32-beam array operating at millimetre wavelengths with InSb MMICs.

---

* At the time of writing, funding for both the Green Bank and Arecibo instruments is under threat.

The One Centimetre Receiver Array (OCRA) project aims to develop a 100-beam array operating at 30 MHz (10 mm) for use on the 32-metre telescope at the Torun observatory in Poland. Currently a prototype, OCRA-F (Figure 1.86; also known as FARADAY), is operating with eight beams using MMICs, whilst the Q/U Imaging Experiment (QUIET) instrument at the Llano de Chajnantor Observatory in Chile completed its observations in 2010. This latter instrument comprised a 1.4-metre Dragonian telescope and used InP MMIC/HEMT arrays for detecting polarisation in the CMB at 43 and 95 GHz (7.0 and 3.2 mm). The arrays comprised 19-element and 90-element assemblies, respectively. SCUBA-2 is also an array device and was discussed earlier.

Very many other systems have been designed to fulfil the same function as a steerable paraboloid but which are easier to construct. The best known of these are the multiple arrays of mixed collinear and broadside type, or similar constructions based upon other aerial types. They are mounted onto a flat plane which is oriented east–west and which is tiltable in altitude to form a transit telescope. Another system such as the 600-metre RATAN telescope uses an off-axis paraboloid that is fixed and which is illuminated by a tiltable flat reflector. Alternatively, the paraboloid may be cylindrical and tiltable itself around its long axis. For all such reflectors some form of feed antenna is required. A parasitic antenna is a common choice at longer wavelengths, whilst a horn antenna, which is essentially a flared end to a waveguide, may be used at higher frequencies.

FIGURE 1.86 The One Centimetre Receiver Array (OCRA-F) during construction. (Reproduced by kind permission of Mike Peel, Jodrell Bank Centre for Astrophysics, University of Manchester.)

A quite different approach is used in the Mills cross type of telescope. These use two collinear arrays oriented north–south and east–west. The first provides a narrow fan beam along the north–south meridian, whilst the second provides a similar beam in an east–west direction. Their intersection is a narrow vertical pencil beam, typically 1° across. The pencil beam may be isolated from the contributions of the remainders of the fan beams by comparing the outputs when the beams are added in phase with when they are added out of phase. The in-phase addition is simply accomplished by connecting the outputs of the two arrays directly together. The out-of-phase addition delays one of the outputs by half a wavelength before the addition and this is most simply done by switching in an extra length of cable to one of the arrays. In the first case, radiation from objects within the pencil beam will interfere constructively, whilst in the second case there will be destructive interference. The signals from objects not within the pencil beam will be mutually incoherent and so will simply add together in both cases. Thus, looking vertically down onto the Mills cross, the beam pattern will alternate between the two cases shown in Figure 1.87. Subtraction of the one from the other will then just leave the pencil beam.

The pencil beam may be displaced by an angle θ from the vertical by introducing a phase shift between each dipole. Again, the simplest method is to switch extra cable into the connections between each dipole, the lengths of the extra portions, L, being given by

$$L = d \sin \theta \qquad (1.94)$$

where d is the dipole separation. The pencil beam may thus be directed around the sky as wished. In practice, the beam is only moved along the north–south plane and the telescope is used as a transit telescope, since the alteration of the cable lengths between the dipoles is a lengthy procedure. The resolution of a Mills cross is the same as that of a parabolic dish whose diameter is equal to the array lengths. The sensitivity, however, is obviously much reduced from that of a dish, since only a tiny fraction of the aperture is filled by the dipoles. As well as being cumbersome to operate, the Mills cross suffers from the disadvantage that it can only operate at a single wavelength unless its dipoles are all changed. Furthermore,

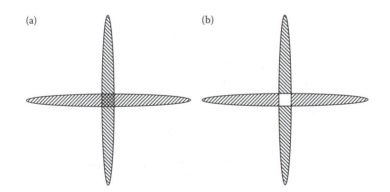

FIGURE 1.87 Beam patterns of a Mills cross radio telescope (a) with the beams added together (i.e. in phase) and (b) with the beams subtracted from each other (i.e. 180° out of phase).

confusion of sources may arise if a strong source is in one of the pencil beams formed by the side lobes, since it will have the same appearance as a weak source in the main beam. An alternative and related system uses a single fan beam and observes the object at many position angles. The structure of the source is then retrievable from the observations in a relatively unambiguous manner.

The phasing of two dipoles, as we have seen, can be used to alter the angle of the beam to their axis. By continuously varying the delay the lobe can be swept across the sky, or accurately positioned on an object and then moved to track that object. This is an important technique for use with interferometers (see Section 2.5) and for solar work. It also forms the basis of a phased array. This latter instrument is basically a Mills cross in which the number of dipoles has been increased until the aperture is filled. It provides great sensitivity since a large area may be covered at a relatively low cost. Small phased arrays are found for example as the planned mid-frequency receivers for the SKA (see below).

A radically different approach to concentrating the radiation is based upon refraction. The Luneburg lens has yet to find much application to radio astronomy although at one time it was being considered for the SKA. The Luneburg lens is a solid sphere within which the refractive index increases linearly inwards from unity at the surface. With a central refractive index of 2, the focus is on the surface of the lens. Since there is no axis of symmetry, the lens can be used to observe in many directions simultaneously, simply by having numerous feeds distributed around it. Many materials potentially can be used for the lens' construction, but to date high-density polystyrene is the one that has been used in practice.

A number of spacecraft carrying microwave detectors have been launched. Thus, Cosmic Background Explorer Satellite (COBE, 1989–1993) carried Dicke radiometers operating at 31.5, 53 and 90 GHz (9.52, 5.66 and 3.33 mm). The Planck spacecraft (2009–2012) was also designed to observe the cosmic microwave background radiation and used HEMTs and MMICs in its low-frequency instrument (30 to 70 GHz, 10 to 4.3 mm) plus bolometers (see Section 1.1) in its high-frequency instrument (100 to 857 GHz, 3 mm to 350 μm). The Wilkinson Microwave Anisotropy Probe (WMAP) (2001–2010) mission observed from 22 to 90 GHz (13.6 to 3.3 mm) using HEMTs. The HIFI instrument on board the Herschel spacecraft has already been mentioned (Figure 1.79).

Few longer wave spacecraft-borne telescopes have been used, since the atmosphere is then transparent. However, the Japanese Halca spacecraft (1997–2003) carried an 8-metre dish operating at centimetre wavelengths as part of an aperture synthesis system (see Section 2.5). In 2011, the Russian-led RadioAstron program launched a 10-metre diameter radio dish (which is currently the largest space telescope). The spacecraft is in an orbit with an apogee of 390,000 km enabling it to undertake mass observations when acting as an interferometer with Earth-based instruments. It operates in several wavebands from 325 MHz to 22 GHz (920 to 13.5 mm) and it has recently made its first successful observations, combining its output with that from several European instruments.

Great improvements in resolution and sensitivity of radio telescopes may be obtained through the use of interferometers and aperture synthesis and these devices and their associated receivers and detectors are discussed in detail in Section 2.5.

1.2.3.1 Construction

The large dishes that we have been discussing pose major problems in their construction. Both the gravitational and wind loads on the structure can be very large and shadowing of parts of the structure can lead to inhomogeneous heating and hence expansion and contraction-induced stresses. The current Green Bank radio telescope (Figure 1.80) replaces an earlier 90-metre dish that collapsed catastrophically in 1988 whilst in use, although fortunately without anyone being injured.

The worst problem is due to wind, since its effect is highly variable. The force can be very large – 1.5×10^6 N (150 tonnes) for a 50-metre dish facing directly into a gale-force wind for example. A rough rule of thumb to allow scaling the degree of wind distortion between dishes of different sizes is that the deflection, Δe, is given by

$$\Delta e \propto \frac{D^3}{A} \tag{1.95}$$

where D is the diameter of the dish and A is the cross-sectional area of its supporting struts. Thus, doubling the diameter of the design of a dish would require the supporting members' sizes to be increased by a factor of 8 if it were to still work at the same wavelength. There are only two solutions to the wind problem: to enclose the dish, or to cease using it when the wind load becomes too great. Some smaller dishes, especially those working at short wavelengths where the physical constraints on the surface accuracy are most stringent, are enclosed in radomes, or space-enclosing structures built from non-conducting materials. But this is not usually practicable for the larger dishes. These must generally cease operating and be parked in their least wind-resistant mode once the wind speed rises above 10 to 15 metres per second (35 to 55 km/h).

The effects of gravity are easier to counteract. The problem arises from the varying directions and magnitudes of the loads placed upon the reflecting surface as the orientation of the telescope changes. Three approaches have been tried successfully for combating the effects of gravity. The first is the brute force approach, whereby the dish is made so rigid that its deformations remain within the surface accuracy limits. Except for very small dishes, this is an impossibly expensive option. The second approach is to compensate for the changing gravity loads as the telescope moves. The surface may be kept in adjustment by systems of weights, levers, guy ropes, springs etc. or more recently by computer-controlled hydraulic jacks. The final approach is much more subtle and is termed the homological transformation system. The dish is allowed to deform, but its supports are designed so that its new shape is still a paraboloid, although, in general, an altered one, when the telescope is in its new position. The only active intervention that may be required is to move the feed antenna to keep it at the foci of the changing paraboloids.

There is little that can be done about inhomogeneous heating other than painting all the surfaces white so that the absorption of the heat is minimised. Fortunately, it is not usually a serious problem in comparison with the first two.

The supporting framework of the dish is generally a complex, cross-braced skeletal structure, whose optimum design requires a computer for its calculation. This framework is then usually placed onto an alt-az mounting (see Section 1.1), since this is generally the cheapest and simplest structure to build and also because it restricts the gravitational load variations to a single plane, so making their compensation much easier. However, a few, usually quite small, radio telescopes do have equatorial mountings.

1.2.3.2 Future

Currently under construction in Guizhou Province (southwestern China) is the Five-hundred metre Aperture Spherical Telescope (FAST). Like the Arecibo instrument this is a fixed dish that lines a natural bowl of suitable size and shape. The reflector is composed of 4600 segments and is actively controlled not only to keep the correct shape, but also to adjust that shape so that it can observe at zenith distances up to 40°, although this latter characteristic reduces the effective aperture to 300 m. Its operating frequencies range from 300 MHz to 5.1 GHz (1 m to 60 mm) and it is expected to be completed around 2016. FAST is expected to be followed at some time in the future with Kilometer-square Area Radio Synthesis Telescope (KARST). This could be a linked array of up to thirty 200-metre versions of FAST with a total collecting area of over a square kilometre.

The Italian Large Scale Polarisation Explorer (LSPE) will be a balloon-borne 0.6-metre telescope observing the CMB at frequencies from 43 to 245 GHz (7 to 1.2 mm). It will use HEMTs at the lower frequencies and bolometers at the higher frequencies. It is planned to fly the balloon in winter from an Arctic site to follow a circumpolar path so that long duration observations are possible, with the first flight in 2015.

The LWA is currently under construction in New Mexico and will operate at 10 to 88 MHz (30 to 3.4 m). It will comprise 13,000 dipoles distributed over a 400 km diameter area. Partial observations commenced in 2011.

The 25-metre Cornel Caltech Atacama Telescope (CCAT) is planned to begin operating in 2017. It will be sited at Cerro Chajnantor and be able to observe from 135 GHz to 1.5 THz (2.2 mm to 200 μm). The high frequencies require a surface that is accurate to 10 μm and so the reflector will be constructed of some 200 segments under active position control.

The SKA is currently under development with a number of preliminary studies and projects already completed. The main sites for the \$2 billion instrument have been chosen as the Murchison Radio Astronomy Observatory in Western Australia and the Karoo desert in South Africa's Northern Cape Province. Additionally, there will also be outlier stations in a number of other southern hemisphere sites. The SKA is scheduled for completion around 2024 and will be a phased array of many small receivers whose total collecting area will be about 1 square kilometre (10^6 m^2). It will cover the spectral region from 70 MHz to 30 GHz (4.3 m to 10 mm) and have a maximum baseline of 3000 km. It is currently planned that three types of instruments will be used to cover this spectral region: dipoles from 70 to 200 MHz (4.3 to 1.5 m), phased array tiles for 200 to 500 MHz (1.5 m to 600 mm) and some three thousand 15-metre dishes for the 500 MHz to 30 GHz (600 to 10 mm) region. Its field of view will range from about 200 square degrees at the lower frequencies to 1 square degree at the high frequencies where its angular resolution will reach 2 mas. The thirty-six 12-metre antennas of the Australian SKA Pathfinder (ASKAP) instrument were completed in 2012 acting both as a major radio telescope in their own right and as a test bed for possible SKA developments. At lower frequencies the Murchison Wide Field Array with 128 phased array tiles is also operational. In South Africa, the seven 13.5-metre dishes of the Karoo Array Telescope Seven (KAT-7) prototype were completed in 2010 and the 64 dishes of the MeerKAT array are expected to be in place by 2016.

Possibly for some time in the 2020s, the North America Array is a concept for the development of the Expanded Very Large Array (EVLA, see Section 2.5) and the Very Long Baseline Array (VLBA, see Section 2.5) and observing at up to 50 GHz (6 mm).

Other developments are likely to be in the area of increasingly sophisticated integrated circuit technology that leads to smaller and less power-hungry detectors, amplifiers, correlators etc. which in turn allows the building of larger array detectors – perhaps up to 1k × 1k size within a decade or so.

In the more distant future, increasing levels of radio noise from artificial terrestrial sources may force radio telescopes to the far side of the moon. However, it seems unlikely that any such project will be possible before there is a substantial and permanently occupied colony (or colonies) on the moon – by which time the radio noise from artificial lunar sources may be just as bad as that on the Earth. One development currently underway that may reduce radio noise is the increasing use of fibre-optic cables for data transmission. The popular demand for faster and faster broadband access seems likely to lead to the installation of fibre-optic lines throughout most of the world within the next decade or two. Once such lines are in place there will be little need for many/most/all of the present-day radio and TV stations etc. and so the radio spectrum may become quieter.

EXERCISES

1.9 Show that the HPBW of a collinear array with n dipoles is given by

$$\text{HPBW} \approx 2\sin^{-1}\left(\frac{6n^2 - 3}{2n^4\pi^2 - \pi^2}\right)^{1/2}$$

when n is large.

1.10 Calculate the dimensions of a Mills cross to observe at a wavelength of 0.3 m with a Rayleigh resolution of 0.25°.

1.11 Show that the maximum number of separable sources using the Rayleigh criterion, for a 60-metre dish working at a wavelength of 0.1 m, is about 3.8 × 10⁵ over a complete hemisphere.

1.3 X-RAY AND GAMMA-RAY DETECTION

1.3.1 Introduction

The electromagnetic spectrum comprises radiation of an infinite range of wavelengths, but the intrinsic nature of the radiation is unvarying. There is, however, a tendency to regard the differing wavelength regions from somewhat parochial points of view and this tends to obscure the underlying unity of the processes that may be involved. The reasons for these attitudes are many; some are historical hangovers from the ways in which the detectors for the various regions were developed, others are more fundamental in that different physical mechanisms predominate in the radiative interactions at different wavelengths. Thus, high-energy gamma rays may interact directly with nuclei, at longer wavelengths we have resonant interactions with atoms and molecules producing electronic, vibrational and rotational transitions, whilst in the radio region currents are induced directly into conductors. But primarily the reason may be traced to the academic backgrounds of the workers involved in each of the regions. Because of the earlier reasons, workers involved in investigations in one spectral region will tend to have different bias in their backgrounds compared with workers in a different spectral region. Thus, there will be different traditions, approaches, systems of notation etc. with a consequent tendency to isolationism and unnecessary failures in communication. The discussion of the point source image in terms of the Airy disc and its fringes by optical astronomers and in terms of the polar diagram by radio astronomers as already mentioned is one good example of this process, and many more exist.

It is impossible to break out from the straitjacket of tradition completely, but an attempt to move towards a more unified approach has been made in this work by dividing the spectrum much more broadly than is the normal case. We only consider three separate spectral regions, within each of which the detection techniques bear at least a familial resemblance to each other. The overlap regions are fairly diffuse with some of the techniques from each of the major regions being applicable. We have already discussed two of the major regions – radio and microwaves and optical and infrared. The third region, X-rays and gamma rays, is the high-energy end of the spectrum and is the most recent area to be explored. This region also overlaps to a very considerable extent, in the nature of its detection techniques, with the cosmic rays that are discussed in the next section. None of the radiation discussed in this section penetrates down to ground level, so its study had to await the availability of observing platforms in space or near the top of the Earth's atmosphere. Thus, significant work on these photons has only been possible since the 1960s and many of the detectors and observing techniques are still under development.

The high-energy electromagnetic spectrum is fairly arbitrarily divided into

- The extreme UV (EUV or XUV region): 100- to 10-nm wavelengths (12- to 120-eV photon energies)

- Soft X-rays: 10 to 1 nm (120 eV to 1.2 keV)

- X-rays: 1 to 0.1 nm (1.2 to 12 keV)

- Hard X-rays: 0.1 to 0.01 nm (12 to 120 keV)

- Soft γ-rays: 0.01 to 0.001 nm (120 keV to 1.2 MeV)

- γ-rays: less than 0.001 nm (greater than 1.2 MeV)

We shall be primarily concerned with the last four regions in this section (i.e. with wavelengths less than 10 nm and photon energies greater than 120 eV). The main production mechanisms for high-energy radiation include electron synchrotron radiation, the inverse Compton effect, free-free radiation and pion decay, whilst the sources include the Sun, supernova remnants, pulsars, bursters, binary systems, cosmic rays, the intergalactic medium, galaxies, Seyfert galaxies and quasars. Absorption of the radiation can be by ionisation with a fluorescence photon or an Auger electron produced in addition to the ion and electron, by Compton scattering, or in the presence of matter, by pair production. This latter process is the production of a particle and its anti-particle and not the pair production process discussed in Section 1.1 which was simply the excitation of an electron from the valence band. The interstellar absorption in this spectral region varies roughly with the cube of the wavelength so that the higher energy radiation can easily pass through the whole galaxy with little chance of being intercepted. At energies under about 2 keV, direct absorption by the heavier atoms and ions can be an important process. The flux of the radiation varies enormously with wavelength. The solar emission alone at the lower energies is sufficient to produce the ionosphere and thermosphere on the Earth. At 1-nm wavelength (1.2 keV) for example, the solar flux is 5×10^9 photons m^{-2} s^{-1}, whilst the total flux from all sources for energies above 10^9 eV is only a few photons per square metre per day.

1.3.2 Detectors

1.3.2.1 Geiger Counters

The earliest detection of high-energy radiation from a source other than the Sun took place in 1962 when soft X-rays from a source that later became known as Sco X-1 were detected by large area Geiger counters flown on a sounding rocket. Geiger counters are no longer used as primary detectors, but variants of them have been developed into much more sophisticated detectors of both high energy e-m radiation and high energy subatomic particles (see Sections 1.4 and 1.5) and are in widespread use today. We start, therefore, with a quick review of the operating principles of the Geiger counter.

Two electrodes inside an enclosure are held at such a potential difference that a discharge in the medium filling the enclosure is on the point of occurring. The entry of ionising radiation triggers this discharge (cf. APDs), resulting in a pulse of current between the electrodes that may then be amplified and detected. The electrodes are usually arranged as the outer wall of the enclosure containing the gas and as a central coaxial wire (Figure 1.88). The medium inside the tube is typically argon at a low pressure with a small amount of an organic gas, such as alcohol vapour, added. The electrons produced in the initial ionisation are accelerated towards the central electrode by the applied potential; as these electrons gain energy they cause further ionisation, producing more electrons, which in turn are accelerated towards the central electrode and so on. The amplification factor can be as high as 10^8 electrons arriving at the central electrode for each one in the initial ionisation trail.

The avalanche of electrons rapidly saturates, so that the detected pulse is independent of the original energy of the photon. Another disadvantage of Geiger counters that also applies to many of the other detectors discussed in this and the following section is that

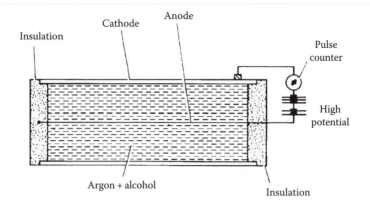

Cathode Anode
Insulation
Pulse counter
High potential
Argon + alcohol
Insulation

FIGURE 1.88 A typical arrangement for a Geiger counter.

a response to one event leaves the detector inoperative for a short interval, known as the dead time. In the Geiger counter the cause of the dead time is that a discharge lowers the potential between the electrodes, so that it is momentarily insufficient to cause a second avalanche of electrons should another X-ray enter the device. The length of the dead time is typically 200 μs.

1.3.2.2 Proportional Counters

Proportional counters are very closely related to Geiger counters and are also known as known as gas-filled ionisation detectors. They are in effect Geiger counters operated at less than the trigger voltage. By using a lower voltage, saturation of the pulse is avoided and its strength is then *proportional* to the energy of the original interaction. The gain of the system operated in this way is reduced to about 10^4 or 10^5, but it is still sufficient for further detection by conventional amplifiers etc. Provided that all the energy of the ionising radiation is absorbed within the detector, its original total energy may be obtained from the strength of the pulse and we have a proportional counter.

At low photon energies, a window must be provided for the radiation. These are typically made from thin sheets of plastic, mica, or beryllium and absorption in the windows limits the detectors to energies above a few hundred electron volts. When a window is used, the gas in the detector has to be continuously replenished because of losses by diffusion through the window.

At high photon energies the detector is limited by the requirement that all the energy of the radiation be absorbed within the detector's enclosure. To this end, proportional counters for high-energy detection may have to be made very large. About 30 electron volts on average are required to produce one ion-electron pair, so that a 1-keV photon produces about 36 electrons and a 10-keV photon about 360 electrons. The spectral energy resolution to two and a half standard deviations from the resulting statistical fluctuations of the electron numbers is thus about 40% at 1 keV and 12% at 10 keV. The quantum efficiencies of proportional counters approach 100% for energies up to 50 keV.

The position of the interaction of the X-ray along the axis of the counter may be obtained through the use of a resistive anode. The pulse is abstracted from both ends of the anode and a comparison of its strength and shape from the two ends then leads to a position of the discharge along the anode. The anode wires are very thin, typically 20 µm across, so that the electric field is most intense very near to the wire. The avalanche of electrons thus develops close to the wire, limiting its spread and giving a precise position. The concept may easily be extended to a two-dimensional grid of anodes to allow genuine imaging. Spatial resolutions of about a tenth of a millimetre are possible. In this form the detector is called a position-sensitive proportional counter. They are also known as multi-wire chambers and as time-projection chambers (TPC) (see Section 1.5), especially in the context of particle physics (see Section 1.4).

Proportional counters and their derivatives are also intrinsically sensitive to the polarisation of the detected radiation. The direction of the initial track of the electrons is (roughly) in the same direction as the electric field of the X-ray photon. Determining the directions of the tracks within the detector therefore reveals the state of polarisation of the incoming X-rays – if the tracks are randomly distributed then the radiation is unpolarised, but a non-random distribution shows that the radiation as a whole is polarised.

Many gases can be used to fill the detector: argon, methane, xenon, carbon dioxide and mixtures thereof at pressures near that of the atmosphere, are amongst the commonest ones. The inert gases are to be preferred since with their single atoms there is then no possibility of the loss of energy into the rotation or vibration of multi-atom molecules.

The Indian Astrosat spacecraft, due for launch in 2013, will carry two instruments based upon proportional counters. The Large Area X-ray Proportional Counter (LAXPC) uses three aligned proportional counters to cover the 3- to 80-keV region with a field of view 1 degree square, whilst the Scanning Sky Monitor (SSM) also uses three proportional counters but each with a coded mask collimator (see below) to cover a $10° \times 90°$ area over the 2- to 10-keV region.

A recent development of the TPC, known as a micro-mesh gaseous structure (Micromegas) (see Section 1.5) detector is to be flown on board the Spectrum-RG spacecraft as the detector for the lobster-eye wide field X-ray telescope (LWFT) (see Section 1.3.4.1). The five modules of the LWFT will instantaneously monitor 10% of the whole sky over the 200-eV to 3-keV spectral region.

1.3.2.3 Scintillation Detectors

The ionising photons do not necessarily knock out only the outermost electrons from the atom or molecule with which they interact. Electrons in lower energy levels may also be removed. When this happens, a hole is left behind into which one of the higher electrons may drop, with a consequent emission of radiation. Should the medium be transparent to this radiation, the photons may be observed and the medium may be used as an X-ray detector. Each interaction produces a flash or scintilla of light, from which the name of the device is obtained. There are many materials that are suitable for this application. Commonly used ones include sodium iodide doped with an impurity such as thallium and caesium iodide doped with sodium or thallium. For these materials the light flashes are detected by a photomultiplier (Figure 1.89). There is no dead time for a scintillation detector and the strength of the flash depends somewhat upon the original photon energy, so

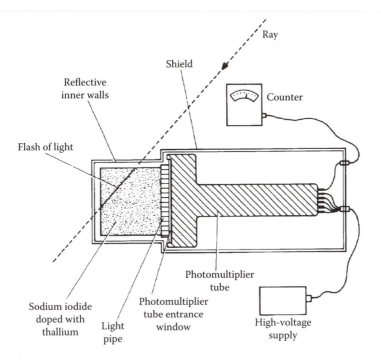

FIGURE 1.89 Schematic experimental arrangement of a scintillation counter.

that some spectral resolution is possible. The noise level, however, is quite high since only about 3% of the X-ray's energy is converted into detectable radiation, with a consequent increase in the statistical fluctuations in their numbers. The spectral resolution is thus about 6% at 1 MeV. Sodium iodide or caesium iodide are useful for X-ray energies up to several hundred kiloelectron volts, organic scintillators such as stilbene ($C_{14}H_{14}N_2$) can be used up to 10 MeV and bismuth germanate (BGO – $Bi_4Ge_3O_{12}$) for energies up to 30 or more MeV. Organically doped plastics are also used and are planned to be used for the Telescope Array cosmic ray observatory in Utah (see Section 1.4).

Scintillator-based detectors have largely been superseded by other devices in recent years for spacecraft-borne detectors. However, the Gamma-ray Burst Monitor (GBM) on board the Fermi gamma-ray space telescope, launched in 2008, employs ten NaI and two BGO scintillators to detect γ-ray bursts over the range from 150 keV to 30 MeV. Gadolinium silicate and BGO scintillators are used on the Suzaku spacecraft (launched 2005) for observing over the 30–600 keV region, whilst the International Gamma-Ray Astrophysics Laboratory (INTEGRAL; launched 2002) uses BGO and plastic scintillators as active shields for two of its instruments.

Discrimination of the X-ray's arrival direction can be obtained by using sodium iodide and caesium iodide in two superimposed layers. The decay time of the pulses differs between the two compounds so that they may be separately identified and the direction of travel of the photon inferred. This arrangement for a scintillation detector is frequently called a phoswich detector. Several gases such as argon, xenon, nitrogen and their mixtures can also be used as scintillators and combined with an optical system to produce another imaging device. X-rays and cosmic rays (Section 1.4) may be distinguished by the differences in their resulting pulses shapes. X-rays are rapidly absorbed and their pulses are sharp and brief. Cosmic ray particles will generally have a much longer path and so their pulses will be comparatively broader and smoother than the X-ray pulse.

The recently completed RXTE mission had two clusters of 4 NaI/CsI phoswich detectors to cover the 15- to 250-keV region, whilst the balloon-borne 2000-kg Polarized Gamma Ray Observer (PoGOLite) instrument uses an array of 217 phoswich detectors to study polarisation of X-ray sources in the 20- to 80-keV region.

1.3.2.4 Gas Scintillation Proportional Counters

A combination of the above two types of detector leads to a significant improvement in the low-energy spectral resolution. Resolutions as good as 8% at 6 keV have been achieved in practice with these devices. The x-radiation produces ion-electron pairs in an argon- or xenon-filled chamber. The electrons are then gently accelerated until they cause scintillations of their own in the gas. These scintillations can then be observed by a conventional scintillation counter system. These have been favoured detectors for launch on several of the more recent X-ray satellites because of their good spectral and positional discrimination, although in some cases their lifetimes in orbit have been rather short.

1.3.2.5 Compton Interaction Detectors

Very-high-energy photons produce electrons in a scintillator through the Compton effect and these electrons can have sufficiently high energies to produce scintillations of their own. Two such detectors separated by a metre or so can provide directional discrimination when used in conjunction with pulse analysers and time-of-flight measurements to eliminate other unwanted interactions. The COMPTEL instrument on board the Compton Gamma Ray Observatory (CGRO) (1991–2000) used an organic liquid scintillator and a NaI scintillator separated by 1.5 metres and both viewed by photomultipliers. γ-rays in the 800 keV to 30 MeV range were detectable over a 1-steradian area of the sky with the photons being scattered in the liquid scintillator before being completely absorbed in the NaI. A plastic scintillator was also used as an active shield. The angular resolution of COMPTEL ranged from 1.7° to 4.4° and its energy resolution from 5% to 8%.

Recently, Compton effect telescopes have been developed using stacks of silicon strip detectors and CdTe strip detectors. The CdTe detector stack lies underneath the SSD stack and CdTe detectors also surround the sides of the stack. The SSDs detect the Compton electrons produced by scattering within them and the CdTe detectors absorb the γ-rays. Japan Aerospace Exploration Agency (JAXA's) ASTRO-H spacecraft's soft γ-ray detector will use six such Compton effect telescopes for observing the 10- to 600-keV region over an area of the sky ranging from half a degree square to 10 degrees square. Its energy resolution will be up to 4000 and an active well shield of BGO will surround each stack of detectors.

1.3.2.6 Solid-State Detectors

There are a number of different varieties of these detectors so that suitable ones may be found for use throughout most of the X-ray and γ-ray regions.

The details of the operating principles for CCDs for optical detection are to be found in Section 1.1. They are also, however, becoming increasingly widely used as primary detectors at EUV and X-ray wavelengths. CCDs become insensitive to radiation in the blue and UV parts of the spectrum because of absorption in the electrode structure on their surfaces. They regain sensitivity at shorter wavelengths as the radiation is again able to penetrate that structure (λ < 10 nm or so, energy > 120 eV). As with optical CCDs, the efficiency of the devices may be improved by using a very thin electrode structure, an electrode structure that only partially covers the surface (virtual phase CCDs), or by illuminating the device from the back. Typically, 3.65 eV of energy is needed to produce a single electron-hole pair in silicon. So, unlike optical photons which can only produce a single such pair, the X-ray photons can each produce many pairs. The CCDs are thus used in a photon-counting mode for detecting X-rays and the resulting pulses are proportional to the X-ray's energy, giving CCDs an intrinsic spectral resolution (see Section 4.1) of around 10 to 50.

Closely related detectors consist of a thick layer (up to 5 mm) of a semiconductor, such as silicon (known as silicon strip detectors [SSDs]). More recently, cadmium (about 90%) and zinc (about 10%) doped with tellurium (known as CZT detectors) and cadmium plus tellurium (CdTe strip detectors) have become popular, since these latter materials have higher stopping powers for the radiation. The semiconductor in each case has a high resistivity by virtue of being back-biased to 100 V or so. The silicon surface is divided into 20 to 25 μm wide strips by thin layers doped with boron. Aluminium electrodes are then deposited on the surface between these layers. When an X-ray, γ-ray or charged particle passes into the material, electron-hole pairs are produced via the photo-electric effect, Compton scattering or collisional ionisation. The pulse of current is detected as the electron-hole pairs are collected at the electrodes on the top and bottom of the semiconductor slice. Processing of the pulses can then proceed by any of the usual methods and a positional accuracy for the interaction of ±10 μm can be achieved.

Silicon drift detectors (SDDs), developed for the Large Hadron Collider, are similar to SSDs and also to deep-depletion CCDs. They comprise a disk of high purity *n*-doped silicon up to 10 mm across and 0.5 mm thick. X-rays penetrating into the disk produce electron-hole pairs. The holes are attracted to negatively charged p-type regions (cathodes) on the flat surfaces of the disk and eliminated. The electrons are attracted to a single central anode and encouraged to move in that direction by a decreasing negative charge on the cathodes from the edge to the centre of the disk. The voltages involved are low (100 V or less) so there is no intrinsic amplification of the cloud of electrons through collisions, but there are sufficient numbers in the original cloud for the pulse to be detected when the electrons reach the anode.

Germanium may be used as a sort of solid proportional counter. A cylinder of germanium cooled by liquid nitrogen is surrounded by a cylindrical cathode and has a central anode (Figure 1.90). A γ-ray scatters off electrons in the atoms until its energy has been consumed in electron-hole pair production. The number of released electrons is proportional

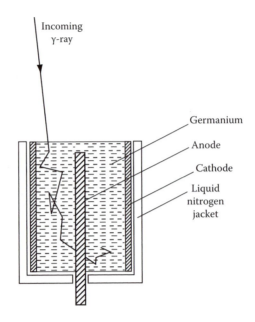

FIGURE 1.90   Germanium γ-ray detector.

to the energy of the γ-ray and these are attracted to the anode where they may be detected. The spectral resolution is very high – 0.2% at 1 MeV – so that detectors of this type are especially suitable for γ-ray line spectroscopy. Other materials that may replace the germanium include germanium doped with lithium, cadmium telluride and mercury-iodine. At lower energies (0.4 to 4 keV) silicon-based solid-state detectors may be used similarly. Their energy resolution ranges from 4% to 30%. The Reuven Ramaty High Energy Solar Spectroscopic Imager (RHESSI) (launched 2002) spacecraft uses germanium detectors to study the spectra of solar flares from 100 keV to 17 MeV.

Solid-state detectors have several advantages that suit them particularly for use in satellite-borne instrumentation – a wide range of photon energies detected (from 1 keV to over 1 MeV), simplicity, reliability, low power consumption, high stopping power for the radiation, room temperature operation (some varieties), no entrance window needed, high counting rates possible, etc. They also have an intrinsic spectral sensitivity since, provided that the photon is absorbed completely, the number of electron-hole pairs produced is proportional to the photon's energy. About 5 eV are needed to produce an electron-hole pair in CZT, so that the spectral sensitivity to two and a half standard deviations is potentially 18% at 1 keV and 0.5% at 1 MeV. The main disadvantages of these detectors are that their size is small compared with many other detectors, so that their collecting area is also small and that unless the photon is stopped within the detector's volume the total energy cannot be determined. This latter disadvantage, however, also applies to most other detectors.

Recent X-ray and γ-ray missions that have utilised CCDs include XMM-Newton which uses three deep-depletion CCDs with an open structure and detects over the 0.15- to 15-keV region, whilst the Chandra spacecraft uses CCDs with 24-μm pixel size giving 0.5″

resolution in its Advanced CCD Imaging Spectrometer (ACIS). Suzaku has four 1 k × 1 k CCDs observing from 0.2 to 12 keV, whilst Swift, launched in 2004 to find GRBs, uses a 600 × 600 pixel CCD to cover from 0.2 to 10 keV in its X-ray telescope. The Spectrum-RG spacecraft, due for launch in 2014, will carry eROSITA, an imager for X-rays from 300 eV to 10 keV that will use seven 384 × 384 pixel deep-depletion frame-store CCDs. The Astro-H spacecraft, also due for launch in 2014, will use four 640 × 640 pixel deep-depletion CCDs in its soft X-ray imaging instrument which will operate over the 400 eV to 12 keV region.

Other types of solid-state detectors have been flown on the Swift space craft, which carries 256 modules each with 128 CZT detectors in its Burst Alert Telescope. The balloon-borne High Energy Focusing Telescope (HEFT) uses six 1000-pixel CZTs. Nuclear Spectroscopic Telescope Array (NuSTAR), launched in 2012, uses eight 32 × 32 pixel CZT detectors, four for each of its two telescopes, whilst INTEGRAL uses 19 germanium detectors within its 20-keV to 8-MeV spectrometer. Astro-H will have silicon strip detectors within its hard X-ray imager alongside a CdTe strip detector.

SSDs are used for the Large Area Telescope (LAT) on board the Fermi spacecraft. These provide positional and directional information on the gamma rays. The silicon strips are arranged orthogonally in two layers. A gamma ray, or the electron-positron-pair produced within a layer of lead by the gamma ray, will be detected within one of the strips in each layer. The position of the ray is then localised to the crossover point of the two strips. By using a number of such pairs of layers of silicon strips piled on top of each other, the direction of the gamma ray can also be determined. The LAT uses 16 such towers, each comprising 16 layers of silicon strip pairs and lead plates, enabling it to determine gamma-ray source positions to within 0.5 to 5 minutes of arc.

It is possible that SDDs will be the detectors chosen for the large area detector on the Large Observatory for X-ray Timing (LOFT) should this ever progress beyond the concept stage.

A rather different type of solid-state detector, the micro-calorimeter, has recently been flown on the Suzaku spacecraft and is planned for use on the Advanced Telescope for High Energy Astrophysics (ATHENA*) mission expected to be launched around 2020. Microcalorimeters have good intrinsic spectral resolution (R between 200 and 1000) (see Section 4.1) and operate by detecting the change in temperature of an absorber when it captures an X-ray photon. So far their principal use has been for low to medium-energy X-ray detection (up to 1 keV). There are three components to a micro-calorimeter – an absorber, a temperature sensor and a thermal sink. The energy of the X-ray photon is converted into heat within the absorber and this change is picked up by the sensor. There is then a weak thermal link to the heat sink so that the temperature of the absorber gradually returns to its original value. Typical response times are a few milliseconds with up to 80 ms recovery time, so micro-calorimeters are not suitable for detecting high fluxes of X-rays. The micro-calorimeter needs to be cooled so that its stored thermal energy is small compared with that released from the X-ray photon. A wide variety of materials can be used for all three components of the device with the main requirement for the absorber being a low thermal

---

* At the time of writing, funding for both the Green Bank and Arecibo instruments is under threat.

capacity and a high stopping rate for the X-rays. Micro-calorimeters can be formed into arrays – Suzaku, for example, carried a 32-pixel micro-calorimeter comprising mercury telluride absorbers combined with silicon thermistors and cooled to 60 mK. Arrays of over a thousand pixels are currently being planned with TES devices as their temperature sensors.

1.3.2.7 Microchannel Plates

For EUV and low-energy X-ray amplification and imaging, there is an ingenious variant of the photomultiplier (see Section 1.1) – the microchannel plate. The devices are also known as MAMAs (see Section 1.1.11) and additionally are used as optical components for imaging X-rays and in these cases they are often called micro-pore optics (see lobster eye collimator, below).

When acting as a detector, the MCP comprises a thin plate pierced by numerous tiny holes, each perhaps only about 10 μm across or less. Its top surface is an electrode with a negative potential of some few thousand volts with respect to its base. The top is also coated with a photoelectron emitter for the X-ray energies of interest. An impinging photon releases one or more electrons that are then accelerated down the tubes. There are inevitably collisions with the walls of the tube during which further electrons are released and these in turn are accelerated down the tube and so on (Figure 1.91). As many as 10^4 electrons can be produced for a single photon and this may be increased to 10^6 electrons

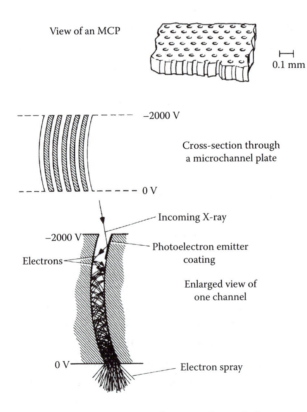

FIGURE 1.91   Schematic view of the operation of a microchannel plate.

in future devices when the problems caused by ion feedback are reduced. The quantum efficiency can be up to 20%. The electrons spray out of the bottom of each tube, where they may be detected by a variety of the more conventional imaging systems, or they may be fed into a second microchannel plate for further amplification.

Early versions of these devices often used curved channels (Figure 1.91). Modern versions now employ two plates with straight channels, the second set of channels being at an angle to the direction of the first set (known as a chevron microchannel plate). The plates are currently manufactured from a billet of glass that has an acid-resisting shell surrounding an acid soluble core. The billet is heated and drawn out until it forms a glass 'wire' about 1 mm across. The wire is cut into short sections and the sections stacked to form a new billet. That billet in turn is heated and drawn out and so on. The holes in the final plate are produced by etching away the acid soluble portions of the glass. Holes down to 6 µm in diameter may be produced in this fashion and the fraction of the plate occupied by the holes can reach 65%. Plates can be up to 0.1 metre square. A single X-ray results in about 30,000,000 electrons that are collected by a grid of wires at the exit holes from the second stage plate.

As examples, the high resolution X-ray camera on board the Chandra spacecraft uses a 93-mm-square chevron microchannel plate detector, with sixty-nine million 10-µm holes and can provide a resolution of 0.5 seconds of arc. The planetary space probe, Bepi Colombo, due to launch in 2015 on a mission to Mercury, will carry microchannel plates as detectors for its far UV spectrometer.

Microchannel plates can also be used in optical and near UV; however, they are then used simply to amplify the photoelectrons produced from a photocathode deposited onto the entrance window of the detector and are thus similar to image intensifiers (see Section 2.3).

1.3.2.8 Čerenkov Detectors

Čerenkov radiation and the resulting detectors are referred to in detail in Section 1.4. For X-ray and gamma radiation their interest lies in the detection of particles produced by the Compton interactions of the very-high-energy photons. If those particles have a velocity higher than the velocity of light in the local medium then Čerenkov radiation is produced. Čerenkov radiation produced in this manner high in the Earth's atmosphere is detectable. Thus, the Collaboration of Australia and Nippon for a Gamma Ray Observatory in the Outback (CANGAROO-III) Čerenkov instrument comprises four 10-metre optical dishes in Australia, each formed from a hundred and fourteen 0.8-metre mirrors. It has a field of view of 2.5° and observes the outer reaches of the Earth's atmosphere with 552 photomultipliers, detecting TeV gamma rays from active galaxies. Similarly, the Very Energetic Radiation Imaging Telescope Array System (VERITAS) instrument, located in Arizona, employs four 12-metre reflectors, each made up from 350 mirror segments whilst Major Atmospheric Gamma Imaging Cherenkov telescope (MAGIC) on La Palma uses two 17-metre multi-segment reflectors and High Energy Stereoscopic System (HESS) in Namibia has four 12-metre dishes plus a single 28-metre dish. The Cherenkov Telescope Array (CTA) is a concept for a much larger installation that is currently under discussion.

It might have several tens of dishes in three different sizes and be located at sites in both the northern and southern hemispheres.

Very-high-energy primary gamma rays (10^{15} eV and more) may also produce cosmic ray showers and so be detectable by the methods discussed in Section 1.4, especially through the fluorescence of atmospheric nitrogen.

1.3.2.9 Future Possibilities

The CCD variant CMOS-APS (also called Hybrid CMOS detectors) (see Section 1.1) which reads out the pixels on an individual basis is likely to replace the CCD in many future X-ray space missions since it is hardened to radiation and can provide high count rates.

The details of the operating principles of STJs and MKIDs may also be found in Section 1.1. They are, however, also sensitive to X-rays and indeed the first applications of STJs were for X-ray detection. Since an X-ray breaks about a thousand times as many Cooper pairs as an optical photon, their spectral discrimination is about 0.1%. The need to cool them to below 1 K and the small sizes of the arrays currently available means that they have yet to be flown on a spacecraft.

Another type of detector altogether is a solid analogue of the cloud or bubble chamber (see Section 1.4). A superconducting solid would be held at a temperature fractionally higher than its critical temperature, or in a non-superconducting state and at a temperature slightly below the critical value. Passage of an ionising ray or particle would then precipitate the changeover to the opposing state and this in turn could be detected by the change in a magnetic field (see TES detectors, Section 1.1).

The currently-under-construction Russian spacecraft, the GAMMA-400, which is scheduled for launch in 2018, will be a γ-ray instrument that combines almost all the individual detectors that we have just discussed. It will observe in the 100-MeV to 1-TeV range with angular resolutions ranging from 2° to 0.1°, with an energy resolution of about 1% and with a 2.5-sr field of view. To accomplish this it will use SSDs interlaced with tungsten strips, time-of-flight scintillators, an imaging microcalorimeter comprising SSDs and tungsten layers, two more scintillators, a BGO microcalorimeter and more scintillators for active shielding.

1.3.3 Shielding

Very few of the detectors that we have just reviewed are used in isolation. Usually, several will be used together in modes that allow the rejection of information on unwanted interactions. This is known as active or anti-coincidence shielding of the detector. The range of possible configurations is very wide and a few examples will suffice to give an indication of those in current use.

1. The germanium solid-state detectors must intercept all the energy of an incoming photon if they are to provide a reliable estimate of its magnitude. If the photon escapes, then the measured energy will be too low. A thick layer of a scintillation crystal therefore surrounds the germanium. A detection in the germanium simultaneously with two in the scintillation crystal is rejected as an escapee.

2. The solid angle viewed by a germanium solid-state detector can be limited by placing the germanium at the bottom of a deep and narrow hole in a scintillation crystal (known as

a well active shield because the hole is shaped like a water well). Only those detections not occurring simultaneously in the germanium and the crystal are used and these are the photons that have entered down the hole. Any required degree of angular resolution can be obtained by sufficiently reducing the size of the entrance aperture.

3. Primary detectors are surrounded by scintillation counters and Čerenkov detectors to discriminate between gamma-ray events and those due to cosmic rays.

4. A sodium iodide scintillation counter may be surrounded, except for an entrance aperture, by one formed from caesium iodide to eliminate photons from the unwanted directions.

5. The Fermi spacecraft uses segmented plastic scintillator tiles for its anti-coincidence shielding. These are read out by photomultipliers.

Recently, nuclear power sources on board other spacecraft have caused interference with gamma-ray observations. These pulses have to be separated from the naturally occurring ones by their spectral signatures.

Passive shields are also used and these are just layers of an absorbing material that screen out unwanted rays. They are especially necessary for the higher-energy photon detectors in order to reduce the much greater flux of lower-energy photons. The mass involved in an adequate passive shield, however, is often a major problem for spacecraft-based experiments with their very tight mass budgets arising from their launcher capabilities.

1.3.4 Imaging

Imaging of high-energy photons is a difficult task because of their extremely penetrating nature. Normal designs of telescope are impossible since in a reflector the photon would just pass straight through the mirror, whilst in a refractor it would be scattered or unaffected rather than refracted by the lenses. At energies under a few kiloelectron volts, forms of reflecting telescope can be made which work adequately, but at higher energies the only options are occultation, collimation and coincidence detection.

1.3.4.1 Collimation

A collimator is simply any device that physically restricts the field of view of the detector without contributing any further to the formation of an image. The image is obtained by scanning the system across the object. Almost all detectors are collimated to some degree, if only by their shielding. The simplest arrangement is a series of baffles (Figure 1.92) that may be formed into a variety of configurations, but all of which follow the same basic principle. Their general appearance leads to their name, honeycomb collimator, even though the cells are usually square rather than hexagonal. They can restrict angles of view down to a few minutes of arc, but they are limited at low energies by reflection off their walls and at high energies by penetration of the radiation through the walls. At high energies, the baffles may be formed from a crystal scintillator and pulses from there used to reject detections of radiation from high inclinations (cf. active shielding).

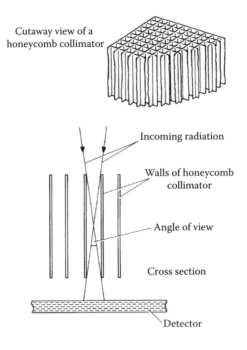

FIGURE 1.92   Honeycomb collimator.

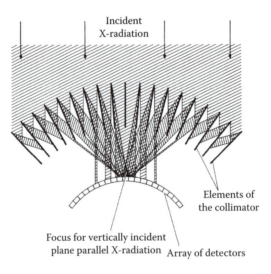

FIGURE 1.93   Cross section through a lobster-eye-focusing wide-angle X-ray collimator.

At the low energies the glancing reflection of the radiation can be used to advantage and a more truly imaging collimator produced. This is called a lobster-eye* focusing collimator (also called micro-pore optics) and is essentially a honeycomb collimator curved into a portion of a sphere (Figure 1.93), with a position-sensitive detector at its focal surface. The imaging is not of very high quality and there is a high background from the unreflected rays and the doubly reflected rays (the latter is not shown in Figure 1.93). But it is a cheap system

---

* The device mimics the way that lobsters' eyes work.

to construct compared with the others that are discussed later in this section and it has the potential for covering very wide fields of view (tens of degrees) at very high resolutions (seconds of arc). In practice, the device is constructed by fusing together thousands of tiny glass rods in a matrix of a different glass. The composite is then heated and shaped by slumping onto a spherical mould and the glass rods etched away to leave the required hollow tubes in the matrix glass. This is the same process as that used to produce microchannel plates and the only difference is that for a lobster-eye collimator, the channels are of square cross section, whereas they are circular in microchannel plates. Multilayer coatings may be applied to enhance reflectivity and increase the waveband covered.

The LWFT for the Spectrum-RG spacecraft will use five lobster-eye telescopes each with a 30° × 30° field of view and covering the 100-eV to 3.5-keV part of the spectrum. Bepi Colombo's Mercury Imaging X-ray Spectrometer (MIXS) which will look for X-ray fluorescence from Mercury's surface in the 500-eV to 7.5-keV region will use two lobster-eye collimators with differing spatial resolutions.

Another system that is known as a modulation collimator, rotation collimator or Fourier transform telescope uses two or more parallel gratings that are separated by a short distance (Figure 1.94). Since the bars of the gratings alternately obscure the radiation and allow it to pass through, the output as the system scans a point source is a sine wave (Figure 1.94). The resolution is given by the angle α

$$\alpha = \frac{d}{s} \qquad (1.96)$$

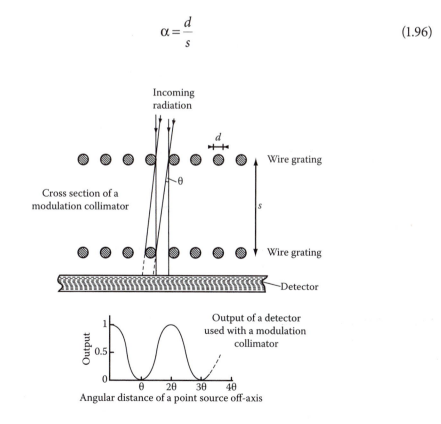

FIGURE 1.94   A modulation collimator.

To obtain unambiguous positions for the sources or for the study of multiple or extended sources, several such gratings of different resolutions are combined. The image may then be retrieved from the Fourier components of the output (cf. aperture synthesis) (see Section 2.5). Two such grating systems at right angles can give a two-dimensional image. With these systems resolutions of a few tens of seconds of arc can be realised even for the high-energy photons.

The RHESSI spacecraft uses modulation collimators to study the X-ray emission from solar flares. There are five pairs of coarse grids and four pairs of fine grids and the detectors are germanium crystals. Spectral resolutions of 1- to 5-keV are obtained over the range 100 keV to 17 MeV with angular resolutions ranging from 2″ to 36″.

A third type of collimating imaging system is a simple pinhole camera. A position-sensitive detector such as a resistive anode proportional counter or a microchannel plate is placed behind a small aperture. The quality of the image is then just a function of the size of the hole and its distance in front of the detector. Unfortunately, the small size of the hole also gives a low flux of radiation so that such systems are limited to the brightest objects. A better system replaces the pinhole with a mask formed from clear and opaque regions. The pattern of the mask is known, so that when sources cast shadows of it onto the detector, their position and structure can be reconstituted in a similar manner to that used for the modulation collimator. The technique is known as coded mask imaging and resolutions of 10 minutes of arc or better can be achieved. Since only half the aperture is obscured, the technique obviously uses the incoming radiation much more efficiently than the pinhole camera. A somewhat related technique known as Hadamard mask imaging is discussed in Section 2.4. The disadvantage of coded mask imaging is that the detector must be nearly the same size as the mask. The Swift space craft, for example, uses a coded mask with an area of 3.2 m², allied with nearly 33,000 CZT detectors covering an area of 0.52 m² in its Burst Alert Telescope (BAT) The BAT detects gamma-ray bursts and it has a field of view of 2 steradians; thus it is able to pinpoint the position of a gamma-ray burst to within 4 arc minutes within 15 seconds of its occurrence. The image from a coded aperture telescope is extracted from the data by a cross correlation between the detected pattern and the mask pattern.

1.3.4.2 Coincidence Detectors
A telescope, in the sense of a device that has directional sensitivity, may be constructed for use at any energy and with any resolution, by using two or more detectors in a line and by rejecting all detections except those which occur in both detectors and separated by the correct flight time. Two separated arrays of detectors can similarly provide a two-dimensional imaging system.

1.3.4.3 Occultation
Although it is not a technique that can be used at will on any source, the occultation of a source by the Moon or another object can be used to give very precise positional and structural information. The use of occultations in this manner is discussed in detail in Section 2.7. The technique is less important now than in the past since other imaging systems are

available, but it was used in 1964, for example, to provide the first indication that the X-ray source associated with the Crab nebula was an extended source.

1.3.4.4 Reflecting Telescopes

At energies below about 100 keV, photons may be reflected with up to 50% efficiency off metal surfaces, when their angle of incidence approaches 90°. Mirrors in which the radiation just grazes the surface (grazing incidence optics) may therefore be built and configured to form reflecting telescopes. The angle between the mirror surface and the radiation has to be very small however: less than 2° for 1-keV X-rays, less than 0.6° for 10-keV X-rays and less than 0.1° for 100-keV X-rays. Several systems have been devised, but the one which has achieved most practical use is formed from a combination of annular sections of very deep paraboloidal and hyperboloidal surfaces (Figure 1.95) and known as a Wolter type I telescope after its inventor. Other designs that are also due to Wolter are shown in Figure 1.96 and have also been used in practice, although less often than the first design. A simple paraboloid can also be used, although the resulting telescope then tends to be excessively long. The aperture of such telescopes is a thin ring, since only the radiation incident onto the paraboloidal annulus is brought to the focus. To increase the effective aperture and hence the sensitivity of the system, several such confocal systems of differing radii may be nested inside each other (Figure 1.97).

The limit of resolution is due to surface irregularities in the mirrors, rather than the diffraction limit of the system. The irregularities are about 0.3 nm in size for the very best of the current production techniques (i.e. comparable with the wavelength of photons of about 1 keV). The mirrors are produced by electro-deposition of nickel onto a mandrel of the inverse shape to that required for the mirror, the mirror shell then being separated from the mandrel by cooling. A coating of a dense metal such as gold, iridium or platinum

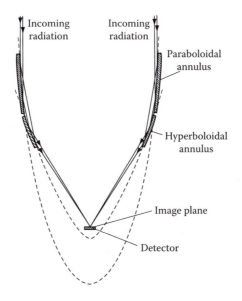

FIGURE 1.95   Cross section through a grazing incidence X-ray telescope.

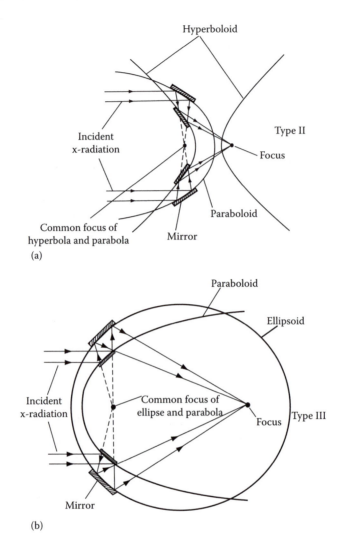

FIGURE 1.96 Cross sections through alternative designs for grazing incidence X-ray telescopes - (a) a Wolter type II design and (b) a Wolter type III design.

may be applied to improve the X-ray reflectivity. At energies of a few kiloelectron volts, glancing incidence telescopes with low but usable resolutions can be made using foil mirrors. The incident angle is less than 1° so a hundred or more mirrors are needed. They may be formed from thin aluminium foil with a lacquer coating to provide the reflecting surface. The angular resolution of around 1′ means that the mirror shapes can be simple cones, rather than the paraboloids and hyperboloids of a true Wolter telescope and so fabrication costs are much reduced.

An alternative glancing incidence system, known as the Kirkpatrick-Baez design, that has fewer production difficulties is based upon cylindrical mirrors. Its collecting efficiency is higher but its angular resolution poorer than the three-dimensional systems discussed above. Two mirrors are used which are orthogonal to each other (Figure 1.98). The surfaces can have a variety of shapes, but two paraboloids are probably the commonest

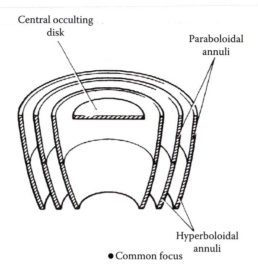

FIGURE 1.97 Section through a nested grazing incidence X-ray telescope.

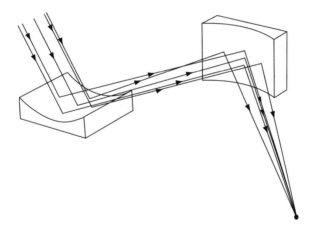

FIGURE 1.98 X-ray imaging by a pair of orthogonal cylindrical mirrors.

arrangement. The mirrors can again be stacked in multiples of the basic configuration to increase the collecting area.

For the XMM-Newton spacecraft a total of 58 such nested telescope shells give a total collecting area of about half a square metre. Chandra has 4 and Swift has 12 nested shells. A position-sensitive detector at the focal plane can then produce images with resolutions as high as a second of arc. There are two identical soft X-ray telescopes to be carried by the ASTRO-H spacecraft. Each is of the Wolter-I design with 203 nested shells, an outer diameter of 450 mm, a focal length of 5.6 metres and a collecting area of 0.056 m². There are to be seven Wolter-I telescopes for the eROSITA instrument on the Spectrum-RG spacecraft; each will have 54 nested shells with an outer diameter of 358 mm and a focal length of 1.6 m. Operating in the 500-eV to 10-keV region, the shells are fabricated from nickel and have a gold reflective coating.

At lower energies, in the EUV and soft X-ray regions, near-normal incidence reflection with efficiencies up to 20% is possible using multilayer coatings. These are formed from tens, hundreds or even thousands of alternate layers of tungsten and carbon, aluminium and gold or magnesium and gold, each about 1 nm thick. The reflection is essentially monochromatic, with the wavelength depending upon the orientation of the crystalline structure of the layers and upon their thickness. Reflection of several wavelengths may be accomplished by changing the thickness of the layers through the stack. The thickest layers are at the top and reflect the longest, least penetrating wavelengths, whilst further down, narrower layers reflect shorter wavelengths. Alternatively, several small telescopes may be used each with a different wavelength response, or as in the Transition Region and Coronal Explorer (TRACE) (1998–2010) spacecraft, different multilayer coatings being applied in each of four quadrants of the optics. Telescopes of relatively conventional design are used with these mirrors and direct images of the Sun at wavelengths down to 4 nm can be obtained.

Multilayer coatings and grazing-incidence optics can be combined to improve reflectivity and extend the usage to higher energies – perhaps to 300 keV in the future. The coatings are alternate layers of high and low atomic number materials such as molybdenum and silicon or tungsten and boron carbide. The layers vary in thickness so that a wide range of X-ray wavelengths are reflected. HEFT uses 72 nested glass multilayer coated mirror shells in each of its three telescopes and NuSTAR uses two grazing incidence telescopes to observe up to 79 keV. The telescopes each have 133 concentric shells and have graded density multilayer reflective coatings. The hard X-ray telescopes on ASTRO-H will have 213 nested Wolter-I shells with potassium-carbon depth graded multilayer reflective coatings.

At energies of tens or hundreds of kiloelectron volts, Bragg reflection (as in the multilayer coatings) and Laue diffraction can be used to concentrate the radiation and to provide some limited imaging. For example, the Laue diffraction pattern of a crystal comprises a number of spots into which the incoming radiation has been concentrated (Figure 1.99).

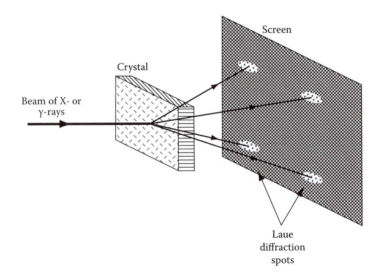

FIGURE 1.99 Laue diffraction of X-rays and gamma rays.

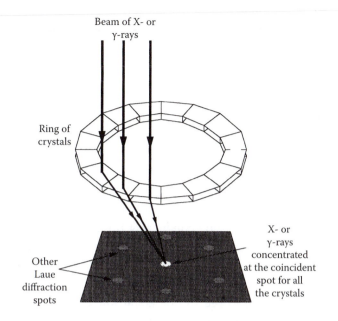

FIGURE 1.100   A Laue diffraction lens for X-rays or gamma rays (only three crystals shown illuminated for clarity).

With some crystals, such as germanium and copper, only a few spots are produced. A number of crystals can be mutually aligned so that one of the spots from each crystal is directed towards the same point (Figure 1.100), resulting in a crude type of lens. With careful design, efficiencies of 25% to 30% can be achieved at some wavelengths. As with the glancing incidence telescopes, several such ring lenses can be nested to increase the effective area of the telescope. A position-sensitive detector such as an array of germanium crystals (Figure 1.90) can then provide direct imaging at wavelengths down to 0.001 nm (1 MeV). A study for a possible future space mission using a Laue lens suggests that the focal length of the lens would be some 500 metres. Thus, two space craft – one to carry the lens and one to carry the detectors – flying in tandem would be needed and γ-rays with energies as high as 1 MeV might be imaged by such a system.

1.3.5 Resolution and Image Identification

At the low-energy end of the X-ray spectrum, resolutions and positional accuracies comparable with those of Earth-based optical systems are possible, as we have seen. There is therefore usually little difficulty in identifying the optical counterpart of an X-ray source, should it be bright enough to be visible. The resolution rapidly worsens, however, as higher energies are approached. The position of a source that has been detected can therefore only be given to within quite broad limits. It is customary to present this uncertainty in the position of a source by specifying an error box. This is an area of the sky, usually rectangular in shape, within which there is a certain specified probability of finding the source. For there to be a reasonable certainty of being able to find the optical counterpart of a source unambiguously, the error box must be smaller than about 1 square minute of arc. Since high-energy-photon imaging systems have resolutions measured in tens of minutes of arc or larger, such identification is not usually possible (although it is hoped that the hard X-ray imager to be launched on the ASTRO-H spacecraft will achieve a 1.7' resolution at 60 keV). The exception to this occurs when an unusual optical object lies within the error box. This object may then, rather riskily, be presumed also to be the X-ray source. Examples of this are the Crab nebula and Vela supernova remnant which lie in the same direction as two strong sources at 100 MeV and are therefore assumed to be those sources, even though the positional uncertainties are measured in many degrees.

1.3.6 Spectroscopy

Many of the detectors that were discussed earlier are intrinsically capable of separating photons of differing energies. The superconducting tunnel junction detector (Section 1.1 and above) has an intrinsic spectral resolution in the X-ray region that potentially could reach 10,000 (i.e. ±0.1 eV at 1 keV). However, the need to operate the devices at lower than 1 K seems likely to prevent their use on spacecraft for some time to come. Microcalorimeters though, which also need to operate at low temperatures, have been flown on the Suzaku mission and are planned to be used on other spacecraft. They have intrinsic spectral resolutions of up to 1000. At photon energies above about 10 keV, it is only this inherent spectral resolution which can provide information on the energy spectrum, although some additional resolution may be gained by using several similar detectors with varying degrees of passive shielding – the lower-energy photons will not penetrate through to the most highly shielded detectors. The hard X-ray imager for the ASTO-H spacecraft employs the detectors themselves in this manner. It will have four stacked silicon strip detectors placed above a single CdTe strip detector. The SSDs absorb (and detect) X-rays of less than 30 keV so that only the higher-energy photons penetrate to the CdTe detector. The instrument will be housed at the bottom of a BGO well active shield.

At low and medium energies, the gaps between the absorption edges of the materials used for the windows of the detectors can form wideband filters. Devices akin to the more conventional idea of a spectroscope, however, can only be used at the low end of the energy spectrum and these are discussed below.

1.3.6.1 Grating Spectrometers

Gratings may be either transmission or grazing incidence reflection. The former may be ruled gratings in a thin metal film which has been deposited onto a substrate transparent to the X-rays, or they may be formed on a pre-ruled substrate by vacuum deposition of a metal from a low angle. The shadow regions where no metal is deposited then form the transmission slits of the grating. The theoretical background for X-ray gratings is identical with that for optical gratings and is discussed in detail in Section 4.1. Typical transmission gratings have around 1000 lines per millimetre. The theoretical spectral resolution (see Section 4.1) is between 10^3 and 10^4, but is generally limited in practice to 50 to 100 by other aberrations.

Reflection gratings are also similar in design to their optical counterparts. Their dispersion differs, however, because of the grazing incidence of the radiation. If the separation of

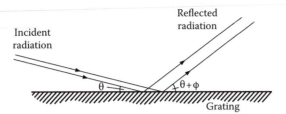

FIGURE 1.101   Optical paths in a grazing incidence reflection grating.

the rulings is d then, from Figure 1.101, we may easily see that the path difference, ΔP, of two rays that are incident onto adjacent rulings is

$$\Delta P = d\left[\cos\theta - \cos(\theta + \phi)\right] \tag{1.97}$$

We may expand this via the Taylor series and neglect powers of θ and ϕ higher than two since they are small angles, to obtain

$$\Delta P = \frac{1}{2}d(\phi^2 - 2\theta\phi) \tag{1.98}$$

In the mth order spectrum, constructive interference occurs for radiation of wavelength λ if

$$m\lambda = \Delta P \tag{1.99}$$

so that

$$\phi = \left(\frac{2m\lambda}{d} + \theta^2\right)^{1/2} - \theta \tag{1.100}$$

and

$$\frac{d\phi}{d\lambda} = \left(\frac{m}{2d\lambda}\right)^{1/2} \tag{1.101}$$

where we have neglected θ^2, since θ is small. The dispersion for a glancing incidence reflection grating is therefore inversely proportional to the square root of the wavelength, unlike the case for near normal incidence, when the dispersion is independent of wavelength. The gratings may be plane or curved and may be incorporated into many different designs of spectrometer (see Section 4.2). Resolutions of up to 10^3 are possible for soft X-rays, but again this tends to be reduced by the effects of other aberrations. Detection may be by scanning the spectrum over a detector (or vice versa), or by placing a position-sensitive detector in the image plane so that the whole spectrum is detected in one go. The XMM-Newton spacecraft for example uses two reflection grating arrays. Each array has 182 individual gratings and they are incorporated into a Rowland-circle (Figure 4.6) spectroscope. CCDs are used as the detectors and spectral resolutions up to 800 are achieved over the 330-eV to 2.5-keV region.

A recent development is the production by photo-lithographic techniques of off-plane grazing incidence gratings. These are placed in the converging beam of X-rays from a Wolter-I telescope (*not* at the focus). The rulings on the grating are parallel to the X-ray beam and converge at the same rate. The result is a spectrum at the focal plane in place of the normal image. Currently, gratings with up to 6000 grooves per mm can be fabricated and potentially lead to soft X-ray spectrometers with spectral resolutions up to 3000 and areas of 0.1 m² – a factor of 10 improvement for both specifications compared with the spectrometers on XMM-Newton and Chandra.

1.3.6.2 Bragg Spectrometers

Distances ranging from 0.1 to 10 nm separate the planes of atoms in a crystal. This is comparable with the wavelengths of X-rays and so a beam of X-rays interacts with a crystal in a complex manner. The details of the interaction were first explained by the Braggs (father and son). Typical radiation paths are shown in Figure 1.102. The path differences for rays such as a, b and c, or d and e, are given by multiples of the path difference, ΔP, for two adjacent layers

$$\Delta P = 2d \sin \theta \tag{1.102}$$

There will be constructive interference for path differences that are whole numbers of wavelengths so that the reflected beam will consist of just those wavelengths, λ, for which this is true

$$M\lambda = 2d \sin \theta \tag{1.103}$$

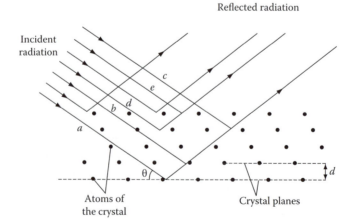

FIGURE 1.102 Bragg reflection.

The Bragg spectrometer uses a crystal to produce monochromatic radiation of known wavelength. If a crystal is illuminated by a beam of X-rays of mixed wavelengths, at an approach angle, θ, then only those X-rays whose wavelength is given by Equation 1.103 will be reflected. The first-order reflection ($m = 1$) is by far the strongest and so the reflected beam will be effectively monochromatic with a wavelength of

$$\lambda_\theta = 2d \sin \theta \qquad (1.104)$$

The intensity of the radiation at that wavelength may then be detected with, say, a proportional counter and the whole spectrum scanned by tilting the crystal to alter θ (Figure 1.103). An improved version of the instrument uses a bent crystal and a collimated beam of X-rays so that the approach angle varies over the crystal. The reflected beam then consists of a spectrum of all the wavelengths (Figure 1.104) and this may then be detected by a single observation using a position-sensitive detector. The latter system has the major advantage for satellite-borne instrumentation of having no moving parts and so has a significantly higher reliability. It also has good time resolution. High spectral resolutions are possible - up to 10^3 at 1 keV – but large crystal areas are necessary for good sensitivity and this may present practical difficulties on a satellite. Many crystals may be used. Amongst the commonest are lithium fluoride, lithium hydride, tungsten disulphide, graphite and potassium acid phthalate (KAP).

Many variants upon the basic spectrometer can be devised. It may be used as a monochromator and combined with a scanning telescope to produce spectroheliograms; for faint sources it may be adapted for use at the focus of a telescope and so on. Designs are rapidly changing and the reader desiring completely up-to-date information must consult the current literature.

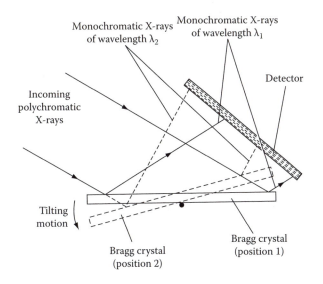

FIGURE 1.103   Scanning Bragg crystal X-ray spectrometer.

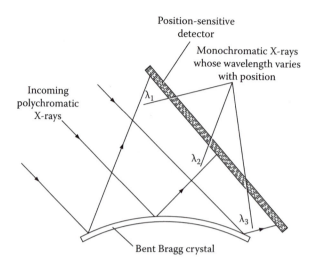

FIGURE 1.104   Bent Bragg crystal X-ray spectrometer.

1.3.7 Polarimetry

Bragg reflection of X-rays is polarisation-dependent. For an angle of incidence of 45°, photons that are polarised perpendicularly to the plane containing the incident and reflected rays will be reflected, whilst those polarised in this plane will not. Thus, a crystal and detector at 45° to the incoming radiation and which may be rotated around the optical axis will function as a polarimeter. The efficiency of such a system would be very low, however, due to the narrow energy bandwidth of Bragg reflection. This is overcome by using many randomly orientated small crystals. The crystal size is too small to absorb radiation significantly. If it should be aligned at the Bragg angle for the particular wavelength concerned, however, that radiation will be reflected. The random orientation of the crystals ensures that overall, many wavelengths are reflected and we have a polarimeter with a broad bandwidth.

A second type of polarimeter looks at the scattered radiation due to Thomson scattering in blocks of lithium or beryllium. If the beam is polarised, then the scattered radiation is asymmetrical and this may be measured by surrounding the block with several pairs of detectors.

Recently X-ray polarimeters have been based upon the intrinsic polarisation sensitivity of proportional counters and their derivatives such as TPCs. The just-cancelled Gravity and Extreme Magnetism Small Explorer (GEMS) spacecraft was due to carry a TPC-based polarimeter for the 2- to 10-keV region as its main instrument and a Bragg reflection polarimeter for 500-eV X-ray observations.

1.3.8 Observing Platforms

The Earth's atmosphere completely absorbs X-ray and γ-ray radiation, so that all the detectors and other apparatus discussed above, with the exception of the Čerenkov detectors, have to be lifted above at least 99% of the atmosphere. The three systems for such lifting equipment are balloons, rockets and spacecraft.

Spacecraft give the best results in terms of their height, stability and the duration of the mission. Their cost is very high, however, and there are weight and space restrictions to be complied with; nonetheless there have been quite a number of satellites, many of which have been mentioned above, launched for exclusive observation of the X-ray and γ-ray region and appropriate detectors have been included on a great many other missions as secondary instrumentation.

Balloons can carry heavy equipment far more cheaply than satellites, but only to a height of about 40 km and this is too low for many of the observational needs of this spectral region. Their mission duration is also comparatively short being a few days to a month for even the most sophisticated of the self-balancing versions. Complex arrangements have to be made for communication and for retrieval of the payload because of the unpredictable drift of the balloon during its mission. The platform upon which the instruments are mounted has to be actively stabilised to provide sufficient pointing accuracy for the telescopes etc. Thus, there are many drawbacks to set against the lower cost of a balloon. The HEFT and PoGOLite missions have already been mentioned as examples of balloon-carried X-ray instruments.

Sounding rockets are even cheaper still and they can reach heights of several hundred kilometres without difficulty. But their flight duration is measured only in minutes and the weight and size restrictions are far tighter even than for spacecraft. Rockets still, however, find uses as rapid response systems for events such as solar flares. A balloon- or spacecraft-borne detector may not be available when the occurs, or may take some time to bear onto the target, whereas a rocket may be held on stand-by on Earth, during which time the cost commitment is minimal and then launched when required at only a few minutes' notice. The High Resolution Coronal Imager (Hi-C) for example was launched in 2011 on a Black Brant rocket to observe the Sun at 64 eV and obtained 165 images with 0.2″ resolution using a CCD camera during its 5 minutes of observing time.

1.4 COSMIC RAY DETECTORS

1.4.1 Background

Cosmic rays comprise two quite separate populations of particles, primary cosmic rays and secondary cosmic rays. The former are the true cosmic rays and consist mainly of atomic nuclei in the proportions 84% hydrogen, 14% helium, 1% other nuclei and 1% electrons and positrons,[*] moving at velocities from a few per cent of the speed of light to 0.999, 999, 999, 999, 999, 999, 9999c. That is, the energies per particle range from 10^6 to a few times 10^{20} eV (10^{-13} to 50 J) with the bulk of the particles having energies near 10^9 eV, or a velocity of 0.9c. These are obviously relativistic velocities for most of the particles, for which the relationship between kinetic energy and velocity is

$$v = \frac{(E^2 + 2Emc^2)^{1/2} c}{E + mc^2} \tag{1.105}$$

where E is the kinetic energy of the particle and m is the rest mass of the particle. The flux in space of the primary cosmic rays of all energies is about 10^4 particles m^{-2} s^{-1}. A limit,

---

[*] The term 'ray' in this context is thus a misnomer arising from before the time that their true nature was understood. However, the usage is unlikely to be corrected now.

the GZK cut-off,* in the cosmic-ray energy spectrum is expected above about 4×10^{19} eV (6 J). At this energy the microwave background photons are blue-shifted to gamma rays of sufficient energy to interact with the cosmic-ray particle to produce pions. This will slow the particle to energies lower than the GZK limit in around 100 million years. However, surprisingly, higher energy cosmic rays *are* observed, implying that there must be one or more sources of ultra-high energy cosmic rays within 100 million light years (30 million pc) of the Earth. The Pierre Auger cosmic-ray observatory (see below) confirmed in 2007 that the most likely sources of cosmic-ray particles with energies above the GZK limit were relatively nearby active galactic nuclei, although the mechanism for accelerating the particles to such enormous energies remains a mystery.

Secondary cosmic rays are produced by the interaction of the primary cosmic rays with nuclei in the Earth's atmosphere. A primary cosmic ray has to travel through about 800 kg m^{-2} of matter on average before it collides with a nucleus. A column of the Earth's atmosphere 1 square metre in cross section contains about 10^4 kg, so the primary cosmic ray usually collides with an atmospheric nucleus at a height of 30 to 60 km. The interaction results in numerous fragments, nucleons, pions, muons, electrons, positrons, neutrinos, gamma rays etc. and these in turn may still have sufficient energy to cause further interactions, producing more fragments and so on. For high-energy primaries (> 10^{11} eV), some of the secondary particles will survive down to sea level† and be observed as secondary cosmic rays. At even higher energies (> 10^{13} eV), large numbers of the secondary particles (> 10^9) survive to sea level and a cosmic-ray shower or extensive air shower (EAS) is produced. At altitudes higher than sea level, the secondaries from the lower-energy primary particles may also be found. Interactions at all energies produce electron and muon neutrinos as a component of the shower. Almost all of these survive to the surface, even those passing through the whole of the Earth, although some convert to tau neutrinos during that passage. The detection of these latter particles is considered in Section 1.5.

We also consider in this section the detection of very-high-energy γ-rays via the Čerenkov radiation that they induce in the Earth's atmosphere since they are detected by the same instruments as those for observing very-high-energy cosmic rays.

1.4.2 Detectors

The methods of detecting cosmic rays may be divided into the following:

Real-time methods. These observe the particles instantaneously and produce information on their direction as well as their energy and composition. They are the standard detectors of nuclear physicists and there is considerable overlap between these detectors and those discussed in Section 1.3. They are generally used in conjunction with passive and/ or active shielding so that directional and spectral information can also be obtained.

---

* Named for Ken Greisen, Timofeevich Zatsepin and Vadim Kuzmin.
† At sea level the radiation exposure coming from secondary cosmic rays is around 0.4 mSv per year. The normal total background exposure is 3 mSv per year. The cosmic ray component will increase though with altitude – reaching perhaps 5 mSv per year for aircraft crews who work full time on high flying aircraft. To put these rates in context, the recommended maximum occupational exposure limit is 20 mSv per year and exposure to 1 Sv in a single short exposure is likely to induce temporary radiation sickness. A 5-Sv dose in a single short exposure is likely to cause 50% fatalities amongst those exposed.

Residual track methods. The path of a particle through a material may be found some time (hours to millions of years) after its passage.

Indirect methods. Information on the flux of cosmic rays at large distances from the Earth, or at considerable times in the past may be obtained by studying the consequent effects of their presence.

1.4.2.1 Real-Time Methods

1.4.2.1.1 Scintillation Detectors   See Section 1.3 for a discussion of scintillation detectors. There is little change required in order for them to detect cosmic-ray particles. Plastic sheet scintillator detectors with areas of 3 square metres and photomultipliers as detectors are used for the 507 surface stations of the Telescope Array Project in Utah. They are also used as triggers and active shields for the PAMELA* experiment launched in 2006.

1.4.2.1.2 Čerenkov Detectors   *Background.* When a charged particle is moving through a medium with a speed greater than the local speed of light in that medium, it causes the atoms of the medium to radiate. This radiation is known as Čerenkov radiation and it arises from the abrupt change in the electric field near the atom as the particle passes by. At subphotic speeds, the change in the field is smoother and little or no radiation results. The radiation is concentrated into a cone spreading outward from the direction of motion of the particle (Figure 1.105), whose half angle, θ, is

$$\theta = \tan^{-1}\left[\left(\mu_v^2 \frac{v^2}{c^2} - 1\right)^{1/2}\right]$$
(1.106)

where μ_v is the refractive index of the material at frequency v and v are the particle's velocity ($v > c/\mu_v$). Its spectrum is given by

$$I_v = \frac{e^2 v}{2\varepsilon_0 c^2}\left(1 - \frac{c^2}{\mu_v^2 v^2}\right)$$
(1.107)

where I_v is the energy radiated at frequency v per unit frequency interval, per unit distance travelled by the particle through the medium. The peak emission depends upon the form of the variation of refractive index with frequency. For a proton in air, peak emission occurs in the visible when the proton's energy is about 2×10^{14} eV.

Detectors. Čerenkov detectors are very similar to scintillation detectors and are sometimes called that, although the flashes of visible radiation are produced by different physical mechanisms in the two detectors. A commonly used system employs a tank of very

---

* Payload for Antimatter Exploration and Light-nuclei Astrophysics.

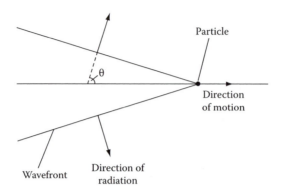

FIGURE 1.105 Čerenkov radiation.

pure water surrounded by photomultipliers for the detection of the heavier cosmic ray particles, whilst high-pressure carbon dioxide is used for the electrons and positrons. With adequate observation and the use of two or more different media, the direction, energy and possibly the type of the particle may be deduced. The Pierre Auger cosmic-ray array in Argentina uses both water-based Čerenkov detectors and atmospheric fluorescence detectors (see below). There are 1600 Čerenkov detectors each containing 12 tons of water and the flashes are detected by photomultipliers. The detectors, which are solar-powered, are spread out in a grid over a 3000-square kilometre area of grazing land. The proposed Pierre Auger North, to be sited in Colorado will have 4400 similar water-based Čerenkov detectors spread over a 20,000 km² area. The data gathered by the Pierre Auger observatory is partially publically available. One per cent of the data is placed into the Public Event Explorer which is at http://auger.colostate.edu/ED/and is accessible freely via the internet.

A related Čerenkov-based detector is called the Ring Imaging Čerenkov (RICH) detector because the Čerenkov radiation spreads out in a cone and thus appears as a ring when intercepted by a flat detector. The Čerenkov radiation is produce within a thin plate of sodium fluoride called a radiator and is detected by an array of photomultipliers. Similar devices that only detect the burst of Čerenkov radiation with a single photomultiplier and so do not image the ring are also in use. The balloon-borne Super Trans-Iron Galactic Element Recorder (Super-TIGER) instrument, for example uses the latter type of Čerenkov detector with acrylic plastic and aerogel as its radiators, alongside scintillation detectors.

As mentioned previously (Čerenkov detectors) (see Section 1.3), the flashes produced in the atmosphere by the primary cosmic rays and high-energy γ-rays can be detected. A large light 'bucket' and a photomultiplier are needed, preferably with two or more similar systems observing the same part of the sky so that non-Čerenkov events can be eliminated by anti-coincidence discrimination. At least two images are also needed so that the direction in space of the incoming particle or photon can be determined. For the high-energy γ-rays, the tracks may be expected to point directly back to their points of origin (interstellar magnetic fields mean that the track directions for charged particles are unrelated to their points of origin). In this way the 10-metre Whipple telescope found tera-electron volt γ-rays coming from the Crab nebula in 1989. The HESS* array in Namibia has four 12-metre and a single 32.6 m × 24.3 m dish. Each of the smaller dishes is made up from three hundred and eighty-two 0.6-metre circular mirrors forming overall spherical reflectors. The larger dish is parabolic and built up from eight hundred and seventy-five 0.9-metre hexagonal segments. The detectors are arrays of 960 photomultipliers. The highest known energy γ-rays (16 TeV) detected to date were found by the High Energy Gamma Ray Astronomy (HEGRA) instrument(the precursor to the MAGIC telescopes) (see below) in 1996 coming from the blazar Mrk 501 in Hercules.

An intriguing aside to Čerenkov detectors arises from the occasional flashes seen by astronauts when in space. These are thought to be Čerenkov radiation from primary cosmic rays passing through the material of the eyeball. Thus, cosmic-ray physicists could observe their subjects directly. It is also just possible that they could listen to them as well! A large extensive air shower hitting a water surface will produce a click sound that is probably detectable by the best of the current hydrophones. Unfortunately, a great many other events produce similar sounds so that the cosmic-ray clicks are likely to be well buried in the noise. Nonetheless, the very highest energy cosmic ray showers might even be audible directly to a skin diver. Recently, several attempts have been made to detect these acoustic pulses and those from ultra-high energy neutrinos, but so far without success. Hydrophones have been included though in several water/ice neutrino detectors including NT200+ (since 2006) and the 2008 ANTARES Modules for the Acoustic Detection Under the Sea (AMADEUS) addition to the Astronomy with a Neutrino Telescope and Abyssal Environmental Research (ANTARES) (see Section 1.5).

1.4.2.1.3 Solid-State Detectors   Solid-state detectors have been discussed in Section 1.3. No change to their operation is needed for cosmic-ray detection. Their main disadvantages are that their size is small compared with many other detectors, so that their collecting area is also small and that unless the particle is stopped within the detector's volume, only the rate of energy loss may be found and not the total energy of the particle. This latter disadvantage, however, also applies to most other detectors. The PAMELA experiment launched in 2006 uses a number of different types of detectors as detectors trackers, active shields and detectors to find primary cosmic-ray anti-protons and positrons with energies up to 270 GeV. Amongst these are silicon detectors, plastic sheet scintillator detectors and tungsten-silicon micro-calorimeters (see Section 1.3). The second Alpha Magnetic Spectrometer (AMS-02) was taken to the International Space Station by the space shuttle in 2011. It also involves a variety of detectors, such as transition radiation detectors, solid-state detectors, calorimeters and RICH detectors. AMS-02, however, also contains a strong permanent magnet. The magnet bends the paths of lower-energy-charged particles enabling them to be identified. It is searching in particular for anti-helium nuclei.

---

* HESS stands for High Energy Stereoscopic System, but also honours Victor Hess who received the 1936 Nobel prize for discovering cosmic radiation during balloon flights between 1911 and 1913.

1.4.2.1.4 Proportional Counters   Proportional counters and their various developments (TPCs, micromegas etc.) were also discussed in Section 1.3. They are commonly called drift chambers when used as cosmic-ray detectors (not to be confused with SDDs, although these can also be used for cosmic-ray detection). A transition radiation detector is a drift chamber which is attached to a slab of material called a radiator. The radiator is a composite of, for example, polypropylene sheets and polymethacrylimide foam. High-energy particles passing through the radiator induce x-radiation within it which is then detected within the drift chamber alongside the electrons produced directly within the chamber by the particle.

A variety of proportional counter called a streamer chamber is also used for cosmic-ray detection. It operates via a short high-voltage pulse that produces a line of ionised particles (the streamer) along the track of the cosmic ray. The track is then imaged directly from the light emitted as the ions and electrons recombine. The device has much in common with the Wilson cloud chamber which shows the paths of charged particles via the drops of liquid that form along the ionised track of the particle.

1.4.2.2 Residual Track Detectors

1.4.2.2.1 Photographic Emulsions   In a sense the use of photographic emulsion for the detection of ionising radiation is the second oldest technique available to the cosmic-ray physicist after the electroscope since Henri Becquerel discovered the existence of ionising particles in 1896 by their effect upon a photographic plate. However, the early emulsions only hinted at the existence of the individual tracks of the particles and were far too crude to be of any real use. For cosmic-ray detection blocks, of photographic emulsion with high silver bromide content were exposed to cosmic rays at the tops of mountains, flown on balloons or sent into space on some early spacecraft. The cosmic rays could be picked up and their nature deduced to some extent by the tracks left in the emulsion after it had been developed. This type of detector is little used now.

1.4.2.2.2 Ionisation Damage Detectors   Ionisation damage detectors provide a selective detector for nuclei with masses above about 150 amu. Their principle of operation is allied to that of the nuclear emulsions in that it is based upon the ionisation produced by the particle along its track through the material. The material is usually a plastic with relatively complex molecules. As an ionising particle passes through it, the large complex molecules are disrupted, leaving behind short, chemically reactive segments, radicals etc. Etching the plastic reveals the higher chemical reactivity along the track of a particle and a conical pit develops along the line of the track. By stacking many thin layers of plastic, the track may be followed to its conclusion. The degree of damage to the molecules and hence the characteristics of the pit which is produced, is a function of the particle's mass, charge and velocity. A particular plastic may be calibrated in the laboratory so that these quantities may be inferred from the pattern of the sizes and shapes of the pits along the particle's track. Cellulose nitrate and polycarbonate plastics are the currently favoured materials. The low weight of the plastic and the ease with which large-area detectors can be formed make this a particularly suitable method for use in space when there is an opportunity for returning the plastic to Earth for processing, as for example, when the flight is a manned one.

Similar tracks may be etched into polished crystals of minerals such as feldspar and rendered visible by infilling with silver. Meteorites and lunar samples can thus be studied and provide data on cosmic rays which have extended back into the past for many millions of years. The majority of such tracks appear to be attributable to iron group nuclei, but the calibration is very uncertain. Because of the uncertainties involved the evidence has so far been of more use to the meteoriticist in dating the meteorite than to the cosmic-ray astronomer.

1.4.2.3 Indirect Detectors

1.4.2.3.1 100-MeV Gamma Rays   Primary cosmic rays occasionally collide with nuclei in the interstellar medium. Even though the chance of this occurring is only about 0.1% if the particle were to cross the galaxy in a straight line, it happens often enough to produce detectable results. In such collisions π° mesons will frequently be produced and these will decay rapidly into two gamma rays, each with an energy of about 100 MeV. π° mesons may also be produced by the interaction of the cosmic ray particles and the 3-K microwave background radiation. This radiation when 'seen' by a 10^{20}-eV proton is Doppler-shifted to a gamma ray of 100-MeV energy and neutral pions result from the reaction

$$p^+ + \gamma \rightarrow p^+ + \pi^\circ \tag{1.108}$$

Inverse Compton scattering of starlight or the microwave background by cosmic-ray particles can produce an underlying continuum around the line emission produced by the pion decay.

γ-rays with energies as high as these are little affected by the interstellar medium so that they may be observed by spacecraft (see Section 1.3) wherever they may have originated within the galaxy. The 100-MeV γ-ray flux thus gives an indication of the cosmic-ray flux throughout the galaxy and beyond.

1.4.2.3.2 Radio Emission   Cosmic-ray electrons are only a small proportion of the total flux and the reason for this is that they lose significant amounts of energy by synchrotron emission as they interact with the galactic magnetic field. This emission lies principally between 1 MHz and 1 GHz (300 m to 300 mm) and is observable as diffuse radio emission from the galaxy. However, the interpretation of the observations into electron energy spectra etc. is not straightforward and is further complicated by the lack of a proper understanding of the galactic magnetic field.

Primary cosmic rays have also recently been detected from their low-frequency emissions as they collide with the Earth's atmosphere. The LOFAR array (see Sections 1.2 and 2.5) has detected bright flashes from cosmic rays that occur about once a day and last for a few tens of nanoseconds.

The charged particles in a cosmic-ray shower emit megahertz synchrotron radiation (sometimes called geosynchrotron radiation) as their paths are affected by the Earth's magnetic field. LOPES (LOFAR Prototype Experimental Station) (see Section 1.2), for example, has been detecting geosynchrotron pulses since 2006 and there are currently proposals for the construction of a radio detector for cosmic rays in the radio-quiet Antarctic continent. Auger Engineering Radio Array (AERA) is currently under construction at the Pierre Auger cosmic-ray observatory and will eventually have 160 radio receivers over a 20-km^2 area observing air showers originating from particles with energies over 10^{17} eV.

Extensive air showers also emit gigahertz radiation from the plasma formed along the particle's track through the atmosphere. Microwave Detection of Air Showers (MIDAS) in Chicago has recently begun operations. It comprises a 4.5-metre dish with a 53-pixel detector. MIDAS' operating frequency lies within the extended C-band (3.4 to 4.2 GHz, 88 to 71 mm) used by satellite TVs, so that almost all the equipment needed is available in inexpensive commercially mass-produced form. The potential for the construction of large arrays relatively cheaply in the future is therefore excellent. Furthermore, these receivers can operate almost all the time – unlike the fluorescence detectors which can only operate during moonless periods.

1.4.2.3.3 Fluorescence   The very-highest-energy extensive air showers are detectable via weak fluorescent light from atmospheric nitrogen. This is produced through the excitation of the nitrogen by the electron-photon component of the cascade of secondary cosmic rays. The equipment required is a light bucket and a detector (cf. Čerenkov detectors and in Section 1.3) and detection rates of a few tens of events per year for particles in the 10^{19}- to 10^{20}-eV range are achieved by several automatic arrays. Fluorescence detectors require very dark sites and moonless and cloudless nights and so are not in continuous operation.

The Fly's Eye detector in Utah used sixty-seven 1.5-metre mirrors feeding photomultipliers on two sites 4 km apart to monitor the whole sky. It operated from 1982 to 1992 and in 1991 it recorded the highest energy cosmic ray yet found: 3×10^{20} eV. This is well above the GZK limit. Fly's Eye was upgraded to HiRes which had a baseline of 12.5 km, with 64 quadruple mirror units feeding 256 photomultipliers. HiRes in turn has now been superseded by the Telescope Array which has thirty-eight 5- to 7-m^2 telescopes each with 256 photomultipliers as their detectors and located in three stations sited at the tops of hills some 25 km apart. The Pierre Auger observatory has four fluorescence detectors separated by about 50 km, each using six 4-metre mirrors and feeding cameras containing 440 photomultipliers.

1.4.2.3.4 Solar Cosmic Rays   Very high fluxes of low-energy cosmic rays can follow the eruption of a large solar flare. The fluxes can be intense enough to lower the Earth's ionosphere and to increase its electron density. This in turn can be detected by direct radar observations, or through long-wave radio communication fade-outs, or through decreased cosmic radio background intensity as the absorption of the ionosphere increases.

1.4.2.3.5 Carbon-14   The radioactive isotope $^{14}_{6}C$ is produced from atmospheric $^{14}_{7}N$ by neutrons from cosmic-ray showers

$$^{14}_{7}N + n \rightarrow {}^{14}_{6}C + p^{+} \tag{1.109}$$

The isotope has a half-life of 5730 years and has been studied intensively as a means of dating archaeological remains. Its existence in ancient organic remains shows that cosmic rays have been present in the Earth's vicinity for at least 20,000 years. The flux seems, however, to have varied markedly at times from its present-day value, particularly between about 4000 and 1000 years BC. But this is probably attributable to increased shielding of the Earth from the low-energy cosmic rays at times of high solar activity, rather than to a true variation in the number of primary cosmic rays.

1.4.3 Arrays

Primary cosmic rays may be studied by single examples of the detectors that we have considered above. The majority of the work on cosmic rays, however, is on the secondary cosmic rays and for these a single detector is not very informative. The reason is that the secondary particles from a single high-energy primary particle have spread over an area of 10 square kilometres or more by the time they have reached ground level from their point of production some 50 km up in the atmosphere. To deduce anything about the primary particle that is meaningful, the secondary shower must hence be sampled over a significant fraction of its area. Thus, arrays of detectors are used rather than single ones (see Section 1.5 for the detection of cosmic-ray-produced neutrinos). Plastic or liquid scintillators (see Section 1.3), water Čerenkov and fluorescence detectors are frequently chosen as the detectors but any of the real-time instruments can be used and these are typically spread out over an area of hundreds to thousands of square kilometres. The resulting several hundred to thousands of individual detectors are all then linked to a central computer for data analysis.

A second reason for using arrays of detectors is that the flux of cosmic rays at the highest energies is very low. Primary cosmic ray particles with energies around 10^{12} eV have a flux of about one particle per square metre per second, at 10^{16} eV the flux is a few particles per square metre per year, whilst at 10^{20} eV the flux falls to less than one particle per square kilometre per century. Thus, to have any chance of catching the highest energy (and perhaps the most interesting) cosmic-ray particles, arrays have to be spread over hundreds or thousands of square kilometres.

The Pierre Auger observatory, for example, is a 3000-square-kilometre array in Argentina and so may catch a few of the highest energy particles per year. It has 1600 water-based Čerenkov detectors distributed over its area and twenty-four 12 m² optical telescopes to detect nitrogen fluorescence at distances of up to 30 km. A second larger array is planned for construction in Colorado to provide coverage of the northern hemisphere. The Telescope Array in Utah uses three fluorescence detectors and 564 two- by three-metre plastic scintillators spread over a thousand-square kilometre area. The High Altitude Water Cherenkov (HAWC) observatory is currently being built in Mexico to detect teraelectron volt γ-rays from their secondary particle products. It will have three hundred tanks, each containing 330 tonnes of water in a dense array covering two hectares. The Čerenkov emissions will be detected by four photomultipliers in each tank.

The analysis of the data from such arrays is difficult. Only a very small sample, typically less than 0.01%, of the total number of particles in the shower is normally caught. The properties of the original particle have then to be inferred from this small sample. However, the nature of the shower varies with the height of the original interaction and

FIGURE 1.106   (**See color insert.**) The two MAGIC telescopes on the Roque de los Muchachos, La Palma. (Reproduced by kind permission of Robert Wagner, Max-Planck-Institut für Physik.)

there are numerous corrections to be applied as discussed below. Thus, normally the observations are fitted to a grid of computer-simulated showers. The original particle's energy can usually be obtained within fairly broad limits by this process, but its further development is limited by the availability of computer time and by our lack of understanding of the precise nature of these extraordinarily high-energy interactions.

Arrays, or at least several telescopes, are also needed for the atmospheric Čerenkov detectors* so that pairs of stereoscopic images may be obtained and used to plot the track's direction in space accurately. The five telescopes of the HESS systems have already been mentioned. Other currently active instruments include MAGIC sited on La Palma in the Canary Islands which, since 2009, has had two 17-metre reflectors each feeding 596 photomultipliers and VERITAS in Arizona with four 12-metre reflectors each feeding 499 photomultipliers, whilst the CANGAROO III system at Woomera has had four 10-metre telescopes operating since 2003. For the future, the planned CTA may have up to a hundred telescopes and possibly sites in both the northern and southern hemispheres. In space, the Japanese Experiment Module – Extreme Universe Space Observatory (JEM-EUSO) is an instrument planned to go onto the International Space Station in 2016. It will look downwards to pick up Čerenkov events over an area up to 10^6 km^2 using a 2.5-metre mirror and some 6000 photomultiplier tubes (Figure 1.106).

1.4.4 Correction Factors

1.4.4.1 Atmospheric Effects

The secondary cosmic rays are produced within the Earth's atmosphere and so its changes may affect the observations. The two most important variations are caused by air mass and temperature.

---

* The possible use of air Čerenkov arrays as intensity interferometers is discussed in Section 2.5.

The air mass depends upon two factors – the zenith angle of the axis of the shower and the barometric pressure. The various components of the shower are affected in different ways by changes in the air mass. The muons are more penetrating than the nucleons and so the effect upon them is comparatively small. The electrons and positrons come largely from the decay of muons and so their variation tends to follow that of the muons. The corrections to an observed intensity are given by

$$I(P_o) = I(P)e^{K(P-P_o)/P_o} \tag{1.110}$$

$$I(0) = I(\theta)e^{K(\sec\theta - 1)} \tag{1.111}$$

where P_o is the standard pressure, P is the instantaneous barometric pressure at the time of the shower, $I(P_o)$ and $I(P)$ are the shower intensities at pressures P_o and P, respectively, $I(0)$ and $I(\theta)$ are the shower intensities at zenith angles of zero and θ, respectively and K is the correction constant for each shower component. K has a value of 2.7 for the muons, electrons and positrons and 7.6 for the nucleons. In practice, a given detector will also have differing sensitivities for different components of the shower and so more precise correction factors must be determined empirically.

The atmospheric temperature changes primarily affect the muon and electron components. The scale height of the atmosphere, which is the height over which the pressure changes by a factor of e^{-1}, is given by

$$H = \frac{kR^2 T}{GMm} \tag{1.112}$$

where R is the distance from the centre of the Earth, T the atmospheric temperature, M the mass of the Earth and m the mean particle mass for the atmosphere. Thus, the scale height increases with temperature and so a given pressure will be reached at a greater altitude if the atmospheric temperature increases. The muons, however, are unstable particles and will have a longer time in which to decay if they are produced at greater heights. Thus, the muon and hence the electron and positron intensity decreases as the atmospheric temperature increases. The relationship is given by

$$I(T_o) = I(T)e^{0.8(T-T_o)/T} \tag{1.113}$$

where T is the atmospheric temperature, T_o the standard temperature and $I(T_o)$ and $I(T)$ are the muon (or electron) intensities at temperatures T_o and T, respectively. Since the temperature of the atmosphere varies with height, Equation 1.113 must be integrated up to the height of the muon formation in order to provide a reliable correction. The temperature profile will, however, only be poorly known, so it is not an easy task to produce accurate results.

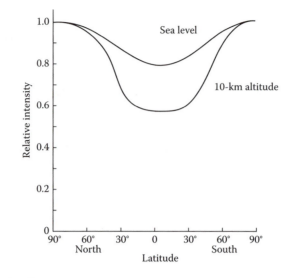

FIGURE 1.107   Latitude effect on cosmic rays.

1.4.4.2 Solar Effects

The Sun affects cosmic rays in two main ways. First, it is itself a source of low-energy cosmic rays whose intensity varies widely. Second, the extended solar magnetic field tends to shield the Earth from the lower energy inter-stellar primaries. Both these effects vary with the sunspot cycle and also on other time scales and are not easily predictable.

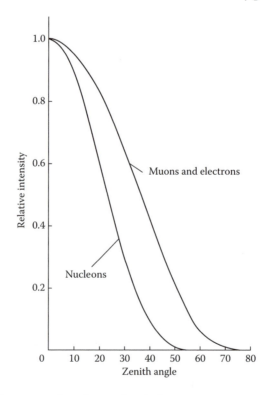

FIGURE 1.108   Zenithal concentration of extensive air shower components as given by Equation 1.111.

1.4.4.3 Terrestrial Magnetic Field

The Earth's magnetic field is essentially dipolar in nature. Lower-energy-charged particles may be reflected by it and so never reach the Earth at all, or particles that are incident near the equator may be channelled towards the poles. There is thus a dependence of cosmic-ray intensity on latitude (Figure 1.107). Furthermore, the primary cosmic rays are almost all positively charged and they are deflected so that there is a slightly greater intensity from the west. The vertical concentration (Equation 1.111 and Figure 1.108) of cosmic rays near sea level makes this latter effect only of importance for high-altitude balloon or satellite observations.

EXERCISES

1.12 Show that the true counting rate, C_t, of a Geiger counter whose dead time is of length Δt, is related to its observed counting rate, C_o, by

$$C_t = \frac{C_o}{1 - \Delta t C_o}$$

(Section 1.3 is also relevant to this problem).

If the effective range of a Geiger counter is limited to $C_t \le 2\, C_o$, calculate the maximum useful volume of a Geiger counter used to detect secondary cosmic rays at sea level if its dead time is 250 μs.

1.13 The minimum particle energy required for a primary cosmic ray to produce a shower observable at ground level when it is incident vertically onto the atmosphere is about 10^{14} eV. Show that the minimum energy required to produce a shower when the primary particle is incident at a zenith angle θ, is given by

$$E_{min}(\theta) = 6.7 \times 10^{12}\, e^{2.7\, \sec\theta}\ eV$$

(Hint: use Equation 1.111 for muons and assume that the number of particles in the shower at its maximum is proportional to the energy of the primary particle.)

The total number of primary particles, $N(E)$, whose energy is greater than or equal to E, is given at high energies by

$$N(E) \approx 10^{22}\, E^{-1.85}\ m^{-2}\, s^{-1}\, str^{-1}$$

for E in electron volts. Hence, show that the number of showers, $N(\theta)$, observable from the ground at a zenith angle of θ is given by

$$N(\theta) \approx 0.019 e^{-5.0\, \sec\theta}\ m^{-2}\, s^{-1}\, str^{-1}$$

1.14 By numerical integration, or otherwise, of the formula derived in Problem 1.13, calculate the total flux of showers of all magnitudes onto a detector array covering 1 square kilometre. (Assume that the primary particle flux is isotropic.)

1.5 NEUTRINO DETECTORS

1.5.1 Background

Neutrino astronomy is in the doldrums at the moment. Only two astronomical sources of neutrinos have ever been observed – the Sun and supernova 1987A in the large Magellanic cloud. Since the solar neutrino problem was solved a couple of decades ago (see below) there are, by comparison, only minor details to be learnt about the Sun from neutrinos. Nonetheless neutrino detectors are still thriving. This is partly in order to study neutrinos in their own right and partly to study cosmic rays via the neutrinos that they produce during their interactions with the Earth's atmosphere. However, as detector sensitivities improve, other types of astronomical objects may become detectable. Thus, IceCube (fully operational since 2011; see below) is expected to be able to pick up neutrinos from AGNs and our own galaxy and may be able to detect GRBs – and, of course, another nearby supernova could happen at any moment. Thus, there is still an astronomical interest in current and forthcoming neutrino detectors.

Wolfgang Pauli postulated the neutrino in 1930 in order to retain the principle of conservation of mass and energy in nuclear reactions. It was necessary in order to provide a mechanism for the removal of residual energy in some beta-decay reactions. From other conservation laws, the neutrino's properties could be defined quite well; zero charge, zero or very small rest mass, zero electric moment, zero magnetic moment and a spin of one half. Over a quarter of a century passed before the existence of this hypothetical particle was confirmed experimentally (in 1956). The reason for the long delay in its confirmation lay in the very low probability of the interaction of neutrinos with other matter (neutrinos interact via the weak nuclear force only). A neutrino originating at the centre of the Sun would have only one chance in 10^{10} of interacting with any other particle during the whole of its 700,000-km journey to the surface of the Sun. The interaction probability for particles is measured by their cross-sectional area for absorption, σ, given by

$$\sigma = \frac{1}{\lambda N} \tag{1.114}$$

where N is the number density of target nuclei and λ is the mean free path of the particle. Even for high-energy neutrinos the cross section for the reaction

$$\nu + {}^{37}_{17}Cl \rightarrow {}^{37}_{18}Ar + e^- \tag{1.115}$$

that was originally used for their detection is only 10^{-46} m^2 (Figure 1.109), so that such a neutrino would have a mean free path of over one parsec even in pure liquid ${}^{37}_{17}Cl$.

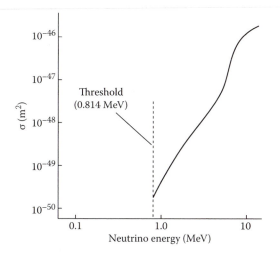

FIGURE 1.109 Neutrino absorption cross sections for the $^{37}_{17}Cl \rightarrow {}^{37}_{18}Ar$ reaction.

Three varieties (or flavours) of neutrino are known, plus their anti-particles. The electron neutrino, ν_e, is the type originally postulated to save the β reactions and which is therefore involved in the archetypal decay: that of a neutron

$$n \rightarrow p^+ + e^- + \tilde{\nu}_e \qquad (1.116)$$

where $\tilde{\nu}_e$ is an anti-electron neutrino. The electron neutrino is also the type commonly produced in nuclear fusion reactions and hence to be expected from the Sun and stars. For example the first stage of the proton-proton cycle is

$$p^+ + p^+ \rightarrow {}^2_1H + e^+ + \nu_e \qquad (1.117)$$

and the second stage of the carbon cycle

$${}^{13}_7N \rightarrow {}^{13}_6C + e^+ + \nu_e \qquad (1.118)$$

and so on.

The other two types of neutrino are the muon neutrino, ν_μ, and the tau neutrino, ν_τ. These are associated with reactions involving the heavy electrons called muons and tauons. For example, the decay of a muon

$$\mu^+ \rightarrow e^+ + \nu_e + \tilde{\nu}_\mu \qquad (1.119)$$

involves an anti-muon neutrino amongst other particles. The muon neutrino was detected experimentally in 1962, but the tau neutrino was not found until 2000.

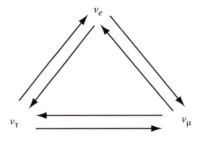

FIGURE 1.110   Neutrino oscillation.

The first neutrino detector started operating in 1968 (see radiochemical detectors) and it quickly determined that the observed flux of solar neutrinos was too low by a factor of 3.3 compared with the theoretical predictions. This deficit was confirmed later by other types of detectors such as Soviet-American Gallium Experiment (SAGE), Gallium Experiment (GALLEX) and Kamiokande and became known as the solar neutrino problem. However, these detectors were only sensitive to electron neutrinos. Then in 1998, the Super Kamiokande detector found a deficit in the number of muon neutrinos produced by cosmic rays within the Earth's atmosphere. Some of the muon neutrinos were converting to tau neutrinos during their flight time to the detector. The Super Kamiokande and Sudbury Neutrino Observatory (SNO) detectors have since confirmed that all three neutrinos can exchange their identities (Figure 1.110). The solar neutrino problem has thus disappeared, since two-thirds of the electron neutrinos produced by the Sun have changed to muon and tau neutrinos by the time that they reach the Earth and so were not detected by the early experiments.

The flux of solar neutrinos (of all types) at the Earth is now measured to be about 6×10^{14} $m^{-2}\ s^{-1}$, in good agreement with the theoretical predictions. The experiments also show that neutrinos have mass, although only upper limits have been determined so far. The tightest constraints on neutrino masses come from cosmology and suggests that the combined masses of all three varieties of neutrino to be no more than 0.3 eV (the mass of an electron for comparison is 511 keV[*]).

1.5.2 Neutrino Detectors

Neutrinos from the Sun were first detected in 1968 by Davis' $^{37}_{17}Cl$-based experiment (see below). Today's main second generation instruments and most of those being considered for the future are based upon observing the Čerenkov radiation in water or ice produced by the superphotic charged subatomic particles resulting from neutrino interactions. The reason for the use of water or ice instead of chlorine is that the former are cheap, or even, in the case of water in the oceans or ice in the Antarctic icecap, free. The neutrino interactions with matter are so weak that vast amounts of material have to be used if they are

---

[*] The mass is given here, as is the conventional practice of sub-atomic physicists, in energy terms through the relationship $e = mc^2$. In more normal units we have that the combined masses of the three varieties of neutrino is no more than 5.4×10^{-37} kg, whilst that of the electron is 9.11×10^{-31} kg.

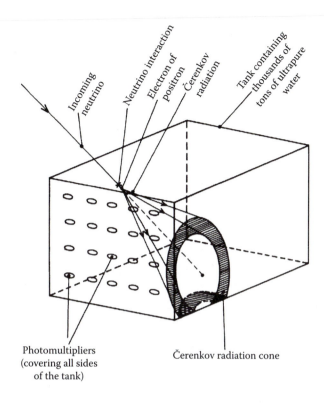

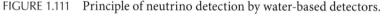

Photomultipliers
(covering all sides
of the tank)

Čerenkov radiation cone

FIGURE 1.111   Principle of neutrino detection by water-based detectors.

to be observed with reasonable frequency. Davis used over 600 tonnes of the chlorine-containing compound tetrachloroethene in his detectors. IceCube (see below), however, monitors a cubic kilometre (10^9 tonnes) of ice and is likely to be regarded by researchers within a decade or so as a 'small' detector.

We examine first, therefore, the water or ice-based Čerenkov detectors before looking more briefly at other ways of detecting neutrinos.

1.5.2.1 Water- and Ice-Based Detectors
After Davis' first neutrino detector, the next working neutrino telescopes did not appear until nearly two decades later. Neither of these instruments, the Kamioka Neutrino Detector (Kamiokande) detector buried 1 km below Mount Ikenoyama in Japan nor the Irvine-Michigan-Brookhaven (IMB) detector, 600 m down a salt mine in Ohio, were built to be neutrino detectors*. Both were originally intended to look for proton decay and were only later converted for use with neutrinos. The design of both detectors was similar (Figure 1.111); they differed primarily only in size (Kamiokande: 3000 tonnes, IMB: 8000 tonnes).

The principle of their operation was the detection of Čerenkov radiation from the products of neutrino interactions. These may directly take two forms, electron scattering and

---

* The large depths of these and other neutrino detectors are so that the overlying material acts as a shield against non-neutrino interactions.

inverse beta-decay. In the former process, a collision between a high-energy neutrino and an electron sends the latter off at a speed in excess of the speed of light in water (225,000 km s^{-1}) and in roughly the same direction as the neutrino. All three types of neutrinos scatter electrons, but the electron neutrinos are 6.5 times more efficient at the process. In inverse beta-decay, an energetic positron is produced via the reaction

$$\tilde{v}_e + p \rightarrow n + e^+ \tag{1.120}$$

Inverse beta decays are about 100 times more probable than the scattering events, but the positron can be emitted in any direction and thus gives no clue to the original direction of the incoming neutrino. Both positron and electron, travelling at superphotic speeds in the water, emit Čerenkov radiation (see Section 1.4) in a cone around their direction of motion. That radiation is then picked up by arrays of photomultipliers that surround the water tank on all sides. The pattern of the detected radiation can be used to infer the energy of the original particle and in the case of a scattering event, also to give some indication of the arrival direction. The minimum detectable energy is around 5 MeV due to background noise.

Both detectors were fortunately in operation in February 1987 and detected the burst of neutrinos from the supernova in the Large Magellanic Cloud (SN1987A). This was the first (and so far only) detection of neutrinos from an astronomical source other than the Sun and it went far towards confirming the theoretical models of supernova explosions.

Both IMB and Kamiokande have now ceased operations,* but a number of detectors operating on the same principle are functioning or are under construction.

Super Kamiokande contains 50,000 tonnes of pure water and uses 13,000 photomultiplier tubes in a tank buried 1 km below Mount Ikenoyama in Japan. Super Kamiokande is able to detect muon neutrinos as well as electron neutrinos. The muon neutrino interacts with a proton to produce a relativistic muon. The Čerenkov radiation from the muon produces a well-defined ring of light. The electron resulting from an electron neutrino scattering event by contrast generates a much fuzzier ring of light. This is because the primary electron produces gamma rays that in turn produce electron-positron pairs. The electron-positron pairs then create additional Čerenkov radiation cones, thus blurring the ring resulting from the primary electron.

This ability to distinguish between electron and muon neutrinos enabled Super Kamiokande to provide the first evidence of neutrino oscillations. Muon neutrinos produced in the Earth's atmosphere by high-energy cosmic rays should be twice as abundant as electron neutrinos.† But in 1998, Super Kamiokande found that although this was the

---

* The IMB photomultipliers were re-used as an active shield for Super Kamiokande.
† The cosmic-ray interaction first produces (amongst other particles) pions. A positively charged pion usually then decays in 2.6×10^{-8} s to a positive muon and a muon neutrino and the muon decays in 2.2×10^{-6} s to a positron, a muon anti-neutrino and an electron neutrino. A negative pion likewise decays to an electron, a muon neutrino, a muon anti-neutrino and an electron anti-neutrino. There should thus be two muon neutrinos to each electron neutrino in an extensive air shower (Section 1.4).

case for neutrinos coming from above (a distance of about 60 km), there were roughly equal numbers of muon neutrinos and electron neutrinos coming from below (i.e. having travelled 12,000 km across the Earth). Since there were no extra electron neutrinos, some of the muon neutrinos would be oscillating to tau neutrinos during the flight time across the Earth to the detector. This result was confirmed in 2000 by a 30% shortfall in the number of muon neutrinos observed from an artificial source at the KEK laboratory at Tsukuba some 250 km away from the detector. Further confirmation of the oscillation of all three types of neutrino has recently come from comparison of the Super Kamiokande and Sudbury Neutrino Observatory (SNO) results and the KamLAND experiment. The designer of the Kamiokande and Super Kamiokande detectors, Matatoshi Koshiba, shared the Nobel physics prize in 2002 for his work.

Man-made water containers much larger than Super Kamiokande are probably impractical. However, much greater quantities of water (or ice, which is equally good) may be monitored using parts of lakes, the sea or the Antarctic ice cap. These types of neutrino telescopes are generally most sensitive to high energy (TeV) neutrinos because of the large spacing (tens of metres) between their photomultipliers.

IceCube, which started operations in 2011 in Antarctica, is currently the largest neutrino detector. It uses photomultipliers to detect Čerenkov radiation that is produced within the ice. Each of its 5160 photomultipliers is housed in a transparent pressure vessel along with its electronics. Sixty of these sensors at a time are suspended from a cable that is lowered down a hole in the ice (Figure 1.112) to a depth between 1.45 and 2.45 km* below the surface. There are 86 such cables spread over an area of about 1 square kilometre. IceCube therefore monitors about a cubic kilometre of ice for Čerenkov radiation. IceCube also monitors a 0.002-km^3 volume with more closely spaced detectors. Called DeepCore, this instrument extends IceCube's sensitivity down to 10 GeV. Additionally, extensive air showers are monitored by IceTop, an array of 160 tanks each containing 2.5 tonnes of ice on the surface above the main IceCube instrument and equipped as Čerenkov detectors.

IceCube is most sensitive to muons and hence to muon neutrinos. Muons, however, are also a large component of extensive air showers and there are around a million cosmic-ray-produced muons for every neutrino-produced muon. IceCube therefore ignores the muons moving downwards that have come from cosmic rays in the atmosphere above the instrument. Instead it selects muons that are moving upwards. These will have been produced by muon neutrinos that have come from the opposite side of the Earth and interacted to produce the muons only when close to or inside the monitored volume (thus in effect using the whole Earth as a filter or shield). The neutrinos coming from the other side of the Earth, however, still mostly originate from cosmic-ray interactions. Neutrinos from more distant astronomical sources can thus only be identified if there is a noticeable change in the flux or the energy distribution or if a particular direction in space suddenly seems to be favoured. Typically, IceCube detects around three upwardly moving neutrinos per hour.

---

* The detectors are placed at depth because the ice there is clear and almost free from air bubbles. Also, the layer of ice above the detectors shields them from most of the charged particles in extensive air showers except for the muons.

FIGURE 1.112 (a) One of the pressure spheres containing the photomultiplier and electronics for the IceCube neutrino telescope ready to be lowered into position. (Reproduced by kind permission of IceCube Collaboration/NSF.) (b) The IceCube pressure sphere on its way down through the hole drilled into the Antarctic icecap. (Reproduced by kind permission of IceCube Collaboration/NSF.)

IceCube has so far detected the shielding effect of the Moon on cosmic-ray protons and a slight anisotropy in the distribution of secondary cosmic-ray muons.

For water-based Čerenkov telescopes the principles are much the same as for IceCube except that the strings of photomultipliers are anchored to the sea or lake bed and float in the water. Thus, Lake Baikal in southeast Siberia is to host the Gigaton Volume Detector (GVD) which is scheduled for completion in 2018. This is expected to comprise 27 clusters of eight strings with 48 photomultipliers on each string. The area covered will be about 2 km^2 and the detectors will range in depth from 600 to 1300 metres below the surface. The total volume to be monitored will thus be 1.5 km^3. GVD will also primarily detect upwardly moving muons. GVD is an upgrade of the Neutrino Telescope (NT)200+ instrument which commenced observations in 2005. NT200+ currently monitors a volume of about 0.01 km^3 using 228 photomultipliers on 11 strings.

Cubic Kilometre Neutrino Telescope (KM3NeT) is also a planned development from existing neutrino detectors, such as ANTARES, to be sited in the Mediterranean. ANTARES was completed in 2008 and is some 40 km off the coast near Toulon. It is 150 metres across by 300 metres high and uses 900 photomultipliers to observe a volume of 0.01 km^3 at a depth of 2.5 km. KM3NeT is unlikely to be in use before 2020 and its design details remain to be confirmed. Currently, plans suggest that it will have around 600 strings each with around 20 detector modules. The detector modules, however, will each contain around 30 photomultipliers giving directional sensitivity over the whole sphere. A monitored volume of 5 km^3 is the aim of the project.

There is the possibility of the detection of extremely-high-energy neutrinos by existing cosmic-ray observatories such as the Pierre Auger Observatory and the Telescope Array (see Section 1.4) via Čerenkov or fluorescence emissions. Extensive air showers resulting from cosmic rays are to be found high in the atmosphere and generally moving in a downward direction. If an EAS were to be found at low altitude or moving horizontally or even in an upward direction, then it almost certainly must have originated from a neutrino interaction. Such detections are currently being sought, but so far without success.

Although the peak Čerenkov emission produced by multi-tera-electron volt particles is in the optical region, radio waves are produced as well (Equation 1.107), mostly in the 100-MHz to 1-GHz (3-m to 300-mm) region as predicted by Gurgen Askaryan (1928–1997) in 1962. At the temperature of the ice cap (about –50°C), the ice is almost completely transparent to metre-wavelength radio waves and so these can be detected by receivers at the surface. A radio receiver borne by a high-altitude balloon would enable the entire Antarctic ice cap to be watched: a volume of around 1 million km^3.

In 2006 and 2008 just such a receiver was flown. Called Antarctic Impulsive Transient Antenna (ANITA), during its flights at a height of 35 km, it picked up millions of events. Most of these events, however, were background non-neutrino pulses. The Askaryan Radio Array (ARA) is currently being deployed on the surface near the South Pole. It is planned to install a total of 37 stations, each with 16 antennas over a 200-km^2 area and so to monitor thousands of cubic kilometres of the icecap. Antarctic Ross Ice-Shelf Antenna Neutrino Array (ARIANNA) could be up to 1000 km^2 in area and use a higher density of antennas than ARA. It is now in the planning stages for possible construction on the Ross ice shelf.

Although not based upon water or ice, this is perhaps the appropriate place to mention another radio-based neutrino detector which potentially will monitor a volume of up to 20 million km^3. This is to look for radio pulses produced by neutrino interactions within the Earth's Moon. The pulses originate via the Askaryan mechanism within the lunar rocks and may be expected to be detectable from any part of the Moon's visible surface. The fourteen 25-metre dishes of the Westerbork synthesis radio telescope have recently conducted a – so far unsuccessful – search for these pulses. It is planned to search for them in the near future with LOFAR and in due course with the SKA.

A neutrino detector based upon 1000 tonnes of heavy water (D_2O) operated at Sudbury in Ontario from 1999 to 2007. Known as SNO, the heavy water was contained in a 12-metre diameter acrylic sphere immersed in an excavated cavity filled with 7000 tonnes of highly purified normal water. The normal water shielded the heavy water from radioactivity in the surrounding rock. It was located 2 km down within the Creighton copper and nickel mine and used 9600 photomultipliers to detect Čerenkov emissions within the heavy water. Neutrinos may be detected through electron scattering as with other water-based detectors, but the deuterium in the heavy water provides two other mechanisms whereby neutrinos may be found. The first of these senses just electron neutrinos. An electron neutrino interacts with a deuterium nucleus to produce two protons and a relativistic electron

$$\nu_e + {}^{2}_{1}H \rightarrow p^+ + p^+ + e^-$$ (1.121)

and the electron is observed via its Čerenkov radiation. This mechanism is called the charged current reaction since the charged W boson mediates it. The second mechanism is mediated by the neutral Z boson and is thus called the neutral current reaction. In it, a neutrino of any type simply splits the deuterium nucleus into its constituent proton and neutron. The neutron is thermalised in the heavy water and eventually combines with another nucleus with the emission of gamma rays. The gamma rays in turn produce relativistic electrons via Compton scattering and these are finally detected from their Čerenkov radiation. Although the neutron can combine with a deuterium nucleus, the capture efficiency is low so that 75% of the neutrons will escape from the detector. The SNO detector could therefore add 2 tonnes of salt to the heavy water to enable the neutron to be captured more easily by a ${}^{35}_{17}Cl$ nucleus (converting it to ${}^{36}_{17}Cl$) and reducing to 17% the number of escaping neutrons. The thermalised neutrons can also be detected using ${}^{3}_{2}He$ proportional counters (see Section 1.3). The neutron combines with the ${}^{3}_{2}He$ to produce a proton and tritium and these then ionise some of the remaining gas to give an output pulse.

Since SNO could detect the number of electron neutrinos separately from the total for all types of neutrino, it provided the definitive data for the solar neutrino problem. The standard solar model predicts that SNO should detect about 30 charged current or neutral current reactions per day and about three electron scattering events. However, the reality of the oscillation of neutrinos between their different type and hence the solution to the solar neutrino problem was demonstrated by the very first results from SNO. It was initially operated in the charged current mode (i.e. without salt being added to the heavy water) to detect just the solar electron neutrinos. This gave a flux that was lower than that measured by Super Kamiokande, since the latter also detects a proportion of the muon and tau neutrinos as well as the electron neutrinos. Some solar electron neutrinos must therefore have oscillated to the other types during their journey from the centre of the Sun. SNO then made measurements of the total solar neutrino flux using the neutral current mode (i.e. with salt added to the heavy water) and in April 2002 finally demonstrated that the total neutrino flux is as predicted by the standard solar model. SNO ceased operations in May 2007 and the heavy water was drained from its tank to be reused elsewhere.

1.5.3 Radiochemical Detectors

Radiochemical detectors operate on a quite different principle from that of the Čerenkov detectors. The neutrino is absorbed by a suitable atomic nucleus, changing it to the nucleus of a radioactive isotope of a different element. The radioactive decay is then separately detected and acts to show the occurrence of the initial neutrino interaction. These detectors are small in volume compared with those just discussed, since the material involved is expensive (sometimes *very* expensive) compared with water or ice. However, they are still in use and also have interest as the first type of detector to register extra-terrestrial neutrinos.

The first neutrino telescope to be built was designed to detect electron neutrinos through the chlorine to argon reaction given in Equation 1.115. The threshold energy of the neutrino for this reaction is 0.814 MeV (see Figure 1.109), so that 80% of the neutrinos from the Sun that are detectable by this experiment arise from the decay of boron in a low probability side chain of the proton-proton reaction

$$\,^{8}_{5}B \rightarrow \,^{8}_{4}Be + e^{+} + \nu_{e} \qquad (1.122)$$

The full predicted neutrino spectrum of the Sun, based upon the conventional astrophysical models, is shown in Figure 1.113.

The chlorine-based detector operated from 1968 to 1998 under the auspices of Ray Davis Jr and his group and so is now largely of historical interest, although many of the operating principles and the precautions that are needed apply to other types of neutrino detector. It consisted of a tank containing over 600 tonnes of tetrachloroethene (C_2Cl_4) located 1.5 km

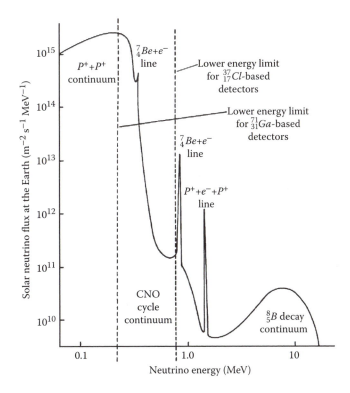

FIGURE 1.113 Postulated solar neutrino spectrum.

down in the Homestake gold mine in South Dakota. About one chlorine atom in four is the $^{37}_{17}Cl$ isotope; so on average each of the molecules contains one of the required atoms, giving a total of about 2×10^{30} $^{37}_{17}Cl$ atoms in the tank. C_2Cl_4 was chosen rather than liquid chlorine for its comparative ease of handling and because of its cheapness: it is a common industrial solvent. The interaction of a neutrino (Equation 1.115) produces a radioactive isotope of argon. The half-life of the argon is 35 days and it decays back to $^{37}_{17}Cl$ by capturing one of its own inner orbital electrons, at the same time ejecting a 2.8-keV electron. The experimental procedure was to allow the argon atoms to accumulate for some time, then to bubble helium through the tank. The argon was swept up by the helium and carried out of the tank. It was separated from the helium by passing the gas stream through a charcoal cold trap. The number of $^{37}_{18}Ar$ atoms was then counted by detecting the 2.8-keV electrons as they were emitted.

The experimental procedure had to be carried out with extraordinary care if the aim of finding a few tens of atoms in 10^{31} was to succeed (similar precautions have to be taken for all neutrino detectors). Pre-eminent amongst the precautions were:

- Siting the tank 1.5 km below the ground to shield it against cosmic rays and natural and artificial radiation sources

- Surrounding the tank with a thick water jacket to shield against neutrons

- Monitoring the efficiency of the argon extraction by introducing a known quantity of $^{36}_{18}Ar$ into the tank before it is swept with helium and determining the percentage of this that was recovered

- Shielding the sample of $^{37}_{18}Ar$ during counting of its radioactive decays and using anti-coincidence techniques to eliminate extraneous pulses

- Use of long integration times to accumulate the argon in the tank

- Measurement of and correction for the remaining noise sources

The Homestake detector typically caught one neutrino every other day. This corresponds to a detection rate of 2.23 ± 0.22 SNU. (A solar neutrino unit, SNU, is 10^{-36} captures per second per target atom.) The expected rate from the Sun is 7.3 ± 1.5 SNU and the apparent disparity between the observed and predicted rates prior to the confirmation of the oscillation of neutrino types gave rise to the previously mentioned solar neutrino problem. In 2002 Ray Davis shared the Nobel physics prize for his work on neutrinos.

There have been two radiochemical neutrino detectors based upon gallium. The SAGE detector used 60 tonnes of liquid metallic gallium buried beneath the Caucasus Mountains at the Baksan laboratory and detected neutrinos via the reaction

$$^{71}_{31}Ga + \nu_e \rightarrow {}^{71}_{32}Ge + e^- \tag{1.123}$$

The germanium product of the reaction is radioactive, with a half-life of 11.4 days. It is separated out by chemical processes on lines analogous to those of the chlorine-based detector. The detection threshold is only 0.236 MeV so that the p-p neutrinos (Figure 1.113) are detected directly. The other gallium-based detector, Gallium Neutrino Observatory (GNO), was based upon the same detection mechanism as SAGE, but used 30 tonnes of gallium in the form of gallium trichloride ($GaCl_3$). It replaced the GALLEX experiment and was located within the Gran Sasso Tunnel under the Apennines, 150 km east of Rome. Both detectors have measured a solar neutrino flux of around 50% of that predicted by the standard solar model.

Other radiochemical bases for past neutrino detectors or possible future neutrino detectors include the conversion of $^{7}_{3}Li$ to $^{7}_{4}Be$, $^{37}_{19}K$ to $^{37}_{18}Ar$, $^{115}_{49}In$ to $^{115}_{50}Sn$, $^{127}_{53}I$ to $^{127}_{54}Xe$ and $^{176}_{70}Yb$ to $^{176}_{71}Lu$.

1.5.4 Scintillator-Based Detectors

High-energy photons and particles resulting from neutrino interactions may be observed using scintillation detectors (see Section 1.3). The Soudan II detector (1989–2001), was located 700 metres below the surface in an old iron mine at Soudan in Minnesota and used 1000 tonnes of alternating sheets of steel and a plastic scintillator (sometimes called a tracking calorimeter). It was originally built to try and detect the decay of protons, but also detected the muons resulting from muon neutrino interactions within the steel plates. Also, in the Soudan mine, are the two tracking calorimeters forming the Main Injector Neutrino Oscillation Search (MINOS) far detector. Each comprises more than 3000 tonnes of alternating 25-mm thick steel and scintillator sheets. Together with the 1000-tonne MINOS near detector at Fermilab it operates to refine the measurements of the parameters of neutrino oscillations by observing a beam of muon neutrinos produced at Fermilab (in Illinois, some 730 km away).

Although not actually a scintillator-based detector, the planned Indian Neutrino Observatory (INO) in Tamil Nadu, is perhaps best mentioned here. It will be a 50,000-tonne tracking calorimeter using iron plates and with resistive plate chambers in place of the scintillator sheets. Resistive plate chambers are a variant on drift chambers (see Section 1.4) using high resistivity plastic plates for the anode and cathode. The electron avalanche is picked up by a separate system of metallic strips.

The KamLAND experiment in Japan uses two nested plastic spheres, 13 and 18 metres in diameter surrounded by 2000 photomultipliers. The spheres are filled with isoparaffin mineral oil and the inner one is also doped with a liquid scintillator. KamLAND detects anti-electron neutrinos from nuclear reactors some 175 km away.

Borexino is an experiment currently aimed at detecting the solar neutrino emission line at 863 keV arising from the decay of $^{7}_{4}Be$ to $^{7}_{3}Li$. It detects all types of neutrinos via their scattering products within 300 tonnes of liquid scintillator observed by 2200 photomultipliers and has a detection threshold of 250 keV. The active scintillator is surrounded by 1000 tonnes of a liquid buffer and that in turn by 2400 tonnes of water, whilst 400 photomultipliers pointing outwards into the water buffer provide an active shield. It is located in the Grand Sasso tunnel in Italy and started operating in 2007.

SNO+ is currently under construction reusing much of the old SNO equipment. It will have 800 tonnes of liquid scintillator surrounded by a water bath and some 10,000 photomultipliers to detect solar and terrestrial neutrinos. The scintillator will be linear alkyl benzene which is used in the manufacture of toilet soap and so available at low cost and in large quantities.

The Low Energy Neutrino Astronomy (LENA) detector is in the early stages of planning and could use 50,000 to 70,000 tonnes of linear alkyl benzene as its scintillator contained in an underground tank 100 metres high and 30 metres in diameter. Possible sites include the Pyhäsalmi mine in Finland and the Laboratoire Souterrain de Modane in the Frejus road tunnel between France and Italy. In either case a chamber for the instrument would need to be specially excavated. LENA is one possible candidate for Large Apparatus studying Grand Unification and neutrino Astrophysics (LAGUNA). The others are Megaton Mass Physics (MEMPHYS), a megaton water Čerenkov detector and Grand Liquid Argon Charge Imaging Experiment (GLACIER), a 100-kiloton concept based upon TPCs (see below) using liquid argon.

1.5.5 Acoustic Detectors

The highest energy cosmic rays (see Section 1.4) have energies measured in several joules. The Pierre Auger cosmic-ray observatory has recently shown that these particles probably originate from AGNs within 30 to 40 mega-parsecs or so of the Earth. It is possible that such AGNs may also produce neutrinos with energies comparable to those of the cosmic ray particles (10^{20} eV). If such a neutrino were to interact with an atom on the Earth, much of its energy would be converted into heating a small volume of the material around the interacting particles. The rapid heat rise would then generate a pulse of sound with 10- to 50-kHz frequencies which would have a bi-polar profile. An array of sensitive microphones could then detect that acoustic pulse and it could be distinguished from the myriad of other pulses by its characteristic shape. Volumes of water, ice or salt of a hundred or more cubic kilometres could thus be monitored for high-energy neutrinos (and also cosmic rays) (see Section 1.4).

No neutrinos or cosmic rays have yet been detected by this technique but hydrophones were added to NT200+ in 2006. In 2008, AMADEUS was added to ANTARES and has 36 hydrophones and South Pole Acoustic Test Setup (SPATS) has been site testing at IceCube using seven transmitters and receivers since 2007.

1.5.6 Indirect Detectors

For neutrinos, at the moment, only one indirect detector exists and that is the influence that neutrinos have upon the CMB. Results from the WMAP spacecraft suggest that when the CMB originated (when the universe was just 380,000 years old), neutrinos formed around 10% of the mass of the universe compared with less than 1% today.

1.5.7 Other Types of Detectors

Detectors based upon $^{37}_{17}Cl$ etc. can only detect the high-energy neutrinos, whilst the bulk of the solar neutrinos are of comparatively low energies (Figure 1.113). Furthermore, no indication of the direction of the neutrinos is possible. Water and gallium-based detectors also look at the high-energy neutrinos. Numerous alternative detectors that are designed to overcome such restrictions are therefore proposed or imagined at the time of writing. A brief survey of the range of possibilities is given below.

Variants on the position-sensitive proportional counters (see Section 1.3) are also used for neutrino and secondary cosmic-ray detection, where they are generally known as multi-wire, TPCs or drift chambers. The TPC is currently widely used by particle physicists and may find application to detecting astronomical neutrinos and cosmic rays in the future. The electron avalanche is confined to a narrow plane by two parallel negatively charged meshes. Sensing wires are spaced at distances of several centimetres. The accelerating voltage is designed so that the electrons gain energy from it at the same rate that they lose it via new ionisations and they thus move between the meshes at a constant velocity. The time of the interaction is determined using scintillation counters around the drift chamber. The exact position of the original interaction between the sensing wires is then found very precisely by the drift time that it takes the electrons to reach the sensing wire. Imaging Cosmic and Rare Underground Signals (ICARUS) T600, for example, currently uses 760 tons of liquid argon and is situated in the Gran Sasso tunnel in Italy. It can detect neutrinos via high energy electrons produced either from electron scattering events or from the conversion of argon to potassium. It has been in operation since 2010.

Other direct interaction detectors which are based upon different principles have recently been proposed, for example, the detection of the change in the nuclear spin upon neutrino absorption by $^{115}_{49}In$ and its conversion to $^{115}_{50}Sn$. Up to 20 solar neutrinos per day might be found by a detector based upon 10 tonnes of superfluid helium held at a temperature below 0.1 K. Neutrinos would deposit their energy into the helium leading to the evaporation of helium atoms from the surface of the liquid. Those evaporated atoms would then be detected from the energy (heat) that they deposit into a thin layer of silicon placed above the helium.

There is also a suggestion to detect neutrinos from the Čerenkov radio radiation produced by their interaction products within natural salt domes in a similar manner to observing the Čerenkov radio emissions within ice. Alternatively, electrons may be stored within a superconducting ring and coherent scattering events detected by the change in the current. With this latter system, detection rates for solar neutrinos of one per day may be achievable with volumes for the detector as small as a few millilitres.

EXERCISES

1.15 Show that if an element and its radioactive reaction product are in an equilibrium state with a steady flux of neutrinos, then the number of decays per second of the reaction product is given by

$$N_p T_{1/2}^{-1} \sum_{n=1}^{\infty} \frac{1}{n} \left(\frac{1}{2} \right)^n$$

when $T_{1/2} \gg$ one second, N_p is the number density of the reaction product and $T_{1/2}$ is its half-life.

Hence, show that the equilibrium ratio of product to original element is given by

$$\frac{N_p}{N_e} = \sigma_v F_v T_{1/2} \left\{ \left[\sum_{n=1}^{\infty} \frac{1}{n} \left(\frac{1}{2} \right)^n \right]^{-1} \right\}$$

where N_e is the number density of the original element, σ_v is the neutrino capture cross section for the reaction and F_v is the neutrino flux.

1.16 Calculate the equilibrium ratio of $^{41}_{20}Ca$ to $^{41}_{19}K$ for $^{8}_{5}B$ solar neutrinos ($T_{1/2} = 80{,}000$ years, $\sigma_v = 1.45 \times 10^{-46}$ m^{-2}, $F_v = 3 \times 10^{10}$ m^{-2} s^{-1}).

1.17 If Davis' chlorine-37 neutrino detector were allowed to reach equilibrium, what would be the total number of $^{37}_{18}Ar$ atoms to be expected? (Hint: see Problem 1.15 and note that $\displaystyle\sum_{n=1}^{\infty} \frac{1}{n} \left(\frac{1}{2} \right)^n = 0.693$).

1.6 GRAVITATIONAL RADIATION

1.6.1 Introduction

When the first edition of this book was written 30 years ago, this section began. "This section differs from the previous ones because none of the techniques that are discussed have yet indisputably detected gravitational radiation". That is still the case. However, there is now quite a widespread expectation that upgrades to some detectors such as LIGO and Virgo (see below) will lead to the first detections within the next decade (i.e. by 2020 to 2025). It has also been stated that a third generation interferometric gravity wave detector (tentatively called the Einstein Telescope) would have a 'guaranteed' discovery rate of thousands of events per year. Thus, we may hope that perhaps a seventh or eighth edition of this book may finally be able to change the opening statement of this section. Whether or not the detection will occur before the centenary of Einstein's prediction of the existence of gravitational waves in 2016, however, still seems unlikely.

However, one change to the introduction to this section from those in previous editions is possible. No gravitational waves may have been detected, but nonetheless LIGO (see below) has produced a result. On 1 February 2007 a short GRB occurred whose position aligned with one of the Andromeda galaxy's (M31) spiral arms. No gravitational wave was picked up from this event, but if the GRB resulted from the merger of two compact objects (neutron stars and/or black holes) as many astronomers surmise, then it should easily have been detected by LIGO. A negative result therefore implies either that the GRB originated via some other process, if it was indeed located within M31, such as being a soft γ-ray repeater, or the alignment with M31 was just due to chance and the

GRB was actually at the usual distance of hundreds to thousands of megaparsecs away from us.

The basic concept of a gravity wave* is simple: if an object with mass changes its position then, in general, its gravitational effect upon another object will change and the information on that changing gravitational field propagates outward through the space–time continuum at the speed of light. The propagation of the changing field obeys equations analogous to those for electromagnetic radiation provided that a suitable adaptation is made for the lack of anything in gravitation that is equivalent to positive and negative electric charges. Hence, we speak of gravitational radiation and gravitational waves. Their frequencies for astronomical sources are anticipated to run from a few kilohertz for collapsing or exploding objects to a few microhertz for binary star systems. A pair of binary stars coalescing into a single object will emit a burst of waves whose frequency rises rapidly with time: a characteristic 'chirp' if we could hear it.

Theoretical difficulties with gravitational radiation arise from the multitudinous metric, curved space-time theories of gravity that are currently extant. The best known of these are due to Einstein (general relativity), Brans and Dicke (scalar-tensor theory) and Hoyle and Narlikar (C-field theory), but there are many others. Einstein's theory forbids dipole radiation, but this is allowed by most of the other theories. Quadrupole radiation is thus the first allowed mode of gravitational radiation in general relativity and will be about two orders of magnitude weaker than the dipole radiation from a binary system that is predicted by the other theories. Furthermore, there are only two polarisation states for the radiation as predicted by general relativity, compared with six for most of the other theories. The possibility of decisive tests for general relativity is thus a strong motive for aspiring gravity wave observers, in addition to the information which may be contained in the waves on such matters as collapsing and colliding stars, supernovae, close binaries, pulsars, early stages of the big bang etc.

The detection problem for gravity waves is best appreciated with the aid of a few order-of-magnitude calculations on their expected intensities. The quadrupole gravitational radiation from a binary system of moderate eccentricity ($e \leq 0.5$), is given in general relativity by

$$L_G \approx \frac{2 \times 10^{-63} M_1^2 M_2^2 (1 + 30 e^3)}{(M_1 + M_2)^{2/3} P^{10/3}} \text{ W} \tag{1.124}$$

where M_1 and M_2 are the masses of the components of the binary system, e is the orbital eccentricity and P is the orbital period. Thus, for a typical dwarf nova system with

$$M_1 = M_2 = 1.5 \times 10^{30} \text{ kg} \tag{1.125}$$

$$e = 0 \tag{1.126}$$

$$P = 10^4 \text{ s} \tag{1.127}$$

---

* The term gravity wave is also used to describe oscillations in the Earth's atmosphere arising from quite different processes. There is not usually much risk of confusion.

we have

$$L_G = 2 \times 10^{24} \text{ W} \tag{1.128}$$

But for an equally typical distance of 250 pc, the flux at the Earth is only

$$F_G = 3 \times 10^{-15} \text{ W m}^{-2} \tag{1.129}$$

This energy will be radiated predominantly at twice the fundamental frequency of the binary with higher harmonics becoming important as the orbital eccentricity increases. Even for a nearby close binary such as ι Boo (distance 23 pc, $M_1 = 2.7 \times 10^{30}$ kg, $M_2 = 1.4 \times 10^{30}$ kg, $e = 0$, $P = 2.3 \times 10^4$ s), the flux only rises to

$$F_G = 5 \times 10^{-14} \text{ W m}^{-2} \tag{1.130}$$

Rotating elliptical objects radiate quadrupole radiation with an intensity given approximately by

$$L_G \approx \frac{GM^2\omega^6 r^4 (A+1)^6 (A-1)^2}{64c^5} \text{ W} \tag{1.131}$$

where M is the mass, ω the angular velocity, r the minor axis radius and A the ratio of the major and minor axis radii. So that for a pulsar with

$$\omega = 100 \text{ rad s}^{-1} \tag{1.132}$$

$$r = 15 \text{ km} \tag{1.133}$$

$$A = 0.99998 \tag{1.134}$$

we obtain

$$L_G = 1.5 \times 10^{26} \text{ W} \tag{1.135}$$

which for a distance of 1000 pc leads to an estimate of the flux at the Earth of

$$F_G = 10^{-14} \text{ W m}^{-2} \tag{1.136}$$

Objects collapsing to form black holes within the galaxy or nearby globular clusters or coalescing binary systems may perhaps produce transient fluxes up to three orders of magnitude higher than these continuous fluxes.

Now these fluxes are relatively large compared with, say, those of interest to radio astronomers, whose faintest sources may have an intensity of 10^{-29} W m^{-2} Hz^{-1}. But the gravitational detectors which have been built to date and those planned for the future have all relied upon detecting the strain ($\delta x/x$) produced in a test object by the *tides* of the gravitational wave rather than the *absolute* gravitational wave flux and this in practice means that measurements of changes in the length of a 1-metre long test object of only 5×10^{-21} m, or about 10^{-12} times the diameter of the hydrogen atom, must be obtained even for the detection of the radiation from ι Boo.

The strain ($\delta x/x$) is conventionally denoted by the symbol h in gravitational wave astronomy. It is a dimensionless quantity. Another parameter,* the strain sensitivity, symbolised by $\tilde{h}$ (pronounced: H-tilde†) is also in use. It is defined as the square root of the input noise that is required to produce the noise observed at the output. It takes account of the bandwidth of the signal and is useful when comparing the performances of various detectors. It has units of (seconds)$^{0.5}$, but customarily Hz$^{-0.5}$ are used. For burst gravitational wave sources, such as supernovae, h and $\tilde{h}$ are related by

$$\tilde{h} = h\Delta t \sqrt{\frac{\pi B}{2}} \text{Hz}^{-0.5} \tag{1.137}$$

where Δt is duration of the burst (typically a millisecond) and B is the detector bandwidth (typically 1 Hz to 1 kHz). Thus, for example, for the NAUTILUS detector (see below) with a 0.001-s burst length and a 5-Hz bandwidth we have $h \approx 3 \times 10^{-19}$ and $\tilde{h} \approx 8 \times 10^{-22}$ Hz$^{-0.5}$. For comparison, Joseph Weber's first detector had $h \approx 10^{-16}$ and $\tilde{h} \approx 4 \times 10^{-19}$ Hz$^{-0.5}$.

In spite of the difficulties that the detection of such small changes must obviously pose, the detector built by Joseph Weber appeared to have detected gravity waves from the centre of the galaxy early in 1969. Unfortunately, other workers did not confirm this and the source of Weber's pulses remains a mystery, although they are not now generally attributed to gravity waves. The pressure to confirm Weber's results, however, has led to a wide variety of gravity wave detectors being proposed and built. The detectors so far built, under construction or proposed are of two types: direct detectors in which there is an attempt to detect the radiation experimentally and indirect detectors in which the existence of the radiation is inferred from the changes that it incidentally produces in other observable properties of the object. The former group may be further subdivided into resonant, or narrow bandwidth detectors and non-resonant or wideband detectors.

---

* A third parameter that is sometimes used is the horizon distance. This is the maximum distance at which the coalescence of two 1.4-solar mass neutron stars could be detected. The basic definition assumes that the coalescence is optimally oriented for detection. To allow for non-optimum situations the basic horizon distance needs dividing by a factor of about two and a quarter. Unfortunately, it is not always apparent which of these figures is being quoted. In this book, therefore, we shall stick with the clearer h and as the detector criteria.

† Other symbols and names for the quantity may be encountered such as hrss (h-root-sum-square), strain amplitude, root-sum-square amplitude, spectral strain sensitivity, strain noise power spectral density etc.

1.6.2 Detectors

1.6.2.1 Direct Resonant Detectors

The vast majority of the early working gravity wave telescopes fell into the category of direct resonant detectors. They are similar to, or improvements on, Weber's original system. This latter used a massive (> 1 tonne) aluminium cylinder which was isolated by all possible means from any external disturbance and whose shape was monitored by piezo-electric crystals attached around its 'equator'. Two such cylinders separated by up to 1000 km were utilised and only coincidental events were regarded as significant in order to eliminate any remaining external interference. With this system, Weber could detect a strain of 10^{-16} (cf. 5×10^{-21} for the detection of ι Boo). Since his cylinders had a natural vibration frequency of 1.6 kHz, this was also the frequency of the gravitational radiation that they would detect most efficiently. Weber detected about three pulses per day at this frequency with an apparent sidereal correlation and a direction corresponding to the galactic centre or a point 180° away from the centre. If originating in the galactic centre, some 500 solar masses would have to be totally converted into gravitational radiation each year in order to provide the energy for the pulses. Unfortunately (or perhaps fortunately from the point of view of the continuing existence of the galaxy), the results were not confirmed even by workers using identical equipment to Weber's and it is now generally agreed that there have been no definite detections of gravitational waves to date.

Although the future seems likely to lie with the Michelson interferometer type of gravitational wave detector (see below) there are still some Weber-type detectors in operation. Thus, the Nautilus (not an acronym) detector near Rome and the Antenna Ultracriogenica Risonante per l'Indagine Gravitazionale Astronomica (AURIGA) detector near Padua both use 2300 kg aluminium bars cooled to around 0.2 K and have strain sensitivities of about 10^{-21} Hz$^{-0.5}$. At Leiden, a detector based upon a spherical mass of copper-aluminium alloy, known as MiniGRAIL, is currently working at a temperature of 50 mK and has a strain sensitivity of 10^{-21} Hz$^{-0.5}$ at the resonant frequency of 3 kHz. It has the advantage of being equally sensitive to gravity waves from any direction. An almost identical instrument, the Mario Schenberg detector is operating in Sao Paulo.

1.6.2.2 Direct, Non-Resonant Detectors

Direct, non-resonant detectors are of three main types and have the potential advantage of being capable of detecting gravity waves with a wide range of frequencies. The first uses a Michelson type interferometer (see Section 2.5) to detect the relative changes in the positions of two or more test masses. A possible layout for the system is shown in Figure 1.114; usually, however, the path length will be amplified by up to a factor of 100 by multiple reflections from the mirrors though these are not shown in the figure in order to preserve simplicity. (The pair of multiply-reflecting mirrors form a Fabry–Perot cavity) (see the discussion of etalons in Section 4.1.) The mirror surfaces have to be flat to 0.5% of the operating wavelength. The light sources are high-stability neodymium-yttrium-garnet lasers with outputs of 100 W or more. The mirrors are mounted on test masses on pendulum suspensions and the whole system, including the light paths, operates in a vacuum. Path length changes are detected by looking at the interference pattern between the two

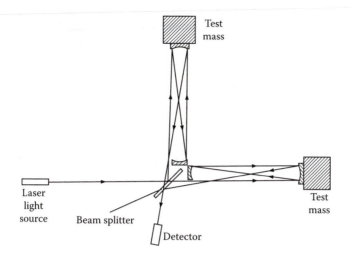

FIGURE 1.114   A possible layout for an interferometric gravity wave detector.

orthogonal beams. When no gravitational disturbance is present there is destructive inter-ference between the two beams at the detector (i.e. no signal) and the light can be returned to the arms of the interferometer, thus increasing the power available – a technique called power or signal recycling. A recent development which may enhance detector sensitivities by a factor of two is the use of quantum entanglement of photons in the beams to reduce the shot noise. 'Squeezing' the light in this way enables phase uncertainty to be reduced whilst amplitude noise is increased. Ultimate strain sensitivities of 10^{-22} Hz$^{-0.5}$ to 10^{-23} Hz$^{-0.5}$ over a bandwidth of 1 kHz are predicted for these systems, enabling detections of collapsing neutron stars, supernovae and coalescing binaries to be made out to distances of 10^7 to 10^8 pc. Terrestrial detectors are limited to frequencies above about 10 Hz because of noise. A proposed space-based system, LISA, which will be able to detect much lower frequencies, is discussed below.

Currently, only one Michelson gravitational wave observatory is in operation. This is the GEO600 German-British instrument near Hanover (Figure 1.115). Its arms are 600 metres in length. It uses a 10-W laser, which, with power recycling increases to 10 kW. It operates in the 50-Hz to 1.5-kHz region with a bandwidth of around 60 Hz and achieves a value of $\tilde{h}$ of 3×10^{-22} Hz$^{-0.5}$.

The two other main Michelson gravitational wave observatories are the Laser Interferometer Gravitational-Wave Observatory (LIGO), and Virgo and both of these are undergoing major upgrades at the time of writing.[*]

LIGO comprised three Michelson interferometers. Two of these have interferom-eter arms, are 4 km long, and they are sited 3000 km apart in Washington state and in Louisiana, so that gravity waves may be distinguished from other disturbances through coincidence techniques. The Washington machine has a second interferometer with 2-km arms that will also help to separate out gravity wave disturbances from other effects. This

---

[*] The Japanese instrument, TAMA300, obtained its last data in 2004.

FIGURE 1.115   The vacuum tube containing one of the arms of GEO600 in its trench. (© Harald Lück. Reproduced by kind permission of H. Lück, Max Planck Institute for Gravitational Physics [Albert Einstein Institute]/Leibniz Universität Hanover.)

is because the gravity wave should induce changes in the 2-km machine that have half the amplitude of those in the 4-km machine, whereas other disturbances are likely to have comparable effects on the two machines. The light beams make 75 round trips along the arms before recombination so that the arms are in effect 600 km long. The upgrade (to Advanced LIGO) will see much more powerful lasers (200 W) and larger mirrors installed. These, together with improvements to the mirror suspensions and seismic isolation should see the sensitivity improved by a factor of 10 over LIGO, perhaps to around 4×10^{-23} Hz$^{-0.5}$. It is possible that the second interferometer at the Washington state site will be relocated to India to give improved triangulation on the positions of gravitational wave sources, but this has still to be confirmed at the time of writing. Completion of the upgrade is likely by 2014.

Anyone with a computer and an Internet link can help with the data processing part of LIGO and GEO600 through the Einstein@home project. This currently has some 180,000 volunteers whose computers process data for the project when they would otherwise be idle. An Internet search for 'Einstein@home' will take anyone interested in joining the project straight to its home page.

The Italian-French Virgo instrument is sited near Pisa and had its first science run in 2007. The light beams undergo 20 multiple reflections in the 3-km-long arms giving a path length of 120 km. The seismic noise is especially low because of very stringent measures taken to isolate the instrument and so it can operate down to a gravitational wave frequency of 10 Hz. Completion of Virgo's upgrade is expected by 2015 and it should have similar levels of strain sensitivity to the Advanced LIGO instruments.

Plans are being considered for the Kamioka gravitational wave telescope (KAGRA) instrument in Japan (previously known as Large-scale Cryogenic Gravity wave Telescope [LCGT]) which will have two sets of 3-km-long arms, be situated in the Kamioka mine and perhaps be operational by 2018. It is expected to have similar levels of performance to Advance LIGO and Advanced Virgo. In addition, there is the Einstein telescope proposal which is currently in the design study phase. This might improve on current detectors' sensitivities by a factor of 100 and comprise three nested detectors with arms up to 10 km long.

The distribution of these detectors over the Earth will not only provide confirmation of detections, but also enable the arrival directions of the waves to be pinpointed to a few minutes of arc (which is why a part of Advanced LIGO may be moved to India).

There is also a European proposal for a space-based interferometer system called New Gravitational Wave Observatory (NGO), previously known as LISA and evolved Laser Interferometer Space Antenna (eLISA). This would employ three drag-free spacecraft (Figure 1.117) at the corners of an equilateral triangle with 5,000,000-km sides. Each spacecraft would be able to operate as the vertex of the interferometer and as the proof mass for the other spacecraft. NGO would thus have three separate interferometer systems. It would orbit at 1 AU from the Sun, but some 20° behind the Earth. Its sensitivity would probably be comparable with the best of the terrestrial gravity wave detectors, but its low noise would enable it to detect the important low frequency waves from binary stars etc. At the time of writing NGO has failed to be chosen as Europe's next large space mission, but a review of its status is possible in 2013. LISA pathfinder, a spacecraft designed to test various technologies for NGO may, however, still be launched, perhaps by 2014. Beyond NGO there may be the Big Bang Observer which is envisaged to comprise four instruments like LISA and be able to detect gravity waves from soon after the Big Bang. Other space-based gravitational wave detector concepts include Astronomical Space Test of Relativity using Optical Devices (ASTROD-GW) which might have up to six spacecraft and arm lengths of hundreds of millions of kilometres and the Japanese Deci-Hertz Interferometer Gravitational Wave Observatory (DECIGO) with three drag-free spacecraft separated by 1000 km.

An ingenious idea underlies the second method in this class. An artificial satellite and the Earth will be independently influenced by gravity waves whose wavelength is less than the physical separation of the two objects. If an accurate frequency source on the Earth is broadcast to the satellite and then returned to the Earth (Figure 1.116), its frequency will be changed by the Doppler shift as the Earth or satellite is moved by a gravity wave. Each gravity wave pulse will be observed three times enabling the reliability of the detection to be improved. The three detections will correspond to the effect of the wave on the transmitter, the satellite and the receiver, although not necessarily in that order. A drag-free satellite (Figure 1.117) would be required in order to reduce external perturbations arising from the solar wind, radiation pressure etc. A lengthy series of X-band observations using

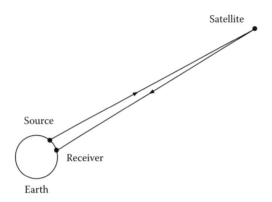

FIGURE 1.116 Arrangement for a satellite-based Doppler tracking gravity wave detector.

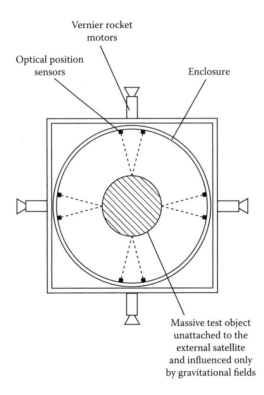

FIGURE 1.117 Schematic cross section through a drag-free satellite. The position of the massive test object is sensed optically and the external satellite driven so that it keeps the test object centred.

the Ulysses spacecraft aimed at detecting gravity waves by this method have been made, but so far without any successful detections.

1.6.2.3 Pulsar Timing Arrays

The third approach is to detect gravitational waves in the nanohertz frequency region from their effects upon the arrival times of pulses from pulsars. A gravitational wave contracts space in some directions and expands it in others as it passes through the solar system. The pulses should therefore arrive earlier or later than normal by a few nanoseconds, respectively. By monitoring the arrival times of the pulses from several tens of pulsars distributed around the whole sky, the passage of the wave should be recordable.

Several projects – the European Pulsar Timing Array (Lovell, Westerbork, Effelsberg and Nançay radio telescopes), the North American Nanohertz Observatory for Gravitational Waves (NANOGrav – Green Bank and Arecibo radio telescopes) and the Parkes Pulsar Timing Array (Parkes radio telescope) are presently searching for gravitational waves in this manner, so far without success although upper limits for h of less than a few times 10^{-15} for the background intensity of these waves have been set.

1.6.2.4 Detectors

Proposals for indirect detectors so far only involve binary star systems. The principle of the method is to attempt to observe the effect on the period of the binary of the loss of angular momentum due to the gravitational radiation. A loss of angular momentum by some process is required to explain the present separations of dwarf novae and other close binary systems with evolved components. These separations are so small that the white dwarf would have completely absorbed its companion during its earlier giant stage, thus the separation at that time must have been greater than it is now. Gravitational radiation can provide an adequate orbital angular momentum loss to explain the observations, but it is not the only possible mechanism. Stellar winds, turbulent mass transfer and tides may also operate to reduce the separation, so that these systems do not provide unequivocal evidence for gravitational radiation.

The prime candidate for study for evidence of gravitational radiation is the binary pulsar (PSR 1913 + 16). Its orbital period is 2.8×10^4 seconds and it was found to be decreasing at a rate of 10^{-4} seconds per year soon after the system's discovery. This was initially attributed to gravitational radiation and the 1993 Nobel physics prize was awarded to Russell Hulse and Joseph Taylor, for their discovery of the pulsar and their interpretation of its orbital decay. However, it now appears that a helium star may be a part of the system and so tides or other interactions could again explain the observations. On the other hand, the effect of a rapid pulsar on the local interstellar medium might mimic the appearance of a helium star when viewed from a distance of 5000 pc. Thus, the detection of gravity waves from the binary pulsar and indeed by all other methods remains not proven at the time of writing.

EXERCISES

1.18 Show that the gravitational radiation of a planet orbiting the Sun is approximately given by

$$L_G \approx 7 \times 10^{21} \, M^2 \, P^{-10/3} \; W$$

where M is the planet's mass in units of the solar mass and P the planet's orbital period in days.

1.19 Calculate the gravitational radiation for each planet in the solar system and show that every planet radiates more energy than the combined total radiated by those planets whose individual gravitational luminosities are lower than its own.

1.7 DARK MATTER AND DARK ENERGY DETECTION

1.7.1 Introduction

Mainstream cosmology currently suggests that the visible portion of the universe (stars, nebulae, planets, galaxies etc.) makes up only about 4% of the total content of the universe. Dark matter makes up about another 22% and dark energy the remaining 74%.

The existence of dark matter was first suggested by Jan Oort in 1932 followed by Fritz Zwicky in 1933. Oort found that more mass was needed to account for the motions of nearby stars than was visible in the Milky Way's disk. Zwicky's observations of the velocities of the galaxies within the Coma galaxy cluster showed that they were too high for the galaxies to be retained by the cluster's gravitational field. Over a few hundred million years the cluster would 'evaporate' and the galaxies would move away as independent entities. However, the existence of many clusters of galaxies suggests that they *are* stable over much longer periods of time. Zwicky therefore theorised that there must be additional material, within the cluster or galaxies or both, whose presence meant that the gravitational field of the cluster was sufficient for the cluster to be stable. Since this matter was not directly visible it became known as dark matter. Zwicky found that the amount of dark matter that was needed to stabilise the Coma cluster was 400 times the amount of visible matter. However, subsequent measurements by many astronomers have brought this figure down to the quantity of dark matter required generally being five to six times that of the visible matter.

Three quarters of a century have passed since Zwicky's observations and although the existence of dark matter is confirmed by many other circumstantial observations – such as the rotation curves of galaxies – the nature of dark matter is still a mystery.* Suggestions for what it might be made up from have included mini black holes, brown dwarfs, large planets together known as massive astrophysical compact halo objects (MACHOs), neutrinos and weakly interacting massive particles (WIMPs). The current front runner is a variety of WIMP called a neutralino that is predicted by the particle physicists' supersymmetry theory to be produced in large numbers at the energies prevalent during the early stages of the Big Bang. Despite the plentiful secondary evidence for the existence of dark matter, three quarters of a century is a long time for no direct evidence of its existence to have been found and some workers now question its existence at all – suggesting that some modification to the way in which the force of gravity operates could explain the observational discrepancies instead. Much effort is therefore currently being put into experiments that might detect the particles, whatever they may be, making up dark matter.

The existence of dark energy, like that of dark matter, has yet to be proven. There are two main circumstantial lines of evidence for its existence – the mean density of the universe

---

* A similar situation though existed for the neutrino – its existence was predicted by Wolfgang Pauli in 1930, but it was not found experimentally until 1956.

and the brightness of Type Ia supernovae. The critical density of the universe is the density that would enable gravity to slow down the expansion of the universe to zero after an infinite length of time. A density lower than the critical value and the universe will expand forever, a higher density and its expansion will eventually halt and it will collapse back again, perhaps ending in the 'Big Crunch'. The actual density of the matter and dark matter in the universe amounts to about 26% of the critical density. For reasons beyond the scope of this book,* if the apparent mean density is as close to the critical density as it appears to be, then the true mean density must almost certainly be exactly equal to the critical density. Thus, something other than matter must comprise 74% of the universe and that 'something' is called dark energy.

The second line of evidence for dark energy emerged in the late 1990s. Observations of distant (≥2,000 Mpc) Type Ia supernovae seemed to suggest that after an initial period during which the expansion of the universe was decelerating, it is now accelerating. One way of causing such acceleration would be the presence within the universe of a large amount of dark energy. If the density of the universe is assumed to be the critical density, then the amount of dark energy needed to provide the observed acceleration turns out to be around 75% of the total mass/energy of the universe.

However, like the case for dark matter, an alternative explanation for dark energy can be found through a modified form of gravity (often called modified Newtonian dynamics [MOND]). A second possibility is if the nature of Type Ia supernovae has changed over time and so they are seen as fainter than expected at large distances because they are fainter than modern supernovae, not because they are more distant than predicted. Thus, observational confirmation of the existence of dark energy is also eagerly being sought.

1.7.2 Dark Matter and Dark Energy Detectors

No new types of detectors will be described in this section – the instruments used in the searches for dark matter and dark energy are amongst those that we have seen in the earlier sections of this chapter. The searches use the existing results and capabilities of those instruments to look for data with the signature of some aspect(s) of dark matter or dark energy. For dark matter the search concentrates on trying to detect the neutralino (if that is what makes up dark matter) via detectors such as those used for gamma rays and neutrinos, whilst evidence is sought for dark energy via its influence on the microwave cosmic background radiation or on galaxy formation and clustering or via improved observations of very distant supernovae.

Dark matter detectors divide into direct and indirect types. The direct types look for interactions of the dark matter particle (DMP) with atoms within the test apparatus. Those interactions are primarily scintillations or ionisations arising from a high-energy normal matter particle produced by a collision with a DMP and acoustic pulses arising from the heat deposited into the material surrounding such a collision. Indirect detectors look for decay products of DMPs or for their influences on aspects of the universe that can be observed.

---

* See *Dark Side of the Universe* by I. Nicolson (Canopus Publishing, 2007) for an excellent explanation of all of this from a layman's point of view.

Recently, the XENON100 experiment has started operations. It is located in the Gran Sasso tunnel so that the Apennines provide some 1.4 km of overlying rock to shield the experiment from unwanted interactions. It is a TPC (see Section 1.5) using 100 kg of liquid xenon and 178 photomultipliers to detect possible DMP scintillations. It has recently completed an observing run of nearly eight months. No definitive detections were made, but the results add to the constraints on the possible properties of DMPs. It is hoped to replace XENON100 with XENON1T by 2015: an instrument using a tonne or more of liquid xenon.

The Dark Matter (DAMA)/Libra experiment is also a scintillation dark matter detector housed in the Gran Sasso tunnel. It uses 250-kg sodium iodide doped with thallium. DEAP1 is a DMP scintillator based upon 7 kg of liquid argon and housed in SNOLab in Canada. It is a prototype for DEAP3600 which is currently in the design stage and will use 3600 kg of liquid argon.

The European Underground Rare Event Calorimeter Array (EURECA) will be the successor to the EDELWEISS instruments. The latter used both ionisation and heat deposition in germanium bolometers cooled to 20 mK. EURECA will have between 100 and 1000 kg of cryogenic calorimeters made from a variety of materials and be housed in the Laboratoire Souterrain de Modane near the French-Italian border. It could be operational by 2015. The Coherent Germanium Neutrino Technology (CoGeNT) experiment in the Soudan underground laboratory in Minnesota uses a 440-g germanium crystal cooled by liquid nitrogen. It senses ionisations arising from nuclei hit by DMPs. It is planned to upgrade the instrument to using 5 kg of germanium in the near future. The Large Underground Xenon experiment (LUX) is due to start operating shortly. It comprises a 370-kg liquid xenon TPC detector that will pick up both scintillations and ionizations. It is located in the Sanford lab in South Dakota, 1.5 km underground.

A possible detection of DMPs was announced in 2011. The Cryogenic Rare Event Search with Superconducting Thermometers (CRESST) II experiment which is in the Gran Sasso tunnel and uses calcium tungstate cryogenic calorimeters and scintillators found 67 signals that could be due to WIMPs. The expected background of such events was only about half this figure. The balance could therefore have arisen from WIMPs with masses in the region of 10 to 25 GeV. However, the data analysis is in the process of being refined and a further run with an upgraded instrument is planned for the near future. At the moment therefore the result is generally regarded as unconfirmed.

Although there have been no definitive detections of DMPs, both the DAMA/Libra and the CoGeNT experiments appear to show an annual variation in their low energy interaction rates at statistically significant levels. If due to DMPs, then this could arise from the changing velocity of the Earth relative to the galaxy as a whole (as so presumably also relative to any dark matter present in the galaxy). The Sun's velocity is roughly towards the star Vega and so the Earth's velocity through space is a maximum in early June – and that is when the detectors' interaction rates also peak. However, other DMP detectors do not confirm this result and although the variation is undoubtedly real, it may arise from a source other than DMPs.

Indirect dark matter detectors search for possible collision or decay products from DMPs such as γ-rays, positrons or anti-protons (although since it is not known what DMPs are, it is also not known what such decay products might be). The searches attempt to find excesses of particles apparently coming from regions where dark matter might be expected to be concentrated, such as the centres of galaxies. Since γ-rays are unaffected by interstellar or inter-galactic magnetic fields, they are the best candidates for the searches, but could also be preferentially produced in the centres of galaxies by other processes. The detectors are identical to the Čerenkov and other γ-ray detectors discussed in Sections 1.3 and 1.4. No significant results have been found to date, although the Fermi spacecraft may have detected a γ-ray emission line originating from near the centre of the Milky Way at an energy of 135 GeV, but this has still to be confirmed.

It is possible that DMPs could accumulate towards the centre of the Sun, or even the centre of the Earth. Their destruction or decay there might produce neutrinos which could be detected by neutrino telescopes (see Section 1.5) on the Earth's surface. In 2010 and 2011, IceCube conducted a search for such neutrinos, although without success. Indirect detection may also be possible via the products of DMP interactions or decays far in space. Amongst these products there may be anti-protons that could be observed directly by some balloon-borne cosmic-ray detectors and positrons that could be found via the 511-keV γ-ray emission line produced when positrons and electrons annihilate each other.

A quite different approach to confirming the existence of dark matter would be to make some ourselves – and that is exactly what the Large Hadron Collider (LHC) may be able to do. When its two beams of protons collide head-on at a combined energy of 1.4×10^{13} eV (14 TeV), amongst the many by-products of the collisions there may be DMPs. These DMPs would, of course, be no easier to detect than those from any other source, but their presence might become apparent if there is 'something' missing when all the other (detectable) products of the interaction have been accounted for.

Confirmation of the existence of dark energy and of its nature (there are several possibilities such as Einstein's cosmological constant, quintessence and phantom energy; see for example 'Dark side of the Universe' for further information) may come through its effect upon the large scale structure of ordinary matter throughout the universe. Galaxies and clusters of galaxies started to form very soon after the Big Bang and their presence produces ripples in the intensity of CMB radiation. Many other processes also produce ripples in the CMB, and although the existence of ripples arising from dark energy has been sought in the data from the WMAP spacecraft, nothing definitive has yet been found. The 10-metre South Pole Telescope may be able to improve on WMAP's results, especially after the recent upgrade to its camera, whilst ESA's Planck mission, launched in 2009, may also provide data of sufficient quality to provide better answers. Planck observed the CMB using an off-axis 1.5 × 1.75-metre Gregorian telescope over the frequency range 27 GHz to 1 THz (11 to 0.3 mm) for around 21 months (the lifetime was limited by the loss of the helium coolant). Its results are expected in 2013.

The large-scale three-dimensional structure of the universe can be studied directly by observing galaxies and clusters of galaxies. To date, surveys such as the Sloan Digital Sky Survey (SDSS) and the AAO's 2dF Galaxy Redshift Survey have suggested that dark energy may take the form of Einstein's cosmological constant. Construction of the LSST has recently begun. This will be an 8.4-metre telescope designed to have a 10-square-degree field of view and using a 3000 mega-pixel camera. It will be able to image all the available sky with 15-second exposures every three nights, covering three-quarters of the whole sky over a year. It is expected to start operating in 2012 and the resulting survey should provide a very detailed three-dimensional map of the large-scale structure of the universe. In the longer term, the SKA, if built, should provide data on at least a billion galaxies.

Studies of distant supernovae provide a history of the acceleration/deceleration of the universe over time and this in turn can put constraints on the nature of dark energy. Such data has already come from surveys like the SDSS and forms the current observational basis for the existence of dark energy. Supernova searches are continuing at several observatories with the eight 2 k × 4 k CCD array MOSAIC camera on the 4-metre telescope at the Cerro Tololo Inter American Observatory and the CFHT's thirty-six 2 k × 4 k CCD MegaCAM being fairly typical. The LSST will be able to detect thousands of supernovae per year out to distances of eleven thousand million light years (3300 Mpc). The recently commissioned Dark Energy Camera on the 4-Metre Blanco telescope is due to survey about an eighth of the sky in the next few years and the Hobby-Eberly Telescope Dark Energy Experiment (HEDTEX) is a similar project using the Hobby-Eberly telescope.

Clues to the large-scale structure of the universe and so to dark energy may also come from studies of the X-ray emission from clusters of galaxies, from weak gravitational lensing and from the frozen-in signatures of acoustic waves (baryon oscillations) in the early universe – now about 500 Mly (160 Mpc) across. The 2.5-metre telescope of the Observatorio Astrofisico de Javalambre in Aragon has recently been equipped with a 1.2 giga-pixel camera specifically to perform a baryon acoustic oscillation survey of the northern sky, whilst the Euclid spacecraft, scheduled for launch in 2020, will attempt to constrain the properties of dark energy by observing both weak gravitational lensing and the baryon oscillations using a 1.2-metre Korsch telescope.

One way or another we may hope that the existence of dark matter and dark energy may be confirmed or refuted within the next decade and at least some clue as to their natures, if they exist, found.

Imaging

2.1 THE INVERSE PROBLEM

A problem that occurs throughout much of astronomy and other remote sensing applications is how best to interpret noisy data so that the resulting deduced quantities are real and not artefacts of the noise. This problem is termed the inverse problem.

For example, stellar magnetic fields may be deduced from the polarisation of the wings of spectrum lines (see Section 5.2). The noise (errors, uncertainties) in the observations, however, will mean that a range of field strengths and orientations will fit the data equally well. The three-dimensional distribution of stars in a globular cluster must be found from observations of its two-dimensional projection onto the plane of the sky. Errors in the measurements will lead to a variety of distributions being equally good fits to the data. Similarly, the images from radio and other interferometers (see Section 2.5) contain spurious features due to the side lobes of the beams. These spurious features may be removed from the image if the effects of the side lobes are known. But since there will be uncertainty in both the data and the measurements of the side lobes, there can remain the possibility that features in the final image are artefacts of the noise, or are incompletely removed side lobe effects.

The latter illustration is an instance of the general problem of instrumental degradation of data. Such degradation occurs from all measurements, since no instrument is perfect: even a faultlessly constructed telescope will spread the image of a true point source into the Airy diffraction pattern (Figure 1.41). If the effect of the instrument and other sources of blurring on a point source or its equivalent are known (the point spread function [PSF] or instrumental profile) then an attempt may be made to remove its effect from the data. The process of removing instrumental effects from data can be necessary for any type of measurement, but is perhaps best studied in relation to imaging, when the process is generally known as deconvolution.

2.1.1 Deconvolution

The removal of the effect of the PSF from an image is known as deconvolution, because the true image *convolves* with the PSF to give the observed (or dirty) image. Inversion of its effect is thus *deconvolution*. Convolution is most easily illustrated in the one-dimensional case (such as the image of a spectrum or the output from a Mills Cross radio array), but applies equally well to two-dimensional images.

A one-dimensional image, such as a spectrum, may be completely represented by a graph of its intensity against the distance along the image (Figure 2.1). The PSF may similarly be plotted and may be found, in the case of a spectrum, by observing the effect of the spectroscope on a monochromatic source.

If we regard the true spectrum as a collection of adjoining monochromatic intensities, then the effect of the spectroscope will be to broaden each monochromatic intensity into the PSF. At a given point (wavelength) in the observed spectrum therefore, some of the original energy will have been displaced out to nearby wavelengths, whilst energy will have been added from the spreading out of nearby wavelengths (Figure 2.2). The process may be expressed mathematically by the convolution integral:

$$O(\lambda_1) = \int_0^\infty T(\lambda_2)I(\lambda_1 - \lambda_2)d\lambda_2 \tag{2.1}$$

where $O(\lambda_1)$ is the intensity in the observed spectrum at wavelength λ_1, $T(\lambda_2)$ is the intensity in the true spectrum at wavelength λ_2 and $I(\lambda_1 - \lambda_2)$ is the response of the instrument (spectroscope) at a distance $(\lambda_1 - \lambda_2)$ from the PSF's centre.

Equation 2.1 is normally abbreviated to

$$O = T * I \tag{2.2}$$

where * is the convolution symbol.

The inversion of Equation 2.1 to give the true spectrum cannot be accomplished directly but involves the use of Fourier transforms.

With the Fourier transform and its inverse in the form

$$F(s) = F(f(x)) = \int_{-\infty}^\infty f(x)e^{-2\pi i x s} dx \tag{2.3}$$

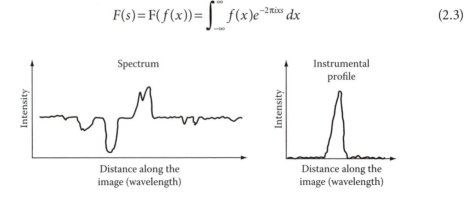

FIGURE 2.1 Representation of a one-dimensional image and the instrumental profile (PSF) plotted as intensity versus distance along image.

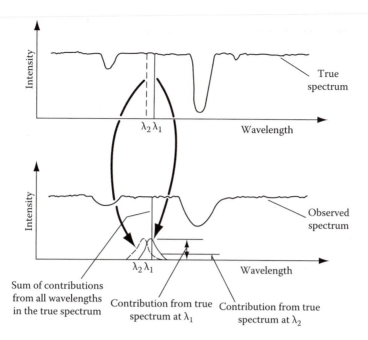

FIGURE 2.2   Convolution of the true spectrum with the PSF to produce the observed spectrum.

$$f(x) = F^{-1}(F(s)) = \int_{-\infty}^{\infty} F(s)e^{2\pi i x s} \, ds \qquad (2.4)$$

then the convolution theorem states that: *convolution* of two functions corresponds to the *multiplication* of their Fourier transforms.

Thus, taking Fourier transforms of Equation 2.2, we have

$$F(O) = F(T * I) \qquad (2.5)$$

$$= F(T) \times F(I) \qquad (2.6)$$

and so the true spectrum may be found from inverting Equation 2.6 and taking its inverse Fourier transform:

$$T = F^{-1}\left[\frac{F(O)}{F(I)}\right] \qquad (2.7)$$

In practice, obtaining the true data (or source function) via Equation 2.7 is complicated by two factors. First, data is sampled at discrete intervals and so is not the continuous function required by Equations 2.3 and 2.4 and also it is not available over the complete range from $-\infty$ to $+\infty$. Second, the presence of noise will produce ambiguities in the calculated values of T.

The first problem may be overcome by using the discrete versions of the Fourier transform and inverse transform:

$$F_D(s_n) = F_D(f(x))_n = \sum_{k=0}^{N-1} f(x_k) e^{-2\pi i k n/N} \Delta \tag{2.8}$$

$$f(x_k) = F_D^{-1}(F_D(s_n)) = \sum_{n=0}^{N-1} F_D(s_n) e^{2\pi i k n/N} \frac{1}{N} \tag{2.9}$$

where $F_D(s_n)$ is the nth value of the discrete Fourier transform of $f(x)$, N is the total number of measurements, $f(x_k)$ is the kth measurement and Δ is the step length between measurements and setting the functions to zero outside the measured range.

Now a function that has a maximum frequency of f is completely determined by sampling at $2f$ (the sampling theorem). Thus, the use of the discrete Fourier transform involves no loss of information providing the sampling frequency $(1/\Delta)$ is twice the highest frequency in the source function. The highest frequency that can be determined for a given sampling interval $(1/2\Delta)$ is known as the Nyquist or critical frequency. If the source function contains frequencies higher than the Nyquist frequency, then these will not be determined by the measurements and the finer detail in the source function will be lost when it is reconstituted via Equation 2.7. Rather more seriously, however, the higher frequency components may beat with the measuring frequency to produce spurious components at frequencies lower than the Nyquist frequency. This phenomenon is known as aliasing and can give rise to major problems in finding the true source function.

The actual evaluation of the transforms and inverse transforms may nowadays be relatively easily accomplished using the fast Fourier transform algorithm on even quite small computers. The details of this algorithm are outside the scope of this book, but may be found in books on numerical computing.

The one-dimensional case just considered may be directly extended to two or more dimensions, though the number of calculations involved then increases dramatically. Thus, for example, the two-dimensional Fourier transform equations are

$$F(s_1, s_2) = F(f(x_1, x_2)) = \int_{-\infty}^{\infty} \int_{-\infty}^{\infty} f(x_1, x_2) e^{-2\pi i x_1 s_1} e^{-2\pi i x_2 s_2} \, dx_1 \, dx_2 \tag{2.10}$$

$$f(x_1, x_2) = F^{-1}(F(s_1, s_2)) = \int_{-\infty}^{\infty} \int_{-\infty}^{\infty} F(s_1, s_2) e^{2\pi i x_1 s_1} e^{2\pi i x_2 s_2} \, ds_1 \, ds_2 \tag{2.11}$$

Some reduction in the noise in the data may be achieved by operating on its Fourier transform. In particular, cutting back on or removing the corresponding frequencies in the transform domain may reduce cyclic noise such as 50- or 60-Hz mains hum or the stripes on scanned images. Random noise may be reduced by using the optimal (or Wiener) filter defined by

$$W = \frac{[F(O)]^2}{[F(O)]^2 + [F(N)]^2} \tag{2.12}$$

where $F(O)$ is the Fourier transform of the observations, without the effect of the random noise and $F(N)$ is the Fourier transform of the random noise. (The noise and the noise-free signal are separated by assuming the high-frequency tail of the power spectrum to be just due to noise and then extrapolating linearly back to the lower frequencies.) Equation 2.7 then becomes

$$T = F^{-1} \left[\frac{F(O)W}{F(I)} \right] \tag{2.13}$$

Whilst processes such as those outlined above may reduce noise in the data, it can never be totally eliminated. The effect of the residual noise, as previously mentioned, is to cause uncertainties in the deduced quantities. The problem is often ill-conditioned; that is to say, the uncertainties in the deduced quantities may, proportionally, be very much greater than those in the original data.

A widely used technique, especially for reducing the instrumental broadening for images obtained via radio telescope, is due to William Richardson and Leon Lucy. The Richardson-Lucy (RL) algorithm is an iterative procedure in which the $n + 1$th approximation to the true image is related to the PSF and nth approximation by

$$T_{n+1} = T_n \int \frac{O}{T_n^* I} I \tag{2.14}$$

The first approximation to start the iterative process is usually taken as the observed data. The RL algorithm has the advantages compared with some other iterative techniques of producing a normalised approximation to the true data without any negative values and it also usually converges quite rapidly.

Recently, several methods have been developed to aid choosing the 'best' version of the deduced quantities from the range of possibilities. Termed 'non-classical' or Bayesian methods, they aim to stabilise the problem by introducing additional information not inherently present in the data as constraints and thus to arrive at a unique solution. The best known of these methods is the maximum entropy method (MEM). The MEM introduces the external constraint that the intensity cannot be negative and finds the solution that has the least structure in it that is consistent with the data. The name derives from the concept of entropy as the inverse of the structure (or information content) of a system. The maximum entropy solution is thus the one with the least structure (the least information content or the smoothest solution) that is consistent with the data. The commonest measure of the entropy is

$$s = -\sum p_i \ln p_i \tag{2.15}$$

where

$$p_i = \frac{d_i}{\sum\limits_j d_j} \tag{2.16}$$

and d_i is the ith datum value, but other measures can be used. A solution obtained by an MEM has the advantage that any features in it must be real and not artefacts of the noise. However, it also has the disadvantages of perhaps throwing away information and that the resolution in the solution is variable, being highest at the strongest values.

Other approaches can be used either separately or in conjunction with MEM to try and improve the solution. The CLEAN method, much used on interferometer images, is discussed in Section 2.5. The method of regularisation stabilises the solution by minimising the norm of the second derivative of the solution as a constraint on the smoothness of the solution. The non-negative least-squares (NNLSs) approach solves for the data algebraically, but subject to the added constraint that there are no negative elements. Myopic deconvolution attempts to correct the image when the PSF is poorly known by determining the PSF as well as the corrected image from the observed image, whilst blind deconvolution is used when the PSF is unknown and requires several images of the same object, preferably with dissimilar PSFs, such as occurs with adaptive optics images (see Section 1.1). Burger-Van Cittert deconvolution firstly convolves the PSF with the observed image and then adds the difference between the convolved and observed images from the observed image. The process is repeated with the new image and iterations continue until the result of convolving the image with the PSF matches the observed image. The $n + 1$th approximation is thus

$$T_{n+1} = T_n + (O - T_n^* I) \tag{2.17}$$

Compared with the RL approach, the Burger-van Cittert algorithm has the disadvantage of producing spurious negative values. The recently developed technique of spectral deconvolution is based upon the relative positions of two real objects (such as a bright star with a much fainter companion) being the same at different wavelengths, whilst artefacts, such as internal reflections will generally change their positions with wavelength, thus enabling real objects to be separated from false ones.

2.2 PHOTOGRAPHY

2.2.1 Introduction

When the first edition of this book appeared in 1984, photography was still the main means of imaging in astronomy throughout the visual region and into the very-NIR (to about 1 μm). CCDs were starting to come on the scene, but were still small and expensive. The situation is now completely reversed. Photography is hardly used at all in professional astronomy – the last photograph on the AAT was taken in 1999 for example. CCDs now dominate imaging from the UV to the NIR, though they are still expensive in the largest sizes and are themselves perhaps on the verge of replacement by wavelength-sensitive imaging detectors such as STJs. Amongst amateur astronomers, small cooled CCDs are now quite common and can be purchased along with sophisticated software for image processing for much less than the cost of a 0.2-metre Schmidt-Cassegrain telescope. Digital still and video cameras produced for the popular market are very cheap can also be used on telescopes, although without the long exposures possible with cooled CCDs specifically manufactured for astronomical use.

However, photography is not *quite* dead yet! Many archives are still in the form of photographs or data derived from photographs so that even recent major surveys such as the Multi mission Archive for the Space Telescope (MAST's) Digitized Sky Survey that produced the Hubble Guide Star Catalogue (GST) and the Anglo-Australian Observatory/UK Schmidt Telescope (AAO/UKST) H-α survey are based upon photographic images. Any Internet search for 'astronomy images' will also show that photography is still widely used amongst amateur astronomers (and many professional astronomers are amateur photographers in their spare time) despite the numbers who use CCDs. Thus, some coverage of photography still needs to be included, if only so that a student working on archive photographs is aware of how they were produced and what their limitations and problems may be.

Photography's pre-eminence as a recording device throughout most of the twentieth century arose primarily from its ability to provide a permanent record of an observation that is (largely) independent of the observer. But it also has many other major advantages including its cheapness, its ability to accumulate light over a long period of time and detecting sources fainter than those visible to the eye, its very high information density and its ease of storage for future reference.

These undoubted advantages coupled with our familiarity with the technique have, however, tended to blind us to its many disadvantages. The most important of these are its slow speed and very low quantum efficiency. The human eye has an effective exposure time of about a tenth of a second, but a photograph would need an exposure of ten to a hundred times longer in order to show the same detail. The impression that photography is far more sensitive than the eye only arises because most photographs of astronomical objects have exposures ranging from minutes to hours and so show far more detail than is seen directly. The processing of a photograph is a complex procedure and there are many possibilities for the introduction of errors. Furthermore, and by no means finally in this list of problems of the photographic method, the final image is recorded as a density and the conversion of this back to intensity is a tedious process. The whole business of photography in fact bears more resemblance to alchemy than to science, especially when it has to be used near the limit of its capabilities.

Colour photography has no place in astronomy, except for taking pretty pictures of the Orion nebula. The reason for this is that the colour balance of both reversal (slide) and print colour films is optimised for exposures of a few seconds or less. Long exposures, such as those needed in astronomy, therefore result in false colours for the object. When a colour image is needed, it is normally obtained by three monochromatic (black and white) images obtained through magenta, yellow and cyan filters that are then printed onto a single final image using the same filters.

2.2.2 Structure of the Photographic Emulsion

Many materials exhibit sensitivity to light that might potentially be usable to produce images. In practice, however, a compound of silver and one or more of the halogens is used almost exclusively. This arises from the relatively high sensitivity of such compounds allied to their ability to retain the latent image for long periods of time. For most purposes the highest sensitivity is realised using silver bromide with a small percentage of iodide ions in solid solution. The silver halogen is in the form of tiny crystals and these are supported in a solid, transparent medium; the whole structure is called the photographic emulsion. The supporting medium is usually gelatine, which has the advantage of adding to the stability of the latent image by absorbing the halogens that are released during its formation. It also allows easy penetration of the processing chemicals and forms the mixing medium during the manufacture of the emulsion.

The size of the silver halide crystals, or grains as they are commonly known, is governed by two conflicting requirements. The first of these is resolution, or the ability of the film to reproduce fine detail. Resolution is affected by both the grain size and by scattering of the light within the emulsion and is measured by the maximum number of lines per millimetre that can be distinguished in images of gratings. It ranges from about 20 to 2000 lines per millimetre, with generally the smaller the grain size the higher the resolution.

The modulation transfer function (MTF) specifies more precisely the resolution of an emulsion. The definition of the MTF is in terms of the recording of an image that varies sinusoidally in intensity with distance. In terms of the quantities shown in Figure 2.3, the modulation transfer, T, is given by

$$T = \frac{(A_R/I_R)}{(A_O/I_O)} \tag{2.18}$$

The MTF is then the manner in which T varies with spatial frequency. Schematic examples of the MTF for some emulsions are shown in Figure 2.4. Using an MTF curve, the resolution may be defined as the spatial frequency at which the recorded spatial variations become imperceptible. For visual observation, this usually occurs for values of T near 0.05.

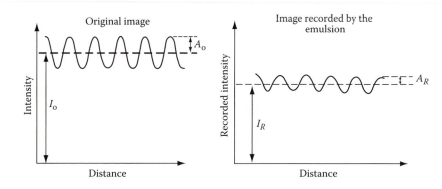

FIGURE 2.3   Schematic representation of the reproduction of an image varying spatially in a sinusoidal manner by an emulsion.

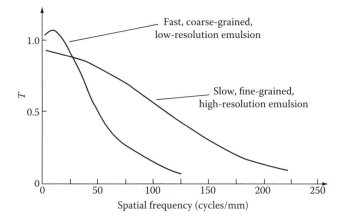

FIGURE 2.4   Schematic MTF curves for typical emulsions.

The second requirement is that of the speed of the emulsion and this is essentially the reciprocal of a measure of the time taken to reach a given image density under some standard illumination. The two systems in common use for measuring the speed of an emulsion are International Standards Organization (ISO), based upon the exposure times for image densities of 0.1 and 0.9 and Deutsche Industrie Norm (DIN), based only on the exposure time to reach an image density of 0.1. ISO is an arithmetic scale, whilst DIN is logarithmic. Their relationship is shown in Figure 2.5, together with an indication of the speeds of normal films. The two numbers are frequently combined into the exposure index (EI) number that is simply the ISO number followed by the DIN number, as in EI 100/21°. The speed of an emulsion is proportional to the volume of a grain for unsensitised emulsions and to the surface area of a grain for dye-sensitised (see below) emulsions. Thus, in either case, higher speed requires larger grains. The conflict between these two requirements means that high-resolution films are slow and that fast films have poor resolution. Grain sizes range from 50 nm for a very high-resolution film, through 800 nm for a normal slow film to 1100 nm for a fast film. Not all the grains are the same size, except in nuclear emulsions

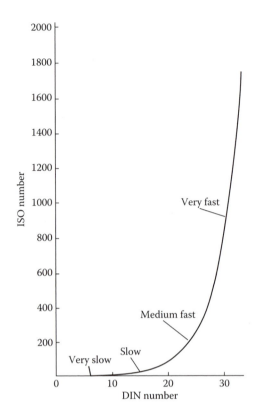

FIGURE 2.5   Emulsion speeds (approximate relationship).

(see Section 1.4) and the standard deviation of the size distribution curve ranges from about 1% of the particle diameter for very high-resolution films to 50% or more for normal commercial films.

A normal silver halide emulsion is sensitive only to short wavelengths (i.e. to the blue, violet and near-UV parts of the spectrum). Hermann Vogel made the fundamental discovery that was to render lifelike photography possible and to extend the sensitivity into the red and infrared in 1873. That discovery was that certain dyes when adsorbed onto the surfaces of the silver halide grains would absorb light at their own characteristic frequencies and then transfer the absorbed energy to the silver halide. The latent image would then be produced in the same manner as normal. The effect of the dye can be very great; a few spectral response curves are shown in Figure 2.6 for unsensitised and sensitised emulsions. Panchromatic or isochromatic emulsions have an almost uniform sensitivity throughout the visible, whilst the orthochromatic emulsions mimic the response of the human eye. Colour film would of course be quite impossible without the use of dyes since it uses three emulsions with differing spectral responses. The advantages of dye sensitisation are very slightly counteracted by the reduction in the sensitivity of the emulsion in its original spectral region, due to the absorption by the dye and more seriously by the introduction of some chemical fogging.

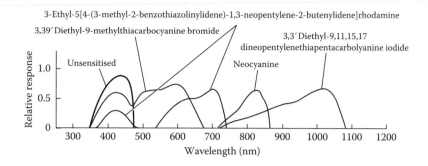

FIGURE 2.6   Effects of dyes upon the spectral sensitivity of photographic emulsion.

In many applications the response of the film can be used to highlight an item of interest; for example comparison of photographs of the same area of sky with blue- and red-sensitive emulsions is a simple way of finding very hot and very cool stars.

2.2.3 The Photographic Image

When an emulsion is exposed to light, the effect is to form the latent image (i.e. an image that requires the further step of development in order to be visible). Its production on silver halide grains is thought to arise in the following manner.

a. Absorption of a photon by an electron occurs in the valence band of the silver halide. The electron thereby acquires sufficient energy to move into the conduction band (see Section 1.1 for a discussion of solid-state energy levels).

b. The removal of the electron leaves a positive hole in the valence band.

c. Both the electron and the hole are mobile. If they meet again, they will recombine and emit a photon. There will then have been no permanent effect. To avoid this recombination, the electron and the hole must be separated by their involvement in alternative reactions. The electrons are immobilised at chemical or physical traps formed from impurities or crystal defects. The positive hole's motion eventually carries it to the surface of the grain. There it can be consumed directly by reaction with the gelatine, or two holes can combine releasing a halogen atom that will then in turn be absorbed by the gelatine.

d. The electron, now immobilised, neutralises a mobile silver ion leaving a silver atom within the crystal structure.

e. The effect of the original electron trap is now enhanced by the presence of the silver atom and it may more easily capture further electrons. In this way, a speck of pure silver containing from a few to a few hundred atoms is formed within the silver halide crystal.

f. Developers are reducing solutions that convert silver halide into silver. They act only very slowly, however, on pure silver halide, but the specks of silver act as catalysts, so that those grains containing silver specks are reduced rapidly to pure silver. Since a 1-μm grain will contain 10^{10} to 10^{11} silver atoms, this represents an amplification of the latent image by a factor of 10^9. Adjacent grains that do not contain an initial silver speck will be unaffected and will continue to react at the normal slow rate.

g. Thus, the latent image consists of those grains that have developable specks of silver on them. For the most sensitive emulsions, three to six silver atoms per grain will suffice, but higher numbers are required for less sensitive emulsions and specks comprising 10 to 12 silver atoms normally represent the minimum necessary for a stable latent image in normal emulsions.

The latent image is turned into a visible one by developing* for a time which is long enough to reduce the grains containing the specks of silver, but which is not long enough to affect those grains without such specks. The occasional reduction of an unexposed grain occurs through other processes and is a component of chemical fog. After the development has stopped, the remaining silver halide grains are removed by a second chemical solution known as the fixer.

The final visible image thus consists of a number of silver grains dispersed in the emulsion. These absorb light and so the image is darkest at those points where it was most brightly illuminated (i.e. the image is the negative of the original). Astronomers customarily work directly with negatives since the production of a positive image normally requires the formation of a photograph of the negative with all its attendant opportunities for the introduction of distortions and errors. This practice may seem a little strange at first, but the student studying photographic images will rapidly become so familiar with negatives that the occasional positive, when encountered, requires considerable thought for its interpretation.

One of the problems encountered in using photographic images to determine the original intensities of sources is that the response of the emulsion is non-linear. The main features of the response curve, or characteristic curve as it is more commonly known, are shown in Figure 2.7. Photographic density is plotted along the vertical axis. This is more properly called the optical density since it is definable for any partially transparent medium and it is given by

$$D = \log_{10}\left(\frac{F_i}{F_t}\right) \tag{2.19}$$

where F_i is the incident flux and F_t is the transmitted flux (Figure 2.8). The transmittance, T, of a medium is defined similarly by

---

* The developer is a chemical that converts silver halide to silver and which is catalysed by the presence of the elemental silver forming the latent image. Further details of how to process photographic emulsions can generally be obtained from the manufacturers of the emulsions.

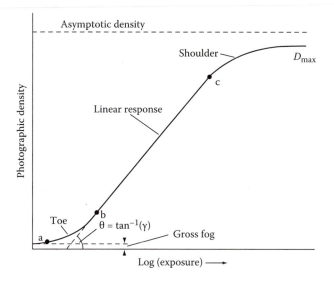
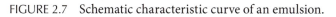

FIGURE 2.7   Schematic characteristic curve of an emulsion.

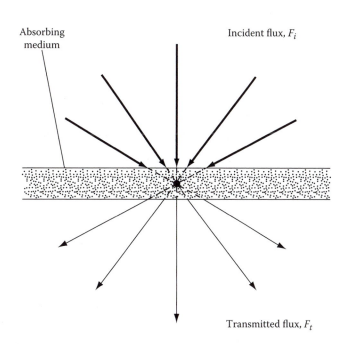

FIGURE 2.8   Quantities used to define photographic density.

$$T = \frac{F_t}{F_i} = 10^{-D} \tag{2.20}$$

and the opacity, A, by

$$A = \frac{F_i}{F_t} = T^{-1} = 10^{D} \tag{2.21}$$

The intensity of the exposure is plotted on the horizontal axis as a logarithm. Since density is also a logarithmic function (Equation 2.19), the linear portion of the characteristic curve (b to c in Figure 2.7) represents a genuinely linear response of the emulsion to illumination. The response is non-linear above and below the linear portion. The gross fog level is the background fog of the unexposed but developed emulsion and of its supporting material. The point marked 'a' on Figure 2.7 is called the threshold and is the minimum exposure required for a detectable image. The asymptotic density is the ultimate density of which the emulsion is capable. The maximum density achieved for a particular developer, D_{max}, will generally be below this value. In practice, the image is ideally formed in the linear portion of the response curve, although this may not always be possible for objects with a very wide range of intensities, or for very faint objects.

The contrast of a film is a measure of its ability to separate regions of differing intensity. It is determined by the steepness of the slope of the linear portion of the characteristic curve. A film with a high contrast has a steep linear section and a low contrast film a shallow one. Two actual measures of the contrast are used: gamma and the contrast index. Gamma is simply the slope of the linear portion of the curve

$$\gamma = \tan \theta \tag{2.22}$$

The contrast index is the slope of that portion of the characteristic curve that is used to form most images. It usually includes some of the toe, but does not extend above a density of 2.0. Contrast and response of emulsions both vary markedly with the type of developer and with the conditions during development. Thus, for most astronomical applications a calibration exposure must be obtained that is developed along with the required photograph and which enables the photographic density to be converted back into intensity.

The resolution of an emulsion is affected by the emulsion structure and by the processing stages. In addition to the grain size and light scattering in the emulsion, there may also be scattering or reflection from the supporting material (Figure 2.9). This is known as halation and it is usually reduced by the addition of an absorbing coating on the back of the film. Processing affects resolution through the change in the concentration of the processing solutions within the emulsion since these are more rapidly consumed in well-exposed regions and almost unused elsewhere. The basic effect of this on the image may best be demonstrated at a sharp boundary between a uniformly exposed region and an unexposed region. The availability of the developer from the unexposed region at the edge

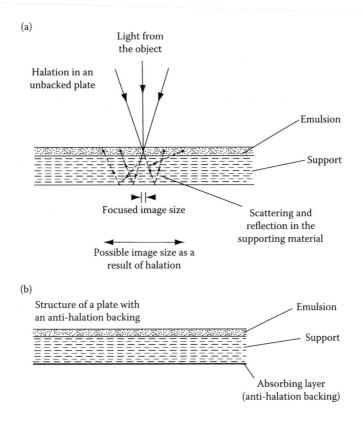

FIGURE 2.9   (a) Halation in an un-backed plate (b) and its suppression by anti-halation backing.

of the exposed region leads to its greater development and so greater density (Figure 2.10). In astronomy this has important consequences for photometry of small images. If two small regions are given exposures of equal length and equal intensity per unit area, then the smaller image will be darker than the larger one by up to 0.3 in density units. This corresponds to an error in the finally estimated intensities by a factor of two, unless account is taken of the effect. A similar phenomenon occurs in the region between two small close images. The developer becomes exhausted by its work within the two images, leaving the region between them underdeveloped. Images of double stars, close spectrum emission lines may therefore appear more widely separated because of this than is the true case.

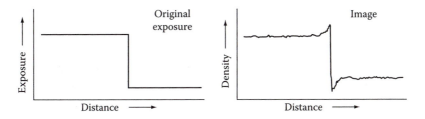

FIGURE 2.10   Edge effect in photographic emulsions.

The quantum efficiency of an emulsion and its speed are two related concepts. We may measure the quantum efficiency by the number of photons required to produce a detectable image compared with an ideal detector (see also Section 1.1). Now all detectors have a certain level of background noise and we may regard a signal as detectable if it exceeds the noise level by one standard deviation of the noise. In a photographic image the noise is due to the granularity of the image and the background fog. For a Poisson variation in granularity, the standard deviation of the noise is given by

$$\sigma = (N\omega)^{1/2} \tag{2.23}$$

where σ is the noise equivalent power (see Section 1.1), N is the number of grains in the exposed region and ω is the equivalent background exposure in terms of photons per grain. (All sources of fogging may be regarded as due to a uniform illumination of a fog-free emulsion. The equivalent background exposure is therefore the exposure required to produce the gross fog level in such an emulsion.) Now for our practical and non-ideal detector let the actual number of photons per grain required for a detectable image be Ω. The detective quantum efficiency (see Section 1.1) is then given by

$$DQE = \left(\frac{\sigma}{N\Omega} \right)^2 \tag{2.24}$$

or

$$DQE = \left(\frac{\omega}{N\Omega^2} \right) \tag{2.25}$$

A given emulsion may further reduce its quantum efficiency by failing to intercept all the photons that fall onto it. Also, between two and 20 photons are required to be absorbed before any given grain reaches developability, as we have already seen. So the quantum efficiency rarely rises above 1% in practice (Figure 2.11).

The speed of an emulsion and its quantum efficiency are related since the speed is a measure of the exposure required to reach a given density in the image. Aspects of the speed have already been mentioned (Figure 2.5 and its discussion) and it is inherent in the position of the characteristic curve along the horizontal axis (Figure 2.7). It can also be affected by processing; the use of more active developers and/or longer developing times usually increases the speed. Other factors that affect the speed include temperature, humidity, the spectral region being observed, the age of the emulsion.

One important factor for astronomical photography that influences the speed very markedly is the length of the exposure. The change in the speed for a given level of total illumination as the exposure length changes is called reciprocity failure. If an emulsion always gave the same density of image for the same number of photons, then the exposure

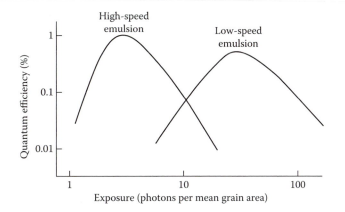

FIGURE 2.11   Quantum efficiency curves for photographic emulsions.

time would be proportional to the reciprocal of the intensity of the illumination. However, for very long and very short exposures this reciprocal relationship breaks down; hence the name reciprocity failure. A typical example of the variation caused by exposure time is shown in Figure 2.12. Here the total exposure (i.e. Intensity × Time, IT) that is required to produce a given density in the final image is plotted against the illumination. Exposure times are also indicated. If the reciprocity law held, then the graph would be a horizontal straight line. The effect of reciprocity failure is drastic; some commonly available roll films for example may have a speed of a few per cent of their normal ratings for exposures of an hour or more.

The photographic emulsion is a weak and easily damaged material and it must therefore have a support. This is usually either a glass plate or a plastic film. Both have been used in astronomy, but plates are preferred for applications requiring high dimensional stability and/or retention for comparative purposes for long periods of time. In both cases great care must be taken in the storage of the material to avoid scratches, high humidity, mould,

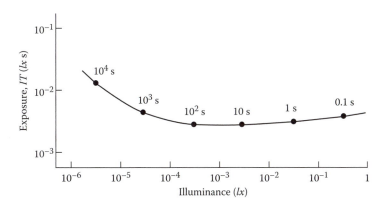

FIGURE 2.12   A typical reciprocity curve for a photographic emulsion.

breakages, distortion that can all too easily destroy the usefulness of the information in the photograph.

2.2.4 Techniques of Astronomical Photography

The prime difference between astronomical photography and more normal types of photography lies in the exposure length. Because of this, not only must the telescope be driven to follow the motion of the stars, but it must also be continually guided in order to correct any errors in the drive. Usually guiding is by means of a second and smaller telescope attached to the main instrument. The astronomer guides the telescope by keeping cross wires in the eyepiece of the secondary telescope centred on the object, or by the use of automatic systems (see Section 1.1). For faint and/or diffuse objects, offset guiding may be necessary. That is, the guide telescope is moved from its alignment with the main telescope until it is viewing a nearby bright star whilst the main telescope continues to point at the required object. Guiding then continues using this bright star. Sometimes the main telescope can act as its own guide by using a guider that inserts a small mirror into the light beam to divert light not producing the main image into the eyepiece.

The determination of the required length of the exposure may be difficult; exposures may be interrupted by cloud, or the star may be a variable whose brightness is not known accurately. Previous experience may provide a guide, and records of all exposures, whether successful or not, should always be kept for this reason. Alternatively, several photographs may be taken with differing exposures. The linear part of the characteristic curve is sufficiently long that the exposures can differ by a factor of two or even four if this approach is tried.

Amongst amateur astronomers, a commonly used technique is to take photographs by eyepiece projection instead of at prime focus. A single-lens reflex (SLR) camera without its normal lens is positioned a little distance behind the eyepiece and the focus adjusted to produce a sharp image on the film. This provides much higher plate scales so that even small objects such as the planets can be resolved with a small telescope. A guide to the extra magnification of eyepiece projection over prime focus photography and to the change in the exposure that may be required is given by the effective focal ratio

$$EFR = \frac{fd}{e} \tag{2.26}$$

where f is the focal ratio of the objective, d is the projection distance (i.e. the distance from the optical centre of the eyepiece to the film plane) and e is the focal length of the eyepiece. If T is the time required for an exposure on an extended object at prime focus, then T', the exposure time for the projected image, is given by

$$T' = T \left(\frac{EFR}{f} \right)^2 \tag{2.27}$$

It is also possible just to point a camera through the eyepiece of a telescope to obtain images. Both the camera and the telescope need to be focused on infinity and an SLR or a digital camera with a real-time display is essential so that the image can be focussed and positioned correctly. This technique is sometimes called magnified imaging and is a very poor second to either direct or eyepiece projection imaging.

Modern commercial digital cameras usually have autofocusing systems which may be either active or passive in nature. Active systems send out a beam of sound or infrared light in order to focus the camera. They are unlikely to work if used for magnified imaging. Passive focussing, however, operates by maximising the contrast in the image and so cameras using this approach (usually the more expensive examples) usually will work if used for magnified imaging.

The observational limit to the length of exposure times is usually imposed by the background sky brightness. This produces additional fog on the emulsion and no improvement in the signal-to-noise ratio will occur once the total fog has reached the linear portion of the characteristic curve. Since the brightness of star images is proportional to the square of the objective's diameter, whilst that of an extended object, including the sky background, is inversely proportional to the square of the focal ratio (see Section 1.1), it is usually possible to obtain adequate photographs of star fields by using a telescope with a large focal ratio. But for extended objects no such improvement is possible since their brightness scales in the same manner as that of the sky background.

Much of the light in the sky background comes from scattered artificial light, so that observatories are normally sited in remote spots well away from built-up areas. Even so, a fast Schmidt camera may have a photographic exposure limit of a few minutes at best. Some further improvement may sometimes then be gained by the use of filters. The scattered artificial light comes largely from sodium and mercury vapour street lighting, so a filter that rejects the most important of the emission lines from these sources can improve the limiting magnitude by two or three stellar magnitudes at badly light-polluted sites. For the study of hot interstellar gas clouds, a narrowband filter centred on the $H\alpha$ line will almost eliminate the sky background whilst still allowing about half the energy from the source to pass through it.

2.2.5 Analysis of Photographic Images

Photographs are used by the astronomer to provide two main types of information: the relative positions of objects and the relative intensities of objects. The analysis for the first of these is straightforward, although care is needed for the highest quality results. The positions of the images are measured by means of a machine that is essentially a highly accurate automated travelling microscope (see Section 5.1).

To determine the relative intensities of sources from their photographs is more of a problem due to the non-linear nature of the emulsion's characteristic curve (Figure 2.7). In order to convert back from the photographic density of the image to the intensity of the source, this curve must be known with some precision. It is obtained by means of a photometric calibration exposure, which is a photograph of a series of sources of differing and known intensities. The characteristic curve is then plotted from the measured densities of the images of these sources. To produce accurate results, great care must be exercised, such as

a. The photometric calibration film must be from the same batch of film as that used for the main exposure

b. The treatment of the two films, including their storage, must be identical

c. Both films must be processed together

d. The exposure lengths, intermittency effects, temperatures during exposures should be as similar as possible for the two plates

e. If the main exposure is for a spectrum, then the calibration exposure should include a range of wavelengths as well as intensities

f. Both plates must be measured under identical conditions

With these precautions, a reasonably accurate characteristic curve may be plotted. It is then a simple although tedious procedure to convert the density of the main image back to intensity. The measurement of a plate is undertaken on a microdensitometer. This is a machine that shines a small spot of light through the image and measures the transmitted intensity. Such machines may readily be connected to a small computer and the characteristic curve also fed into it, so that the conversion of the image to intensity may be undertaken automatically.

Photometry of the highest accuracy is no longer attempted from photographs for stars, but a rough guide to their brightnesses may be obtained from the diameters of their images, as well as by direct microphotometry. Scattering, halation, diffraction rings around star images mean that bright stars have larger images than faint stars. If a sequence of stars of known brightnesses is on the plate, then a calibration curve may be plotted and estimates of the magnitudes of large numbers of stars made very rapidly.

2.3 ELECTRONIC IMAGING

2.3.1 Introduction

The alternatives to photography for recording images directly are almost all electronic in nature. They have two great advantages over the photographic emulsion. First, the image is usually produced as an electrical signal and can therefore be relayed or transmitted to a remote observer, which is vital for satellite-borne instrumentation and useful in many other circumstances and also enables the data to be fed directly into a computer. Second, with many of the systems the quantum efficiency is up to a factor of a hundred or so higher than that of the photographic plate. Subsidiary advantages can include intrinsic amplification of the signal, linear response and long or short wavelength sensitivity.

The most basic form of electronic imaging simply consists of an array of point-source detecting elements. The method is most appropriate for the intrinsically small detectors such as photoconductive cells (see Section 1.1). Hundreds or thousands or even more of the individual elements (usually termed pixels) can then be used to give high spatial resolution. The array is simply placed at the focus of the telescope or spectroscope in place of the photographic plate. Other arrays such as CCDs, infrared arrays and STJs were reviewed in detail in Section 1.1. Millimetre-wave and radio arrays were also considered in Sections 1.1 and 1.2. Here, therefore, we are concerned with other electronic approaches to imaging, most of which have been superseded by array detectors, but which have historical and archival interest.

2.3.2 Television and Related Systems

Low-light-level television systems, combined with image intensifiers and perhaps using a slower scan rate than normal, have been used in the past. They were particularly found on the guide systems of large telescopes where they enable the operator at a remote-control console to have a viewing/finding/guiding display. Also, many of the early planetary probes and some other spacecraft such as the International Ultra-violet Observer used TV cameras of various designs. These systems have now been replaced by solid-state arrays.

2.3.3 Image Intensifiers

Microchannel plates are a form of image intensifier (see Section 1.3). The term is usually reserved for the types of devices that produce an intensified image visible to the eye. Although used in the past for astronomical purposes, they are now rarely encountered and are mostly found as night sights for the military and for wildlife observers. Details may be found in earlier editions of this book if required.

2.3.4 Photon Counting Imaging Systems

A combination of a high-gain image intensifier and TV camera led in the 1970s to the development by Alec Boksenberg and others of the image photon counting system (IPCS). In the original IPCS, the image intensifier was placed at the telescope's (or spectroscope's) focal plane and produced a blip of some 10^7 photons at its output stage for each incoming photon in the original image. A relatively conventional TV camera then viewed this output and the video signal fed to a computer for storage. Again, this type of device is now superseded by solid-state detectors.

More recently, a number of variants on Boksenberg's original IPCS have been developed. These include microchannel plate image intensifiers linked to CCDs and other varieties of image intensifiers linked to CCDs or other array-type detectors. The principles of the devices remain unchanged however.

2.4 SCANNING

This is such an obvious technique as scarcely to appear to deserve a separate section and indeed at its simplest it hardly does so. A two-dimensional image may be built up using any point source detector if that detector is scanned over the image or vice versa. Many images at all wavelengths are obtained in this way. Sometimes the detector moves, sometimes the whole telescope or satellite. Occasionally it may be the movement of the secondary mirror of the telescope or of some *ad hoc* component in the apparatus which allows the image to be scanned. Scanning patterns are normally raster or spiral (Figure 2.13). Other patterns may be encountered, however, such as the continuously nutating roll employed by some of the early artificial satellites. The only precautions advisable are the matching of the scan rate to the response of the detector or to the integration time being used and the matching of the separation of the scanning lines to the resolution of the system. Scanning by radio telescopes (see Section 1.2) used to be by observing individual points within the scan pattern for a set interval of time. More recently, the scanning has been continuous, or 'on the fly', since this enables slow changes due to the atmosphere or the instrument to be eliminated. Specialised applications of scanning occur in the spectrohelioscope (see Section 5.3) and the scanning spectrometer (see Section 4.2). Many Earth observation satellites use push-broom scanning. In this system a linear array of detectors is aligned at right angles to the spacecraft's ground track. The image is then built up as the spacecraft's motion moves the array to look at successive slices of the swathe of ground over which the satellite is passing.

A more sophisticated approach to scanning is to modulate the output of the detector by interposing a mask of some type in the light beam. Examples of this technique are discussed elsewhere and include the modulation collimator and the coded array mask used in x-ray imaging (Section 1.3). A very great improvement over either these methods or the basic approach discussed above may be made by using a series of differing masks, or by scanning a single mask through the light beam so that differing portions of it are utilised. This improved method is known as Hadamard mask imaging. We may best illustrate its principles by considering one-dimensional images such as might be required for spectrophotometry. The optical arrangement is shown in Figure 2.14. The mask is placed in the image plane of the telescope and the Fabry lens directs all the light that passes through the mask onto the single detector. Thus, the output from the system consists of a simple measurement of the intensity passed by the mask. If a different mask is now substituted for the first, then a new and in general different intensity reading will be obtained. If the image is to be resolved into N elements, then N such different masks must be used and N intensity readings determined. If $\mathbf{D}$ is the vector formed from the detector output readings, $\mathbf{I}$ is the vector of the intensities of the elements of the image and $\mathbf{M}$ the $N \times N$ matrix whose

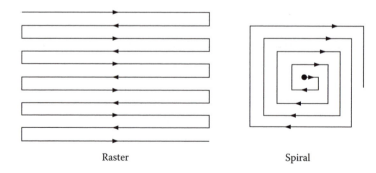

Raster   Spiral

FIGURE 2.13   Scanning patterns.

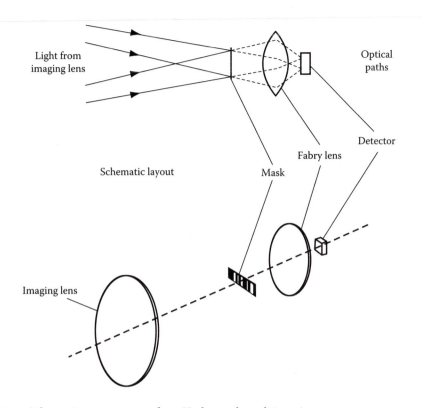

FIGURE 2.14　Schematic arrangement for a Hadamard mask imaging system.

columns each represent one of the masks, with the individual elements of the matrix corresponding to the transmissions of the individual segments of the mask; that is

$$\mathbf{D} = [D_1, D_2, D_3, \ldots D_N] \tag{2.28}$$

$$\mathbf{I} = [I_1, I_2, I_3, \ldots I_N] \tag{2.29}$$

$$M = \begin{bmatrix} m_{11} & m_{12} & m_{13} & \ldots & m_{1N} \\ m_{21} & & & & \\ m_{31} & & & & \\ \cdot & & & & \\ \cdot & & & & \\ \cdot & & & & \\ m_{N1} & & & m_{NN} \end{bmatrix} \tag{2.30}$$

Then, ignoring any contribution arising from noise, we have

$$\mathbf{D} = \mathbf{I}\,\mathbf{M} \tag{2.31}$$

and so

$$I = D \, M^{-1} \tag{2.32}$$

Thus, the original image is simply obtained by inverting the matrix representing the masks. The improvement of the method over a simple scan lies in its multiplex advantage (cf. the Fourier transform spectrometer, Section 4.1). The masks usually comprise segments that either transmit or obscure the radiation completely; that is

$$m_{ij} = 0 \text{ or } 1 \tag{2.33}$$

and on average, about half of the total image is obscured by a mask. Thus, $N/2$ image segments contribute to the intensity falling onto the detector at any one time. Hence, if a given signal-to-noise ratio is reached in a time T, when the detector observes a single image element, then the total time required to detect the whole image with a simple scan is

$$N \times T \tag{2.34}$$

and is approximately

$$\sqrt{2NT} \tag{2.35}$$

for the Hadamard masking system. Thus, the multiplex advantage is approximately a factor of

$$\sqrt{\frac{N}{2}} \tag{2.36}$$

improvement in the exposure length.

In practice, moving a larger mask across the image generates the different masks. The matrix representing the masks must then be cyclic, or in other words, each successive column is related to the previous one by being moved down a row. The use of a single mask in this way can lead to a considerable saving in the construction costs, since $2N - 1$ segments are needed for it compared with N^2 if separate masks are used. An additional constraint on the matrix **M** arises from the presence of noise in the output. The errors introduced by this are minimised when

$$Tr \, [M^{-1} \, (M^{-1})^T] \tag{2.37}$$

is minimised. There are many possible matrices that will satisfy these two constraints, but there is no completely general method for their generation. One method of finding suitable matrices and so of specifying the mask is based upon the group of matrices known as the

Hadamard matrices (hence the name for this scanning method). These are matrices whose elements are ±1s and which have the property

$$\mathbf{H}\,\mathbf{H}^T = N\,\mathbf{I} \tag{2.38}$$

where $\mathbf{H}$ is the Hadamard matrix and $\mathbf{I}$ is the identity matrix. A typical example of the mask matrix, $\mathbf{M}$, obtained in this way might be

$$\mathbf{M} = \begin{bmatrix} 0 & 1 & 0 & 1 & 1 & 1 & 0 \\ 0 & 0 & 1 & 0 & 1 & 1 & 1 \\ 1 & 0 & 0 & 1 & 0 & 1 & 1 \\ 1 & 1 & 0 & 0 & 1 & 0 & 1 \\ 1 & 1 & 1 & 0 & 0 & 1 & 0 \\ 0 & 1 & 1 & 1 & 0 & 0 & 1 \\ 1 & 0 & 1 & 1 & 1 & 0 & 0 \end{bmatrix} \tag{2.39}$$

for N having a value of 7.

Variations to this scheme can include making the obscuring segments out of mirrors and using the reflected signal as well as the transmitted signal, or arranging for the opaque segments to have some standard intensity rather than zero intensity. The latter device is particularly useful for infrared work. The scheme can also easily be extended to more than one dimension, although it then rapidly becomes very complex. A two-dimensional mask may be used in a straightforward extension of the one-dimensional system to provide two-dimensional imaging. A two-dimensional mask combined suitably with another one-dimensional mask can provide data on three independent variables – for example, spectrophotometry of two-dimensional images. Two two-dimensional masks could then add polarimetry to this and so on, except in the one-dimensional case, the use of a computer to unravel the data is obviously essential, and even in the one-dimensional case it will be helpful as soon as the number of resolution elements rises above five or six.

2.5 INTERFEROMETRY

2.5.1 Introduction

Interferometry is the technique of using constructive and destructive addition of radiation to determine information about the source of that radiation. Interferometers are also used to obtain high-precision positions for optical and radio sources and this application is considered in Section 5.1. Two principal types of interferometer exist: the Michelson stellar interferometer, first proposed by Armand Fizeau in 1868 and the intensity interferometer proposed by Robert Hanbury Brown in 1949. There are also a number of subsidiary techniques such as amplitude interferometry, nulling interferometry, speckle interferometry, Fourier spectroscopy, intensity interferometry. The latter until recently has only been of historical interest; however, intensity interferometry may soon be revived to map stellar surfaces using γ-ray air Čerenkov telescopes.

2.5.2 Michelson Optical Stellar Interferometer

The Michelson stellar interferometer is so called in order to distinguish it from the type of interferometer used in the Michelson-Morley experiment, gravitational wave interferometers and in Fourier spectroscopy. Since the latter type of interferometer is discussed primarily in Sections 1.6 and 4.1, we shall drop the 'stellar' qualification for this section when referring to the first of the two main types of interferometer.

In practice, many interferometers use the outputs from numerous telescopes; however, this is just to reduce the time taken for the observations and it is actually the outputs from pairs of telescopes that are combined to produce the interference effects. We may start therefore by considering the interference effects arising from two beams of radiation.

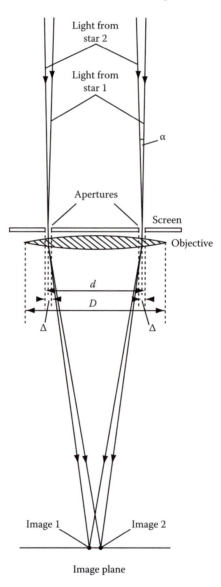

FIGURE 2.15   Optical arrangement of a Michelson interferometer.

The complete optical system that we shall study is shown in Figure 2.15. The telescope objective, which for simplicity is shown as a lens (but exactly the same considerations apply to mirrors), is covered by an opaque screen with two small apertures and is illuminated by two monochromatic equally bright point sources at infinity. The objective diameter is D, its focal length F, the separation of the apertures is d, the width of the apertures is Δ, the angular separation of the sources is α and their emitted wavelength is λ.

Let us first consider the objective by itself, without the screen and with just one of the sources. The image structure is then just the well-known diffraction pattern of a circular lens (Figure 2.16). With both sources viewed by the whole objective, two such patterns are superimposed. There are no interference effects between these two images, for their radiation is not mutually coherent. When the main maxima are superimposed upon the first minimum of the other pattern, we have Rayleigh's criterion for the resolution of a lens (Figures 1.41 and 2.17). The Rayleigh criterion for the minimum angle between two separable sources α', as we saw in Section 1.1, is

$$\alpha' = \frac{1.22\lambda}{D} \tag{2.40}$$

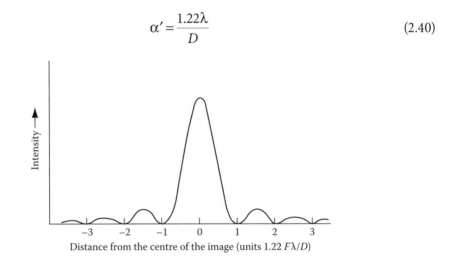

Distance from the centre of the image (units 1.22 $F\lambda/D$)

FIGURE 2.16   Image structure for one point source and the whole objective.

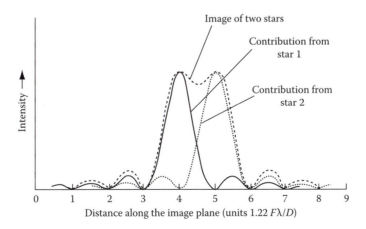

Distance along the image plane (units 1.22 $F\lambda/D$)

FIGURE 2.17   Image structure for two point sources and the whole objective.

Now let us consider the situation with the screen in front of the objective and first consider just one aperture looking at just one of the sources. The situation is then the same as that illustrated in Figure 2.16, but the total intensity and the resolution are both reduced because of the smaller size of the aperture compared with the objective (Figure 2.18). Although the image for the whole objective is also shown for comparison in Figure 2.18, if it were truly to scale for the situation illustrated in Figure 2.15, then it would be one seventh of the width that is shown and 1800 times higher.

Now consider what happens when one of the sources is viewed simultaneously through both small apertures. If the apertures were infinitely small, then ignoring the inconvenient fact that no light would get through anyway, we would obtain a simple interference pattern (Figure 2.19). The effect of the finite width of the apertures is to modulate the straightforward variation of Figure 2.19 by the shape of the image for a single aperture (Figure 2.18). Since two apertures are now contributing to the intensity and the energy 'lost' at the minima reappears at the maxima, the overall envelope of the image peaks at four times the intensity for a single aperture (Figure 2.20). Again, for the actual situation shown in Figure 2.15, there should be a total of 33 fringes inside the major maximum of the envelope.

Finally, let us consider the case of two equally bright point sources viewed through two apertures. Each source has an image whose structure is that shown in Figure 2.20 and these simply add together in the same manner as the case illustrated in Figure 2.17 to form the combined image. The structure of this combined image will depend upon the separation of the two sources. When the sources are almost superimposed, the image structure will be identical with that shown in Figure 2.20, except that all the intensities will have doubled. As the sources move apart, the two fringe patterns will also separate, until when the sources are separated by an angle α'', given by

$$\alpha'' = \frac{\lambda}{2d} \tag{2.41}$$

the maxima of one fringe pattern will be superimposed upon the minima of the other and vice versa. The fringes will then disappear and the image will be given by their envelope (Figure 2.21). There may still be a very slight ripple on this image structure due to the incomplete filling of the minima by the maxima, but it is unlikely to be noticeable. The

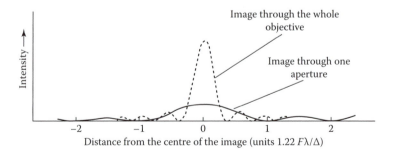

FIGURE 2.18 Image structure for one point source and one aperture.

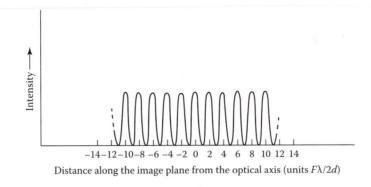

FIGURE 2.19 Image structure for a single point source viewed through two infinitely small apertures.

fringes will reappear as the sources continue to separate until the pattern is almost double that of Figure 2.20 once again. The image patterns are then separated by a whole fringe width and the sources by $2\alpha''$. The fringes disappear again for a source separation of $3\alpha''$ and reach yet another maximum for a source separation of $4\alpha''$ and so on. Thus, the fringes are most clearly visible when the source's angular separation is given by $2n\alpha''$ (where n is an integer) and they disappear or reach a minimum in clarity for separations of $(2n + 1)\alpha''$. Applying the Rayleigh criterion for resolution, we see that the resolution of two apertures

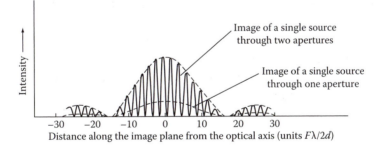

FIGURE 2.20 Image structure for one point source viewed through two apertures.

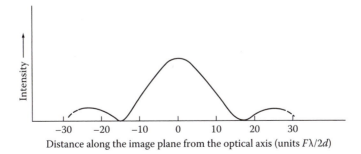

FIGURE 2.21 Image structures for two point sources separated by an angle, $\lambda/2d$, viewed through two apertures.

is given by the separation of the sources for which the two fringe patterns are mutually displaced by half a fringe width. This as we have seen is simply the angle α″ and so we have

$$\frac{\text{Resolution through two apertures}}{\text{Resolution of the objective}} = \frac{\alpha''}{\alpha'} \tag{2.42}$$

$$= \frac{2.44d}{D} \tag{2.43}$$

Imagining the two apertures placed at the edge of the objective (i.e. $d = D$), we see the quite remarkable result that the resolution of an objective may be increased by almost a factor of two and a half by screening it down to two small apertures at opposite ends of one of its diameters. The improvement in the resolution is only along the relevant axis – perpendicular to this, the resolution is just that of one of the apertures. The resolution in both axes may, however, be improved by using a central occulting disc which leaves the rim of the objective clear. This also enables us to get a feeling for the physical basis for this improvement in resolution. We may regard the objective as composed of a series of concentric narrow rings, each of which has a resolution given by Equation 2.41, with d being the diameter of the ring. Since the objective's resolution is the average of all these individual resolutions, it is naturally less than their maximum. In practice, some blurring of the image may be detected for separations of the sources that are smaller than the Rayleigh limit (see Section 1.1). This blurring is easier to detect for the fringes produced by the two apertures than it is for the images through the whole objective. The effective improvement in the resolution may therefore be even larger than that given by Equation 2.43.

Now for the situation illustrated in Figure 2.15, the path difference between the two light beams on arriving at the apertures is likely to be small, since the screen will be perpendicular to the line of sight to the objects. In order to produce fringes with the maximum clarity however, that path difference must be close to zero. This is because radiation is never completely monochromatic; there is always a certain range of wavelengths present known as the bandwidth of the signal. Now for a zero path difference at the apertures, all wavelengths will be in phase when they are combined and will interfere constructively. However, if the path difference is not zero, some wavelengths will be in phase but others will be out of phase to a greater or lesser extent, since the path difference will equal different numbers of cycles or fractions of cycles at the different wavelengths. There will thus be a mix of constructive and destructive interference and the observed fringes will have reduced contrast. The path difference for which the contrast in the fringes reduces to zero (i.e. no fringes are seen) is called the coherence length, l, and is given by

$$l = \frac{c}{\Delta \nu} = \frac{\lambda^2}{\Delta \lambda} \tag{2.44}$$

where $\Delta \nu$ and $\Delta \lambda$ are the frequency and wavelength bandwidths of the radiation. So that for $\lambda = 500$ nm and $\Delta \lambda = 1$ nm, we have a coherence length of 0.25 mm, for white light

($\Delta\lambda \approx 300$ nm) this reduces to less than a micron, but in the radio region it can be large; 30 metres, for example, at $\nu = 1.5$ GHz and $\Delta\nu = 10$ MHz. However, as we shall, the main output from an interferometer is the fringe contrast (usually known as the fringe visibility, V; Equation 2.45), so that in order for this not to be degraded, the path difference to the apertures, or their equivalent, must be kept to a small fraction of the coherence length. For interferometers such as Michelson's 1921 stellar interferometer (Figure 2.24), the correction to zero path difference is small and in that case was accomplished through the use of adjustable glass wedges. However, most interferometers, whether operating in the optical or radio regions, now use separate telescopes on the ground (Figures 2.27 and 2.35 for example), so the path difference to the telescopes can be over 100 metres for optical interferometers and up to thousands of kilometres for very-long-baseline radio interferometry. These path differences have to corrected either during the observation by hardware, or afterwards during the data processing.

Michelson's original interferometer was essentially identical to that shown in Figure 2.15, except that the apertures were replaced by two movable parallel slits to increase the amount of light available and a narrowband filter was included to give nearly monochromatic light (earlier we had assumed that the *sources* were monochromatic). To use the instrument to measure the separation of a double star with equally bright components, the slits are aligned perpendicularly to the line joining the stars and moved close together, so that the image fringe patterns are undisplaced and the image structure is that of Figure 2.20. The actual appearance of the image in the eyepiece at this stage is shown in Figure 2.22. The slits are then moved apart until the fringes disappear (Figures 2.21 and 2.23). The distance between the slits is then such that the fringe pattern from one star is filling-in the fringe pattern from that other and their separation is given by α'' (Equation 2.41, with d given by the distance between the slits). If the two stars are of differing brightnesses, then the fringes will not disappear completely, but the fringe visibility, V, is given by

$$V = \frac{(I_{max} - I_{min})}{(I_{max} + I_{min})} \tag{2.45}$$

FIGURE 2.22 Appearance of the image of a source seen in a Michelson interferometer at the instant of maximum fringe visibility.

FIGURE 2.23   Appearance of the image of a source seen in a Michelson interferometer at the time of zero fringe visibility.

where I_{max} is the intensity of a fringe maximum and I_{min} is the intensity of a fringe minimum, and will reach a minimum at the same slit distance.

To measure the diameter of a symmetrical object like a star or the satellite of a planet, the same procedure is used. However, the two sources are now the two halves of the star's disk, represented by point sources at the optical centres* of the two semicircles. The diameter of the star is then given by 2.44 α'', so that in determining the diameters of objects, an interferometer has the same resolution as a conventional telescope with an objective diameter equal to the separation of the slits.

Very stringent requirements on the stability and accuracy of the apparatus are required for the success of this technique. As we have seen, the path difference to the slits must not be greater than a small fraction of the coherence length of the radiation (Equation 2.44). Furthermore, that path difference must remain constant to considerably better than the wavelength of the radiation that is being used, as the slits are separated. Vibrations and scintillation are additional limiting factors. In general, the paths will not be identical when the slits are close together and the fringes are at their most visible. But the path difference will then be some integral number of wavelengths and provided that it is significantly less than the coherence length the interferometer may still be used. Thus, in practice, the value that is required for d is the difference between the slit separations at minimum and maximum fringe visibilities. A system such as has just been described was used by Michelson in 1891 to measure the diameters of the Galilean satellites. Their diameters are in the region of 1 second of arc, so that d has a value of a few tens of millimetres.

Now the angular diameters of stellar disks are very much smaller than those of the Galilean satellites; 0.047″ for α Orionis (Betelgeuse) for example, one of the largest angular diameter stars in the sky. A separation for the apertures of several metres is thus needed if stellar diameters are to be measured. This led to the improved version of the system that is rather better known than the one used for the Galilean satellites and which is the type of interferometer that is usually intended when the Michelson stellar interferometer is mentioned. It was used by Albert Michelson and Francis Pease in 1921 to measure the diameters of half a dozen or so of the larger stars. The slits were replaced by movable mirrors on a

---

* The same as their centres of gravity if they are thin, flat, uniform semicircles.

rigid 6-metre-long bar mounted on top of the 2.5-metre Hooker telescope (Figure 2.24). By this means, *d* could be increased beyond the diameter of the telescope. A system of mirrors then reflected the light into the telescope. The practical difficulties with the system were very great and although larger systems have since been attempted none has been successful and so the design is now largely of historical interest only.

Modern optical interferometers are based upon separate telescopes rather than two apertures feeding a single instrument. They are still primarily used to measure stellar diameters, although some aperture synthesis systems are now working and others are under construction. There is no difference in the principle of operation of these interferometers from that of Michelson, but there are many practical changes. The main change arises because the path difference to the two telescopes can now be many metres. A system for compensating for the path difference is thus required. In most designs the path compensation is via a delay line; the light from one telescope is sent on a longer path than that from the other one before they are mixed together (Figure 2.25). The extra path length is arranged to be equal to the path difference between the telescopes. Since the path difference will change as the object moves across the sky, the delay must also change. This is usually accomplished by having one or more of the mirrors directing the light into the longer path mounted on a moveable carriage. Even using narrow bandwidths, the small value of the coherence length (Equation 2.44) at optical wavelengths means that the carriage

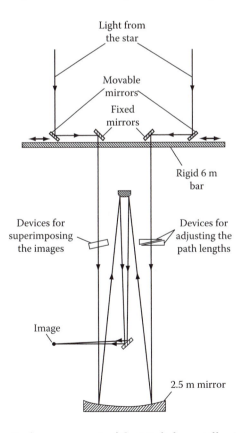

FIGURE 2.24 Schematic optical arrangement of the Michelson stellar interferometer.

FIGURE 2.25   One of the retro-reflectors making up a delay line for the VLTI. The retro-reflector
is essentially a small Cassegrain telescope in which the incoming light beam is reflected from the
secondary mirror back to be reflected from the primary mirror a second time. The light beam is
thus returned back to the other end of the delay line. The retro-reflector is mounted on a wheeled
carriage which can move on rails over a 60-metre distance and so provide a maximum delay line
length of 120 metres. The required positioning accuracy of the reflector is about 50 nm and this is
achieved by a combination of movement of the carriage and piezoelectric transducers monitored
by a laser measuring system. (Reproduced by kind permission of ESO.)

must be capable of being positioned to submicron accuracy and yet be able to travel tens
of metres.

The second major difference is that the telescopes are usually fixed in position or moveable only slowly between observations. It is thus not possible to change their separation
in order to observe maximum and minimum fringe visibilities. A change in the effective
separation occurs as the angle between the baseline and the source in the sky alters, but
this is generally insufficient to range from maximum to minimum fringe visibility. Thus,
instead a measurement of the fringe visibility is made for just one, or a small number, of
separations. This measurement is then fitted to a theoretical plot of the variation of the
fringe visibility with the telescope separation and the size of the source in order to determine the latter.

The third difference arises from the atmospheric turbulence. We have seen that there
is an upper limit to the size of a telescope aperture, given by Fried's coherence length
(Equation 1.79). In the visible this is only about 120 mm, though in the infrared it can
rise to 300 mm. Telescopes with apertures larger than Fried's limit will thus be receiving
radiation through several, perhaps many, atmospheric cells, each of which will have different phase shifts and wavefront distortions (Figure 1.67). When the beams are combined,

the desired fringes will be washed out by all these differing contributions. Furthermore, the atmosphere changes on a time scale of around 5 ms, so that even if small apertures are used, exposures have to be short enough to freeze the atmospheric motion (see also speckle interferometry, Section 2.6). Thus, many separate observations have to be added together to get a value with a sufficiently high signal-to-noise ratio. In order to use large telescopes as a part of an interferometer, their images have to be corrected through adaptive optics (see Section 1.1). Even a simple tip–tilt correction will allow the useable aperture to be increased by a factor of three. However, in order to use 8- and 10-metre telescopes, such as ESO's VLT and the Keck telescopes, full atmospheric correction is needed.

When only two telescopes form the interferometer, the interruption of the phase of the signal by the atmosphere results in the loss of that information. Full reconstruction of the morphology of the original object is then not possible; only the fringe visibility can be determined (see below for a discussion of multielement interferometers, closure phase and aperture synthesis). However, the fringe visibility still enables stellar diameters and double star separations to be found and so several two-element interferometers either have been until recently or are currently in operation.

Nulling interferometry can be undertaken with two-element interferometers as well as those with more elements. It has recently come into prominence as a way of detecting exoplanets. The interferometer is adjusted so that the stellar fringe pattern is at a maximum (Figure 2.22). The exoplanet's fringe pattern will not then, in general, simultaneously be at a fringe maximum since the exoplanet is separated by a few microarc to milliarc seconds from the star in the sky. The exoplanet's fringe pattern can therefore be studied separately from that of the star.

Sydney University Stellar Interferometer (SUSI) in Australia operates in the visible region with two 0.20-metre mirrors producing 0.14-metre light beams. It has recently been upgraded and can operate with baselines ranging from 5 to 160 metres over the 400- to 950-nm region giving it a best resolution of about 1 milliarc second. It has EMCCDs as detectors and uses monochromators to narrow the bandwidth down to a few nanometres. The LBT interferometer has recently been commissioned and it operates as a nulling instrument in the MIR with an effective 14.4-metre baseline giving it a resolution of 100 milliarc seconds. The two 10-metre Keck telescopes have operated as a two-element NIR nulling interferometer in the past. However, the additional four 1.8-metre outrigger telescopes designed to make the instruments into a full interferometer array have never been installed and the two main telescopes now work independently.

A variation on the system that also has some analogy with intensity interferometry is known as amplitude interferometry. This has been used recently in a successful attempt to measure stellar diameters. At its heart is a device known as a Köster prism that splits and combines the two separate light beams (Figure 2.26). The interfering beams are detected in a straightforward manner using photomultipliers. Their outputs may then be compared. The fringe visibility of the Michelson interferometer appears as the anticorrelated component of the two signals, whilst the atmospheric scintillation affects the correlated component. Thus, the atmospheric effects can be screened out to a large extent and the stability of the system is vastly improved.

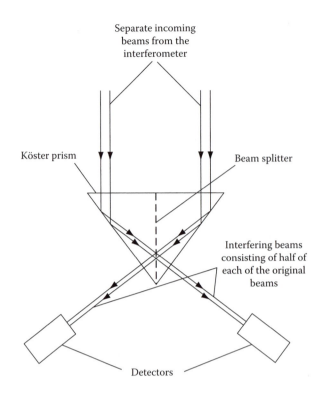

FIGURE 2.26   The Köster prism.

2.5.3 Michelson Radio Interferometer

Radio aerials are generally only a small number of wavelengths in diameter (see Section 1.2) and so their intrinsic resolution (Equation 1.34) is poor. The use of interferometers to obtain higher resolution therefore dates back to soon after the start of radio astronomy. There is no difference in principle between a radio and an optical interferometer when they are used to measure the separation of double sources or the diameters of uniform objects. However, the output from a radio telescope contains both the amplitude and phase of the signal, so that complete imaging of the source is possible. Radio interferometers generally use more than two antennae and these may be in a two-dimensional array. The basic principle, however, is the same as for just two antennae, although the calculations become considerably more involved.

The individual elements of a radio interferometer are usually fairly conventional radio telescopes (see Section 1.2) and the electrical signal that they output varies in phase with the received signal. In fact with most radio interferometers it is the electrical signals that are mixed to produce the interference effect rather than the radio signals themselves. Their signals may then be combined in two quite different ways to provide the interferometer output. In the simplest version, the signals are simply added together before the square law detector (Figure 1.78) and the output will then vary with the path difference between the two signals. Such an arrangement, however, suffers from instability problems, particularly due to variations in the voltage gain.

A system that is now preferred to the simple adding interferometer is the correlation or multiplying interferometer. In this, as the name suggests, the IF signals from the receivers are multiplied together. The output of an ideal correlation interferometer will contain only the signals from the source (which are correlated); the other components of the outputs of the telescopes will be zero as shown below.

If we take the output voltages from the two elements of the interferometer to be $(V_1 + V_1')$ and $(V_2 + V_2')$ where V_1 and V_2 are the correlated components (i.e. from the source in the sky) and V_1' and V_2' are the uncorrelated components, the product of the signals is then

$$V = (V_1 + V_1') \times (V_2 + V_2') \tag{2.46}$$

$$= V_1 V_2 + V_1' V_2 + V_1 V_2' + V_1' V_2' \tag{2.47}$$

If we average the output over time, then any component of Equation 2.47 containing an uncorrelated component will tend to zero. Thus

$$\bar{V} = \overline{V_1 V_2} \tag{2.48}$$

In other words, the time-averaged output of a correlation interferometer is the product of the correlated voltages. Since most noise sources contribute primarily to the uncorrelated components, the correlation interferometer is inherently much more stable than the adding interferometer. The phase-switched interferometer is an early example of a correlation interferometer, though direct multiplying interferometers are now more common.

The systematics of radio astronomy differ from those of optical work (see Section 1.2), so that some translation is required to relate optical interferometers to radio interferometers. If we take the polar diagram of a single radio aerial (Figure 1.84 for example), then this is the radio analogue of our image structure for a single source and a single aperture (Figures 2.16 and 2.18). The detector in a radio telescope accepts energy from only a small fraction of this image at any given instant (though array detectors with small numbers of pixels are now coming into use) (see Section 1.2). Thus, scanning the radio telescope across the sky corresponds to scanning this energy-accepting region through the optical image structure. The main lobe of the polar diagram is therefore the equivalent of the central maximum of the optical image and the side lobes are the equivalent of the diffraction fringes. Since an aerial is normally directed towards a source, the signal from it corresponds to a measurement of the central peak intensity of the optical image (Figures 2.16 and 2.18).

If two stationary aerials are now considered and their outputs combined, then when the signals arrive without any path differences, the final output from the radio system as a whole corresponds to the central peak intensity of Figure 2.20. If a path difference does exist however, then provided that it is less than the coherence length, the final output will correspond to some other point within that image. In particular when the path difference is a whole number of wavelengths, the output will correspond to the peak intensity of one of the fringes and when it is a whole number plus half a wavelength, it will correspond to one of the minima. Now the path differences arise in two main ways: from the angle of inclination of the source to the line joining the two antennae and from delays in the electronics and cables between the antennae and the central processing station (Figure 2.27). The latter will normally be small and constant and may be ignored or corrected. The former will alter as the rotation of the Earth changes the angle of inclination. Thus, the output of the interferometer will vary with time as the value of P changes. The output over a period of time, however, will not follow precisely the shape of Figure 2.20 because the rate of change of path difference varies throughout the day, as we may see from the equation for P

$$P = s \cos \phi \cos(\psi - E) \tag{2.49}$$

where μ is the altitude of the object and is given by

$$\mu = \sin^{-1}[\sin \delta \sin \phi + \cos \delta \cos \phi \cos(T - \alpha)] \tag{2.50}$$

ψ is the azimuth of the object and is given by

$$\psi = \cot^{-1}[\sin \phi \cot(T - \alpha) - \cos \phi \tan \delta \operatorname{cosec}(T - \alpha)] \tag{2.51}$$

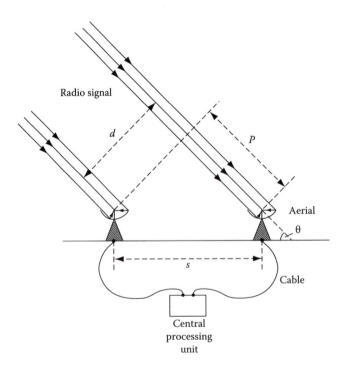

FIGURE 2.27 Schematic arrangement of a radio interferometer with fixed aerials.

In these equations, E is the azimuth of the line joining the aerials, α and δ are the right ascension and declination of the object, T is the local sidereal time at the instant of observation and ϕ is the latitude of the interferometer. The path difference also varies because the effective separation of the aerials, d, where

$$d = s\,[\sin^2 \mu + \cos^2 \mu \, \sin^2(\psi - E)]^{1/2} \qquad (2.52)$$

also changes with time, thus the resolution and fringe spacing (Equation 2.41) are altered. Hence, the output over a period of time from a radio interferometer with fixed antennae is a set of fringes whose spacing varies, with maxima and minima corresponding to those of Figure 2.20 for the instantaneous values of the path difference and effective aerial separation (Figure 2.28).

An improved type of interferometer, which has increased sensitivity and stability, is the phase switched interferometer. The phase of the signal from one aerial is periodically changed by 180° by, for example, switching an extra piece of cable half a wavelength long into or out of the circuit. This has the effect of oscillating the beam pattern of the interferometer through half a fringe width. The difference in the signal for the two positions is then recorded. The phase switching is generally undertaken in the latter stages of the receiver (see Section 1.2), so that any transient effects of the switching are not amplified. The output fluctuates on either side of the zero position as the object moves across the sky (Figure 2.29).

When the aerials are driven so that they track the object across the sky (Figure 2.30), then the output of each aerial corresponds to the central peak intensity of each image (Figures

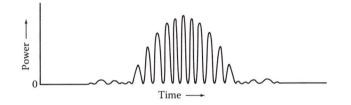

FIGURE 2.28 Output from a radio interferometer with stationary aerials, viewing a single source.

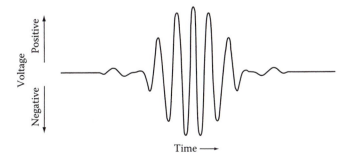

FIGURE 2.29 Output from a radio interferometer with stationary aerials and phase switching, viewing a single point source.

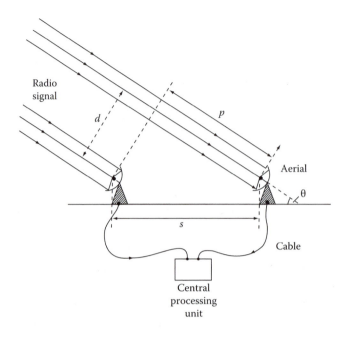

FIGURE 2.30   Schematic arrangement of a radio interferometer with tracking aerials.

2.16 and 2.18). The path difference then causes a simple interference pattern (Figure 2.31 and cf. Figure 2.19) whose fringe spacing alters due to the varying rate of change of path difference and aerial effective spacing as in the previous case. The maxima are now of constant amplitude since the aerials' projected effective areas are constant. In reality, many more fringes would occur than are shown in Figure 2.31. As with the optical interferometer however, the path differences at the aerials arising from the inclination of the source to the baseline have to be compensated to a fraction of the coherence length. This is much easier at the longer wavelengths since the coherence length is large – 30 metres for a 10-MHz bandwidth and 300 metres for a 1-MHz bandwidth. The path difference can therefore be corrected by switching in extra lengths of cable or by shifting the recorded signals with respect to each other during data processing.

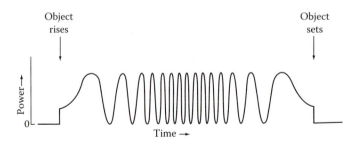

FIGURE 2.31   Output from a radio interferometer with tracking aerials viewing a single point source.

2.5.4 Aperture Synthesis

An interferometer works most efficiently, in the sense of returning the most information, for sources whose separation is comparable with its resolution. For objects substantially larger than the resolution, little or no useful information may be obtained. We may, however, obtain information about a *larger* source by using an interferometer with a *smaller* separation of its elements. The resolution is thereby degraded until it is comparable with the angular size of the source. By combining the results of two interferometers of differing separations, one might thus obtain information on both the large- and small-scale structure of the source. Following this idea to its logical conclusion led Sir Martin Ryle in the early 1960s to the invention and development of the technique of aperture synthesis. He was awarded the Nobel physics prize in 1974 for this work.

By this technique, which also goes under the name of Earth-rotation synthesis and is closely related to synthetic aperture radar (see Section 2.8), observations of a stable source by a number of interferometers are combined to give the effect of an observation using a single very large telescope. The simplest way to understand how aperture synthesis works is to take an alternative view of the operation of an interferometer. The output from a two-aperture interferometer viewing a monochromatic point source is shown in Figure 2.19 and is a simple sine wave. This output function is just the Fourier transform (Equation 2.3) of the source. Such a relationship between source and interferometer output is no coincidence, but is just a special case of the van Cittert-Zernicke theorem:

> The instantaneous output of a two-element interferometer is a measure of one component of the two-dimensional Fourier transform (Equation 2.10) of the objects in the field of view of the telescopes.

Thus, if a large number of two-element interferometers were available, so that all the components of the Fourier transform could be measured, then the inverse two-dimensional Fourier transform (Equation 2.11) would immediately give an image of the portion of the sky under observation.

Now the complete determination of the Fourier transform of even a small field of view would require an infinite number of interferometers. In practice, therefore, the technique of aperture synthesis is modified in several ways. The most important of these is relaxing the requirement to measure all the Fourier components *at the same instant*. However, once the measurements are spread over time, the source(s) being observed must remain unvarying over the length of time required for those measurements.

Given then, a source that is stable at least over the measurement time, we may use one or more interferometer pairs to measure the Fourier components using different separations and angles. Of course, it is still not possible to make an infinite number of such measurements, but we may use the discrete versions of Equations 2.10 and 2.11 to bring the required measurements down to a finite number (cf. the one-dimensional analogues, Equations 2.3, 2.4, 2.8 and 2.9) though at the expense of losing the high-frequency components of the transform and hence the finer details of the image (see Section 2.1). The

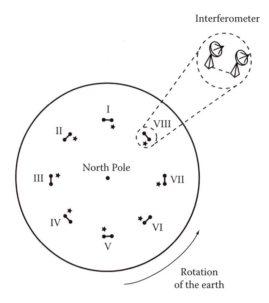

FIGURE 2.32 Changing orientation of an interferometer. The Earth is viewed from above the North Pole and successive positions of the interferometer at three-hour intervals are shown. Notice how the orientation of the starred aerial changes through 360° with respect to the other aerial during a day.

problem of observing with many different baselines is eased because we are observing from the rotating Earth. Thus, if a single pair of telescopes observes a source over 24 hours, the orientation of the baseline revolves through 360° (Figure 2.32). If the object is not at the North (or South) Pole then the projected spacing of the interferometer will also vary and they will seem to trace out an ellipse. The continuous output of such an interferometer is a complex function whose amplitude is proportional to the amplitude of the Fourier transform and whose phase is the phase shift in the fringe pattern.

The requirement for 24 hours of observation would limit the technique to circumpolar objects. Fortunately, however, only 12 hours are actually required; the other 12 hours can then be calculated from the conjugates of the first set of observations. Hence, aperture synthesis can be applied to any object in the same hemisphere as the interferometer.

Two elements arranged upon an East–West line follow a circular track perpendicular to the Earth's rotational axis. It is thus convenient to choose the plane perpendicular to the Earth's axis to work in and this is usually termed the u-v plane. If the interferometer is not aligned East–West, then the paths of its elements will occupy a volume in the u-v-w space and additional delays will have to be incorporated into the signals to reduce them to the u-v plane. The paths of the elements of an interferometer in the u-v plane range from circles for an object at a declination of ±90° through increasingly narrower ellipses to a straight line for an object with a declination of 0° (Figure 2.33).

A single 12-hour observation by a two-element interferometer thus samples all the components of the Fourier transform of the field of view covered by its track in the u-v plane. We require, however, the whole of the u-v plane to be sampled. Thus, a series of 12-hour observations must be made, with the interferometer baseline changed by the diameter of one of its

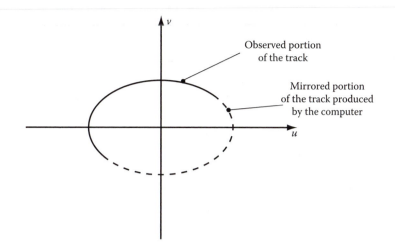

FIGURE 2.33 Track of one element of an interferometer with respect to the other in the *u-v* plane for an object at a declination of ±35°.

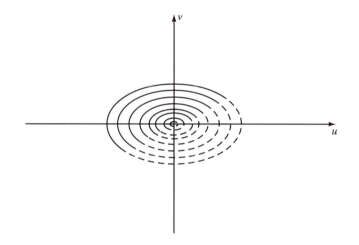

FIGURE 2.34 Successive tracks in the *u-v* plane of a two-element interferometer as its baseline is varied.

elements each time (Figure 2.34). The *u-v* plane is then sampled completely out to the maximum baseline possible for the interferometer and the inverse Fourier transform will give an image equivalent to that from a single telescope with a diameter equal to the maximum baseline.

A two-element radio interferometer with 20-metre diameter aerials and a maximum baseline of 1 km would need to make fifty 12-hour observations in order to synthesise a 1-km diameter telescope. By using more than two aerials, however, the time required for the observations can be much reduced. Thus, with six elements, there are 15 different pairings.* If the spacings of these pairs of elements are all different (non-redundant spacing, something that it is not always possible to achieve) then the 50 visibility functions required by the previous example can be obtained in just four 12-hour observing sessions. In 1997,

---

* For *N* elements there are [*N*(*N*-1)/2] possible pairs.

Eric Keto showed that the optimum (in the sense of sampling the Fourier transform best) layout for the elements of an interferometer was in the shape of a curve of constant width. The circle is the best known such curve, but is the least satisfactory choice for this purpose. Other curves, known as Reuleaux polygons, are better. A Reuleaux polygon is just a straight-sided polygon with an odd number of sides in which the straight sides are replaced by arcs of circles whose centres are the vertices opposite to each of the sides.* Of these curves, the Reuleaux triangle is the best of all for an interferometer.

If the source has a reasonably smooth spectrum, the *u-v* plane may be sampled even more rapidly by observing simultaneously at several nearby frequencies (multifrequency synthesis). Since the path differences between the aerials are dependent upon the operating wavelength, this effectively multiplies the number of interferometer pairs with different separations by the number of frequencies being observed. The analysis of the results from such observations, however, will be considerably complicated by any variability of the source over the frequencies being used.

A radio aperture synthesis system such as we have just been considering is a filled-aperture system. That is, the observations synthesise the sensitivity *and* resolution of a telescope with a diameter equal to the maximum available baseline. Although some aperture synthesis systems are of this type, it becomes increasingly impractical and expensive to sample the whole *u-v* plane as the baseline extends beyond a kilometre or so. For many observations, even arrays that are capable of synthesising a filled aperture may not need to do so.

Most aperture synthesis arrays in practice therefore, synthesise sparse apertures in which only selected annuli within the whole aperture are completed. Sparse apertures can range from systems which are almost filled to very-long baseline interferometry (VLBI) where a few to a few tens of annuli, each only 10 to 100 metres in width may be completed within a baseline 10,000 km long. In such cases, the Fourier transform of the source is not fully sampled. Special techniques, known as hybrid mapping are then required to produce the final maps of radio sources.

The Westerbork Radio Synthesis Telescope (WRST) in the Netherlands is an example of an array that can come close to synthesising a filled aperture. It uses 10 fixed and four movable 25-metre dishes over a maximum baseline of 2.7 km. The WRST is due to be upgraded shortly to using array detectors, so increasing its field of view by a factor of 25. In the United Kingdom, e-MERLIN[†] is an array of seven fixed radio telescopes with a maximum baseline of 217 km sited across central England.[‡] It has recently been upgraded to using fibre optics to link the telescopes and now can reach a resolution of 50 milliarc seconds at 5 GHz (60 mm). The Karl G. Jansky Very Large Array ([VLA]; previously known as EVLA) in New

---

* The Reuleaux heptagon will be familiar to UK readers, at least, since it is the shape of the 20-p and 50-p coins.
† Multi-Element Radio Linked Interferometer – although the radio links have recently been replaced by fibre optics that have increased MERLIN's sensitivity by a factor of 30.
‡ An eighth telescope may be added in the future and sited at Goonhilly Downs in Cornwall. This will double the maximum baseline to 450 km.

FIGURE 2.35   The central part of the VLA. (Reproduced by kind permission of NRAO/AUI and NRAO.)

Mexico (Figure 2.35) has also recently been refurbished and uses twenty-seven* 25-metre dishes arranged in a 'Y' pattern with a maximum baseline of 36 km. Details, insofar as they are determined at the time of writing, of the planned SKA have been discussed in Section 1.2.

Beyond e-MERLIN, the VLA and similar radio systems, we have VLBI. For VLBI, the elements of the interferometer may be separated by thousands of kilometres over several continents and provide milliarc second resolutions or better. In VLBI, the signals from each element are separately recorded along with timing pulses from an atomic clock. The recordings are then physically brought together and processed by a computer that uses the time signals to ensure the correct registration of one radio signal with respect to another. Recently, real-time analysis of the data from a VLBI system comprising the Arecibo radio dish and telescopes in the United Kingdom, Sweden, the Netherlands and Poland became possible when they were linked via Internet research networks – earning the system the name of e-VLBI.

The Very Long Baseline Array (VLBA) VLBI system uses ten 25-metre telescopes spread out over the United States from Hawaii to the US Virgin islands. Its maximum baseline is 8600 km and at a wavelength of 7 mm its resolution can be 150 microarc seconds. It has, for example, recently been able to measure the parallax for the Orion nebula to an accuracy of 100 microarc seconds, reducing the previously accepted distance for the nebula from 475 to 385 pc. It has also recently had new receivers installed and with upgraded computers it is now improved by a factor of around 5000 compared with its performance when it first started operating in 1993. Along with the Arecibo, Effelsberg and GBT dishes and the VLA it also forms the High Sensitivity Array. This can improve upon the sensitivity of the VLBA by a factor of 10 and has a maximum baseline of 10,300 km.

For even longer baselines than the HSA, it is necessary to put one or more receivers into space. The Japanese Halca spacecraft which carried an 8-metre radio telescope was linked with up to 40 ground-based dishes, to give baselines up to 30,000 km in length. The spacecraft lost altitude control in 2003, however, and has now ceased to operate. Also, as mentioned in Section 1.2, the Russian RadioAstron spacecraft is currently carrying a 10-metre dish and by linking with ground-based instruments offers baselines up to 390,000 km.

The fields of view in VLBI have generally been only a few arc seconds. However, the computational problems previously encountered arising from the available computer capacities are now being overcome as computer power increases. It seems likely that fields of view of up to half a degree may become possible in the near future, greatly increasing the utility of VLBI. VLBI is also potentially extendable to submillimetre wavelengths with more or less the existing equipment and telescopes. Already APEX, SMA and the 10-metre Submillimeter Telescope in Arizona have conducted VLBI at a wavelength of 1.3 mm (230 GHz) and achieved 28 microarc second resolutions. It seems likely that submillimetre observations will be made in the near future.

At millimetre and submillimetre wavelengths the Owens Valley Radio Observatory in California is operating the Combined Array for Research in Millimeter-wave Astronomy

---

* Twenty-seven dishes implies 351 baselines – and this is sufficient for the VLA to be able to obtain images in the time it takes to make a single set of measurements of the source, and so not have to wait for the Earth's rotation to 'do any of the work' for it. This is known as the VLA's snapshot mode of operation. However, the snapshot images are much noisier than those obtained by conventional aperture synthesis.

(CARMA) array. This is actually two independent arrays. One has 15 dishes with 6.1- or 10.4-metre diameters. The second has eight 3.5-metre dishes. The former observes in the 85- to 115-GHz and 215- to 270-GHz bands (3.5 to 2.6 mm and 1.4 to 1.1 mm). The latter covers 26 to 36 GHz and 80 to 115 GHz (11.5 to 8.3 mm and 3.8 to 2.6 mm). The maximum baseline is 2 km and the resolutions are 150 milliarc seconds and 1 arc minute, respectively.

On the Plateau de Bure in France, the Institut de Radio Astronomie Millimétrique (IRAM) operates a system with six 15-metre telescopes with a maximum baseline of 760 m. It observes in the 100- to 300-GHz band (3 to 1 mm) with a best resolution of 500 milliarc seconds.

The Submillimeter Array (SMA) on Mauna Kea observes between 180 and 700 GHz (1.7 mm and 430 μm) with eight 6-metre dishes. It can also link with the JCMT and the 10.4-metre Caltech Submillimeter Observatory telescope (both also on Mauna Kea) via fibre-optic cables as a 10- element interferometer (when it is called the extended SMA or eSMA). Baselines range up to 500 metres for SMA and 780 metres for eSMA. It can achieve a resolution of a few tenths of a second of arc. The SMA has four configurations that are based upon Reuleaux triangles insofar as is practicable.

ALMA is due to be completed by 2013 although observations are already being made using the partially completed instrument. It will have fifty 12-metre antennas plus a central smaller array of twelve 7-metre and four 12-metre dishes for observing extended sources (Figure 2.36a). It will operate between 84 and 720 GHz (3.6 mm and 420 μm) and have angular resolutions down to 5 milliarc seconds at its maximum aerial separation of 16 km. The individual radio telescopes weigh around 100 tonnes each but can be moved in one piece from one position to another within the array so that the configuration of the array can be changed (Figure 2.36b). There are a total of 196 available positions for the 66 antennas. Correction for seeing at millimetre wavelengths will be needed if ALMA is to achieve its intended resolution (see also optical real-time atmospheric compensation, Section 1.1). This will be accomplished using 183-GHz radiometers to measure the atmospheric properties along the line of sight and by observing point sources (quasars) near to the field of view.

At visible and NIR wavelengths, aperture synthesis has only recently been successfully attempted because of the very stringent requirements on the stability and accuracy of construction of the instruments imposed by the small wavelengths involved.

The Center for High Angular Resolution Astronomy (CHARA) array on Mount Wilson can reach a resolution of 200 microarc seconds in the visible and NIR using six 1-metre telescopes in a Y-shaped array with a maximum baseline of 330 metres.

The four 8.2-metre telescopes of ESO's VLTI can be combined with four 1.8-metre auxiliary telescopes for aperture synthesis and the latter can be moved to 30 different positions offering baselines of up to 200 metres (see Figure 2.25 and cover image). The system operates in the NIR and MIR with angular resolutions down to 2 milliarc seconds for imaging and 10 microarc seconds for astrometry. There are several second generation instruments either just commissioned or about to be commissioned for the VLTI at the time of writing. These include Gravity, Multi Aperture mid-Infrared SpectroScopic Experiment (MATISSE), Precision Integrated-Optics Near-infrared Imaging Experiment (PIONIER) and Phase-Referenced Imaging and Microarcsecond Astrometry (PRIMA).

FIGURE 2.36 (**See color insert.**) (a) Some of the individual radio telescopes making up ALMA, photographed at night. (Reproduced by kind permission of ALMA [ESO/NAOJ/NRAO], C. Padilla.) (b) Moving one of ALMA's radio telescopes to a new site. (Reproduced by kind permission of ALMA [ESO/NAOJ/NRAO], S. Rossi [ESO].)

The recently decommissioned PIONIER has been operating as an imaging interferometer in the NIR since 2010. It was a visitor instrument built in France and combined the outputs from the four 8.2-metre telescopes or the four 1.8-metre telescopes. PRIMA will be able to observe two objects separated by up to 1 arc minute simultaneously. One of the objects will be a guide star allowing additional corrections to be made to the image of the other object. It is intended for NIR astrometry with a capability in the 10-microarc-second region and is currently being commissioned. Gravity will be a NIR instrument designed for imaging (4-milliarc-second resolution) and astrometry (10-microarc-second accuracy) using the four 8.2-metre unit telescopes. It is expected to become available around 2014 to 2015. MATISSE is also expected to become available around 2014 to 2015. It will operate in the MIR combining the outputs from up to four of the telescopes as a spectro-interferometer.

The Kenneth J. Johnston Navy Precision Optical Interferometer (NPOI) has been operational since 1994. It currently has six telescopes with effective apertures of 0.14 metres in a Y-shaped distribution and with baselines up to 79 metres. It operates in the visible for imaging and low-resolution spectroscopy. The four 1.8-metre outrigger telescopes originally intended for the Keck aperture synthesis system are currently being incorporated into the NPOI instrument. Its primary purpose is astrometry as an aid to the navigation of spacecraft.

The Magdalena Ridge Observatory Interferometer (MROI) in New Mexico is currently under construction. It will operate from 600 nm to 2.4 μm with up to ten 1.4-metre telescopes in a Y-shaped configuration and with baselines ranging from 7.8 to 340 metres. It is hoped eventually to achieve a resolution of 600 microarc seconds at a wavelength of 1 μm.

There have been a number of proposals for space-based interferometer systems. Most of these have been cancelled at some stage. The Halca and RadioAstron spacecraft-based radio interferometers though have made it into space and were discussed earlier. One future ground-based advance that currently has a prototype called Dragonfly being commissioned is based upon photonic optical components (see Section 1.1). It has been tried out successfully on the AAT. It remains to be seen, however, whether the approach can improve upon existing instruments.

2.5.5 Data Processing

The Common Astronomy Software Applications (CASA) computing package has been developed from the earlier Astronomical Image Processing System (AIPS and AIPS++) produced by the National Radio Astronomy Observatory (NRAO). It is used at many interferometric observatories and provides for many of the processing steps outlined in the list below.

The extraction of the image of the sky from an aperture synthesis system is complicated by the presence of noise and errors. Overcoming the effects of these adds additional stages to the data reduction process. In summary it becomes

- Data calibration

- Inverse Fourier transform

- Deconvolution of instrumental effects

- Self-calibration

Data calibration is required in order to compensate for problems such as errors in the locations of the elements of the interferometer and variations in the atmosphere. It is carried out by comparing the theoretical instrumental response function with the observed response to an isolated stable point source. These responses should be the same and if there is any difference then the data calibration process attempts to correct it by adjusting the amplitudes and phases of the Fourier components. Often ideal calibration sources cannot be found close to the observed field. Then, calibration can be attempted using special calibration signals, but these are insensitive to the atmospheric variations and the result is much inferior to that obtained using a celestial source.

Applying the inverse Fourier transform to the data has already been discussed and further details are given in Section 2.1.

After the inverse Fourier transformation has been completed, we are left with the 'dirty' map of the sky. This is the map contaminated by artefacts introduced by the PSF of the interferometer. The primary components of the PSF, apart from the central response, are the side lobes. These appear on the dirty map as a series of rings that extend widely over the sky and which are centred on the central response of the PSF. The PSF can be calculated from interferometry theory to a high degree of precision. The deconvolution can then proceed as outlined in Section 2.1.

The maximum entropy method discussed in Section 2.1 has recently become widely used for determining the best source function to fit the dirty map. Another method of deconvolving the PSF, however, has long been in use and is still used by many workers and that is the method known as CLEAN.

The CLEAN algorithm was introduced by Jan Högbom in 1974. It involves the following stages:

a. Normalise the PSF (instrumental profile or dirty beam) to ($g\,I_{MAX}$), where I_{MAX} is the intensity of the point of maximum intensity in the dirty map and g is the loop gain with a value between 0 and 1.

b. Subtract the normalised PSF from the dirty map.

c. Find the point of maximum intensity in the new map – this may or may not be the same point as before – and repeat the first two steps.

d. Continue the process iteratively until I_{MAX} is comparable with the noise level.

e. Produce a final clear map by returning all the components removed in the previous stages in the form of clean beams with appropriate positions and amplitudes. The clean beams are typically chosen to be Gaussian with similar widths to the central response of the dirty beam.

CLEAN has proved to be a useful method for images made up of point sources despite its lack of a substantial theoretical basis. For extended sources MEMs are better because CLEAN may then require thousands of iterations. However, as pointed out in Section 2.1, MEMs also suffer from problems, especially the variation of resolution over the image.

The final stage of self-calibration is required for optical systems and for radio systems when the baselines become more than a few kilometres in length, since the atmospheric effects then differ from one telescope to another. Under such circumstances with three or more elements we may use the closure phase that is independent of the atmospheric phase delays. The closure phase is defined as the sum of the observed phases for the three baselines made by three elements of the interferometer. It is independent of the atmospheric phase delays as we may see by defining the phases for the three baselines in the absence of an atmosphere to be ϕ_{12}, ϕ_{23} and ϕ_{31} and the atmospheric phase delays at each element as a_1, a_2 and a_3. The observed phases are then

$$\phi_{12} + a_1 - a_2$$

$$\phi_{23} + a_2 - a_3$$

$$\phi_{31} + a_3 - a_1$$

The closure phase is then given by the sum of the phases around the triangle of the baselines:

$$\phi_{123} = \phi_{12} + a_1 - a_2 + \phi_{23} + a_2 - a_3 + \phi_{31} + a_3 - a_1 \tag{2.53}$$

$$= \phi_{12} + \phi_{23} + \phi_{31} \tag{2.54}$$

From Equation 2.54 we may see that the closure phase is independent of the atmospheric effects and is equal to the sum of the phases in the absence of an atmosphere. In a similar way an atmosphere-independent closure amplitude can be defined for four elements:

$$G_{1234} = \frac{A_{12} A_{34}}{A_{13} A_{24}} \tag{2.55}$$

where A_{12} is the amplitude for the baseline between elements 1 and 2. Neither of these closure quantities are actually used to form the image, but they are used to reduce the number of unknowns in the procedure. At millimetre wavelengths, the principal atmospheric effects are due to water vapour. Since this absorbs the radiation, as well as leading to the phase delays, monitoring the sky brightness can provide information on the amount of water vapour along the line of sight and so provide additional information for correcting the phase delays.

For VLBI, hybrid mapping is required since there is insufficient information in the visibility functions to produce a map directly. Hybrid mapping is an iterative technique that uses a mixture of measurements and guesswork. An initial guess is made at the form of the required map. This may well actually be a lower resolution map from a smaller interferometer. From this map the visibility functions are predicted and the true phases estimated. A new map is then generated by Fourier inversion. This map is then improved and used to provide a new start to the cycle. The iteration is continued until the hybrid map is in satisfactory agreement with the observations.

2.5.6 Intensity Interferometer

Until the last couple of decades, the technical difficulties of an optical Michelson interferometer severely limited its usefulness. Most of these problems, however, may be reduced in a device that correlates intensity fluctuations. Robert Hanbury-Brown originally invented the device in 1949 as a radio interferometer, but it has found its main application in the optical region. The disadvantage of the system compared with the Michelson interferometer is that phase information is lost and so the structure of a complex source cannot be reconstituted. The far greater ease of operation of the intensity interferometer led to it being able to measure some hundred stellar diameters between 1965 and 1972. Hanbury-Brown's instrument has long been decommissioned, but there is now a revival of interest in the technique since existing γ-ray Čerenkov arrays such as MAGIC and VERITAS (see Sections 1.3 and 1.4) have the potential to utilise the approach to image stellar surfaces.

The principle of the operation of the interferometer relies upon phase differences in the low-frequency beat signals from different mutually incoherent sources at each aerial, combined with electrical filters to reject the high-frequency components of the signals. The schematic arrangement of the system is shown in Figure 2.37. We may imagine the signal from a source resolved into its Fourier components and consider the interaction of one such component from one source with another component from the other source. Let the frequencies of these two components be ν_1 and ν_2, then when they are mixed, there will be two additional frequencies – the upper and lower beat frequencies, $(\nu_1 + \nu_2)$ and $(\nu_1 - \nu_2)$ involved. For lightwaves, the lower beat frequency will be in the radio region (typically 10 to 100 MHz) and this component may easily be filtered out within the electronics from the much higher original and upper beat frequencies. The low frequency outputs from the two telescopes are multiplied and integrated over a short time interval and the signal bandwidth to produce the correlation function, Robert Hanbury-Brown and Richard Twiss were able to show that K was simply the square of the fringe visibility (Equation 2.45) for the Michelson interferometer.

The intensity interferometer was used to measure stellar diameters and $K(d)$ reaches its first zero when the angular stellar diameter, θ', is given by

$$\theta' = \frac{1.22\lambda}{d} \tag{2.56}$$

where d is the separation of the receivers. So the resolution of an intensity interferometer (and also of a Michelson interferometer) for stellar discs is the same as that of a telescope (Equation 2.40) whose diameter is equal to the separation of the receivers (Figure 2.38).

The greater ease of construction and operation of the intensity interferometer over the Michelson interferometer arises from its dependence upon the beat frequency of two light beams of similar wavelengths, rather than upon the actual frequency of the light. A

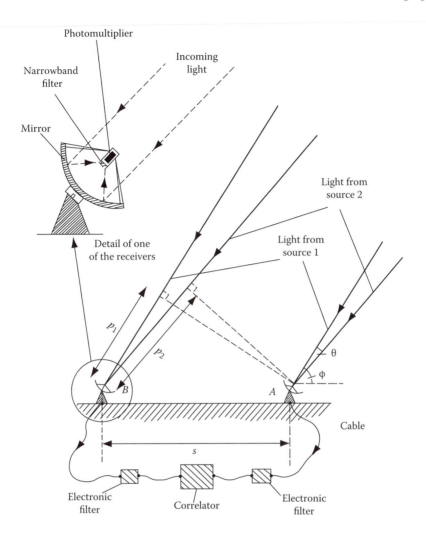

FIGURE 2.37 Schematic arrangement of an intensity interferometer.

typical value of the lower beat frequency is 100 MHz, which corresponds to a wavelength of 3 metres. Thus, the path differences for the two receivers may vary by up to about 0.3 metres during an observing sequence without ill effects. Scintillation, by the same argument, is also negligible.

Only one working intensity interferometer has been constructed. It was built by Hanbury-Brown at Narrabri in Australia. It has now been decommissioned. It used two 6.5-metre reflectors that were formed from several hundred smaller mirrors. There was no need for very high optical quality since the reflectors simply acted as light buckets and only the brightest stars could be observed. The reflectors were mounted on trolleys on a circular track 94 metres in radius. The line between the reflectors could therefore always be kept perpendicular to the line of sight to the source and their separation could be varied from 0 to 196 metres (Figure 2.39). It operated at a wavelength of 433 nm, giving it a maximum resolution of about 500 microarc seconds.

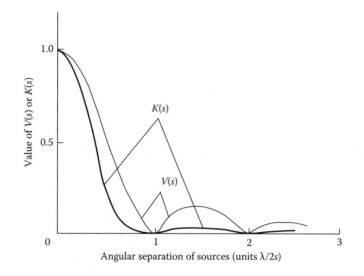

FIGURE 2.38   Comparison of the fringe visibility of a Michelson interferometer with the correlation function of an intensity interferometer.

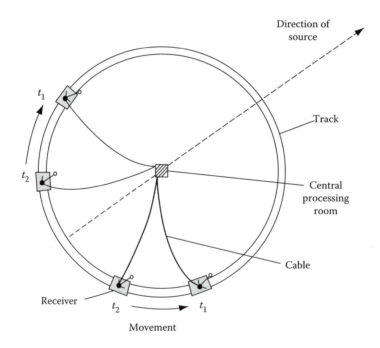

FIGURE 2.39   Schematic layout of the intensity interferometer at Narrabri, showing the positions of the receivers for observing in the same direction with different baselines. Note there were only *two* receivers; the diagram shows the arrangement at two separate baselines, superimposed.

High optical quality is not needed for the telescopes involved in intensity interferometry.* The telescopes in Hanbury-Brown's instrument for example had individual resolutions of 6 minutes of arc. Furthermore, the detectors operated only in an analogue fashion. γ-Ray air Čerenkov arrays (see Sections 1.3 and 1.4) are therefore well suited to being used as intensity interferometers. Since they can have up to five telescopes (and may have more in the future) with sizes up to 28 metres and since photon-counting detectors can be used, the performances of such arrays will be far better than that of the original instrument. Computer modelling suggests that at a wavelength of 400 nm resolutions of 60 microarc seconds could be obtainable. This would be sufficient for maps of the angularly larger stars to be produced with several tens or even a few hundred resolution elements – more than adequate to monitor the emissions from larger active regions. It seems likely that this application of γ-ray Čerenkov arrays will be tried out in practice in the near future.

EXERCISES

2.1 Calculate the separation of the slits over a telescope objective in the first type of Michelson stellar interferometer, which would be required to measure the diameter of Ganymede (*a*) at opposition and (*b*) near conjunction of Jupiter. Take Ganymede's diameter to be 5000 km and the eye's sensitivity to peak at 550 nm.

2.2 Observations of a quasar that is thought to be at a distance of 1500 Mpc just reveal structure to a VLBI working at 50 GHz. If its maximum baseline is 9000 km, what is the linear scale of this structure?

2.3 Calculate the maximum distance at which the Narrabri intensity interferometer would be able to measure the diameter of a solar-type star.

2.6 SPECKLE INTERFEROMETRY

This technique is a kind of poor man's space telescope since it provides near diffraction-limited performance from Earth-based telescopes. It works by obtaining images of the object sufficiently rapidly to freeze the blurring of the image that arises from atmospheric scintillation. The total image then consists of a large number of small dots or speckles, each of which is a diffraction-limited interference fringe for some objective diameter up to and including the diameter of the actual objective (Figure 1.67). An alternative to this technique is the adaptive optics telescope (see Section 1.1), where adjusting the telescope optics to compensate for the atmospheric effects recombines the speckles. Adaptive optics, especially with artificial laser guide stars, has replaced speckle interferometry for most of the largest instruments, but the technique is still to be encountered in use on smaller

---

* Of course, better resolution does not do any harm – so there is no reason why the VLT or the Keck instruments could not be used this way as well.

telescopes – such as, for example, the U.S. Naval Observatory's 0.66-metre refractor and the 2.5-metre Hooker telescope on Mount Wilson. A modern adaptation of the technique is used on the VLT and will be available on the JWST (see later in this section).

We may see how this speckled image structure arises by considering the effect of scintillation upon the incoming wavefront from the object. If we assume that above the atmosphere the wavefront is planar and coherent, then the main effect of scintillation is to introduce differential phase delays across it. The delays arise because the atmosphere is non-uniform with different cells within it having slightly different refractive indices. A typical cell size is 0.1 m and the scintillation frequencies usually lie in the range 1 to 100 Hz. Thus, some 100 atmospheric cells will affect an average image from a 1-metre telescope at any given instant. These will be rapidly changing and over a normal exposure time, which can range from seconds to hours, they will form an integrated image that will be large and blurred compared with the diffraction-limited image. Even under the best seeing conditions, the image is rarely less than 1 second of arc across. An exposure of a few milliseconds, however, is sufficiently rapid to freeze the image motion and the observed image is then just the resultant of the contributions from the atmospheric cells across the telescope objective at that moment.

Now the large number of these cells renders it highly probable that some of the phase delays will be similar to each other and so some of the contributions to the image will be in phase with each other. These particular contributions will have been distributed over the objective in a random manner. Considering two such contributions, we have in fact a simple interferometer and the two beams of radiation will combine in the image plane to produce results identical with those of an interferometer whose baseline is equal to the separation of the contributions on the objective. We have already seen what this image structure might be (Figure 2.20). If several collinear contributions are in phase, then the image structure will approach that shown in Figure 4.2, modulated by the intensity variation due to the aperture of a single cell. The resolution of the images is then given by the maximum separation of the cells. When the in-phase cells are distributed in two dimensions over the objective, the images have resolutions in both axes given by the maximum separations along those axes at the objective. The smallest speckles in the total image therefore have the diffraction-limited resolution of the whole objective, assuming always of course that the optical quality of the telescope is sufficient to reach this limit. Similar results will be obtained for those contributions to the image that are delayed by an integral number of wavelengths with respect to each other.

Intermediate phase delays will cause destructive interference to a greater or lesser extent at the point in the image plane that is symmetrical between the two contributing beams of radiation, but will again interfere constructively at other points, to produce an interference pattern that has been shifted with respect to its normal position. Thus, all pairs of contributions to the final image interfere with each other to produce one or more speckles.

The true image is the Fourier transform of the interference pattern, just as for any other type of interference pattern. To obtain the true image from a speckled image, it must thus be Fourier analysed and the power spectrum obtained. This is the square of the modulus of the Fourier transform of the image intensity (Equation 2.3). Today, the Fourier transform is mostly obtained directly using quite small computers. However, when the technique was first developed in the 1970s and 1980s even the largest computers were inadequate for the task. The Fourier transform was therefore obtained optically by illuminating the image with collimated coherent light (Figure 2.40). The image (then a photographic negative) was placed at one focus of an objective and its Fourier transform imaged at the back focus of the objective. A spatial filter at this point can be used to remove unwanted frequencies if required and then the Fourier transform reimaged.

However, it may be obtained, the power spectrum can then be inverted to give centro-symmetric information such as diameters, limb darkening, oblateness and binarity. Non-centrosymmetric structure can only be obtained if there is a point source close enough for its image to be obtained simultaneously with that of the object and for it to be affected by the atmosphere in the same manner as the image of the object (i.e. the point source is within the same isoplanatic patch of sky as the object of interest) (Figure 1.68). Then decon-volution (see Section 2.1) can be used to retrieve the image structure.

An alternative way of processing speckle images that requires little in the way of com-puting power is called shift and add. It is the same process that is mentioned in Section 1.1 as a means of (somewhat) correcting atmospheric image degradation for small telescopes. With a number of speckle images of the same object, the brightest speckle in each image is found and all the images moved until those brightest speckles are aligned with each other. The images are then just added together to produce a generally much less noisy and a sharper image.

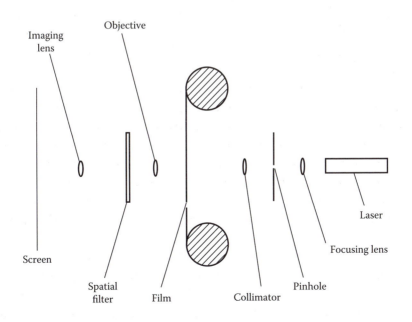

FIGURE 2.40   Arrangement for obtaining the Fourier transform of a speckle photograph by optical means.

The practical application of speckle interferometry requires the use of large telescopes since not only are very short exposures required (0.001 to 0.1 seconds), but very large plate scales (0.1 to 1 seconds of arc per millimetre) are also needed in order to separate the individual speckles. Furthermore, the wavelength range must be restricted by a narrowband filter to 20 to 30 nm. Even with such a restricted wavelength range, it may still be necessary to correct any remaining atmospheric dispersion using a low-power direct vision spectroscope. Thus, the limiting magnitude of the technique is currently about $+18^m$ and it has found its major applications in the study of red supergiants, the orbits of close binary stars, Seyfert galaxies, asteroids and fine details of solar structure.

A recent improvement to the technique is obtained, rather counter-intuitively, by masking the whole telescope objective down to a few, much smaller, holes. The sizes of the holes are typically similar to those of an individual atmospheric cell (100 to 300 mm depending upon the wavelength at which the observations are being made). The mask may be placed directly over the telescope objective. More conveniently however, a lens after the telescope's focal point may be used to produce a parallel beam of light with a much smaller diameter than that of the main mirror. A suitably scaled mask is then placed across the beam of parallel light and the light is reimaged. The holes are usually arranged over the mask so that the distances between them are not duplicated along either the x- or y-axes of a plane perpendicular to the light beam – although the x separations *should* be duplicated along the y direction to give equal resolutions in both. In effect the original single telescope mirror is now an interferometer array (see Section 2.5) directly analogous to systems like the VLTI and the VLA.

Masking the main telescope mirror in this way is called speckle masking or sparse aperture masking. When, as described, the separations of the holes (baselines in interferometric parlance) are not duplicated, it is termed a non-redundant mask. Sometimes, in order to increase the amount of light available, separations may be duplicated and then it is termed a partially redundant mask.

The image obtained through a sparse aperture mask is a collection of speckles from each pair of holes. Because the holes are comparable in size with the atmospheric cells, many will only be affected by one such cell (i.e. no wavefront distortion for that particular hole). The speckle from two such holes will have no noise contribution from the atmosphere – just a displacement across the field of view if there is a phase difference between the light beams. The speckles will thus be much cleaner than those obtained with an unmasked mirror. Analysis of the data is based upon the calculation of an average closure phase (Equations 2.53 and 2.54 – the closure phase is sometimes called the bispectrum in this context) which is inverted to obtain the image.

Sparse aperture masking is currently being undertaken on several telescopes – for example with the Nasmyth Adaptive Optics System-Near-Infrared Imager and Spectrograph (NAOS-CONICA) camera of the VLT where 340 milliarc second resolutions have been reached. It will also be an option on the JWST where Near-InfraRed Imager and Slitless Spectrograph (NIRISS) will provide the telescope's highest resolution imaging at around 75 milliarc seconds for a 4.6-μm wavelength.

2.7 OCCULTATIONS

2.7.1 Background

Three astronomical objects whose movements though space have put them into a straight line is a phenomenon termed a syzygy. With the Earth as one of the three objects involved, there are three commonly encountered such alignments: eclipses (when the two objects seen from the Earth have comparable angular sizes*), transits (when the nearer of the two objects has a much smaller angular size as seen from the Earth compared with the more distant one) and occultations (when the nearer of the two objects has a much larger angular size as seen from the Earth compared with the more distant one).

Occultations of more distant objects by the Moon occur frequently because of the large angular size of the Moon. Observation of such occultations has a long history, with records of lunar occultations stretching back some two and a half millennia. Occultations by planets, their satellites and asteroids occur much less often and are usually only to be seen from a restricted part of the Earth's surface. Recently, interest in the events has been revived for their ability to give precise positions for objects observed at low angular resolution in, say, the x-ray region and for their ability to give structural information about objects at better than the normal diffraction-limited resolution of a telescope. To see how the latter effect is possible, we must consider what happens during an occultation.

First, let us consider Fresnel diffraction at a knife edge (Figure 2.41) for radiation from a monochromatic point source. The phase difference, δ, at a point P between the direct and the diffracted rays is then

$$\delta = \frac{2\pi}{\lambda}\left\{ d_1 + \left[d_2^2 + (d_2 \tan\theta)^2 \right]^{1/2} - \left[(d_1 + d_2)^2 + (d_2 \tan\theta)^2 \right]^{1/2} \right\} \tag{2.57}$$

which, since θ is very small, simplifies to

$$\delta = \frac{\pi d_1 d_2}{\lambda(d_1 + d_2)}\theta^2 \tag{2.58}$$

The intensity at a point in a diffraction pattern is obtainable from the Cornu spiral (Figure 2.42), by the square of the length of the vector, **A**. P' is the point whose distance along the curve from the origin, l, is given by

$$l = \left(\frac{2d_1 d_2}{\lambda(d_1 + d_2)} \right)^{1/2} \theta \tag{2.59}$$

---

* A lunar eclipse is not therefore a true eclipse, although it is a syzygy. If observed from the Moon by an astronaut it would properly be called a transit of the Sun by the Earth. It also seems most appropriate to consider stellar coronagraphs under this heading since they operate by producing artificial eclipses. Solar coronagraphs are considered in Section 5.3.

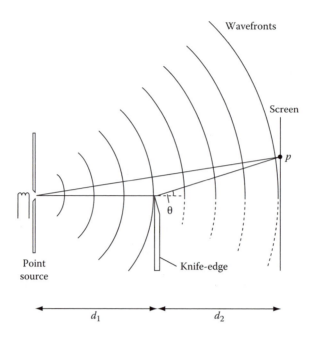

FIGURE 2.41   Fresnel diffraction at a knife edge.

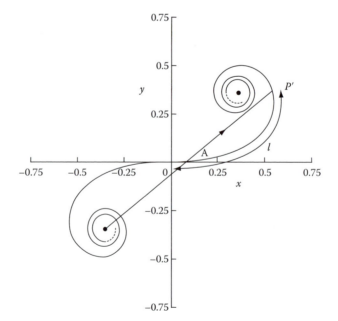

FIGURE 2.42   The Cornu spiral.

and the phase difference at P' is the angle that the tangent to the curve at that point makes with the x-axis, or from Equation 2.58

$$\delta = \frac{1}{2}\pi l^2 \tag{2.60}$$

The coordinates of P', which is the point on the Cornu spiral giving the intensity at P, x and y, are obtainable from the Fresnel integrals

$$x = \frac{1}{\sqrt{2}}\int_0^l \cos\left(\frac{1}{2}\pi l^2\right)dl \tag{2.61}$$

$$y = \frac{1}{\sqrt{2}}\int_0^l \sin\left(\frac{1}{2}\pi l^2\right)dl \tag{2.62}$$

whose pattern of behaviour is shown in Figure 2.43.
    If we now consider a star occulted by the Moon, then we have

$$d_1 \gg d_2 \tag{2.63}$$

so that

$$l \approx \left(\frac{2d_2}{\lambda}\right)\theta \tag{2.64}$$

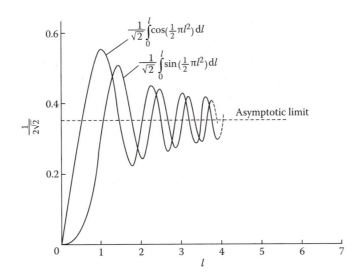

FIGURE 2.43   The Fresnel integrals.

but otherwise the situation is unchanged from the one we have just discussed. The edge of the Moon of course is not a sharp knife-edge, but since even a sharp knife-edge is many wavelengths thick, the two situations are not in practice any different. The shadow of the Moon cast by the star onto the Earth therefore has a standard set of diffraction fringes around its edge. The intensities of the fringes are obtainable from the Cornu spiral and Equations 2.61 and 2.62 and are shown in Figure 2.44. The first minimum occurs for

$$l = 1.22 \tag{2.65}$$

so that for the mean Earth–Moon distance

$$d_2 = 3.84 \times 10^8 \text{ m} \tag{2.66}$$

we obtain

$$\theta = 6.4 \text{ milliarc seconds} \tag{2.67}$$

at a wavelength of 500 nm. Therefore, the fringes have a linear width of about 12 metres when the shadow is projected onto a part of the Earth's surface that is perpendicular to the line of sight. The mean rate of angular motion of the Moon, $\dot{\theta}$, is

$$\dot{\theta} = 0.55''s^{-1} \tag{2.68}$$

so that at a given spot on the Earth, the fringes will be observed as intensity variations of the star as it is occulted, with a basic frequency of up to 85 Hz. Usually the basic frequency

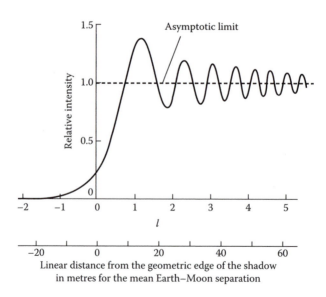

FIGURE 2.44 Fringes at the edge of the lunar shadow.

is lower than this since the Earth's rotation can partially offset the Moon's motion and because the section of the lunar limb that occults the star will generally be inclined to the direction of the motion.

If the star is not a point source, then the fringe pattern shown in Figure 2.44 becomes modified. We may see how this happens by imagining two point sources separated by an angle of about 3.7 milliarc seconds in a direction perpendicular to the limb of the Moon. At a wavelength of 500 nm the first maximum of one star is then superimposed upon the first minimum of the other star and the amplitude of the resultant fringes is much reduced compared with the single point source case. The separation of the sources parallel to the lunar limb is, within reason, unimportant in terms of its effect on the fringes. Thus, an extended source can be divided into strips parallel to the lunar limb (Figure 2.45) and each strip then behaves during the occultation as though it were a centred point source of the relevant intensity. The fringe patterns from all these point sources are then superimposed in the final tracing of the intensity variations. The precise nature of the alteration from the point source case will depend upon the size of the star and upon its surface intensity variations through such factors as limb darkening and gravity darkening as discussed later in this section. A rough guide, however, is that changes from the point source pattern are just detectable for stellar diameters of 2 milliarc seconds, whilst the fringes disappear completely for object diameters of about 200 milliarc seconds. In the latter situation, the diameter may be recoverable from the total length of time that is required for the object to fade from sight. Double stars may also be distinguished from point sources when their separation exceeds about 2 milliarc seconds.

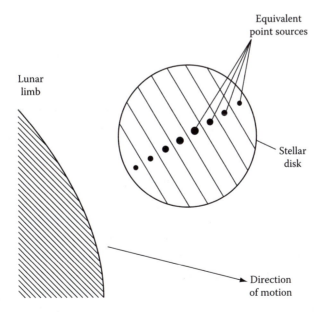

FIGURE 2.45 Schematic decomposition of a non-point source into equivalent point sources for occultation purposes.

2.7.2 Techniques

In complete contrast to almost all other areas of astronomy, the detection of occultations in the optical region is best undertaken using comparatively small telescopes. This arises from the linear size of the fringes which we have seen and is about 12 metres. Telescopes larger than about 1 metre will therefore simultaneously sample widely differing parts of the fringe so that the detected luminosity variations become smeared. Since the signal-to-noise ratio decreases with decreasing size of telescope, the optimum size for observing occultations usually lies between 0.5 and 2 metres.

Most of the photometers described in Section 3.2 can be used to detect the star. However, we have seen that the observed intensity variations can have basic frequencies of 85 Hz with harmonics of several hundred hertz. The photometer and its associated electronics must therefore be capable of responding sufficiently rapidly to pick up these frequencies. The photometers used to detect occultations thus require response times of one to a few milliseconds.

Fringes are also blurred by the waveband over which the photometer is operating. We have seen from Equation 2.64 that the monochromatic fringe pattern is wavelength-dependent. A bichromatic source with wavelengths differing by a factor of 2.37 would have the first maximum for one wavelength superimposed upon the first minimum for the other, so that the fringes would almost disappear. Smaller bandwidths will still degrade the visibility of the fringes although to a lesser extent. Using the standard UBV filters (Section 3.1) only five or six fringes will be detectable even in the absence of all other noise sources. A bandwidth of about 20 nm must be used if the fringes found using a medium-sized telescope are not to deteriorate through this effect.

CCDs can also be used to determine the diffraction pattern of an occulted star. The diffraction pattern moves over the CCD at a calculable velocity and the charges in the pixels are moved through the device at the same rate (cf. image tracking for liquid mirrors, Section 1.1). By this process of TDI, high signal-to-noise ratios can be reached because each portion of the diffraction pattern is observed for a longer time than when a single element detector is used.

Since the observations are obviously always carried out within a few minutes of arc of the brightly illuminated portion of the Moon, the scattered background light is a serious problem. It can be minimised by using clean, dust-free optics, precisely made light baffles and a very small entrance aperture for the photometer, but it can never be completely eliminated. The total intensity of the background can easily be many times that of the star even when all these precautions have been taken. Thus, electronic methods must be used to compensate for its effects. The simplest system is to use a differential amplifier that subtracts a preset voltage from the photometer's signal before amplification. It must be set manually a few seconds before the occultation since the background is continually changing. A more sophisticated approach uses a feedback to the differential amplifier that continually adjusts the preset voltage so that a zero output results. The time constant of this adjustment, however, is arranged to be long, so that whilst the slowly changing background is counteracted, the rapid changes during the occultation are unaffected. The latter system is particularly important for observing disappearances, since it does not require last-minute movements of the telescope onto and off the star.

An occultation can be observed at the leading edge of the Moon (a disappearance event) or at the trailing edge (when it is a reappearance) providing that the limb in each case near the star is not illuminated. Disappearances are by far the easiest events to observe since the star is visible and can be accurately tracked right up to the instant of the occultation. Reappearances can be observed only if a very precise offset is available for the photometer or telescope, or if the telescope can be set onto the star before its disappearance and then kept tracking precisely on its position for tens of minutes until it reappears. The same information content is available, however, from either event. The Moon often has quite a large motion in declination, so that it is occasionally possible for both a disappearance and the following reappearance to occur at a dark limb. This is a particularly important situation since the limb angles will be different between the two events and a complete two-dimensional map of the object can be found.

Scintillation is another major problem in observing an occultation. The frequencies involved in scintillation are similar to those of the occultation so that they cannot be filtered out. The noise level is therefore largely due to the effects of scintillation and it is rare for more than four fringes to be detectable in practice. Little can be done to reduce the problems caused by scintillation since the usual precautions of observing only near the zenith and on the steadiest nights cannot be followed; occultations have to be observed when and where they occur.

Occultations can be used to provide information other than whether the star is a double, or to determine its diameter. Precise timing of the event can give the lunar position to 50 milliarc seconds or better and can be used to calibrate ephemeris time. Until recently, when GPS satellites became available, lunar occultations were used by surveyors to determine the positions on the Earth of isolated remote sites, such as oceanic islands.

Occultations of stars by planets are quite rare but are of considerable interest. Their main use is as a probe for the upper atmosphere of the planet, since it may be observed through the upper layers for some time before it is completely obscured. There has also been the serendipitous discovery of the rings of Uranus in 1977 through their occultation of a star. In spectral regions other than the visible, lunar occultations have in the past found application in the determination of precise positions of objects.

The rapid recent improvement in the angular resolution of the observational techniques available at most wavelengths (Sections 1.1, 1.2, and 1.3) has somewhat reduced the interest in occultation measurements. However, there are research occultation programmes still being undertaken such as that at Williams College in Massachusetts using the Portable Occultation, Eclipse, and Transit System (POETS) photometer on a 0.4-metre telescope. High Speed Imaging Photometer for Occultations (HIPO) is one of the instruments available on the SOFIA telescope and Infrared Spectrometer And Array Camera (ISAAC) on the VLT has been used for lunar occultation work until recently, although the instrument is soon to be decommissioned. Additionally, there are few finer ways of honing the skills of future observational astronomers than observing and analysing occultations of any kind.

2.7.3 Analysis

The analysis of the observations to determine the diameters and/or duplicity of the sources is simple to describe, but very time-consuming to carry out. It involves the comparison of the observed curve with synthesised curves for various models. The synthetic curves are obtained in the manner indicated in Figure 2.45, taking account also of the bandwidth and the size of the telescope. One additional factor to be taken into account is the inclination of the lunar limb to the direction of motion of the Moon. If the Moon were a smooth sphere this would be a simple calculation, but the smooth limb is distorted by lunar surface features. Sometimes the actual inclination at the point of contact may be determinable from artificial satellite photographs of the Moon. On other occasions, the lunar slope must appear as an additional unknown to be determined from the analysis. Very rarely, a steep slope may be encountered and the star may reappear briefly before its final extinction. More often the surface may be too rough on a scale of tens of metres to provide a good straight-edge. Both these cases can usually be recognised from the data and have to be discarded, at least for the purposes of determination of diameters or duplicity.

The analysis for positions of objects simply comprises determining the precise position of the edge of the Moon at the instant of occultation. One observation limits the position of the source to a semicircle corresponding to the leading (or trailing) edge of the Moon. Two observations from separate sites or from the same site at different lunations fix the object's position to the one or two intersections of two such semicircles. This will usually be sufficient to allow an unambiguous optical identification for the object, but if not, further observations can continue to be added to reduce the ambiguity in its position and to increase the accuracy for as long as the occultations continue. Generally, a source that is occulted one month as observed from the Earth, or from a close Earth-orbiting satellite, will continue to be occulted for several succeeding lunations. Eventually, however, the rotation of the lunar orbit in the Saros cycle of 18 years will move the Moon away from the object's position and no further occultations will occur for several years.

2.7.4 Stellar Coronagraphs

A coronagraph is an instrument that is designed to enable a faint object to be studied when that object is so close to a much brighter object that it would normally be swamped by the scattered light from that brighter object. The solar corona is typically less than 0.0001% as bright as the solar photosphere and the first coronagraphs were built to enable the solar corona to be studied outside those occasions when there was a total solar eclipse – hence the name given to the instruments. Solar coronagraphs are discussed in Section 5.3; here we are concerned with high-contrast situations involving other objects.

There are many high-contrast situations that are of interest – such as stars with circumstellar gas and dust envelopes and faint stellar companions to brighter stars –but much of the recent work in this area has been motivated by the desire to study and/or to image exoplanets directly. Although exoplanets can be up to 10 times (or more) the mass of Jupiter, the latter is close to the maximum physical size attained by exoplanets.* To put the problem of directly observing exoplanets in perspective, if alien astronomers were to observe

---

* The material towards the centre of exoplanets that are higher in mass than Jupiter is much more compressed than that at the centre of Jupiter due to the more massive planets' higher central pressures and so their sizes are not much more than that of Jupiter.

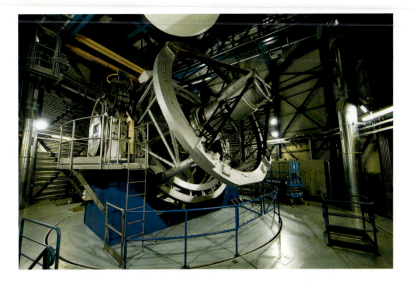

FIGURE 1.26   ESO's 4.1-metre VISTA telescope. The large metal cylinder mounted at and project-ing beyond the top ring of the telescope is the infrared camera containing the cryostat, baffles, filters and, of course, the 67.1-mega-pixel detector array. (Reproduced by kind permission of ESO.)

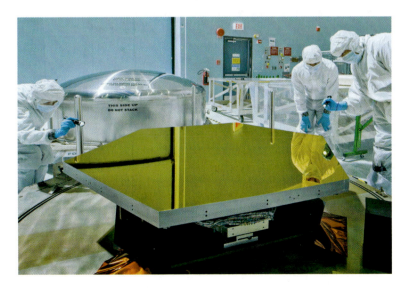

FIGURE 1.53   One of the 18 mirror segments for the JWST. The segment is 1.32 metres in diameter (between the flat edges) and weighs around 20 kg. The gold reflective coating can be clearly seen. (Reproduced by kind permission of NASA/Chris Gunn.)

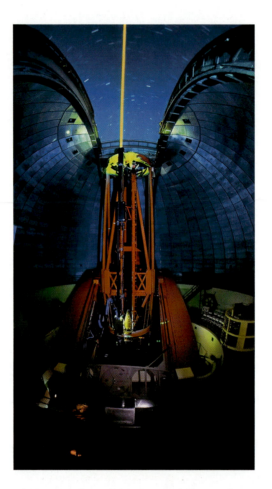

FIGURE 1.69   A laser beam being sent from the 3.05-metre Shane telescope at the Lick Observatory to produce an artificial guide star for its adaptive optics system. The exposure is five minutes. (Reproduced by kind permission of L.Hatch. © 2002 Laurie Hatch. See: www.lauriehatch.com for many more of her beautiful astro-images.)

FIGURE 1.80   The world's largest fully steerable radio telescope, the 100 m × 110 m Robert C. Byrd Green Bank Telescope. (Reproduced by kind permission of NRAO/AUI.)

FIGURE 1.106   The two MAGIC telescopes on the Roque de los Muchachos, La Palma. (Reproduced by kind permission of Robert Wagner, Max-Planck-Institut für Physik.)

FIGURE 2.36 (a) Some of the individual radio telescopes making up ALMA, photographed at night. (Reproduced by kind permission of ALMA [ESO/NAOJ/NRAO], C. Padilla.) (b) Moving one of ALMA's radio telescopes to a new site. (Reproduced by kind permission of ALMA [ESO/NAOJ/NRAO], S. Rossi [ESO].)

FIGURE 4.33 A solar spectrum obtained using the BACHES echelle spectrograph. The Hα line is near the centre of the second segment from the top and the sodium D lines near the centre of the sixth segment from the top. (Reproduced by kind permission of Burwitz/Club of Aficionados in Optical Spectroscopy [CAOS]. For more information on astronomical spectroscopy using small telescopes and by amateur astronomers see http://spectroscopy.wordpress.com/.)

the solar system from a distance of (say) 100 pc, then Jupiter's maximum brightness would be just 0.0000002% that of the Sun and its image would be separated from that of the Sun by just 50 milliarc seconds – and most exoplanets are physically *much* closer to their host stars than Jupiter is to the Sun.

Exoplanets hosted by cool, faint stars or brown dwarfs and in large orbits may, however, reduce the contrast problem considerably – the brown dwarf 2M1207* in the TW Hydrae association is just 100 times brighter than its ≥4 $M_{Jupiter}$ exoplanet.[†] The exoplanet, 2M1207 b is 45 AU out from its host brown dwarf (800 milliarc minutes as seen from Earth) and was first imaged in 2004 using the VLT's Nasmyth Adaptive Optics System (NAOS) and Coudé Near Infrared Camera (CONICA) NaCo instrument but without using the coronagraph option. Since many exoplanets are intrinsically hot (whether because they are still condensing, have internal tidal heating or are heated by their host star) observations in the NIR or MIR may also reduce contrast since the exoplanet's own emissions will be added to its reflected light.

We have already seen that the nulling interferometer (see Section 2.5) can be used to reduce the luminosity of a bright object whilst leaving the fainter object unaffected. This is also essentially what a coronagraph does, except that the latter does so by producing an artificial eclipse. Stellar coronagraphs can be ground- or space-based and the eclipse can be external or internal to the telescope. In all of the variants, careful precautions must be taken to minimise all forms of scattered light and it is in this respect that the coronagraph principally differs from a normal telescope. For the ground-based varieties, the use of real-time atmospheric compensation is essential (see Section 1.1) and quasi-static speckles need suppressing by differential observations of some type.

Quasi-static speckles have a similar appearance to the speckles produced by atmospheric disturbances (Figure 1.67) but change on a time scale of tens of minutes to hours. They may easily be mistaken for faint companions to the main star. These speckles arise from diffraction and interference effects caused by the mirror mountings and supports and other parts of the structure of the telescope and possibly also slight imperfections and misalignments of the mirrors. They alter due to the changing gravitational loading on the telescope structure and consequent sagging as it tracks an object across the sky and the expansion or contraction of the structure due to temperature changes. The quasi-static speckles, however, can often be distinguished from faint companions by obtaining two or more images under circumstances where the speckles change but the companion's image does not. The three current approaches to undertaking this process are called angular differential imaging (ADI), polarimetric differential imaging (PDI) and spectral differential imaging (SDI).

ADI relies upon the usually unwanted phenomenon of field rotation (see Section 1.1) for images at Nasmyth foci. With a telescope on an alt-az mounting, the telescope structure maintains a fixed orientation to the vertical. The speckles therefore also maintain a fixed orientation to the vertical. However, without the field de-rotator in action, the field of view

---

* Its full name is 2MASSWJ1207334-393254.
[†] The mass of an exoplanet is usually only a lower limit obtained by assuming that the plane of the exoplanet's orbit lies along the line of sight. For other orientations of the orbit, the exoplanetary mass will have a larger value.

will rotate as the telescope tracks the image across the sky. Several exposures obtained at intervals of an hour or so (the image de-rotator will need to operate during each exposure) will thus change the mutual orientation of the real objects in the sky and the quasi-static speckles. Subtracting the average frame from each of the raw frames will leave just the planet's image on each frame, plus any uncompensated noise. De-rotating the processed frames and then adding them together will reinforce the exoplanet's image whilst further averaging the background noise. With PDI, two images of the field of view are obtained simultaneously through orthogonally oriented linear polarisers. The star light and the quasi-static speckle light will be almost unpolarised, but the reflected light from the exoplanet will be polarised to some degree or other. Subtracting the two images will then reduce the brightnesses of the stellar and speckle images by proportionally more than the reduction in the brightness of the exoplanet's image, so improving the contrast of the latter. SDI is essentially similar to PDI except that the images are obtained through narrowband filters whose wavelengths are closely adjacent to each other. SDI also differs from PDI in that more than two images can be used by employing more filters so improving on the level of noise reduction.

The Lyot coronagraph is the basic instrument for producing the artificial eclipse internally (shown in Figure 5.29 and described in Section 5.3). The first image of an exoplanet obtained with a coronagraph of this design was that of Fomalhaut b announced in 2008. The HST's Advanced Camera for Surveys (ACS) operating in its coronagraphic mode had obtained the images in 2004 and 2006. More recently, HiCIAO on the 8.2-metre Subaru telescope and which is also uses a Lyot-pattern coronagraph has detected a companion to Gliese 878, a sixth-magnitude star in Lyra. The companion has a mass estimated to lie between 10 and 40 $M_{Jupiter}$ and so could be an exoplanet but is more likely to be a brown dwarf.

Recently, Project 1640* began observations on the 5-metre Hale telescope. The coronagraph uses an apodised image (see Section 5.3) and is combined with an integral field spectrograph and the Palm-3000 adaptive optics system. The Gemini Planet Imager (GPI), currently being commissioned for the 8.1-metre Gemini South telescope, will operate in the NIR using a coronagraph based upon apodised masks. Also being commissioned is SPHERE for the VLT which will incorporate an apodised coronagraph. In space, the JWST will carry three coronagraphs operating from 2.1 to longer than 10 μm, whilst for SPICA it is planned to incorporate a coronagraph operating from 3.5 to 27 μm. A concept currently being discussed is for a space-based stellar coronagraph using an external occulter (or star shade). Provisionally called the New Worlds Observer it might comprise two spacecraft flying in formation with a separation of several tens of thousands of kilometres near the second Lagrangian point of the Sun–Earth system. One spacecraft would carry a 4-metre telescope, the other a 50-metre diameter star shade with a deeply serrated edge. Its aim would be to discover and study at least one habitable Earth-like exoplanet. At the time of writing however, the New Worlds Observer has not been funded beyond its initial study.

A recent development for NIR coronagraphs uses a phase plate. The phase plate is a disk of zinc selenide with annular zones of differing thicknesses. Interference effects arising

---

* Named for its optimum operating wavelength of 1640 nm.

from the different times that it takes the infrared light to pass through the different thicknesses of the material again lead to the suppression of the star's light. A phase plate commissioned for the VLT's NaCo instrument has recently been used to image β Pic b.

Another recent development is to apodise the coronagraph using two aspherical mirrors in place of the apodising masks. The technique, known as Phase-Induced Amplitude Apodization (PIAA) retains the full throughput of the instrument and its angular resolution and is achromatic. It can observe contrasts between two objects at the 10^{10} level and promises to enable exoplanet imaging using 1- to 2-metre-class telescopes. A PIAA coronagraph is under consideration for the planned Exoplanetary Circumstellar Environments and Disk Explorer (EXCEDE) spacecraft which has a possible launch date of 2019.

A completely different approach to producing a stellar coronagraph has recently been suggested and this is to use an optical vortex. The optical vortex looks like a 360° turn of the steps of a spiral staircase. It is a helical mask with steps of progressively increasing thickness constructed from a transparent material. When the phase delay of the radiation passing through the thinnest step is two wavelengths less than that through the thickest step, destructive interference occurs such that the radiation passing through the centre of the mask is eliminated. The operating wavelength of the optical vortex depends upon the step heights and so it is essentially a monochromatic device; nonetheless it may provide the possibility for future stellar coronagraphs that are much more compact than existing devices.

The Simultaneous Differential Imager (SDI) is a device that fulfils a similar function to a coronagraph by a different approach. The image is split into four and passed through narrowband filters centred on and just outside the strong methane absorption band in the near infrared. Jovian-type exoplanets contain methane and will be fainter when seen through the filters within the absorption region than when compared with the image through filters outside that region. The host star though will have the same brightness in all images. Subtracting one image from another will eliminate the star's image, but leave that of the planet. Exoplanets have yet to be observed using the SDI, but brown dwarfs have been detected.

2.8 RADAR

2.8.1 Introduction

Radar astronomy and radio astronomy are very closely linked because the same equipment is often used for both purposes. Radar astronomy is less familiar to the astrophysicist however, for only one star, the Sun, has ever been studied with it and this is likely to remain the case until other stars are visited by spacecraft, which will be a while yet.* Other

---

* A recent study suggests that this comment might be (very slightly) pessimistic. Several studies have shown that sending a spacecraft to a nearby star is, as a conservative estimate, likely to cost in 2012, US dollars between $100 billion for a minimal non-return fly-by mission with results transmitted back to the Earth within a century and $200 trillion for an unmanned sampling and return mission (see the author's *Exoplanets – Finding, Exploring and Understanding Alien Worlds* for a more detailed discussion of interstellar mission possibilities). The new study suggests that a radar system with a receiver/transmitter area of 10,000 km² might detect Earth-sized planets up to 6 pc (20 ly) away at a cost of $20 trillion. The results would be received within 9 years (α Centauri) and 40 years (stars and planets 20 ly away). To put these figures in perspective, the current US gross domestic product is around $20 trillion and in 2012 dollars the Apollo Moon landing programme cost $200 billion.

aspects of astronomy benefit from the use of radar to a much greater extent, so the technique is included, despite being almost outside the limits of this book, for completeness and because its results may find applications in some areas of astrophysics.

Radar, as used in astronomy, can be fairly conventional equipment for use on board spacecraft, or for studying meteor trails, or it can be of highly specialised design and construction for use in detecting the Moon, planets, Sun from the Earth. Its results potentially contain information on three aspects of the object being observed – distance, surface fine scale structure and relative velocity.

2.8.2 Theoretical Principles

Consider a radar system as shown in Figure 2.46. The transmitter has a power P and is an isotropic emitter. The radar cross section of the object is α, defined as the cross-sectional area of a perfectly isotropically scattering sphere which would return the same amount of energy to the receiver as the object. Then the flux at the receiver, f, is given by

$$f = \frac{P\alpha}{4\pi R_1^2 \, 4\pi R_2^2} \qquad (2.69)$$

Normally, we have the transmitter and receiver close together, if they are not actually the same antenna, so that

$$R_1 = R_2 = R \qquad (2.70)$$

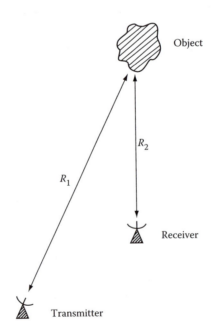

FIGURE 2.46   Schematic radar system.

Furthermore, the transmitter would not, in practice, be isotropic but would have a gain, g (see Section 1.2). If the receiver has an effective collecting area of A_e then the received signal, F, is given by

$$F = \frac{A_e \alpha P g}{16 \pi^2 R^4} \tag{2.71}$$

For an antenna, we have the gain from Equation 1.89

$$g = \frac{4 \pi \nu^2 A_e'}{c^2} \tag{2.72}$$

where A_e' is the effective area of the transmitting antenna. Much of the time in radar astronomy, the transmitting and receiving dishes are the same, so that

$$A_e' = A_e \tag{2.73}$$

and

$$F = \frac{P \alpha A_e^2 \nu^2}{4 \pi c^2 R^4} \tag{2.74}$$

This last equation is valid for objects that are not angularly resolved by the radar. For targets that are comparable with the beam width or larger, the radar cross section, α, must be replaced by an appropriate integral. Thus, for a spherical target which has the radar beam directed towards its centre, we have

$$F \approx \frac{P A_e^2 \nu^2}{4 \pi c^2 R^4} \int_0^{\pi/2} 2 \pi r \alpha(\phi) \sin\phi \left[s\left(\frac{r \sin\phi}{R} \right) \right]^2 d\phi \tag{2.75}$$

where r is the radius of the target and is assumed to be small compared with the distance, R, ϕ is the angle at the centre of the target to a point on its surface illuminated by the radar beam, $\alpha(\phi)$ is the radar cross section for the surface of the target when the incident and returned beams make an angle ϕ to the normal to the surface. The function is normalised so that the integral tends to α as the beam width increases. $s(\theta)$ is the sensitivity of the transmitter/receiver at an angle θ to its optical axis (cf. Equation 1.90).

The amount of flux that is received is not the only criterion for the detection of objects as we saw in Section 1.2 (Equation 1.83). The flux must also be sufficiently stronger than the noise level of the whole system if the returned pulse is to be distinguishable. Now we have seen in Section 1.2 that the receiver noise may be characterised by comparing it with the noise generated in a resistor at some temperature. We may similarly characterise all the other noise sources and obtain a temperature, T_s, for the whole system, which includes the effects of the target and of the transmission paths as well as the transmitter and receiver. Then from Equation 1.84, we have the noise in power terms, N

$$N = 4\,k\,T_s\,\Delta\nu \tag{2.76}$$

where $\Delta\nu$ is the bandwidth of the receiver in frequency terms. The signal-to-noise ratio is therefore

$$\frac{F}{N} = \frac{P\alpha A_e^2 \nu^2}{16\pi kc^2 R^4 T_s \Delta\nu} \tag{2.77}$$

and must be unity or larger if a single pulse is to be detected. Since the target that is selected fixes α and R, the signal-to-noise ratio may be increased by increasing the power of the transmitter, the effective area of the antenna, or the frequency, or by decreasing the system temperature or the bandwidth. Only over the last of these items does the experimenter have any real control; all the others will be fixed by the choice of antenna. However, the bandwidth is related to the length of the radar pulse. Even if the transmitter emits monochromatic radiation, the pulse will have a spread of frequencies given by the Fourier transform of the pulse shape. To a first approximation

$$\Delta\nu \approx \frac{1}{\tau}\,Hz \tag{2.78}$$

where τ is the length of the transmitted pulse in seconds. Thus, increasing the pulse length can increase the signal-to-noise ratio. Unfortunately, for accurate ranging, the pulse needs to be as short as possible, and so an optimum value which minimises the conflict between these two requirements must be sought and this will vary from radar system to radar system and from target to target.

When ranging is not required, so that information is only being sought on the surface structure and the velocity, the pulse length may be increased considerably. Such a system is then known as a continuous wave (CW) radar and its useful pulse length is only limited by the stability of the transmitter frequency and by the spread in frequencies introduced into the returned pulse through Doppler shifts arising from the movement and rotation of the target. In practice, CW radars work continuously even though the signal-to-noise ratio remains that for the optimum pulse. The alternative to CW radar is pulsed radar. Astronomical pulsed radar systems have a pulse length typically between 10 μs and 10 ms, with peak powers of several tens of megawatts. For both CW and pulsed systems, the receiver pass band must be matched to the returned pulse in both frequency and bandwidth. The receiver must therefore be tunable over a short range to allow for the Doppler shifting of the emitted frequency.

The signal-to-noise ratio may be improved by integration. With CW radar, samples are taken at intervals given by the optimum pulse length. For pulsed radar, the pulses are simply combined. In either case the noise is reduced by a factor of $N^{1/2}$, where N is the number of samples or pulses that are combined. If the radiation in separate pulses from a pulsed radar system is coherent, then the signal-to-noise ratio improves directly with N. Coherence, however, may only be retained over the time interval given by the optimum pulse length for the system operating in the CW mode. Thus, if integration is continued over several times the coherence interval, the signal-to-noise ratio for a pulsed system decreases as $(NN'^{1/2})$, where N is the number of pulses in a coherence interval and N' the number of coherence intervals over which the integration extends.

For most astronomical targets, whether they are angularly resolved or not, there is some depth resolution with pulsed systems. The physical length of a pulse is $(c\tau)$ and this ranges from about 3 to 3000 km for most practical astronomical radar systems. Thus, if the depth of the target in the sense of the distance along the line of sight between the nearest and furthest points returning detectable echoes is larger than the physical pulse length, the echo will be spread over a greater time interval then the original pulse. This has both advantages and disadvantages. First, it provides information on the structure of the target with depth. If the depth information can be combined with Doppler shifts arising through the target's rotation, then the whole surface may be mappable, although there may still remain a twofold ambiguity about the equator. Second, and on the debit side, the signal-to-noise ratio is reduced in proportion approximately to the ratio of the lengths of the emitted and returned pulses.

Equation 2.77 and its variants that take account of resolved targets is often called the radar equation. Of great practical importance is its R^{-4} dependence. Thus, if other things are equal, the power required to detect an object increases by a factor of 16 when the distance to the object doubles. Thus, Venus at greatest elongation needs 37 times the power required for its detection at inferior conjunction. The current state-of-the-art for Earth-based planetary radar studies is given by the Arecibo radio telescope's detection of Saturn's rings and its satellite, Titan.

The high dependence upon R can be made to work in our favour, however, when using spacecraft-borne radar systems. The value of R can then be reduced from tens of millions of kilometres to a few hundred. The enormous gain in the sensitivity that results means that the small low-power radar systems that are suitable for spacecraft use are sufficient to map the target in detail. The prime example of this to date is, of course, Venus. Most of the data we have on the surface features have come from radars carried by Magellan and earlier spacecraft. Magellan (1989–1994) was the last spacecraft to study Venus using radar. Two Venus missions are currently in the planning stages and may continue Magellan's radar work. These are the Indian Venus Orbiter mission and the Russian Venera-D. Both are currently scheduled for 2016 launches, but this seems rather optimistic. The radar on board Magellan was of a different type from those so far considered. It is known as synthetic aperture radar (SAR) and has much in common with the technique of aperture synthesis (see Section 2.5). Such radars have also been used on board several remote sensing satellites, such as Seasat and ERS 1 for studying the Earth.

Synthetic aperture radar uses a single antenna which, since it is on board a spacecraft, is moving with respect to the surface of the planet below. The single antenna therefore successively occupies the positions of the multiple antennae of an array (Figure 2.47). To simulate the observation of such an array, all that is necessary, therefore, is to record the output from the radar when it occupies the position of one of the elements of the array and then to add the successive recordings with appropriate shifts to counteract their various time delays. In this way the radar can build up an image whose resolution, along the line-of-flight of the spacecraft, is many times better than that which the simple antenna would provide by itself. In practice, of course, the SAR operates continually, not just at the positions of the elements of the synthesised array.

The maximum array length that can be synthesised in this way is limited by the period over which a single point on the planet's surface can be kept under observation. The resolution of a parabolic dish used as a radar is approximately λ/D. If the spacecraft is at a height, h, the radar footprint of the planetary surface is thus about

$$L = \frac{2\lambda h}{D} \tag{2.79}$$

in diameter (Figure 2.48). Since h, in general, will be small compared with the size of the planet, we may ignore the curvature of the orbit. We may then easily see (Figure 2.49) that a given point on the surface only remains observable whilst the spacecraft moves a distance, L.

Now, the resolution of an interferometer (see Section 2.5) is given by Equation 2.41. Thus, the angular resolution of an SAR of length L is

$$\text{Resolution} = \frac{\lambda}{2L} = \frac{D}{4H} \text{ radians} \tag{2.80}$$

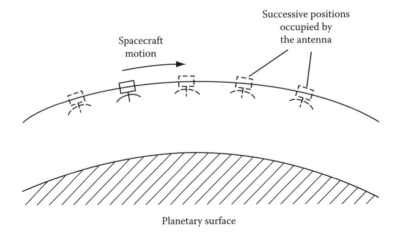

FIGURE 2.47 Synthesis of an interferometer array by a moving single antenna.

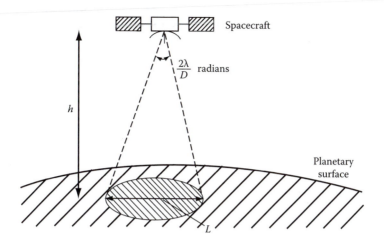

FIGURE 2.48   Radar footprint.

The diameter of the synthesised radar footprint is then

$$\text{Footprint diameter} = 2\left(\frac{D}{4h}\right)h = \frac{D}{2} \qquad (2.81)$$

Equation 2.81 shows the remarkable result that the linear resolution of an SAR is improved by *decreasing* the size of the radar dish! The linear resolution is just half the dish diameter. Other considerations, such as signal-to-noise ratio (Equation 2.77) require the dish to be as large as possible. Thus, an actual SAR system has a dish size that is a compromise and existing systems typically have 5- to 10-metre diameter dishes.

The above analysis applies to a focused SAR; that is to say, an SAR in which the changing distance of the spacecraft from the point of observation is compensated by phase-shifting

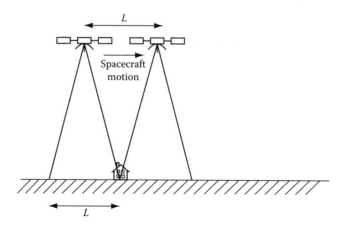

FIGURE 2.49   Maximum length of an SAR.

the echoes appropriately. In an unfocused SAR the point at which the uncorrelated phase shifts due to the changing distance to the observed object reach about $\lambda/4$ limits the synthesised array length. The diameter of the array footprint is then about

$$\text{Unfocused SAR footprint} = \sqrt{2\lambda h} \qquad (2.82)$$

2.8.3 Equipment

Radar systems for use in studying planets have much in common with radio telescopes. Almost any filled aperture radio antenna is usable as a radar antenna. Steerable paraboloids are particularly favoured because of their convenience in use. Unfilled apertures, such as collinear arrays (see Section 1.2) are not used since most of the transmitted power is wasted.

The receiver is also generally similar to those in use in radio astronomy, except that it must be very accurately tunable to provide compensation for Doppler shifts and the stability of its frequency must be very high. Its band pass should also be adjustable so that it may be matched accurately to the expected profile of the returned pulse in order to minimise noise.

Radar systems, however, do require three additional components that are not found in radio telescopes. First and obviously, there must be a transmitter, second a very high stability master oscillator and third an accurate timing system. The transmitter is usually second only to the antenna as a proportion of the cost of the whole system. Different targets or purposes will generally require different powers, pulse lengths, frequencies from the transmitter. Thus, it must be sufficiently flexible to cope with demands that might typically range from pulse emission at several megawatts for a millisecond burst to continuous wave generation at several hundred kilowatts. Separate pulses must be coherent, at least over a time interval equal to the optimum pulse length for the CW mode, if full advantage of integration to improve signal-to-noise ratio is to be gained. The master oscillator is in many ways the heart of the system. In a typical observing situation, the returned pulse must be resolved to 0.1 Hz for a frequency in the region of 1 GHz and this must be accomplished after a delay between the emitted pulse and the echo that can range from minutes to hours. Thus, the frequency must be stable to one part in 10^{12} or 10^{13} and it is the master oscillator that provides this stability. It may be a temperature-controlled crystal oscillator or an atomic clock. In the former case, corrections for ageing will be necessary. The frequency of the master oscillator is then stepped up or down as required and used to drive the exciter for the transmitter, the local oscillators for the heterodyne receivers, the Doppler compensation system and the timing system. The final item, the timing system, is essential if the range of the target is to be found. It is also essential to know with quite a high accuracy when to expect the echo. This arises because the pulses are not broadcast at constant intervals. If this were to be the case then there would be no certain way of relating a given echo to a given emitted pulse. Thus, instead, the emitted pulses are sent out at varying and coded intervals and the same pattern sought amongst the echoes. Since the distance between the Earth and the target also varies, an accurate timing system is vital to allow the resulting echoes with their changing intervals to be found amongst all the noise.

2.8.4 Data Analysis

The analysis of the data in order to obtain the distance to the target is simple in principle, being merely half the delay time multiplied by the speed of light. In practice, the process is much more complicated. Corrections for the atmospheres of the Earth and the target (should it have one) are of particular importance. The radar pulse will be refracted and delayed by the Earth's atmosphere and there will be similar effects in the target's atmosphere with in addition the possibility of reflection from an ionosphere as well as, or instead of, reflection from the surface. This applies especially to solar detection. Then reflection can often occur from layers high above the visible surface; 10.000 km above the photosphere, for example, for a radar frequency of 1 GHz. Furthermore, if the target is a deep one (i.e. it is resolved in depth), then the returned pulse will be spread over a greater time interval than the emitted one so that the delay length becomes ambiguous. Other effects such as refraction or delay in the interplanetary medium may also need to be taken into account.

For a deep, rotating target, the pulse is spread out in time and frequency and this can be used to plot maps of the surface of the object (Figure 2.50). A cross section through the returned pulse at a given instant of time will be formed from echoes from all the points over an annular region such as that shown in Figure 2.51. For a rotating target, which for simplicity we assume has its rotational axis perpendicular to the line of sight, points such as A and A' will have the same approach velocity and their echoes will be shifted to the same higher frequency. Similarly, the echoes from points such as B and B' will be shifted to the same lower frequency. Thus, the intensity of the echo at a particular instant and frequency is related to the radar properties of two points that are equidistant from the equator and are north and south of it. Hence, a map of the radar reflectivity of the target's surface may be produced, although it will have a twofold ambiguity about the equator. If the rotational axis is not perpendicular to the line of sight, then the principle is similar, but the calculations are more complex. However, if such a target can be observed at two different angles of inclination of its rotational axis, then the twofold ambiguity may be removed and a genuine map of the surface produced. Again, in practice, the process needs many

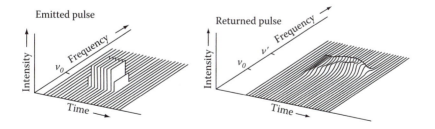

FIGURE 2.50 Schematic change in pulse shape for radar pulses reflected from a deep rotating target. The emitted pulse would normally have a Gaussian profile in frequency, but it is shown here as a square wave in order to enhance the clarity of the changes between it and the echo.

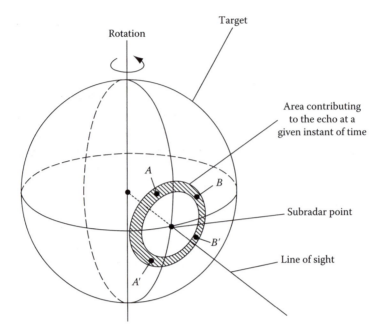

FIGURE 2.51   Region of a deep radar target contributing to the echo at a particular instant of time.

corrections and must have the frequency and time profiles of the emitted pulse removed from the echo by deconvolution (see Section 2.1). The features on radar maps are often difficult to relate to more normal surface features, since they are a conflation of the effects of surface reflectivity, small scale structure, height, gradient.

The relative velocity of the Earth and the target is much simpler to obtain. Provided that the transmitter is sufficiently stable in frequency, then it may be found from the shift in the mean frequency of the echo compared with the mean frequency of the emitted pulse, by the application of the Doppler formula. Corrections for the Earth's orbital and rotational velocities are required. The former is given by Equation 4.40, whilst the latter may be obtained from

$$v_c = v_o + 462 \cos\delta \, \sin(\alpha - \text{LST}) \, \sin(\text{LAT}) \text{ m s}^{-1} \tag{2.83}$$

where v_c is the corrected velocity (m s^{-1}), v_o is the observed velocity (m s^{-1}), α and δ are the right ascension and declination of the target, LST is the local sidereal time at the radar station when the observation is made and LAT is the latitude of the radar station.

The combination of accurate knowledge of both distance and velocity for a planet on several occasions enables its orbit to be calculated to a very high degree of precision. This then provides information on basic quantities such as the astronomical unit and enables tests of metrical gravitational theories such as general relativity to be made and so is of considerable interest and importance even outside the area of astronomy. Producing a precise orbit for a planet by this method requires, however, that the Earth's and the planet's orbits both be known precisely so that the various corrections can be applied. Thus, the procedure is a bootstrap one, with the initial data leading to improved orbits and these improved orbits then allowing better interpretation of the data and so on over many iterations.

The radar reflection contains information on the surface structure of the target. Changes in the phase, polarisation and of the manner in which the radar reflectivity changes with angle can in principle be used to determine rock types, roughness of the surface and so on. In practice, the data are rarely sufficient to allow anything more than vague possibilities to be indicated. The range of unfixed parameters describing the surface is so large that very many different models for the surface can fit the same data. This aspect of radar astronomy has therefore been of little use to date.

2.8.5 Ground Penetrating Radar

Ground penetrating radar (also known as microwave tomography) is frequently used on the Earth as a non-destructive means of studying the top 100 metres of so of the Earth's crust. The radar operates in the 10-MHz to 1-GHz region (30 m to 300 mm) with the radar antenna usually in contact with the ground, although airborne instruments are also used. It seems likely that such radars will be installed on planetary landers in the future. In the meantime, ESA's Mars Express carries Mars Advanced Radar for Subsurface and Ionosphere Sounding (MARSIS) to study the top few kilometres of Mars' crust from orbit. It operates in the 2- to 5-MHz region (150 to 60 m) looking in particular for water/ice interfaces. The instruments and data interpretation for ground-penetrating radar are little different from those of other radars except for the low frequencies at which they operate.

2.8.6 Meteors

The ionised vapour along the track of a meteor reflects radar transmissions admirably. Over the last few decades, therefore, numerous stations have been set up to detect meteors by such means. Since meteors can be observed equally well during the day as at night, many previously unknown daytime meteor showers have been discovered. Either pulsed or CW radar is suitable and a wavelength of a few metres is usually used.

The echoes from a meteor trail are modulated into a Fresnel diffraction pattern (Figure 2.44), with the echoed power expressible in terms of the Fresnel integrals (Equations 2.61 and 2.62) as

$$F(t) = F(x^2 + y^2) \tag{2.84}$$

where F is the echo strength for the fully formed trail, $F(t)$ is the echo strength at time t during the formation of the trail. The parameter l of the Fresnel integrals in this case is the distance of the meteor from the minimum range point. The modulation of the echo arises because of the phase differences between echoes returning from different parts of the trail. The distance of the meteor and hence its height may be found from the delay time as usual. In addition to the distance however, the cross-range velocity can also be obtained from the echo modulation. Several stations can combine, or a single transmitter can have several well-separated receivers, in order to provide further information such as the position of the radiant, atmospheric deceleration of the meteor.

EXERCISE

2.4 If a 10-MW peak power radar is used to observe Venus from the Earth at inferior conjunction, using a 300-metre dish aerial, calculate the peak power required for a satellite-borne radar situated 1000 km above Venus' surface which has a 1-metre dish aerial if the same signal-to-noise ratio is to be achieved. Assume that all the other parameters of the radar systems are similar and take the distance of Venus from the Earth at inferior conjunction to be 4.14×10^7 km.

If the linear resolution of the Earth-based radar at Venus is 10^4 km, what is that of the satellite-based system?

2.9 ELECTRONIC IMAGES

2.9.1 Image Formats

There are various ways of storing electronic images. One currently in widespread use for astronomical images is the Flexible Image Transport System (FITS). There are various versions of FITS, but all have a header, the image in binary form and an end section. The header must be a multiple of 2880 bytes (this number originated in the days of punched card input to computers and represents 36 cards each of 80 bytes). The header contains information on the number of bits representing the data in the image, the number of dimensions of the image (one for a spectrum, two for a monochromatic image, three for a colour image), the number of elements along each dimension, the object observed, the telescope used, details of the observation and other comments. The header is then padded out to a multiple of 2880 with zeros. The image section of a FITS file is simply the image data in the form specified by the header. The end section then pads the image section with zeros until it too is an integer multiple of 2880 bytes.

Other formats may be encountered such as Joint Photographic Experts Group (JPEG), Graphic Interchange Format (GIF) and Tagged Image File Format (TIFF) that are more useful for processed images than raw data.

2.9.2 Image Compression

Most images, howsoever they may be produced, and at whatever wavelength they were obtained, are nowadays stored in electronic form. This is even true of many photographic images that may be scanned using microdensitometers (see Section 2.2) and fed into a computer. Electronic images may be stored by simply recording the precise value of the intensity at each pixel in an appropriate two-dimensional matrix. However, this uses large amounts of memory; for example 400 megabytes for a 32-bit $10,000 \times 10,000$ pixel image. Means whereby storage space for electronic images may be utilised most efficiently are therefore at a premium. The most widely used approach to the problem is to compress the images.

Image compression relies for its success upon some of the information in the image being redundant. That information may therefore be omitted or stored in shortened form without losing information from the image as a whole. Image compression that conserves all the information in the original image is known as lossless compression. For some purposes a reduction in the amount of information in the image may be acceptable (for example, if one is only interested in certain objects or intensity ranges within the image) and this is the basis of lossy compression.

There are three main approaches to lossless image compression. The first is differential compression. This stores not the intensity values for each pixel, but the differences between the values for adjacent pixels. For an image with smooth variations, the differences will normally require fewer bits for their storage than the absolute intensity values. The second approach, known as run compression, may be used when many sequential pixels have the same intensity. This is often true for the sky background in astronomical images. Each run of constant intensity values may be stored as a single intensity combined with the number of pixels in the run, rather than storing the value for every pixel. The third approach, which is widely used on the Internet and elsewhere, where it goes by various proprietary names such as ZIP and DoubleSpace, is similar to run compression except that it stores repeated sequences whatever their nature as a single code, not just constant runs. For some images, lossless compressions by a factor of 5 or 10 are possible.

Sometimes not all the information in the original images is required. Lossy compression can then be by factors of 20, 50 or more. For example, a 32-bit image to be displayed only on a computer monitor which has 256 brightness levels need only be stored to 8-bit accuracy, or only the (say) 64 × 64 pixels covering the image of a planet, galaxy need be stored from a 1000 × 1000 pixel image if that is all that is of interest within the image. Most workers, however, will prefer to store the original data in a lossless form.

Lossless compression techniques do not work effectively on noisy images since the noise is incompressible. Then a lossy compression may be of use since it will primarily be the noise that is lost. An increasingly widely used lossy compression for such images is based on the wavelet transform

$$T = \frac{1}{\sqrt{|a|}} \int f(t)\psi\left(\frac{t-b}{a}\right) dt \tag{2.85}$$

This can give high compression ratios for noisy images. The details of the technique are, however, beyond the scope of this book.

2.9.3 Image Processing

Image processing is the means whereby images are produced in an optimum form to provide the information that is required from them.

Image processing divides into data reduction and image optimisation. Data reduction is the relatively mechanical process of correcting for known problems and faults. Many aspects of it are discussed elsewhere, such as dark frame subtraction, flat fielding and cosmic ray spike elimination on CCD images (see Section 1.1), deconvolution and maximum entropy processing (see Section 2.1) and CLEANing aperture synthesis images (see Section

2.5). Much, if not all, of the data reduction needed may be done automatically with a suitable computer program.

Image optimisation is more of an art than a science. It is the further processing of an image with the aim of displaying those parts of the image that are of interest to their best effect for the objective for which they are required. The same image may therefore be optimised in quite different ways if it is to be used for different purposes. For example, an image of Jupiter and its satellites would be processed quite differently if the surface features on Jupiter were to be studied, compared with if the positions of the satellites with respect to the planet were of interest. There are dozens of techniques used in this aspect of image processing and there are few rules other than experience to suggest which will work best for a particular application. The main techniques are outlined below, but the interested reader will need to look to other sources and to obtain practical experience before being able to apply them with consistent success.

2.9.3.1 Grey Scaling*

Grey scaling is probably the technique in most widespread use. It arises because many detectors have dynamic ranges far greater than those of the computer monitors or hard copy devices that are used to display their images. In many cases the interesting part of the image will cover only a restricted part of the dynamic range of the detector. Grey scaling then consists of stretching that section of the dynamic range over the available intensity levels of the display device.

For example, the spiral arms of a galaxy on a 16-bit CCD image (with a dynamic range from 0 to 65,535) might have CCD level values ranging from 20,100 to 20,862. On a computer monitor with 256 intensity levels, this would be grey scaled so that CCD levels 0 to 20,099 corresponded to intensity 0 (i.e. black) on the monitor, levels 20,100 to 20,103 to intensity 1, levels 20,104 to 20,106 to intensity 2 and so on up to levels 20,859 to 20,862 corresponding to intensity 254. CCD levels 20,863 to 65,535 would then all be lumped together into intensity 255 (i.e. white) on the monitor.

Other types of grey scaling may aim at producing the most visually acceptable version of an image. There are many variations of this technique whose operation may best be envisaged by its effect on a histogram of the display pixel intensities. In histogram equalisation, for example, the image pixels are mapped to the display pixels in such a way that there are equal numbers of image pixels in each of the display levels. For many astronomical images, a mapping which results in an exponentially decreasing histogram for the numbers of image pixels in each display level often gives a pleasing image since it produces a good dark background for the image. The number of other mappings available is limited only by the imagination of the operator.

2.9.3.2 Image Combination

Several images of the same object may be added together to provide a resultant image with lower noise. More powerful combination techniques, such as the subtraction of two

---

* This term is also used for the conversion of a colour image to a monochromatic image.

images to highlight changes, or false-colour displays, are, however, probably in more widespread use.

2.9.3.3 Spatial Filtering

This technique has many variations that can have quite different effects upon the image. The basic process is to combine the intensity of one pixel with those of its surrounding pixels in some manner. The filter may best be displayed by a 3 × 3 matrix (or 5 × 5, 7 × 7 matrices) with the pixel of interest in the centre. The elements of the matrix are the weightings given to each pixel intensity when they are combined and used to replace the original value for the central pixel (Figure 2.52). Some commonly used filters are shown in Figure 2.52.

Various commercial image processing packages are available and enable anyone to undertake extensive image processing. Much of this is biased towards commercial artwork and the production of images for advertising. There are, however, packages specifically designed for astronomical image processing and many are available free of charge.

The most widespread of these packages for optical images is Image Reduction and Analysis Facility (IRAF), produced and maintained by National Optical Astronomy Observatories (NOAO). IRAF is available free for downloading via the Internet (see http://iraf.noao.edu/) or for a cost of a few tens of pounds/dollars on CD-ROM (see previous web site for further details and also for details of IRAF's image processing packages). You will also need your computer to have a UNIX or LINUX operating system. The latter is available, also without charge for downloading onto PCs, at http://www.linux.org/. You are warned however, that although available free of charge, these packages are not easy

Strong smoothing filter

+1/9	+1/9	+1/9
+1/9	+1/9	+1/9
+1/9	+1/9	+1/9

Sharpening filter

0	−1/4	0
−1/4	+2	−1/4
0	−1/4	0

Weak smoothing filter

0	+1/8	0
+1/8	+1/2	+1/8
0	+1/8	0

Edge enhancement

−1	−1	−1
−1	+8	−1
−1	−1	−1

FIGURE 2.52   Examples of commonly used spatial filters.

for an inexperienced computer user to implement successfully. If you have purchased a CCD camera designed for amateur astronomical use, then the manufacturer will almost certainly have provided at least a basic image processing package as a part of the deal. Most such manufacturers also have Internet sites where other software produced by themselves and by users of the cameras may be found. For radio astronomical data and interferometric data, CARA (previously AIPS and AIPS++) is the most widely used data reduction package (see Section 2.5).

CHAPTER 3

Photometry

3.1 PHOTOMETRY

3.1.1 Background

3.1.1.1 Introduction

The measurement of the intensity of electromagnetic energy coming to us from an astronomical object has traditionally been called photometry. However, physics distinguishes between 'photometry' and 'radiometry'. The former is the brightness of an object as perceived by the eye, whilst the latter is the brightness measured in absolute terms.* Thus, when the eye was the only means of determining the brightness of stars, nebulae then astronomers were, indeed, making photometric measurements. By definition however, photometry can only be undertaken within the visible part of the spectrum – typically from 390 to 700 nm (see Section 1.1). Thus, in 1800 when Sir William Herschel discovered infrared radiation from the Sun and in 1801 when Johann Ritter discovered solar ultraviolet radiation, they were making the first radiometric measurements.

Today, no one would consider making eye estimates of the brightnesses of stars for any work that had the least claim to being serious. Basic photometers based upon pin photodiodes are available cheaply and can be used with the smallest of astronomical telescopes. CCDs are also used extensively by amateur as well as professional astronomers and both CCDs and pin diodes fundamentally measure brightness in absolute terms. Thus, on a strict definition, when astronomers nowadays measure the brightness of objects in the sky they are undertaking radiometry, not photometry. However, the use of the term photometry in this context is so well established amongst astronomers that a change to the accurate terminology seems unlikely to occur and so we will continue to use photometry in the astronomers' sense in this book.

---

* The difference is reflected in the units used. For radiometry (for example), radiant flux or radiant power is the total power emitted as EM radiation by an EM source in all directions and over all wavelengths and is measured in watts. For photometry, the equivalent quantity is the luminous flux or luminous power and is the total visible light power emitted by an EM source in all directions as *perceived* by the eye. It is measured in lumens. At a wavelength of 555 nm:
  1 lm(555) = 0.001464 W = 4.1×10^{15} 555 nm photons but the relationship varies with wavelength, so that at 500 nm
  1 lm(500) = 0.004043 W = 1.0×10^{16} 500 nm photons and so on.

Photometry is the measurement of the energy coming from an astronomical object within a restricted range of wavelengths. On this definition, just about every observation that is made is a part of photometry. It is customary and convenient, however, to regard spectroscopy and imaging as separate from photometry, although the differences are blurred in the overlap regions and outside the optical part of the spectrum. Spectroscopy (see Chapter 4) is the measurement of the energy coming from an object within several (usually hundreds or thousands) of adjacent wavebands and with the widths of those wavebands a hundredth of the operating wavelength or less (usually a thousandth to a hundred-thousandth of the operating wavelength). However, some narrowband photometric systems on this definition could be regarded as low-resolution spectroscopy. It is harder to separate imaging (see Chapters 1 and 2) from photometry, especially with array detectors, since much of photometry is now just imaging through an appropriate filter or filters. So it is perhaps best to regard photometry as imaging with the primary purpose of measuring the brightnesses of some or all of the objects within the image and imaging as having the primary purpose of determining the spatial structure or appearance of the object. The situation is further confused by the recent development of integral field spectroscopy (see Section 4.2) where spectra are simultaneously obtained for some or all of the objects within an image.

Since most radio receivers are narrowband instruments and many of the detectors used for FIR and microwave observations operate over restricted wavebands, essentially all their observations come under photometry and have therefore been covered in Sections 1.1 and 1.2. Likewise EUV, X-ray and gamma-ray detectors often have some intrinsic sensitivity to wavelength (i.e. photon energy) and so automatically operate as photometers or low-resolution spectroscopes. Their properties have been covered in Section 1.3. In this section we are thus concerned with photometry as practiced over the near UV, visible, NIR and MIR regions of the spectrum (roughly 100 nm to 100 μm).

3.1.1.2 Magnitudes

The system used by astronomers to measure the visible brightnesses of stars is a very ancient one and for that reason is a very awkward one. It originated with the earliest of the known astronomical catalogues: that due to Hipparchos in the late second century BC. The stars in this catalogue were divided into six classes of brightness, the brightest being of the first class and the faintest of the sixth class. With the invention and development of the telescope such a system proved to be inadequate and it had to be refined and extended. William Herschel initially undertook this, but our present-day system is based upon the work of Norman Pogson in 1856. He suggested a logarithmic scale that approximately agreed with the earlier measures. As we have seen in Section 1.1, the response of the eye is nearly logarithmic, so that the ancient system due to Hipparchos was logarithmic in terms of intensity. Hipparchos' class five stars were about 2.5 times brighter than his class six stars and were about 2.5 times fainter than class four stars and so on. Pogson expressed this as a precise mathematical law in which the magnitude was related to the logarithm of the diameter of the minimum size of telescope required to see the star (i.e. the limiting magnitude) (Equation 3.3). Soon after this, the proposal was adapted to express the magnitude directly in terms of the energy coming from the star and the equation is now usually used in the form:

$$m_1 - m_2 = -2.5 \log_{10}\left(\frac{E_1}{E_2}\right) \tag{3.1}$$

where m_1 and m_2 are the magnitudes of stars 1 and 2 and E_1 and E_2 are the energies per unit area at the surface of the Earth for stars 1 and 2. The scale is awkward to use because of its peculiar base, leading to the ratio of the energies of two stars whose brightnesses differ by one magnitude being 2.512.* Also, the brighter stars have magnitudes whose actual values are smaller than those of the fainter stars. However (like photometry), it seems unlikely that astronomers will change the system, so the student must perforce become familiar with it as it stands.

One aspect of the scale is immediately apparent from Equation 3.1 and that is that the scale is a relative one – the magnitude of one star is expressed in terms of that of another, there is no direct reference to absolute energy or intensity units. The zero of the scale is therefore fixed by reference to standard stars. On Hipparchos' scale, those stars that were just visible to the naked eye were of class six. Pogson's scale is thus arranged so that stars of magnitude six are just visible to a normally sensitive eye from a good observing site on a clear, moonless night. The standard stars are chosen to be non-variables and their magnitudes are assigned so that the above criterion is fulfilled. The primary stars are known as the North Polar Sequence† and comprise stars within 2° of the pole star, so that they may be observed from almost all northern hemisphere observatories throughout the year. Secondary standards may be set up from these to allow observations to be made in any part of the sky. A commonly used standard for the Johnson and Morgan UBV filter system (see below) is the star Vega (α Lyrae) which has a magnitude of +0.03^m in the visible region. Several recent large photometric surveys have started measuring the brightness of objects in terms of absolute units (see below). The data, however, are still presented as a magnitude scale and made as consistent with earlier work as possible.

The faintest stars visible to the naked eye, from the definition of the scale, are of magnitude six; this is termed the limiting magnitude of the eye. For point sources the brightnesses are increased by the use of a telescope by a factor, G, which is called the light grasp of the telescope (see Section 1.1). Since the dark-adapted human eye has a pupil diameter of about 7 mm, G is given by

$$G \approx 2 \times 10^4 \, d^2 \tag{3.2}$$

---

* $=10^{0.4}$. For a difference of two magnitudes, the energies differ by ×6.31 (2.512²), for three magnitudes by ×15.85 (2.512³), four magnitudes by ×39.81 (2.512⁴) and by five magnitudes by exactly ×100 (2.512⁵).

† The stars of the North Polar Sequence were listed in editions 1 to 4 of this book. If anyone should need this information, then it may be found in those earlier editions of this book, or in the primary source, Leavitt H. S., *Annals of the Harvard College Observatory* **71**(3): 49–52.

where d is the telescope's objective diameter in metres. Thus, the limiting magnitude through a visually used telescope, m_L, is

$$m_L = 16.8 + 5 \log_{10} d \qquad (3.3)$$

CCDs and other detection techniques will generally improve on this by some 5 to 10 stellar magnitudes.

The magnitudes so far discussed are all apparent magnitudes and so result from a combination of the intrinsic brightness of the object and its distance. This is obviously an essential measurement for an observer, since it is the criterion determining exposure times; however, it is of little intrinsic significance for the object in question. A second type of magnitude is therefore defined that is related to the actual brightness of the object. This is called the absolute magnitude and it is defined as the apparent magnitude of the object if its distance from the Earth were 10 parsecs. It is usually denoted by M, whilst apparent magnitude uses the lowercase, m. The relation between apparent and absolute magnitudes may easily be obtained from Equation 3.1. Imagine the object moved from its real distance to 10 parsecs from the Earth. Its energy per unit area at the surface of the Earth will then change by a factor $(D/10)^2$, where D is the object's actual distance in parsecs. Thus

$$M - m = -2.5 \log_{10} \left(\frac{D}{10} \right)^2 \qquad (3.4)$$

$$M = m + 5 - 5 \log_{10} D \qquad (3.5)$$

The difference between apparent and absolute magnitudes is called the distance modulus and is occasionally used in place of the distance itself

$$\text{Distance modulus} = m - M = 5 \log_{10} D - 5 \qquad (3.6)$$

Equations 3.5 and 3.6 are valid so long as the only factors affecting the apparent magnitude are distance and intrinsic brightness. However, light from the object often has to pass through interstellar gas and dust clouds, where it may be absorbed. A more complete form of Equation 3.5 is therefore

$$M = m + 5 - 5 \log_{10} D - AD \qquad (3.7)$$

where A is the interstellar absorption in magnitudes per parsec. A typical value for A for lines of sight within the galactic plane is 0.002 mag pc^{-1}. Thus, we may determine the absolute magnitude of an object via Equations 3.5 or 3.7 once its distance is known. More frequently, however, the equations are used in the reverse sense in order to determine the distance. Often the absolute magnitude may be estimated by some independent method and then

$$D = 10^{[(m-M+5)/5]} \, pc \qquad (3.8)$$

Such methods for determining distance are known as standard candle methods, since the object is in effect acting as a standard of some known luminosity. The best known examples are the classical Cepheids with their period-luminosity relationship:

$$M = -1.9 - 2.8 \log_{10} P \qquad (3.9)$$

where P is the period of the variable in days. Many other types of stars such as dwarf Cepheids, RR Lyrae stars, W Virginis stars and β Cepheids are also suitable. Also, Type Ia supernovae, or the brightest novae, or the brightest globular clusters around a galaxy, or even the brightest galaxy in a cluster of galaxies can provide rough standard candles. Yet another method is due to Wilson and Bappu who found a relationship between the width of the emission core of the ionised calcium line at 393.3 nm (the Ca II K line) in late-type stars and their absolute magnitudes. The luminosity is then proportional to the sixth power of the line width.

Both absolute and apparent magnitudes are normally measured over some well-defined spectral region. Whilst the above discussion is quite general, the equations only have validity within a given spectral region. Since the spectra of two objects may be quite different, their relationship at one wavelength may be very different from that at another. The next section discusses the definitions of these spectral regions and their interrelationships.

3.1.2 Filter Systems

Numerous filter systems and filter-detector combinations have been devised. They may be grouped into wideband, intermediate band and narrowband systems according to the bandwidths of their transmission curves. In the visible, wideband filters typically have bandwidths of around 100 nm, intermediate band filters range from 10 to 50 nm, whilst narrowband filters range from 0.05 to 10 nm. The division is convenient for the purposes of discussion here, but is not of any real physical significance.

The filters used in photometry are of two main types based upon either absorption/transmission or on interference. The absorption/transmission filters use salts such as nickel or cobalt oxides dissolved in glass or gelatine, or a suspension of colloid particles. These filters typically transmit over a 100-nm wide region. They are thus mostly used for the wideband photometric systems. Many of these filters will also transmit in the red and infrared and so must be used with a red blocking filter made with copper sulphate. Shortwave blocking filters may be made using cadmium sulphide or selenide, sulphur or gold. Although not normally used for photometry, two other types of filters may usefully be mentioned here: dichroic mirrors and neutral density filters. Neutral density filters absorb by a constant amount over a wide range of wavelengths and may be needed when observing very bright objects like the Sun (see Section 5.3). They are normally a thin deposit of a reflecting metal such as aluminium or stainless steel on a glass or very thin plastic substrate. Dichroic mirrors reflect over one wavelength range and transmit over another. For example a mirror thinly plated with a gold coating is transparent to visual radiation, but reflects the infrared. They may also be produced using multilayer interference coatings. Dichroic mirrors may be used in spectroscopy to feed two spectroscopes operating over different wavelength ranges (see Section 4.2).

Interference filters are mostly Fabry–Perot etalons (see Section 4.1) with a very small separation of the mirrors. The transmission wavelength and the bandwidth of the filter can be tuned by changing the mirror separation and/or their reflectivities (see also tunable filters, Section 4.1 and solar H-α filters, Section 5.3). Combining several Fabry–Perot cavities can alter the shape of the transmission band. Interference filters are usually used for the narrower filters of a photometric system since they can be made with bandwidths ranging from a few tens of nanometres to a hundredth of a nanometre. The recently developed rugate filter can transmit or reflect simultaneously at several wavelengths. It uses interference coatings in which the refractive index varies continuously throughout the depth of the layer.

The earliest filter system was given by the response of the human eye (Figure 1.3) and peaks around 510 nm with a bandwidth of 400 nm or so. Since all normal (i.e. not colour-blind) eyes have roughly comparable responses, the importance of the spectral region within which a magnitude was measured was not realised until the application of the photographic plate to astronomical detection. Then it was discovered that the magnitudes of stars that had been determined from such plates often differed significantly from those found visually. The discrepancy arose from the differing sensitivities of the eye and the photographic emulsion. Early emulsions and today's unsensitised emulsions (see Section 2.2) have a response that peaks near 450 nm; little contribution is made to the image by radiation of a wavelength longer than about 500 nm. The magnitudes determined with the two methods became known as visual and photographic magnitudes, respectively, and are denoted by m_v and m_p or M_v and M_p. These magnitudes may be encountered when reading older texts or working on archive material, but have now largely been replaced by more precise filter systems.

With the development of photoelectric methods of detection of light and their application to astronomy by Joel Stebbins and others in the last century, other spectral regions became available and their more precise definitions were required. There are now a very great number of different photometric systems. Whilst it is probably not quite true that there is one system for every observer, there is certainly at least one separate system for every slightly differing purpose to which photometric measurements may be put. Many of these are developed for highly specialised purposes and so are unlikely to be encountered outside their particular area of application. Other systems have a wider application and so are worth knowing about in more detail.

For a long while, the most widespread of these more general systems was the UBV system, defined by Harold Johnson and William Morgan in 1953. This is a wideband system with the B and V regions corresponding approximately to the photographic and visual responses and with the U region in the violet and UV. The precise definition requires a reflecting telescope with aluminised mirrors and uses an RCA 1P21 photomultiplier.

The filters are

Corning 3384	for the V region
Corning 5030 plus Schott GG 13	for the B region
Corning 9863	for the U region

Filter Name	U	B	V
Central wavelength (nm)	365	440	550
Bandwidth (nm)	70	100	90

The response curves for the filter-photomultiplier combination are shown in Figure 3.1. The absorption in the Earth's atmosphere normally cuts off the short wavelength edge of the U response. At sea level, roughly half of the incident energy at a wavelength of 360 nm is absorbed, even for small zenith angles. More importantly, the absorption can vary with changes in the atmosphere. Ideal U response curves can therefore only be approached at very high-altitude, high-quality observing sites. The effect of the atmosphere on the U response curve is shown in Figure 3.2. The B and V response curves are comparatively unaffected by differential atmospheric absorption across their wavebands; there is only a total reduction of the flux through these filters (see Section 3.2). The scales are arranged so that the magnitudes through all three filters are equal to each other for A0 V stars such as α Lyrae (Vega, see Table 3.1). The choice of filters for the UBV system was made on the basis

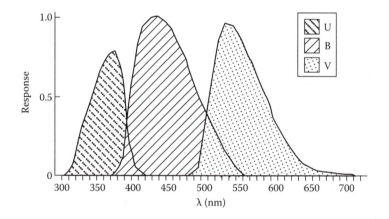

FIGURE 3.1   UBV response curves, excluding the effects of atmospheric absorption.

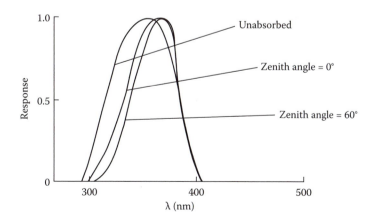

FIGURE 3.2   Effect of atmospheric absorption at sea level upon the U response curve (normalised).

TABLE 3.1 UBV Primary Standard Stars

		Photometric Standards				
Star	Spectral Type	U	B	V	B–V	U–B
α Ari	K2III	4.28	3.16	1.996	1.164	1.12
β Lib	B8V	2.176	2.535	2.605	−0.07	−0.359
α Ser	K2III	5.08	3.80	2.63	1.117	1.28
β Cnc	K4III	6.75	5.00	3.52	1.48	1.75
τ Her	B5IV	3.19	3.75	3.90	−0.15	−0.56
ε CrB	K2III	6.65	5.36	4.13	1.23	1.29
η Hya	B3V	3.38	4.10	4.30	−0.20	−0.72
10 Lac	O9V	3.65	4.67	4.88	−0.21	−1.02
HD 18331	A3V	5.28	5.25	5.16	0.09	0.03
HD 219134	K3V	7.46	6.56	5.57	0.99	0.90
		Colour Index[a] Standards				
α Lyr	A0V	0.02	0.03	0.03	0.00	−0.01
α CrB	A0V	2.214	2.244	2.214	0.03	−0.03
γ UMa	A0V	2.44	2.45	2.44	0.01	−0.01
γ Oph	A0V	3.83	3.79	3.75	0.04	0.04
109 Vir	A0V	3.68	3.71	3.72	−0.01	−0.03
HD 71155	A0V	3.86	3.88	3.90	−0.02	−0.02

[a] The colour index is the difference between two magnitudes through two different filters – in this section. It is also, rather ambiguously, sometimes just called the colour.

of what was easily available at the time. The advent of CCDs has led to several new photometric systems being devised (Section 3.2), with filters more related to the requirements of astrophysics, but the UBV system and its extensions are still in use, if only to enable comparisons to be made with measurements taken some time ago.

The relationship between UBV magnitudes and absolute physical quantities has been determined using laboratory sources such as platinum furnaces. It is sometimes called the Vega system. It is defined (in the units usually used for this purpose*) by the flux for a zero magnitude star of spectral type, A0, at 5500 Å wavelength being

$$F(5500 \text{ Å}) = 3.63 \times 10^{-9} \text{ erg s}^{-1} \text{ cm}^{-2} \text{ Å}^{-1} \tag{3.10}$$

or using frequencies

$$F(545 \text{ THz}) = 3.63 \times 10^{-20} \text{ erg s}^{-1} \text{ cm}^{-2} \text{ Hz}^{-1} \tag{3.11}$$

Any measurement of an object's brightness expressed in the units used in Equation 3.10 can be converted into its V magnitude through the equation

$$V = -2.5 \log_{10} \frac{\int R(\lambda) F_{Object}(\lambda) d\lambda}{\int R(\lambda) F_{Vega}(\lambda) d\lambda} + 0.03 \tag{3.12}$$

---

* In more normal units these would be $F(5500 \text{ Å}) = 3.63 \times 10^{-11} \text{ W m}^{-2} \text{ nm}^{-1}$ or $F(545 \text{ THz}) = 3.63 \times 10^{-22} \text{ W m}^{-2} \text{ Hz}^{-1}$.

(or its equivalent in frequency terms), where $R(\lambda)$ is the V filter response at wavelength λ and the 0.03 is Vega's V magnitude. The relationships for the other filters are based upon the colour indices being zero (i.e. $B - V = U - B = 0$).

There are several extensions to the UBV system for use at longer wavelengths. A widely used system was introduced by Johnson in the 1960s for lead sulphide infrared detectors. With the photographic R and I bands this adds another eight regions:

Filter Name	R	I	J	K	L	M	N	Q
Central wavelength (nm)	700	900	1250	2200	3400	4900	10,200	20,000
Bandwidth (nm)	220	240	380	480	700	300	5000	5000

The longer wavelength regions in this system are defined by atmospheric transmission and so are very variable. Recently, therefore, filters have been used to define all the pass bands and the system has been adapted to make better use of the CCD's response and the improvements in infrared detectors. This has resulted in slight differences to the transmission regions in what is sometimes called the Johnson, Alan Cousins and Ian Glass (JCG) system:

Filter Name	U	B	V	R	I	Z[a]	J
Central wavelength (nm)[b]	367	436	545	638	797	908	1220
Bandwidth (nm)	66	94	85	160	149	96	213

[a] Added for use with CCDs.
[b] Data from *The Encyclopaedia of Astronomy and Astrophysics*, P. Murdin, ed. IoP Press 2001, p. 1642.

Filter Name	H	K	L	M
Central wavelength (nm)	1630	2190	3450	4750
Bandwidth (nm)	307	390	472	460

Two other wideband filter systems are now used because of the large data sets that are available. These are from the Hubble Space Telescope (HST) and the SDSS. The HST Wide field and planetary camera 2 (WFPC2) system had six filters:

Central wavelength (nm)	336	439	450	555	675	814
Bandwidth (nm)	47	71	107	147	127	147

The Wide field camera 3 (WFC3) has 12 filters:

Central wavelength (nm)	222.4	235.9	270.4	335.53	92.1	432.5
Bandwidth (nm)	32.2	46.7	39.8	51.1	89.6	61.8

Central wavelength (nm)	477.3	530.8	588.7	624.2	764.7	802.4
Bandwidth (nm)	134.4	156.2	218.2	146.3	117.1	153.6

The magnitudes obtained using the HST (known as STMag) are related to the flux from the object (cf. Equation 3.10) by

$$m(\lambda) = -2.5 \log_{10} F(\lambda) - 21.1 \tag{3.13}$$

where the constant normalises the magnitudes to that for Vega in the Johnson and Morgan V band.

The SDSS is based upon the 2.5-metre telescope at the Apache Point Observatory in New Mexico and uses five filters that cover the whole range of sensitivity of CCDs:

Filter Name	u'	g'	r'	i'	z'
Central wavelength (nm)	358	490	626	767	907
Bandwidth (nm)	64	135	137	154	147

The SDSS magnitudes are based upon the brightnesses of spectral type, F, subdwarf stars and known as the Gunn system after James Gunn whom proposed the idea along with Trinh Thuan in 1976. The choice of these stars is due to their spectra being much smoother than those of Vega and similar A0 stars. The star BD +17 4708 (spectral type sdF8, $U = 9.724$, $B = 9.886$, $V = 9.45$) is the main standard and has all its SDSS magnitudes equal to 9.50 (thus again more or less normalising the system to the Johnson and Morgan V band). The SDSS measurements are based upon the flux per unit frequency interval (cf. Equation 3.11) and the magnitudes* obtained are related to the flux from the object by

$$m(\nu) = -2.5 \log_{10} F(\nu) - 48.6 \tag{3.14}$$

Amongst the intermediate pass band systems, the most widespread is the uvby or Strömgren system. This was proposed by Bengt Strömgren in the late 1960s and is now in fairly common use. Its transmission curves are shown in Figure 3.3. It is often used with two additional filters centred on the Hβ line, 3 and 15 nm wide, respectively, to provide better temperature discrimination for the hot stars.

Filter Name	u	v	b	y	β_n	β_w
Central wavelength (nm)	349	411	467	547	486	486
Bandwidth (nm)	30	19	18	23	3	15

Narrowband work mostly concentrates on isolating spectral lines. Hα and Hβ are common choices and their variations can be determined by measurements through a pair of filters which are centred on the line but have different bandwidths. No single system is in general use, so that a more detailed discussion is not very profitable. Other spectral features can be studied with narrowband systems where one of the filters isolates the spectral

---

* Magnitudes based upon the flux per unit frequency interval are sometimes called AB magnitudes, the AB being derived from 'absolute'. Unfortunately, the possibility of confusion with absolute magnitude (Equation 3.5) is great, and thus, the term should be avoided.

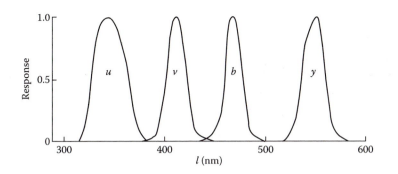

FIGURE 3.3 Normalised transmission curves for the uvby system, not including the effects of atmospheric absorption.

feature and the other is centred on a nearby section of the continuum. The reason for the invention of these and other photometric systems lies in the information that may be obtained from the comparison of the brightness of the object at different wavelengths. Many of the processes contributing to the final spectrum of an object as it is received on Earth preferentially affect one or more spectral regions. Thus, estimates of their importance may be obtained simply and rapidly by measurements through a few filters. Some of these processes are discussed in more detail in the next subsection and in Section 3.2. The most usual features studied in this way are the Balmer discontinuity and other ionisation edges, interstellar absorption and strong emission or absorption lines or bands.

There is one additional photometric system that has not yet been mentioned. This is the bolometric system, or rather bolometric magnitude, since it has only one pass band. The bolometric magnitude is based upon the total energy emitted by the object at all wavelengths. Since it is not possible to observe this in practice, its value is determined by modelling calculations based upon the object's intensity through one or more of the filters of a photometric system. Although X-ray, UV, infrared and radio data are now available to assist these calculations, many uncertainties remain and the bolometric magnitude is still rather imprecise especially for high-temperature stars. The calculations are expressed as the difference between the bolometric magnitude and an observed magnitude. Any filter of any photometric system could be chosen as the observational basis, but in practice, the V filter of the standard UBV system is normally used. The difference is then known as the bolometric correction, BC,

$$BC = m_{bol} - V \tag{3.15}$$

$$= M_{bol} - M_V \tag{3.16}$$

and its scale is chosen so that it is zero for main sequence stars with a temperature of about 6500 K (i.e. about spectral class F5 V). The luminosity of a star is directly related to its absolute bolometric magnitude

$$L_* = 3 \times 10^{28} \times 10^{-0.4 M_{bol}} \quad W \tag{3.17}$$

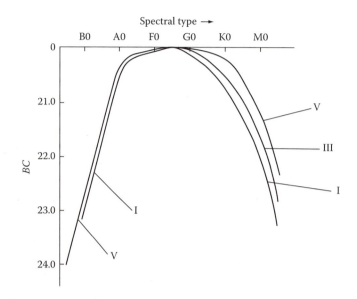

FIGURE 3.4   Bolometric corrections for main sequence stars (type V), giants (type III) and supergiants (type I).

Similarly, the flux of the stellar radiation just above the Earth's atmosphere, f_*, is related to the apparent bolometric magnitude

$$f_* = 2.5 \times 10^{-8} \times 10^{-0.4 m_{bol}} \quad Wm^{-2} \tag{3.18}$$

The bolometric corrections are plotted in Figure 3.4.

Measurements in one photometric system can sometimes be converted to another system. This must be based upon extensive observational calibrations or upon detailed calculations using specimen spectral distributions. In either case, the procedure is very much second best to obtaining the data directly in the required photometric system and requires great care in its application. Its commonest occurrence is for the conversion of data obtained with a slightly nonstandard UBV system to the standard system.

3.1.3 Stellar Parameters

The usual purpose of making measurements of stars in the various photometric systems is to determine some aspect of the star's spectral behaviour by a simpler and more rapid method than that of actually obtaining the spectrum. The simplest approach to highlight the desired information is to calculate one or more colour indices. The colour index is just the difference between the star's magnitudes through two different filters, such as the B and V filters of the standard UBV system. The colour index, C, is then just

$$C = B - V \tag{3.19}$$

where B and V are the magnitudes through the B and V filters, respectively. Similar colour indices may be obtained for other photometric systems such as 439 − 555 for the HST system and $g' - r'$ for the SDSS. Colour indices for the uvby intermediate band system are discussed below. It should be noted, however, that a fairly common alternative usage is for C to denote the so-called international colour index, which is based upon the photographic and photovisual magnitudes. The interrelationship is

$$m_p - m_{pv} = C = B - V - 0.11 \tag{3.20}$$

The $B - V$ colour index is closely related to the spectral type (Figure 3.5) with an almost linear relationship for main sequence stars. This arises from the dependence of both spectral type and colour index upon temperature. For most stars, the B and V regions are located on the long wavelength side of the maximum spectral intensity. In this part of the spectrum the intensity then varies approximately in a black body fashion for many stars. If we assume that the B and V filters effectively transmit at 440 and 550 nm, respectively, then using the Planck equation

$$F_\lambda = \frac{2\pi hc^2}{\lambda^5 \left[e^{hc/\lambda kT} - 1 \right]} \tag{3.21}$$

we obtain

$$B - V = -2.5 \log_{10} \left\{ \frac{(5.5 \times 10^{-7})^5 \left[\exp\left(\dfrac{6.62 \times 10^{-34} \times 3 \times 10^8}{5.5 \times 10^{-7} \times 1.38 \times 10^{-23} \times T} \right) - 1 \right]}{(4.4 \times 10^{-7})^5 \left[\exp\left(\dfrac{6.62 \times 10^{-34} \times 3 \times 10^8}{4.4 \times 10^{-7} \times 1.38 \times 10^{-23} \times T} \right) - 1 \right]} \right\} \tag{3.22}$$

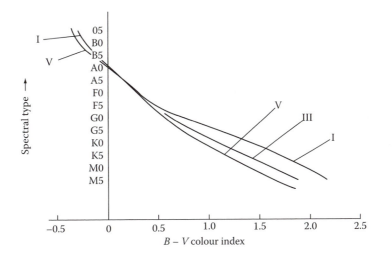

FIGURE 3.5   Relationship between spectral type and $B - V$ colour index.

which simplifies to

$$B - V = -2.5 \log_{10} \left(3.05 \frac{\left[\exp(2.617 \times 10^4 / T) - 1 \right]}{\left[\exp(3.27 \times 10^4 / T) - 1 \right]} \right) \tag{3.23}$$

which for $T < 10{,}000$ K is approximately

$$B - V \approx -2.5 \log_{10} \left(3.05 \frac{\exp(2.617 \times 10^4 / T)}{\exp(3.27 \times 10^4 / T)} \right) \tag{3.24}$$

$$= -1.21 + \frac{7090}{T} \tag{3.25}$$

Now the magnitude scale is an arbitrary one as we have seen and it is defined in terms of standard stars, with the relationship between the B and the V magnitude scale such that $B - V$ is zero for stars of spectral type A0 (Figure 3.5). Such stars have a surface temperature of about 10,000 K and so a correction term of +0.5 must be added to Equation 3.20 to bring it into line with the observed relationship. Thus, we get

$$B - V = -0.71 + \frac{7090}{T} \tag{3.26}$$

giving

$$T = \frac{7090}{(B - V) + 0.71} K \tag{3.27}$$

Equations 3.21 and 3.22 are still only poor approximations because the filters are very broad and so the monochromatic approximation used in obtaining the equations is a very crude one. Furthermore, the effective wavelengths (i.e. the average wavelength of the filter, taking account of the energy distribution within the spectrum) of the filters change with the stellar temperature (Figure 3.6). However, an equation of a similar nature may be fitted empirically to the observed variation (Figure 3.7), with great success for the temperature range 4000 to 10,000 K and this is

$$B - V = -0.865 + \frac{8540}{T} \tag{3.28}$$

$$T = \frac{8540}{(B - V) + 0.865} K \tag{3.29}$$

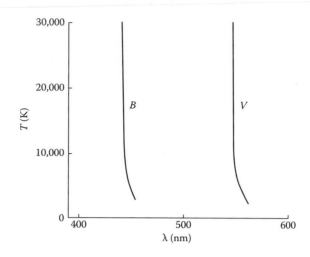

FIGURE 3.6 Change in the effective wavelengths of the standard B and V filters with changing black body temperature.

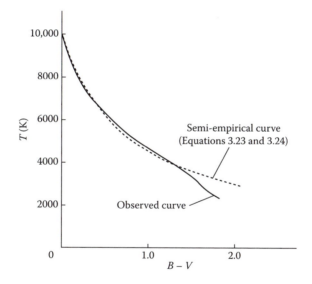

FIGURE 3.7 Observed and semi-empirical $(B - V)/T$ relationships over the lower part of the main sequence.

At higher temperatures, the relationship breaks down, as would be expected from the approximations made to obtain Equation 3.19. The complete observed relationship is shown in Figure 3.8. Similar relationships may be found for other photometric systems that have filters near the wavelengths of the standard B and V filters. For example, the relationship between spectral type and the $b - y$ colour index of the uvby system is shown in Figure 3.9.

For filters that differ from the B and V filters, the relationship of colour index with spectral type or temperature may be much more complex. In many cases the colour index is a measure of some other feature of the spectrum and the temperature relation is of little use or interest. The $U - B$ index for the standard system is an example of such a case, since the

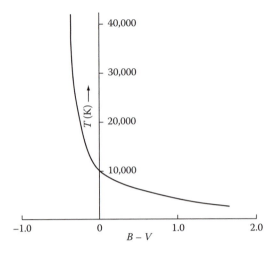

FIGURE 3.8   Observed $B - V$ versus T relationship for the whole of the main sequence.

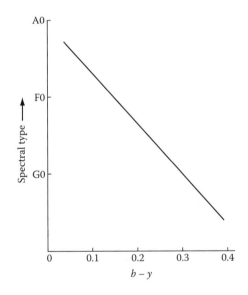

FIGURE 3.9   Relationship between spectral type and $b - y$ colour index.

U and B responses bracket the Balmer discontinuity (Figure 3.10). The extent of the Balmer discontinuity is measured by a parameter, D, defined by

$$D = \log_{10}\left(\frac{I_{364+}}{I_{364-}}\right) \tag{3.30}$$

where I_{364+} is the spectral intensity at wavelengths just longer than the Balmer discontinuity (which is at or near 364 nm) and I_{364-} is the spectral intensity at wavelengths just shorter than the Balmer discontinuity. The variation of D with the $U - B$ colour index is

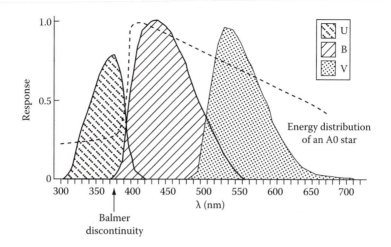

FIGURE 3.10   Position of the standard UBV filters with respect to the Balmer discontinuity.

shown in Figure 3.11. The relationship is complicated by the overlap of the B filter with the Balmer discontinuity and the variation of the effective position of that discontinuity with spectral and luminosity class. The discontinuity reaches a maximum at about A0 for main sequence stars and at about F0 for supergiants. It almost vanishes for very hot stars and for stars cooler than about G0, corresponding to the cases of too high and too low temperatures for the existence of significant populations with electrons in the hydrogen $n = 2$ level. The colour index is also affected by the changing absorption coefficient of the negative hydrogen ion (Figure 3.12) and by line blanketing in the later spectral types. A similar relationship may be obtained for any pair of filters that bracket the Balmer discontinuity and another commonly used index is the c_1 index of the uvby system

$$c_1 = u + b - 2v \tag{3.31}$$

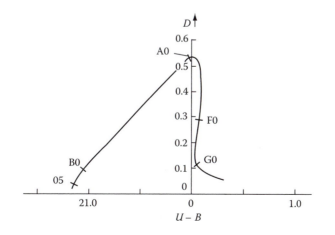

FIGURE 3.11   Variation of D with $U - B$ for main sequence stars.

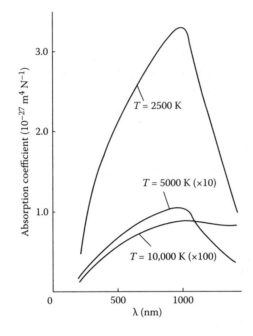

FIGURE 3.12   Change in the absorption coefficient of the negative hydrogen ion with temperature.

The filters are narrower and do not overlap the discontinuity, leading to a simpler relationship (Figure 3.13), but the effects of line blanketing must still be taken into account.

Thus, the $B - V$ colour index is primarily a measure of stellar temperature, whilst the $U - B$ index is a more complex function of both luminosity and temperature. A plot of one against the other provides a useful classification system analogous to the Hertzsprung-Russell diagram. It is commonly called the colour–colour diagram and is shown in Figure 3.14. The deviations of the curves from that of a black body arise from the effects just mentioned: Balmer discontinuity, line blanketing, negative hydrogen ion absorption coefficient

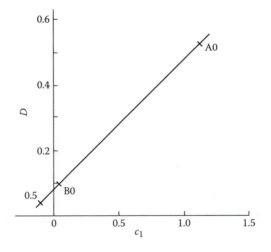

FIGURE 3.13   Variation of D with the c_1 index of the uvby system for main sequence stars.

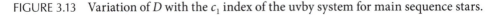

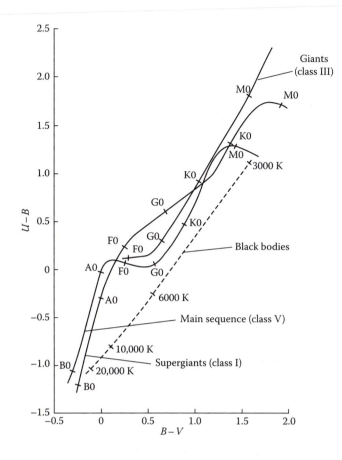

FIGURE 3.14 $(U - B)$ versus $(B - V)$ colour–colour diagram.

variation and also because the radiation originates over a region of the stellar atmosphere rather than in a single layer characterised by a single temperature. The latter effect is due to limb darkening and because the thermalisation length (the distance between successive absorptions and re-emissions) normally corresponds to a scattering optical depth many times unity. The final spectral distribution therefore contains contributions from regions ranging from perhaps many thousands of optical depths below the visible surface to well above the photosphere and so cannot be assigned a single temperature.

Figure 3.14 is based upon measurements of nearby stars. More distant stars are affected by interstellar absorption and since this is strongly inversely dependent upon wavelength (Figure 3.15), the $U - B$ and the $B - V$ colour indices are altered. The star's spectrum is progressively weakened at shorter wavelengths and so the process is often called interstellar reddening.* The colour excesses measure the degree to which the spectrum is reddened

$$E_{U-B} = (U - B) - (U - B)_0 \tag{3.32}$$

$$E_{B-V} = (B - V) - (B - V)_0 \tag{3.33}$$

---

* This is not the same as the redshift observed for distant galaxies. That results from the galaxies' velocities away from us and changes the observed wavelengths of the spectrum lines. With interstellar reddening the spectrum lines' wavelengths are unchanged.

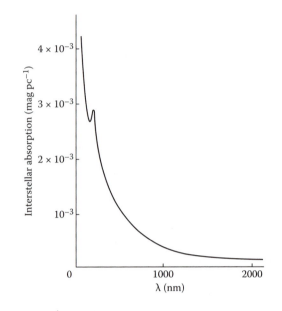

FIGURE 3.15 Average interstellar absorption.

where the subscript 0 denotes the unreddened quantities. These are called the intrinsic colour indices and may be obtained from the spectral type and Figures 3.5 and 3.16. Interstellar absorption (Figure 3.15) varies over most of the optical spectrum in a manner that may be described by the semi-empirical relationship

$$A_\lambda = \frac{6.5\times10^{-10}}{\lambda} - 2.0\times10^{-4} \qquad \text{mag pc}^{-1} \tag{3.34}$$

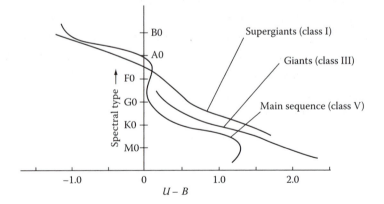

FIGURE 3.16 Relationship between spectral type and $U - B$ colour index.

Hence, approximating U, B and V filters by monochromatic responses at their central wavelengths, we have, for a distance of D parsecs,

$$\frac{E_{U-B}}{E_{B-V}} = \frac{(U-U_0)-(B-B_0)}{(B-B_0)-(V-V_0)} \tag{3.35}$$

$$= \frac{\left[\dfrac{6.5\times10^{-10}}{3.65\times10^{-7}}-2.0\times10^{-4}\right]D - \left[\dfrac{6.5\times10^{-10}}{4.4\times10^{-7}}-2.0\times10^{-4}\right]D}{\left[\dfrac{6.5\times10^{-10}}{4.4\times10^{-7}}-2.0\times10^{-4}\right]D - \left[\dfrac{6.5\times10^{-10}}{5.5\times10^{-7}}-2.0\times10^{-4}\right]D} \tag{3.36}$$

$$= \frac{\left[\dfrac{1}{365}-\dfrac{1}{440}\right]}{\left[\dfrac{1}{440}-\dfrac{1}{550}\right]} \tag{3.37}$$

$$= 1.027 \tag{3.38}$$

Thus, the ratio of the colour excesses is independent of the reddening. This is an important result and the ratio is called the reddening ratio. Its actual value is somewhat different from that given in Equation 3.33, because of our monochromatic approximation and furthermore it is slightly temperature-dependent and not quite independent of the reddening. Its best empirical values are

$$\frac{E_{U-B}}{E_{B-V}} = (0.70\pm0.10)+(0.045\pm0.015)E_{B-V} \qquad \text{at } 30{,}000\,\text{K} \tag{3.39}$$

$$\frac{E_{U-B}}{E_{B-V}} = (0.72\pm0.06)+(0.05\pm0.01)E_{B-V} \qquad \text{at } 10{,}000\,\text{K} \tag{3.40}$$

$$\frac{E_{U-B}}{E_{B-V}} = (0.82\pm0.12)+(0.065\pm0.015)E_{B-V} \qquad \text{at } 5000\,\text{K} \tag{3.41}$$

The dependence upon temperature and reddening is weak, so that for many purposes we may use a weighted average reddening ratio

$$\overline{\left(\frac{E_{U-B}}{E_{B-V}}\right)} = 0.72\pm0.03 \tag{3.42}$$

The colour factor, Q, defined by

$$Q = (U - B) - \overline{\left(\frac{E_{U-B}}{E_{B-V}}\right)}(B - V) \tag{3.43}$$

is then independent of reddening, as we may see from its expansion

$$Q = (U - B)_0 + E_{U-B} - \overline{\left(\frac{E_{U-B}}{E_{B-V}}\right)}\left[(B-V)_0 + E_{B-V}\right] \tag{3.44}$$

$$= (U - B)_0 - \overline{\left(\frac{E_{U-B}}{E_{B-V}}\right)}(B - V)_0 \tag{3.45}$$

$$= (U - B)_0 - 0.72(B - V)_0 \tag{3.46}$$

and provides a precise measure of the spectral class of the B type stars (Figure 3.17). Since the intrinsic $B - V$ colour index is closely related to the spectral type (Figure 3.5), we therefore also have the empirical relation for the early spectral type

$$(B - V)_0 = 0.332\, Q \tag{3.47}$$

shown in Figure 3.18. Hence, we have

$$E_{B-V} = (B - V) - 0.332\, Q = 1.4\, E_{U-B} \tag{3.48}$$

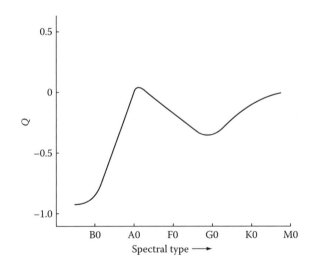

FIGURE 3.17   Variation of colour factor with spectral type.

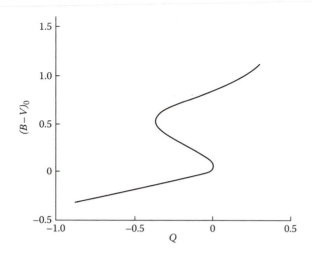

FIGURE 3.18   Relationship between colour factor and the $B - V$ intrinsic colour index.

Thus, simple UBV photometry for hot stars results in determinations of temperature, Balmer discontinuity, spectral type and reddening. The latter may also be translated into distance when the interstellar absorption in the star's direction is known. Thus, we have potentially a very high return of information for a small amount of observational effort and the reader may see why the relatively crude methods of UBV and other wideband filter photometry still remain popular.

Similar or analogous relationships may be set up for other filter systems. Only one further example is briefly discussed here. The reader is referred to the bibliography for further information on these and other systems. The uvby system plus filters centred on Hβ (often called the uvbyβ system) has several indices. We have already seen that $b - y$ is a good temperature indicator (Figure 3.9) and that c_1 is a measure of the Balmer discontinuity for hot stars (Figure 3.13). A third index is labelled m_1 and is given by

$$m_1 = v + y - 2b \tag{3.49}$$

It is sometimes called the metallicity index, for it provides a measure of the number of absorption lines in the spectrum and hence, for a given temperature the abundance of the metals* (Figure 3.19). The fourth commonly used index of the system uses the wide and narrow Hβ filters and is given by

$$\beta = m_n - m_w \tag{3.50}$$

where m_n and m_w are the magnitudes through the narrow and wide Hβ filters, respectively. β is directly proportional to the equivalent width of Hβ (Figure 3.20) and therefore acts

---

* Astronomically speaking all elements heavier than helium are called metals – even the nonmetallic ones. The reason for this is that some metals such as iron have so many spectrum lines (iron has over 20,000 in the visible part of the solar spectrum alone) that their lines dominate most stellar optical spectra.

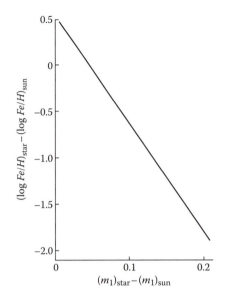

FIGURE 3.19   Relationship between metallicity index and iron abundance for solar-type stars.

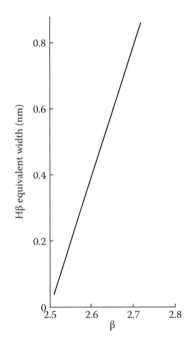

FIGURE 3.20   Relationship between β and the equivalent width of Hβ.

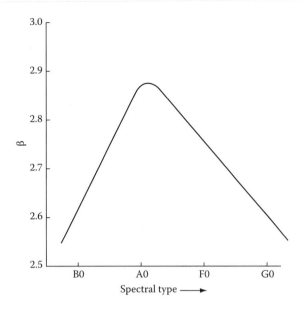

FIGURE 3.21 Relationship of β with spectral type.

as a guide to luminosity and temperature amongst the hotter stars (Figure 3.21). Colour excesses may be defined as before

$$E_{b-y} = (b - y) - (b - y)_0 \tag{3.51}$$

and so on. Correction for interstellar absorption may also be accomplished in a similar manner to the UBV case and results in

$$(c_1)_0 = c_1 - 0.20 \ (b - y) \tag{3.52}$$

$$(m_1)_0 = m_1 + 0.18 \ (b - y) \tag{3.53}$$

$$(u - b)_0 = (c_1)_0 + 2 \ (m_1)_0 \tag{3.54}$$

By combining the photometry with model atmosphere work, much additional information such as effective temperatures and surface gravities may be deduced.

EXERCISES

3.1 Calculate the absolute magnitude of Jupiter from its apparent magnitude at opposition, −2.6 and the absolute magnitude of M 31, from its apparent magnitude, +3.5 (assuming it to be a point source). Their distances are, respectively, 4.2 AU and 670 kpc.

3.2 Calculate the distance of a Cepheid whose apparent magnitude is +13^m on average and whose absolute magnitude is −4^m on average, assuming that it is affected by interstellar absorption at a rate of 1.5×10^{-3} mag pc^{-1}.

3.3 Standard UBV measures of a main sequence star give the following results: $U = 3.19$; $B = 4.03$; $V = 4.19$. Hence, calculate or find $(U - B)$, $(B - V)$; Q; $(B - V)_0$, E_{B-V}; E_{U-B}, $(U - B)_0$; spectral type; temperature; distance (assuming that the interstellar absorption shown in Figure 3.15 may be applied); U_0, B_0, V_0 and M_V.

3.2 PHOTOMETERS

3.2.1 Instruments

3.2.1.1 Introduction

We may distinguish three basic methods of measuring stellar brightness, based upon the eye, the photographic plate, or upon a variety of photoelectric devices. Only the latter is now of any significance, although a few amateur astronomers may still make eye estimates of brightness as part of long-term monitoring of variable stars. Forthcoming developments like the LSST (see Section 1.7), which will monitor tens of millions of stars every few days, will soon make even this usage obsolete. Photographic magnitudes may be encountered when working with archive material. The Hubble space telescope second GSC 2 containing details of 500 million objects, for example, is based upon photographic plates obtained by the 1.2-metre Schmidt cameras at Mount Palomar and the Anglo-Australian Observatory. Precision photometry, however, is now exclusively undertaken by means of photoelectric devices, with the CCD predominating in the optical region and solid-state devices in the infrared.

3.2.1.2 Photographic Photometry

Apart from being used with Schmidt cameras and automatic plate measuring machines to produce relatively low accuracy ($\pm 0.01^m$) measurements of the brightnesses of large numbers of stars and other objects, photography has been replaced by the CCD, even for most amateur work. Only a brief overview of the method is therefore included here. The emulsion's characteristic curve (Figure 2.7) must first be found. This may be via a separate calibration exposure on the edge of the image, or if sufficient standard stars of known magnitudes are on the image, as is likely to be the case with large area Schmidt plates, then they may be measured and used instead.

The images of stars of different magnitudes will be of different sizes. This arises through the image of the star not being an ideal point. It is spread out by diffraction, scintillation, scattering within the emulsion and by reflection off the emulsion's support (halation). The actual image structure will therefore have an intensity distribution that probably approaches a Gaussian pattern. However, the recorded image will have a size governed by the point at which the illumination increases the photographic density above the gross fog level. Images of stars of differing magnitudes will then be recorded as shown schematically in Figure 3.22.

The measurement that is made of the stellar images may simply be that of its diameter, when a semi-empirical relationship of the form

$$D = A + B \log_{10} I \tag{3.55}$$

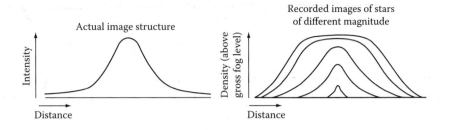

FIGURE 3.22 Schematic variation in image size and density for stars of differing magnitudes recorded on photographic emulsion.

(where D is the image diameter, I the intensity of the star and A and B are constants to be determined from the data) can often be fitted to the calibration curve. A more precise measurement than the actual diameter would be the distance between points in the image which are some specified density greater than that of the gross fog, or between the half-density points. Alternatively, a combination of diameter and density may be measured. Automatic measuring machines scan the image (see Section 5.1), so that the total density may be found by integration. Another system is based upon a variable diaphragm. This is set over the image and adjusted until it just contains that image. The transmitted intensity may then be measured and a calibration curve plotted from known standard stars as before.

The magnitudes obtained from a photograph are of course not the same as those of the standard photometric systems (see Section 3.1). Usually, the magnitudes are photographic or photovisual. However, with careful selection of the emulsion and the use of filters, a response that is close to that of some filter systems may be obtained (Figure 3.23). Alternatively, a combination of emulsion types and filters may be used that gives a useful set of bands, but without trying to imitate any of the standard systems. The Palomar Digital Sky Survey (DPOSS) for example used blue, red and NIR bands, obtained using the now discontinued IIIaJ, IIIaF and IVN plates combined with GG395, RG610 and RG9 filters, respectively.

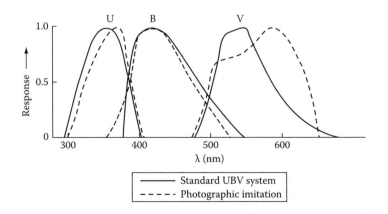

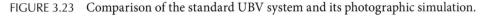

FIGURE 3.23 Comparison of the standard UBV system and its photographic simulation.

3.2.1.3 CCD and Photoelectric Photometers

The most precise photometry relies upon photoelectric devices of various types. These are primarily the CCD in the visual and NIR regions and infrared array detectors (see Section 1.1.15) at longer wavelengths. Photomultipliers continue to be used occasionally, as do p-i-n photo-diodes in small photometers aimed at the amateur market. Photometry with a CCD differs little from ordinary imaging, except that several images of the same area need to be obtained through appropriate filters (see Section 3.1). Anti-blooming CCDs should be avoided since their response is non-linear for objects nearing saturation. The normal data reduction techniques such as dark signal subtraction and flat fielding (see Section 1.1) need to be applied to the images.

The brightness of a star is obtained directly by adding together the intensities of the pixels covered by the star image and subtracting the intensities in a similar number of pixels covering a nearby area of sky background. Conversion to stellar magnitudes then requires the brightnesses of several standard stars also to be measured. With mosaics of several large area CCDs, there are likely to be several standard stars on any image. However, if this is not the case, or with detectors covering smaller areas of the sky, one or more separate calibration images will be needed. These are as identical to the main image as possible – as close in time as possible, as close in altitude and azimuth as possible, identical exposures, filters and data processing Small CCDs sold for the amateur market normally have software supplied by the manufacturer to enable stars' brightnesses to be found. This is accomplished by summing the intensities in a square of pixels that the user centres onto the stars' images one at a time. The size of the square is usually adjustable to take account of changes to the image sizes under different seeing conditions. Clearly the same size of square is needed to obtain the background reading.

Images from larger CCDs are usually processed automatically using software such as IRAF (see Section 2.9) or specialised programs written for the individual CCD. Only when stars' images overlap may the observer need to intervene in this process. Integrated magnitudes for extended objects such as galaxies are obtained in a similar fashion, though more input from the observer may be required to ensure that the object is delineated correctly before the pixel intensities are added together. From a good site under good observing conditions, photometry to an accuracy of $\pm 0.001^m$ is now achievable with CCDs.

At NIR and MIR wavelengths the procedure is similar to that for CCDs. But in the long wavelength infrared regions, array detectors are still small (see Section 1.1) and so only one or two objects are likely to be on each image, making separate calibration exposures essential. In many cases, the object being observed is much fainter than the background noise. For the MIR and FIR regions, special observing techniques such as chopping rapidly between the source and the background, subtracting the background from the signal and integrating the result over a long period need to be used. Some telescopes designed specifically for infrared work have secondary mirrors that can be oscillated to achieve this switching.

Photomultipliers (see Section 1.1) continue to be used when individual photons need to be detected as in the neutrino and cosmic ray Čerenkov detectors (see Sections 1.3 and 1.4) or when very rapid responses are required as in the observation of occultations (see Section

2.7). They may also be used on board spacecraft for UV measurements in the 10- to 300-nm region where CCDs are insensitive.

3.2.2 Observing Techniques

Probably the single most important technique for successful photometry lies in the selection of the observing site. Only the clearest and most consistent of skies are suitable for precision photometry. Haze, dust, clouds, excessive scintillation, light pollution and the variations in these, all render a site unsuitable for photometry. Furthermore, for infrared work, the amount of water vapour above the site should be as low as possible. For these reasons, good photometric observing sites are rare. Generally, they are at high altitudes and are located where the weather is particularly stable. Oceanic islands and mountain ranges with a prevailing wind from an ocean, with the site above the inversion layer, are fairly typical of the best choices. Sites that are less than ideal can still be used for photometry, but their number of good nights will be reduced. Restricting the observations to near the zenith is likely to improve the results obtained at a mediocre observing site.

The second most vital part of photometry lies in the selection of the comparison star(s). This must be non-variable, close to the star of interest, of similar apparent magnitude and spectral class and have its own magnitude known reliably on the photometric system that is in use. Amongst the brighter stars, there is usually some difficulty in finding a suitable standard that is close enough. With fainter stars, the likelihood that a suitable comparison star exists is higher, but the chances of its details being known are much lower. Thus, ideal comparison stars are found only rarely. The best-known variables already have lists of good comparison stars, which may generally be found from the literature. However, for less well studied stars and those being investigated for the first time, there is no such useful information to hand. It may even be necessary to undertake extensive prior investigations, such as checking back through archive records to find non-variable stars in the region and then measuring these to obtain their magnitudes before studying the star itself. An additional star may need to be observed several times throughout an observing session to supply the data for the correction of atmospheric absorption (see below).

A single observatory can clearly only observe objects for a fraction of the time, even when there are no clouds. Rapidly changing objects, however, may need to be monitored continuously over 24 hours. A number of observatories distributed around the globe have therefore instituted various cooperative programmes so that such continuous photometric observations can be made. The Whole Earth Blazar Telescope (WEBT) for example is a consortium of (currently) over 30 observatories studying compact quasars (see also BISON and GONG, Section 5.3). Similarly, the Liverpool telescope on La Palma and the two Faulkes telescopes in Hawaii and Australia are 2-metre robotic telescopes that can respond to GRB alerts in less than 5 minutes.

3.2.3 Data Reduction and Analysis

The reduction of the data is performed in three stages – correction for the effects of the Earth's atmosphere, correction to a standard photometric system and correction to heliocentric time.

The atmosphere absorbs and reddens the star's light and its effects are expressed by Bouguer's law

$$m_{\lambda,0} = m_{\lambda,z} - a_\lambda \sec z \tag{3.56}$$

where $m_{\lambda,z}$ is the magnitude at wavelength λ and at zenith distance z and a_λ is a constant that depends on λ. The law is accurate for zenith distances up to about 60°, which is usually sufficient since photometry is rarely carried out on stars whose zenith distances are greater than 45°. Correction for atmospheric extinction may be simply carried out once the value of the extinction coefficient a_λ is known. Unfortunately, a_λ varies from one observing site to another, with the time of year, from day to day and even throughout the night. Thus, its value must be found on every observing occasion. This is done by observing a standard star at several different zenith distances and by plotting its observed brightness against $\sec z$. Now for zenith angles less than 60° or 70°, to a good approximation $\sec z$ is just a measure of the air mass along the line of sight, so that we may reduce the observations to above the atmosphere by extrapolating them back to an air mass of zero (ignoring the question of the meaning of a value of $\sec z$ of zero; see Figure 3.24). For the same standard star, this brightness should always be the same on all nights, so that an additional point, which is the average of all the previous determinations of the above-atmosphere magnitude of the star, is available to add to the observations on any given night. The extinction coefficient is simply obtained from the slope of the line in Figure 3.24. The coefficient is strongly wavelength dependent (Figure 3.25) and so it must be determined separately for every filter that is being used. Once the extinction coefficient has been found, the above-atmosphere magnitude of the star, m_λ, is given by

$$m_\lambda = m_{\lambda,z} - a_\lambda (1 + \sec z) \tag{3.57}$$

Thus, the observations of the star of interest and its standards must all be multiplied by a factor k_λ

$$k_\lambda = 10^{0.4 a_\lambda (1+\sec z)} \tag{3.58}$$

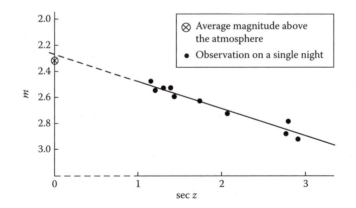

FIGURE 3.24 Schematic variation in magnitude of a standard star with zenith distance.

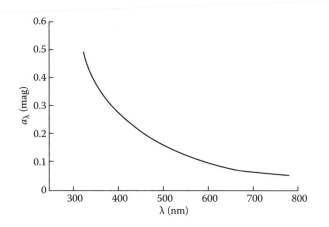

FIGURE 3.25   A typical dependence of the extinction coefficient with wavelength for a good observing site.

to correct them to their unabsorbed values. When the unknown star and its comparisons are very close together in the sky, the differential extinction will be negligible and this correction need not be applied. But the separation must be very small; for example at a zenith distance of 45°, the star and its comparisons must be within 10 minutes of arc of each other if the differential extinction is to be less than a thousandth of a magnitude. If E_λ and E'_λ are the original average signals for the star and its comparison, respectively, through the filter centred on λ then the corrected magnitude is given by

$$m_\lambda = m'_\lambda - 2.5 \log_{10} \left(\frac{E_\lambda 10^{0.4a_\lambda(1+\sec z)}}{E'_\lambda 10^{0.4a_\lambda(1+\sec z')}} \right) \qquad (3.59)$$

$$= m'_\lambda + a_\lambda(\sec z' - \sec z) - 2.5 \log_{10} \left(\frac{E_\lambda}{E'_\lambda} \right) \qquad (3.60)$$

where m_λ and m'_λ are the magnitudes of the unknown star and its comparison, respectively, and z and z' are similarly their zenith distances. The zenith angle is given by

$$\cos z = \sin \phi \sin \delta + \cos \phi \cos \delta \cos(\text{LST} - \alpha) \qquad (3.61)$$

where ϕ is the latitude of the observatory, α and δ are the right ascension and declination of the star and LST is the local sidereal time of the observation.

If the photometer is working in one of the standard photometric systems, then the magnitudes obtained through Equation 3.60 may be used directly and colour indices, colour excesses obtained as discussed in Section 3.1. Often, however, the filters or the detector may not be of the standard type. Then the magnitudes must be corrected to a standard system. This can only be attempted if the difference is small, since absorption lines, ionisation edges will make any large correction exceedingly complex. The required correction is best determined empirically by observing a wide range of the standard stars of the photometric system. Suitable correction curves may then be plotted from the known magnitudes of these stars.

Finally, and especially for variable stars, the time of the observation should be expressed in heliocentric Julian days. The geocentric Julian date is tabulated in Appendix A and if the time of the observation is expressed on this scale, then the heliocentric time is obtained by correcting for the travel time of the light to the Sun

$$T_{Sun} = T + 5.757 \times 10^{-3} \left[\sin \delta_* \sin \delta_{Sun} - \cos \delta_* \cos \delta_{Sun} \cos(\alpha_{Sun} - \alpha_*) \right] \qquad (3.62)$$

where T is the actual time of observation in geocentric Julian days, T_{Sun} is the time referred to the Sun, α_{Sun} and δ_{Sun} are the right ascension and declination of the Sun at the time of the observation and α_* and δ_* are the right ascension and declination of the star. For very precise work, the varying distance of the Earth from the Sun may need to be taken into account as well.

Further items in the reduction and analysis of the data such as the corrections for interstellar reddening, calculation of colour index, temperature were covered in Section 3.1.

3.2.4 High-Speed Photometry

Measurements of objects' brightnesses at intervals of a millisecond or so are necessary for observing occultations (see Section 2.7) and for real-time atmospheric compensation (see Section 1.1). There are also many astrophysical processes, such as phenomena within pulsars and X-ray binary systems, surface oscillations on white dwarfs, solar and stellar flares where observations at millisecond to microsecond intervals may be used profitably.

Almost any photometer based upon a detector with a short response time can be used to observe phenomena needing time resolutions in the 0.1- to 10-s range. As the required time resolution decreases below 0.1 s, however, the read-out times of array-type detectors start to limit photometers' responses. Also, the number of photons available to the detector decreases directly with the time resolution and so shot and other noise sources become more significant. Thus, for time resolutions in the region of 1 μs to 1 ms, special adaptations and/or specially built instruments are needed. In particular, only the few pixels containing the image of interest within an array detector may be read out and large telescopes are needed to gather sufficient light. A recent proposal for high-speed photometry incorporates both these requirements. It is to use the dishes of γ-ray air Čerenkov detectors, such as MAGIC and VERITAS (see Sections 1.3 and 1.4) for the detection of optical transients. With VERITAS, using one of the 12-metre dishes and reading out only from the central (of 499) photomultiplier, a prototype photometer capable of operating at around 5-μs intervals is currently being tested.

3.2.5 Exoplanets

The first planet orbiting a star other than the Sun (an exoplanet) was discovered in 1995.[*] The discovery was made by detecting changes in the radial velocity of the star as the planet

---

[*] Or 1992 if you accept that the incinerated remnants of 'something' (probably a small star) orbiting the unvapourized/recondensed remnants of a star that had exploded as a supernova constitute a normal planetary system.

orbited around it (see Chapter 4). Since then some 900 exoplanets have been found, the majority by the same method as the first – stellar radial velocity variations. However, a substantial minority of the discoveries (about 28%) have come about through photometry of the exoplanets' host stars. Exoplanets may be detected via their host stars' brightness variations through three distinct processes: transits, transit timings and gravitational microlensing.

When an exoplanet transits its host star as seen from the Earth, the brightness of that star decreases slightly because a small portion of its (bright) surface is blocked by the (dark) silhouette of the planet. When a star is already known to host an exoplanet, timings of the transits of that planet may sometimes be early and sometimes late due to another planet's gravitational field pulling the observed exoplanet forwards or backwards. The second exoplanet is thus discovered through timings of the transits of the first exoplanet.

Gravitational microlensing arises because gravity affects the paths of light beams passing near to the surface of a star. For a light (or radio, microwave, IR, UV, X-ray or γ-ray) beam just skimming the surface of the Sun, the deflection is about 1.7 seconds of arc and this deflection decreases as the distance of the light beam from the Sun's surface increases. Thus, when we have a star hosting an exoplanet that, as seen from the Earth, is in front of a more distant star, the light from the latter is bent by the gravitational field of the former. Since both stars will be in relative motion with respect to the Earth, the gravitational lensing effect changes on a time scale of a few days to a few tens of days. The observed effect is that the brightness of the more distant star increases (sometimes by a factor of 10 or more), as the nearer star passes between it and the Earth. If the nearer star has an exoplanet, then the gravitational field of the exoplanet will distort the light variations of the distant star; adding brief sharp brightenings or dimmings to the normal light curve.*

In order to detect an exoplanet through the brightness variations arising from any of these processes, stars' brightnesses must be measured to an accuracy of around one part in a thousand (±0.001^m). Furthermore, since the chances of two stars lining up, as seen from the Earth, sufficiently closely for gravitational lensing to be significant are very small, thousands of stars must be monitored for the brightness changes for months or years at a time. However, some 250 to 300 exoplanets have been discovered though photometric measurements, so clearly these highly stringent observational requirements are being met.

Foremost amongst the (larger number of) exoplanet photometric programmes is NASA's spacecraft, Kepler (Figure 3.26). This was launched in 2009 into a Sun-centred orbit and it monitors the brightnesses of around 150,000 stars in a patch of the sky between Deneb (α Cyg) and Vega (α Lyr) to an accuracy of ±0.002^m. At the time of writing, Kepler has had confirmed discoveries of over 60 exoplanets and has nearly 3000 candidates. A candidate is a star whose photometric variations suggest the presence of an exoplanet but which requires further observations for this to be confirmed. Unfortunately, recent problems with the gyroscopic stabilisation of the spacecraft have cast doubt on whether it will be able to complete its planned 6-year mission.

---

* For further details of these processes see the author's *Exoplanets: Finding, Exploring and Understanding Alien Worlds*, Springer Verlag, 2012.

FIGURE 3.26 The photometric exoplanet discoverer spacecraft, Kepler; an artist's impression of its appearance in space. Note that the background exoplanetary system is completely imaginary; Kepler remains in orbit around the Sun; it does not travel to other planetary systems. (Reproduced by kind permission of NASA, the Kepler Mission and Wendy Stenzel.)

There are many ongoing ground-based programmes aimed at discovering and studying exoplanets. The author's previously mentioned book (see footnote) gives much more information on most of these programmes. Here there is only room to choose (probably unfairly!) a couple to discuss to give an idea of what is being done.

Super Wide Angle Search for Planets (SuperWASP) is based upon 110-mm cameras (yes – that is 110 mm!). There are two instruments, each with eight cameras and based in the northern and southern hemispheres, respectively. The cameras use commercial telephoto lenses feeding CCD detectors and image up to a million stars every minute. Some 80 exoplanets have been discovered from their transits to date.

Optical Gravitational Lensing Experiment (OGLE) uses the 1.3-metre Warsaw telescope feeding thirty-two 2k × 4k pixel CCDs giving it a 1.4-square-degree field of view. It monitors the Milky Way's galactic bulge and the Magellanic clouds for gravitational lensing events. Seventeen exoplanets have been found by the programme at the time of writing.

EXERCISE

3.4 Show that if the differential extinction correction is to be less than Δm magnitudes, then the zenith distances of the star and its comparison must differ by less than Δz, where

$$\Delta z = \frac{\Delta m}{a_\lambda} \cos z \operatorname{cosec} z \qquad radians$$

CHAPTER 4

Spectroscopy

4.1 SPECTROSCOPY

4.1.1 Introduction

Practical spectroscopes are usually based upon one or other of two quite separate optical principles – interference and differential refraction. The former produces instruments based upon diffraction gratings or interferometers, whilst the latter results in prism-based spectroscopes. There are also some hybrid designs. The details of the spectroscopes themselves are considered in the next section; here we discuss the basic optical principles that underlie their designs.

4.1.2 Diffraction Gratings

The operating principle of diffraction gratings relies upon the effects of diffraction and interference of light waves. Logically therefore it might seem that they should be included within the section on interference-based spectroscopes below. Diffraction gratings, however, are in such widespread and common use in astronomy, that they merit a section to themselves.

We have already seen in Section 2.5 the structure of the image of a single source viewed through two apertures (Figure 2.20). The angular distance of the first fringe from the central maximum is λ/d, where d is the separation of the apertures. Hence, the position of the first fringe and of course the positions of all the other fringes, is a function of wavelength. If such a pair of apertures were illuminated with white light, all the fringes apart from the central maximum would thus be short spectra with the longer wavelengths furthest from the central maximum. In an image such as that of Figure 2.20, the spectra would be of little use since the fringes are so broad that they would overlap each other long before a useful dispersion could be obtained. However, if we add a third aperture in line with the first two and separated from the nearer of the original apertures by a distance, d, again, then we find that the fringes remain stationary, but become narrower and more intense. Weak secondary maxima also appear between the main fringes (Figure 4.1). The peak intensities are of course modulated by the pattern from a single slit when looked at on a larger scale, in the manner of Figure 2.20. If further apertures are added in line with the first three and with the same separations, then the principal fringes continue to narrow and intensify and

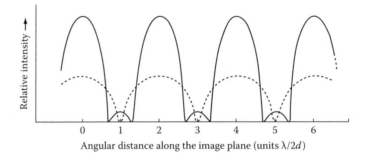

FIGURE 4.1   A small portion of the image structure for a single point source viewed through two apertures (broken curve) and three apertures (full curve).

further weak maxima appear between them (Figure 4.2). The intensity of the pattern at some angle, θ, to the optical axis is given by (cf. Equation 1.12)

$$I(\theta)=I(0)\left[\frac{\sin^2\left(\dfrac{\pi D\sin\theta}{\lambda}\right)}{\left(\dfrac{\pi D\sin\theta}{\lambda}\right)^2}\right]\left[\frac{\sin^2\left(\dfrac{N\pi d\sin\theta}{\lambda}\right)}{\sin^2\left(\dfrac{\pi d\sin\theta}{\lambda}\right)}\right]\tag{4.1}$$

where D is the width of the aperture and N the number of apertures. The term

$$\left[\frac{\sin^2\left(\dfrac{\pi D\sin\theta}{\lambda}\right)}{\left(\dfrac{\pi D\sin\theta}{\lambda}\right)^2}\right]\tag{4.2}$$

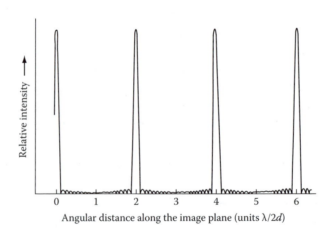

FIGURE 4.2   A small portion of the image structure for a single point source viewed through 20 apertures.

represents the modulation of the image by the intensity structure for a single aperture, whilst the term

$$\left[\frac{\sin^2\left(\dfrac{N\pi d \sin\theta}{\lambda}\right)}{\sin^2\left(\dfrac{\pi d \sin\theta}{\lambda}\right)} \right]$$

(4.3)

represents the result of the interference between N apertures. We may write

$$\Delta = \left(\frac{\pi D \sin\theta}{\lambda} \right)$$

(4.4)

and

$$\delta = \left(\frac{\pi d \sin\theta}{\lambda} \right)$$

(4.5)

and Equation 4.1 then becomes

$$I(\theta) = I(0)\frac{\sin^2\Delta}{\Delta^2}\frac{\sin^2(N\delta)}{\sin^2\delta}$$

(4.6)

Now consider the interference component as δ tends to $m\pi$, where m is an integer. Putting

$$P = \delta - m\pi$$

(4.7)

we have

$$\lim_{\delta\to m\pi}\left(\frac{\sin(N\delta)}{\sin\delta} \right) = \lim_{P\to 0}\left(\frac{\sin[N(P+m\pi)]}{\sin(P+m\pi)} \right)f$$

(4.8)

$$= \lim_{P\to 0}\left(\frac{\sin(NP)\cos(Nm\pi) + \cos(NP)\sin(Nm\pi)}{\sin P\cos(m\pi) + \cos P\sin(m\pi)} \right)$$

(4.9)

$$= \lim_{P\to 0}\left(\pm\frac{\sin(NP)}{\sin P} \right)$$

(4.10)

$$= \pm N \lim_{P \to 0} \left(\frac{\sin(NP)}{NP} \frac{P}{\sin P} \right) \tag{4.11}$$

$$= \pm N \tag{4.12}$$

Hence, integer multiples of π give the values of δ for which we have a principal fringe maximum. The angular positions of the principal maxima are given by

$$\theta = \sin^{-1} \left(\frac{m\lambda}{d} \right) \tag{4.13}$$

and m is usually called the order of the fringe. The zero intensities in the fringe pattern will be given by

$$N\delta = m' \pi \tag{4.14}$$

where m' is an integer, but excluding the cases where $m' = mN$ that are the principal fringe maxima. Their positions are given by

$$\theta = \sin^{-1} \left(\frac{m'\lambda}{Nd} \right) \tag{4.15}$$

The angular width of a principal maximum, W, between the first zeros on either side of it is thus given by

$$W = \frac{2\lambda}{Nd\cos\theta} \tag{4.16}$$

The width of a fringe is therefore inversely proportional to the number of apertures, whilst its peak intensity, from Equations 4.6 and 4.12, is proportional to the square of the number of apertures. Thus, for a bichromatic source observed through a number of apertures we obtain the type of image structure shown in Figure 4.3. The angular separation of fringes of the same order for the two wavelengths, for small values of θ, can be seen from Equation 4.13 to be proportional to both the wavelength and to the order of the fringe, whilst the fringe width is independent of the order (Equation 4.16). For a white light source, by a simple extension of Figure 4.3, we may see that the image will consist of a series of spectra on either side of a white central image. The Rayleigh resolution within this image is obtained from Equation 4.16:

$$W' = \frac{\lambda}{Nd\cos\theta} \tag{4.17}$$

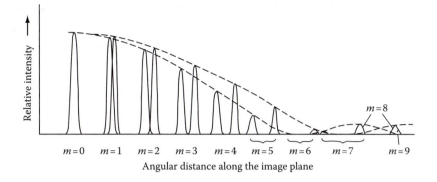

FIGURE 4.3 A portion of the image structure for a single bichromatic point source viewed through several apertures.

and is independent of the fringe order. The ability of a spectroscope to separate two wavelengths is called the spectral resolution and is denoted by W_λ and may now be obtained from Equation 4.17:

$$W_\lambda = W' \frac{d\lambda}{d\theta} \tag{4.18}$$

but from Equation 4.13

$$\frac{d\lambda}{d\theta} = \frac{d}{m} \cos\theta \tag{4.19}$$

so that

$$W_\lambda = \frac{\lambda}{Nm} \tag{4.20}$$

The spectral resolution thus improves directly with the fringe order because of the increasing dispersion of the spectra.

More commonly the resolution is expressed as the ratio of the operating wavelength to the spectral resolution and denoted by R (often and confusingly also called the spectral resolution)

$$R = \frac{\lambda}{W_\lambda} = Nm \tag{4.21}$$

The resolution for a series of apertures is thus just the product of the number of apertures and the order of the spectrum. It is independent of the width and spacing of the apertures.

From Figure 4.3, we may see that at higher orders the spectra are overlapping. This occurs at all orders when white light is used. The difference in wavelength between two superimposed wavelengths from adjacent spectral orders is called the free spectral range, Σ. From Equation 4.13 we may see that if λ_1 and λ_2 are two such superimposed wavelengths, then

$$\sin^{-1}\left(\frac{m\lambda_1}{d}\right) = \sin^{-1}\left(\frac{(m+1)\lambda_2}{d}\right) \tag{4.22}$$

that is, for small angles

$$\Sigma = \lambda_1 - \lambda_2 \approx \frac{\lambda_2}{m} \tag{4.23}$$

For small values of m, Σ is therefore large and the unwanted wavelengths in a practical spectroscope may be rejected by the use of filters. Some spectroscopes, such as those based on Fabry–Perot etalons and echelle gratings, however, operate at very high spectral orders and both of the overlapping wavelengths may be needed. Then it is necessary to use a cross disperser so that the final spectrum consists of a two-dimensional (2D) array of short sections of the spectrum (see Figure 4.32).

A practical device for producing spectra by diffraction uses a large number of closely spaced, parallel, narrow slits or grooves and is called a diffraction grating. Typical gratings for astronomical use have between 100 and 1000 grooves per millimetre and 1000 to 50,000 grooves in total. They are used at orders ranging from 1 up to 200 or so. Thus, the resolutions range from 10^3 to 10^5. Although the earlier discussion was based upon the use of clear apertures, a narrow plane mirror can replace each aperture without altering the results. Thus, diffraction gratings can be used either in transmission or reflection modes. Most astronomical spectroscopes are in fact based upon reflection gratings. Often the grating is inclined to the incoming beam of light, but this changes the discussion only marginally. There is a constant term, $d \sin i$, added to the path differences, where i is the angle made by the incoming beam with the normal to the grating. The whole image (Figure 4.3) is shifted an angular distance i along the image plane. Equation 4.13 then becomes

$$\theta = \sin^{-1}\left[\left(\frac{m\lambda}{d}\right) - \sin i\right] \tag{4.24}$$

and in this form is often called the grating equation.

Volume phase holographic gratings (VPHGs) are currently starting to be used within astronomical spectroscopes. These have a grating in the form of a layer of gelatine within which the refractive index changes (Section 4.2), with the lines of the grating produced by regions of differing refractive indices. VPHGs operate through Bragg diffraction

(Section 1.3, Figure 1.102 and Equation 1.104). Their efficiencies can thus be up to 95% in the first order. They can be used either as transmission or reflection gratings and replace conventional gratings in spectroscopes at the appropriate Bragg angle for the operating wavelength.

To form a part of a spectroscope, a grating must be combined with other optical elements. The basic layout is shown in Figure 4.4; practical designs are discussed in Section 4.2. The grating is illuminated by parallel light that is usually obtained by placing a slit at the focus of a collimating lens, but sometimes may be obtained simply by allowing the light from a very distant object to fall directly onto the grating. After reflection from the grating the light is focused by the imaging lens to form the required spectrum and this may then be recorded, observed through an eyepiece, projected onto a screen as desired. The collimator and imaging lenses may be simple lenses as shown, in which case the spectrum will be tilted with respect to the optical axis because of chromatic aberration, or they may be achromats or mirrors.

The angular dispersion of a grating is not normally used as a parameter of a spectroscopic system. Instead it is combined with the focal length of the imaging element to give either the linear dispersion or the reciprocal linear dispersion. If x is the linear distance along the spectrum from some reference point, then we have for an achromatic imaging element of focal length f_2,

$$\frac{dx}{d\lambda} = f_2 \frac{d\theta}{d\lambda} \tag{4.25}$$

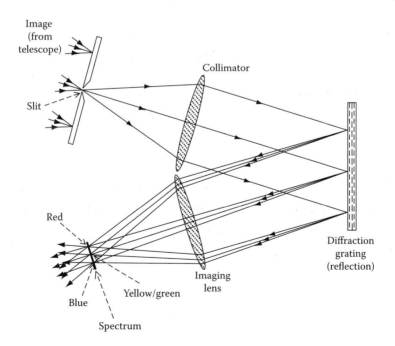

FIGURE 4.4 Basic optical arrangement of a reflection grating spectroscope.

where θ is small and is measured in radians. From Equation 4.24, the linear dispersion within each spectrum is thus given by

$$\frac{dx}{d\lambda} = \pm \frac{mf_2}{d\cos\theta} \tag{4.26}$$

Now since θ varies very little over an individual spectrum, we may write

$$\frac{dx}{d\lambda} \approx \text{constant} \tag{4.27}$$

The dispersion of a grating spectroscope is thus roughly constant compared with the strong wavelength dependence of a prism spectroscope (Equation 4.69). More commonly, the reciprocal linear dispersion, $\frac{d\lambda}{dx}$, is quoted and used. For practical astronomical spectrometers, this usually has values in the range

$$10^{-7} < \frac{d\lambda}{dx} < 5 \times 10^{-5} \tag{4.28}$$

The commonly used units are nanometres change of wavelength per millimetre along the spectrum so that the above range is from 0.1 to 50 nm mm^{-1}. The use of Å mm^{-1} is still fairly common practice amongst astronomers; the magnitude of the dispersion is then a factor of 10 larger than the standard measure. The advent of electronic detectors has also made the use of nanometres or angstroms per pixel a measure of dispersion.

The resolving power of a spectroscope is limited by the spectral resolution of the grating, the resolving power of its optics (see Section 1.1) and by the projected slit width. The spectrum is formed from an infinite number of monochromatic images of the entrance slit. It is easy to see that the width of one of these images, S, is given by

$$S = s\frac{f_2}{f_1} \tag{4.29}$$

where s is the slit width, f_1 is the collimator's focal length and f_2 is the imaging element's focal length. In wavelength terms, the slit width, $S\frac{d\lambda}{dx}$, is sometimes called the spectral purity of the spectroscope. The entrance slit must have a physical width of $s_{\max}$ or less, if it is not to degrade the spectral resolution, where

$$s_{\max} = \frac{\lambda f_1}{Nd\cos\theta} \tag{4.30}$$

(cf. Equation 4.77).

If the optics of the spectroscope are well corrected then we may ignore their aberrations and consider only the diffraction limit of the system. When the grating is fully illuminated, the imaging element will intercept a rectangular beam of light. The height of the beam is just the height of the grating and has no effect upon the spectral resolution. The width of the beam, D, is given by

$$D = L \cos\theta \tag{4.31}$$

where L is the length of the grating and θ is the angle of the exit beam to the normal to the plane of the grating. The diffraction limit is just that of a rectangular slit of width D. So that from Figure 1.37, the linear Rayleigh limit of resolution, W'', is given by

$$W'' = \frac{f_2 \lambda}{D} \tag{4.32}$$

$$= \frac{f_2 \lambda}{L \cos\theta} \tag{4.33}$$

If the beam is limited by some other element of the optical system and/or is of circular cross section, then D must be evaluated as may be appropriate, or the Rayleigh criterion for the resolution through a circular aperture (Equation 1.62) used in place of that for a rectangular aperture. Optimum resolution occurs when

$$S = W'' \tag{4.34}$$

i.e.

$$s = \frac{f_1 \lambda}{D} \tag{4.35}$$

$$= \frac{f_1 \lambda}{L \cos\theta} \tag{4.36}$$

The major disadvantage of a grating as a dispersing element is immediately obvious from Figure 4.3; the light from the original source is spread over a large number of spectra. The grating's efficiency in terms of the fraction of light concentrated into the spectrum of interest is therefore very low. This disadvantage, however, may be largely overcome with reflection gratings through the use of the technique of blazing the grating. In this technique, the individual mirrors that comprise the grating are angled so that they concentrate the light into a single narrow solid angle (Figure 4.5). For instruments based upon the use

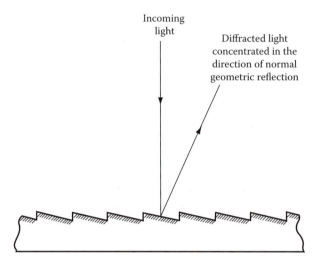

FIGURE 4.5   Enlarged section through a blazed reflection grating.

of gratings at low orders, the angle of the mirrors is arranged so that the light is concentrated into the spectrum to be used and by this means up to 90% efficiency can be achieved. In terms of the interference patterns, the grating is designed so that central peak due to an individual aperture just extends over the width of the desired spectrum. The blaze angle then shifts that peak along the array of spectra until it coincides with the desired order.

For those spectroscopes that use gratings at high orders, the grating can still be blazed, but then the light is concentrated into short segments of many different orders of spectra. By a careful choice of parameters, these short segments can be arranged so that they overlap slightly at their ends and so coverage of a much wider spectral region may be obtained by producing a montage of the segments. Transmission gratings can also be blazed although this is less common. Each of the grooves then has the cross section of a small prism, the apex angle of which defines the blaze angle. Blazed transmission gratings for use at infrared wavelengths can be produced by etching the surface of a block of silicon in a similar manner to the way in which integrated circuits are produced.

Another problem that is intrinsically less serious but which is harder to counteract is that of shadowing. If the incident and/or reflected light makes a large angle to the normal to the grating, then the step-like nature of the surface (Figure 4.5) will cause a significant fraction of the light to be intercepted by the vertical portions of the grooves and so lost to the final spectrum. There is little that can be done to eliminate this problem except either to accept the light loss, or to design the system so that large angles of incidence or reflection are not needed.

Curved reflection gratings are frequently produced. By making the curve that of an optical surface, the grating itself can be made to fulfil the function of the collimator and/or the imaging element of the spectroscope, thus reducing light losses and making for greater simplicity of design and reduced costs. The grooves should be ruled so that they appear parallel and equally spaced when viewed from infinity. The simplest optical principle employing a curved grating is that due to Henry Rowland. The slit, grating and spectrum all lie on a single circle that is called the Rowland circle (Figure 4.6). This has a diameter

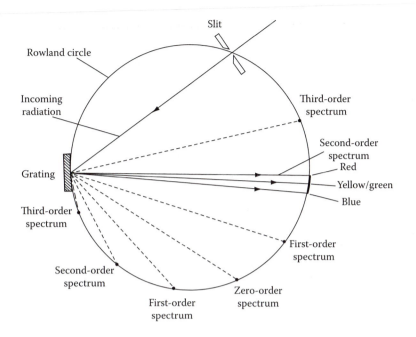

FIGURE 4.6   Schematic diagram of a spectroscope based upon a Rowland circle, using a curved grating blazed for the second order.

equal to the radius of curvature of the grating. The use of a curved grating at large angle to its optical axis introduces astigmatism and spectral lines may also be curved due to the varying angles of incidence for rays from the centre and ends of the slit (cf. Figure 4.14). Careful design, however, can reduce or eliminate these defects and there are several practical designs for spectroscopes based upon the Rowland circle (see Section 4.2). Aspherical curved gratings are also possible and can be used to provide very highly corrected designs with few optical components.

Higher dispersion can be obtained by using immersed reflection gratings. The light in these interacts with the grating within a medium other than air (or a vacuum). The refractive index of the medium shortens the wavelength so that the groove spacing is effectively increased relative to the wavelength and so the dispersion is augmented (Equation 4.19). One approach to producing an immersion grating is simply to illuminate the grating from the back (i.e. through the transparent substrate). A second approach is to flood the grating with a thin layer of oil kept in place by a cover sheet.

A grating spectrum generally suffers from unwanted additional features superimposed upon the desired spectrum. Such features are usually much fainter than the main spectrum and are called ghosts. They arise from a variety of causes. They may be due to overlapping spectra from higher or lower orders, or to the secondary maxima associated with each principal maximum (Figure 4.2). The first of these is usually simple to eliminate by the use of filters since the overlapping ghosts are of different wavelengths from the overlapped main spectrum. The second source is usually unimportant since the secondary maxima are very faint when more than a few tens of apertures are used, though they still contribute to the wings of the PSF. Of more general importance are the ghosts that arise through errors in the grating. Such errors most commonly take the form of a periodic variation in the groove spacing. A variation with a single period gives rise to Rowland ghosts that appear as faint lines close to and on either side of strong spectrum lines. Their intensity is proportional to the square of the order of the spectrum. Thus, echelle gratings (see below) must be of very high quality since they may use spectral orders of several hundred. If the error is multi-periodic, then Lyman ghosts of strong lines may appear. These are similar to the Rowland ghosts, except that they can be formed at large distances from the line that is producing them. Some compensation for these errors can be obtained through deconvolution of the PSF (see Section 2.1), but for critical work, the only real solution is to use a grating without periodic errors, such as a holographically produced grating.

Wood's anomalies may also sometimes occur. These do not arise through grating faults, but are due to light that should go into spectral orders behind the grating (were that to be possible) reappearing within lower-order spectra. The anomalies have a sudden onset and a slower decline towards longer wavelengths and are almost 100% plane polarised. They are rarely important in efficiently blazed gratings.

By increasing the angle of a blazed grating, we obtain an echelle grating (Figure 4.7). This is illuminated more or less normally to the groove surfaces and therefore at a large angle to the normal to the grating. It is usually a very coarse grating – 10 lines per millimetre is not uncommon – so that the separation of the apertures, d, is very large. The reciprocal linear dispersion

$$\frac{d\lambda}{dx} = \pm \frac{d \cos^3 \theta}{m f_2} \tag{4.37}$$

is therefore also very large. Such gratings concentrate the light into many overlapping high-order spectra and so from Equation 4.21, the resolution is very high. A spectroscope that is based upon an echelle grating requires a second low dispersion grating or prism whose dispersion is perpendicular to that of the echelle and so is called a cross-disperser in order to separate out each of the orders (see Section 4.2).

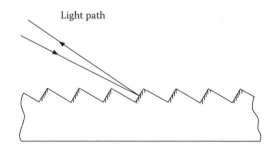

FIGURE 4.7   Enlarged view of an echelle grating.

A quantity known variously as throughput, etendu, or light gathering power, is useful as a measure of the efficiency of the optical system. It is the amount of energy passed by the system when its entrance aperture is illuminated by unit intensity per unit area per unit solid angle and it is denoted by u

$$u = \tau A \Omega \qquad (4.38)$$

where Ω is the solid angle accepted by the instrument, A is the area of its aperture and τ is the fractional transmission of its optics, incorporating losses due to scattering, absorption, imperfect reflection. For a spectroscope, Ω is the solid angle subtended by the entrance slit at the collimator or, for slitless spectroscopes, the solid angle accepted by the telescope-spectroscope combination. A is the area of the collimator or the effective area of the dispersing element, whichever is the smaller, τ will depend critically upon the design of the system, but as a reasonably general rule it may be taken to be the product of the transmissions for all the surfaces. These will usually be in the region of 0.8 to 0.9 for each surface, so that τ for the design illustrated in Figure 4.4 will have a value of about 0.4. Older spectroscope designs often had very low throughputs – less than 10% was not uncommon. Even today a throughput of 40% such as has been achieved by the Bench-mounted High Resolution Optical Spectrometer (bHROS) instrument for the Gemini South telescope is considered excellent. Thus, much light is lost in spectroscopy compared with direct imaging and since the remaining light is then spread out over the spectrum, the exposures needed for spectroscopy are typically a hundred or more times those required for imaging.

The product of resolution and throughput, P, is a useful figure for comparing the performances of different spectroscope systems

$$P = R u \qquad (4.39)$$

Normally it will be found that, other things being equal, P will be largest for Fabry–Perot spectroscopes (see below), of intermediate values for grating-based spectroscopes and lowest for prism-based spectroscopes.

4.1.3 Prisms

Pure prism-based spectroscopes will rarely be encountered today, except within instruments constructed some time ago, although as an exception to that, a low-resolution prism-based MIR spectroscope is currently planned for use on the JWST. However, prisms are commonly used in conjunction with gratings in some modern instruments. The combination is known as a 'grism' and the deviation of the light beam by the prism is used to counteract that of the grating, so that the light passes straight through the instrument. The spectroscope can then be used for direct imaging just by removing the grism and without having to move the camera. Prisms are also often used as cross-dispersers for high spectral order spectroscopes based upon echelle gratings or etalons and may be used non-spectroscopically for folding light beams.

When monochromatic light passes through an interface between two transparent isotropic media at a fixed temperature, then we can apply the well-known Snell's law relating the angle of incidence, i, to the angle of refraction, r, at that interface

$$\mu_1 \sin i = \mu_2 \sin r \qquad (4.40)$$

where μ_1 and μ_2 are constants that are characteristic of the two media. When $\mu_1 = 1$, which strictly only occurs for a vacuum, but which holds to a good approximation for most gases, including air, we have

$$\frac{\sin i}{\sin r} = \mu_2 \qquad (4.41)$$

and μ_2 is known as the refractive index of the second medium. Now we have already seen that the refractive index varies with wavelength for many media (see Section 1.1, Equation 1.38). The manner of this variation may, over a restricted wavelength interval, be approximated by the Hartmann dispersion formula, in which A, B and C are known as the Hartmann constants (see also the Cauchy formula, Equation 4.147):

$$\mu_\lambda = A + \frac{B}{\lambda - C} \qquad (4.42)$$

If the refractive index is known at three different wavelengths, then we can obtain three simultaneous equations for the constants, from Equation 4.42, giving

$$C = \frac{\left[\left(\frac{\mu_1 - \mu_2}{\mu_2 - \mu_3}\right)\lambda_1(\lambda_2 - \lambda_3) - \lambda_3(\lambda_1 - \lambda_2)\right]}{\left[\left(\frac{\mu_1 - \mu_2}{\mu_2 - \mu_3}\right)(\lambda_2 - \lambda_3) - (\lambda_2 - \lambda_3)\right]} \qquad (4.43)$$

$$B = \frac{\mu_1 - \mu_2}{\left(\dfrac{1}{\lambda_1 - C} - \dfrac{1}{\lambda_2 - C}\right)} \qquad (4.44)$$

$$A = \mu_1 - \frac{B}{\lambda_1 - C} \qquad (4.45)$$

The values for the constants for the optical region for typical optical glasses are

	A	B	C
Crown glass	1.477	3.2×10^{-8}	-2.1×10^{-7}
Dense flint glass	1.603	2.08×10^{-8}	1.43×10^{-7}

Thus, white light refracted at an interface is spread out into a spectrum with the longer wavelengths refracted less than the shorter ones. This phenomenon was encountered in Section 1.1 as chromatic aberration and there we were concerned with eliminating or minimising its effects; for spectroscopy by contrast we are interested in maximising the dispersion.

Consider, then, a prism with light incident upon it as shown in Figure 4.8. For light of wavelength λ, the deviation, θ, is given by

$$\theta = i_1 + r_2 - \alpha \tag{4.46}$$

So that using Equation 4.42 and using the relations

$$\mu_\lambda = \frac{\sin i_1}{\sin r_1} = \frac{\sin r_2}{\sin i_2} \tag{4.47}$$

and

$$\alpha = r_1 + i_2 \tag{4.48}$$

we get

$$\theta = i_1 - \alpha + \sin^{-1}\left\{\left(A + \frac{B}{\lambda - C}\right)\sin\left[\alpha - \sin^{-1}\left(\frac{\sin i_1}{\left(A + \frac{B}{\lambda - C}\right)}\right)\right]\right\} \tag{4.49}$$

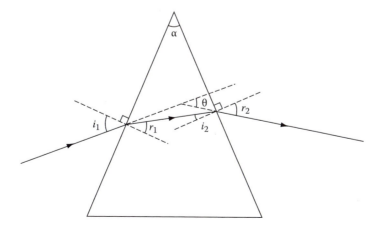

FIGURE 4.8 Optical path in a prism.

Now we wish to maximise $\dfrac{\partial \theta}{\partial \lambda}$, which we could study by differentiating Equation 4.49, but which is easier to obtain from

$$\frac{\Delta \theta}{\Delta \lambda} = \frac{\theta_{\lambda_1} - \theta_{\lambda_2}}{\lambda_1 - \lambda_2} \tag{4.50}$$

so that

$$\frac{\Delta \theta}{\Delta \lambda} = \frac{\left\{ \sin^{-1}\left[\mu_{\lambda_1} \sin\left(\alpha - \sin^{-1}\left\{ \dfrac{\sin i_1}{\mu_{\lambda_1}} \right\} \right) \right] - \sin^{-1}\left[\mu_{\lambda_2} \sin\left(\alpha - \sin^{-1}\left\{ \dfrac{\sin i_1}{\mu_{\lambda_2}} \right\} \right) \right] \right\}}{(\lambda_1 - \lambda_2)} \tag{4.51}$$

The effect of altering the angle of incidence or the prism angle is now most simply followed by an example. Consider a dense flint prism for which

$$\lambda_1 = 4.86 \times 10^{-7} \text{ m} \quad \mu_{\lambda_1} = 1.664 \tag{4.52}$$

$$\lambda_2 = 5.89 \times 10^{-7} \text{ m} \quad \mu_{\lambda_2} = 1.650 \tag{4.53}$$

and

$$\frac{\Delta \theta}{\Delta \lambda} = \frac{\left\{ \sin^{-1}\left[1.664 \sin\left(\alpha - \sin^{-1}\left\{ \dfrac{\sin i_1}{1.664} \right\} \right) \right] - \sin^{-1}\left[1.650 \sin\left(\alpha - \sin^{-1}\left\{ \dfrac{\sin i_1}{1.650} \right\} \right) \right] \right\}}{(1.03 \times 10^{-7})} \tag{4.54}$$

(in° m^{-1}). Figure 4.9 shows the variation of $\dfrac{\Delta \theta}{\Delta \lambda}$ with angle of incidence for a variety of apex angles as given by Equation 4.54. From this figure it is reasonably convincing to see that the maximum dispersion of $1.02 \times 10^{8°}$ m^{-1} occurs for an angle of incidence of 90° and an apex angle of 73.8776°. This represents the condition of glancing incidence and exit from the prism (Figure 4.10) and the ray passes symmetrically through the prism.

The symmetrical passage of the ray through the prism is of importance apart from being one of the requirements for maximum $\dfrac{\Delta \theta}{\Delta \lambda}$. It is only when this condition applies that the astigmatism introduced by the prism is minimised. The condition of symmetrical ray

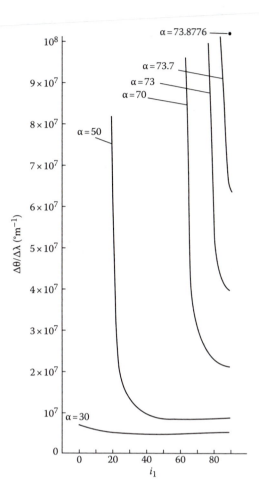

FIGURE 4.9 Variation of $\dfrac{\Delta\theta}{\Delta\lambda}F$ with angle of incidence for a dense flint prism, for various apex angles, α.

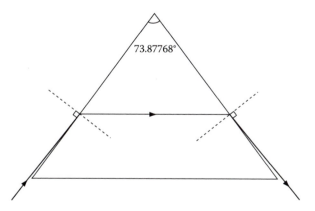

FIGURE 4.10 Optical path for maximum dispersion in a dense flint prism.

passage for any prism is more normally called the position of minimum deviation. Again an example quickly illustrates why this is so. For a dense flint prism with an apex angle of 30° and at a wavelength of 500 nm, Figure 4.11 shows the variation of the deviation with angle of incidence. The minimum value of θ occurs for $i_1 = 25.46°$, from which we rapidly find that $r_1 = 15°$, $i_2 = 15°$ and $r_2 = 25.46°$ and so the ray is passing through the prism symmetrically when its deviation is a minimum. More generally, we may write θ in terms of r_1,

$$\theta = \sin^{-1}(\mu_\lambda \sin r_1) + \sin^{-1}[\mu_\lambda \sin(\alpha - r_1)] - \alpha \tag{4.55}$$

so that

$$\frac{\partial \theta}{\partial r_1} = \frac{\mu_\lambda \cos r_1}{\sqrt{1 - \mu_\lambda^2 \sin^2 r_1}} - \frac{\mu_\lambda \cos(\alpha - r_1)}{\sqrt{1 - \mu_\lambda^2 \sin^2(\alpha - r_1)}} \tag{4.56}$$

Since $\dfrac{\partial \theta}{\partial r_1} = 0$ for θ to be an extremum, we obtain at this point after some manipulation

$$\left[\cos(2r_1) - \cos(2\alpha - 2r_1) \right](1 - 2\mu_\lambda^2) = 0 \tag{4.57}$$

giving

$$r_1 = \alpha/2 \tag{4.58}$$

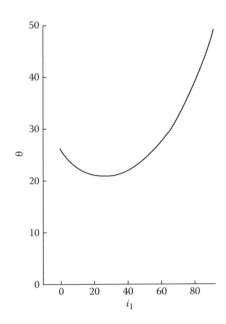

FIGURE 4.11 Deviation of dense flint prism with an apex angle of 30° at a wavelength of 500 nm.

and so also

$$i_2 = \alpha/2 \qquad (4.59)$$

and the minimum deviation always occurs for a symmetrical passage of the ray through the prism.

In practice, the maximum dispersion conditions of glancing incidence and exit are unusable because most of the light will be reflected at the two interactions and not refracted through the prism. Anti-reflection coatings (see Section 1.1) cannot be employed because these can only be optimised for a few wavelengths and so the spectral distribution would be disturbed. Thus, the apex angle must be less than the optimum, but the minimum deviation condition must be retained in order to minimise astigmatism. A fairly common compromise, then, is to use a prism with an apex angle of 60°, which has advantages from the manufacturer's point of view in that it reduces the amount of waste material if the initial blank is cut from a billet of glass. For an apex angle of 60°, the dense flint prism considered earlier has a dispersion of $1.39 \times 10^{7\circ}$ m^{-1} for the angle of incidence of 55.9° that is required to give minimum deviation ray passage. This is almost a factor of 10 lower than the maximum possible value.

With white light, it is obviously impossible to obtain the minimum deviation condition for all the wavelengths and the prism is usually adjusted so that this condition is preserved for the central wavelength of the region of interest. To see how the deviation varies with wavelength, we consider the case of a prism with a normal apex angle of 60°. At minimum deviation, we then have

$$r_1 = 30° \qquad (4.60)$$

Putting these values into Equation 4.55 and using Equation 4.42, we get

$$\theta = 2\sin^{-1}\left\{\frac{1}{2}\left[A + \frac{B}{\lambda - C}\right]\right\} - 60° \qquad (4.61)$$

so that

$$\frac{d\theta}{d\lambda} = \frac{-180B}{\pi(\lambda - C)^2 \left\{1 - \frac{1}{4}\left[A + \frac{B}{\lambda - C}\right]^2\right\}^{1/2}} \,^\circ\,\text{m}^{-1} \qquad (4.62)$$

Now

$$A + \frac{B}{\lambda - C} = \mu_\lambda \approx 1.5 \qquad (4.63)$$

and so

$$\left\{1-\frac{1}{4}\left[A+\frac{B}{\lambda-C}\right]^2\right\}^{-1/2} \approx A+\frac{B}{\lambda-C} \tag{4.64}$$

Thus

$$\frac{d\theta}{d\lambda} \approx \frac{-180AB}{\pi(\lambda-C)^2} - \frac{180B^2}{(\lambda-C)^3} \tag{4.65}$$

Now the first term on the right-hand side of Equation 4.65 has a magnitude about 30 times larger than that of the second term for typical values of λ, A, B and C. Hence

$$\frac{d\theta}{d\lambda} \approx \frac{-180AB}{\pi(\lambda-C)^2} \quad °\mathrm{m}^{-1} \tag{4.66}$$

and hence

$$\frac{d\theta}{d\lambda} \propto (\lambda-C)^{-2} \tag{4.67}$$

and so the dispersion of a prism increases rapidly towards shorter wavelengths. For the example we have been considering involving a dense flint prism, the dispersion is nearly five times larger at 400 nm than at 700 nm.

To form a part of a spectrograph a prism, like a diffraction grating, must be combined with other elements. The basic layout is shown in Figure 4.12; practical designs are discussed in Section 4.2.

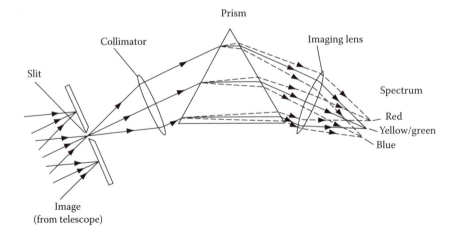

FIGURE 4.12 Basic optical arrangement of a prism spectroscope.

The linear dispersion of a prism is obtained in a similar way to that for the grating. If x is the linear distance along the spectrum from some reference point, then we have for an achromatic imaging element of focal length f_2,

$$\frac{dx}{d\lambda} = f_2 \frac{d\theta}{d\lambda} \tag{4.68}$$

where θ is small and is measured in radians. Thus, from Equation 4.66,

$$\frac{dx}{d\lambda} = \frac{-180 ABf_2}{\pi(\lambda - C)^2} \tag{4.69}$$

The reciprocal linear dispersion, $\frac{d\lambda}{dx}$, is thus

$$\frac{d\lambda}{dx} = \frac{\pi(\lambda - C)^2}{-180 ABf} \tag{4.70}$$

In a similar way to the grating spectroscope, the resolving power of a prism spectroscope is limited by the spectral resolution of the prism, the resolving power of its optics (see Section 1.1) and by the projected slit width. The spectrum is formed from an infinite number of monochromatic images of the entrance slit. The width of one of these images, S, is again given by

$$S = s \frac{f_2}{f_1} \tag{4.71}$$

where s is the slit width, f_1 is the collimator's focal length and f_2 is the imaging element's focal length. If the optics of the spectroscope are well corrected then we may ignore their aberrations and consider only the diffraction limit of the system. When the prism is fully illuminated, the imaging element will intercept a rectangular beam of light. The height of the beam is just the height of the prism and has no effect upon the spectral resolution. The width of the beam, D, is given by

$$D = L \left[1 - \mu_\lambda^2 \sin^2 \left(\frac{\alpha}{2} \right) \right]^{1/2} \tag{4.72}$$

where L is the length of a prism face and the diffraction limit is just that of a rectangular slit of width D. So that from Figure 1.37, the linear Rayleigh limit of resolution, W'', is given by

$$W'' = \frac{f_2 \lambda}{D} \tag{4.73}$$

$$= \frac{f_2 \lambda}{L\left[1 - \mu_\lambda^2 \sin^2\left(\dfrac{\alpha}{2}\right)\right]^{1/2}} \tag{4.74}$$

If the beam is limited by some other element of the optical system and/or is of circular cross section, then D must be evaluated as may be appropriate, or the Rayleigh criterion for the resolution through a circular aperture (Equation 1.62) used in place of that for a rectangular aperture. Optimum resolution occurs when

$$S = W'' \tag{4.75}$$

that is

$$s = \frac{f_1 \lambda}{D} \tag{4.76}$$

$$= \frac{f_1 \lambda}{L\left[1 - \mu_\lambda^2 \sin^2\left(\dfrac{\alpha}{2}\right)\right]^{1/2}} \tag{4.77}$$

The spectral resolution may now be obtained from Equations 4.68 and 4.73:

$$W_\lambda = W'' \frac{d\lambda}{dx} \tag{4.78}$$

$$= \frac{\lambda}{d} \frac{d\lambda}{d\theta} \tag{4.79}$$

$$\approx \frac{\lambda(\lambda - C)^2}{ABL\left\{1 - \left[A + \dfrac{B}{\lambda - C}\right]^2 \sin^2\left(\dfrac{\alpha}{2}\right)\right\}^{1/2}} \tag{4.80}$$

and the resolution is

$$R = \frac{\lambda}{W_\lambda} \tag{4.81}$$

$$\approx \frac{ABL\left\{1-\left[A+\dfrac{B}{\lambda-C}\right]^2 \sin^2\left(\dfrac{\alpha}{2}\right)\right\}^{1/2}}{(\lambda-C)^2} \tag{4.82}$$

For a dense flint prism with an apex angle of 60° and a side length of 0.1 m, we then obtain in the visible

$$R \approx 1.5 \times 10^4 \tag{4.83}$$

and this is a fairly typical value for the resolution of a prism-based spectroscope. We may now see another reason why the maximum dispersion (Figure 4.9) is not used in practice. Working back through the equations, we find that the term $L\left\{1-\left[A+\dfrac{B}{\lambda-C}\right]^2 \sin^2\left(\dfrac{\alpha}{2}\right)\right\}^{1/2}$ that is involved in the numerator of the right-hand side of Equation 4.82 is just the width of the emergent beam from the prism. Now for maximum dispersion, the beam emerges at 90° to the normal to the last surface of the prism. Thus, however, as large as the prism may be, the emergent beam width is zero and thus R is zero as well. The full variation of R with α is shown in Figure 4.13. A 60° apex angle still preserves 60% of the maximum resolution

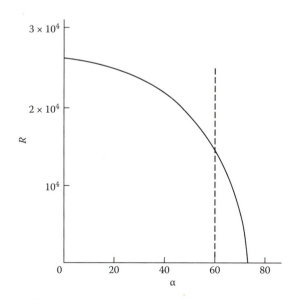

FIGURE 4.13   Resolution of a dense flint prism with a side length of 0.1 m at a wavelength of 500 nm, with changing apex angle α.

and so is a reasonable compromise in terms of resolution as well as dispersion. The truly optimum apex angle for a given type of material will generally be close to but not exactly 60°. Its calculation will involve a complex trade-off between resolution, dispersion and throughput of light, assessed in terms of the final amount of information available in the spectrum and is not usually attempted unless absolutely necessary.

The resolution varies slightly across the width of the spectrum, unless cylindrical lenses or mirrors are used for the collimator and these have severe disadvantages of their own. The variation arises because the light rays from the ends of the slit impinge on the first surface of the prism at an angle to the optical axis of the collimator (Figure 4.14). The peripheral rays therefore encounter a prism whose effective apex angle is larger than that for the paraxial rays. The deviation is also increased for such rays and so the ends of the spectrum lines are curved towards shorter wavelengths. Fortunately, astronomical spectra are mostly so narrow that both these effects can be neglected, although the wider beams encountered in integral field and multi-object spectroscopes (see Section 4.2) may be affected in this way.

The material used to form prisms depends upon the spectral region that is to be studied. In the visual region, the normal types of optical glass may be used, but these mostly start to absorb in the near UV. Fused silica and crystalline quartz can be formed into prisms to extend the limit down to 200 nm. Crystalline quartz, however, is optically active (see Section 5.2) and therefore must be used in the form of a Cornu prism. This has the optical axis parallel to the base of the prism so that the ordinary and extraordinary rays coincide and is made in two halves cemented together. The first half is formed from a right-handed crystal and the other half from a left-handed crystal, and the deviations of left- and right-hand circularly polarised beams are then similar. If required, calcium fluoride or lithium fluoride can extend the limit down to 140 nm or so, but astronomical spectroscopes working at such short wavelengths are normally based upon gratings. In the infrared, quartz can again be used for wavelengths out to about 3.5 μm. Rock salt can be used at even longer wavelengths, but it is extremely hygroscopic which makes it difficult to use. More

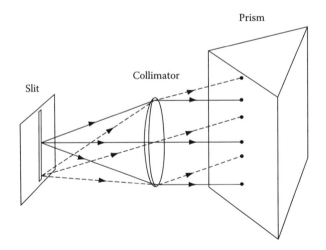

FIGURE 4.14 Light paths of rays from the centre and edge of the slit of a spectroscope.

commonly Fourier spectroscopy is applied when high-resolution spectroscopy is required in the far infrared.

4.1.4 Interferometers

We only consider here in any detail two main types of spectroscopic interferometry: the Fabry–Perot interferometer or etalon and the Michelson interferometer or Fourier-transform spectrometer. Other systems exist but at present are of little importance for astronomy.

4.1.4.1 Fabry–Perot Interferometer

Two parallel, flat, partially reflecting surfaces are illuminated at an angle θ (Figure 4.15). The light undergoes a series of transmissions and reflections as shown, and pairs of adjoining emergent rays differ in their path lengths by ΔP, where

$$\Delta P = 2t \cos \theta \qquad (4.84)$$

Constructive interference between the emerging rays will then occur at those wavelengths for which

$$\mu \, \Delta P = m \, \lambda \qquad (4.85)$$

where m is an integer; that is

$$\lambda = \frac{2t\mu \cos\theta}{m} \qquad (4.86)$$

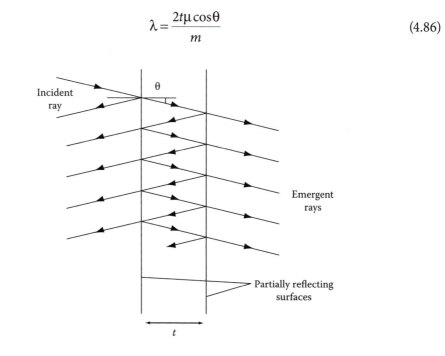

FIGURE 4.15 Optical paths in a Fabry–Perot interferometer.

If such an interferometer is used in a spectroscope in place of the prism or grating (Figure 4.16), then the image of a point source is still a point. However, the image is formed from only those wavelengths for which Equation 4.86 holds. If the image is then fed into another spectroscope it will be broken up into a series of monochromatic images. If a slit is now used as the source, the rays from the different points along the slit will meet the etalon at differing angles and the image will then consist of a series of superimposed short spectra. If the second spectroscope is then set so that its dispersion is perpendicular to that of the etalon (a cross-disperser), then the final image will be a rectangular array of short parallel spectra. The widths of these spectra depend upon the reflectivity of the surfaces. For a high reflectivity, we get many multiple reflections, whilst with a low reflectivity the intensity becomes negligible after only a few reflections. The monochromatic images of a point source are therefore not truly monochromatic but are spread over a small wavelength range in a similar, but not identical, manner to the intensity distributions for several collinear apertures (Figures 4.1 and 4.2). The intensity distribution varies from that of the multiple apertures since the intensities of the emerging beams decrease as the number of reflections required for their production increases. Examples of the intensity distribution with wavelength are shown in Figure 4.17. In practice, reflectivities of about 90% are usually used.

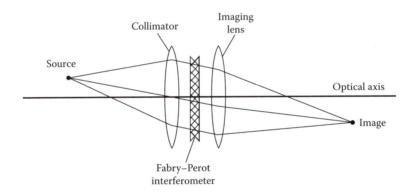

FIGURE 4.16 Optical paths in a Fabry–Perot spectroscope.

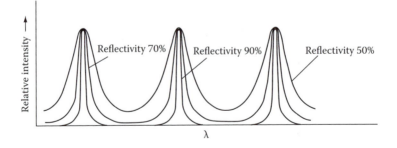

FIGURE 4.17 Intensity versus wavelength in the image of a white-light point source in a Fabry–Perot spectroscope, assuming negligible absorption.

In the absence of absorption, the emergent intensity at a fringe peak is equal to the incident intensity at that wavelength. This often seems a puzzle to readers, who on inspection of Figure 4.15, might expect there to be radiation emerging from the left as well as on the right, so that at the fringe peak the total emergent intensity would appear to be greater than the incident intensity. If we examine the situation more closely, however, we find that when at a fringe peak for the light emerging on the right, there is zero intensity in the beam emerging on the left. If the incident beam has intensity I and amplitude a ($I = a^2$), then the amplitudes of the successive beams on the left in Figure 4.15 are

$$-aR^{1/2},\ aR^{1/2}T,\ aR^{3/2}T,\ aR^{5/2}T,\ aR^{7/2}T, \ldots$$

where the first amplitude is negative because it results from an internal reflection. It has therefore an additional phase delay of 180° compared with the other reflected beams. T is the fractional intensity transmitted and R the fractional intensity reflected by the reflecting surfaces (note that $T + R = 1$ in the absence of absorption). Summing these terms (assumed to go on to infinity) gives zero amplitude and therefore zero intensity for the left-hand emergent beam. Similarly, the beams emerging on the right have amplitudes

$$aT,\ aTR,\ aTR^2,\ aTR^3,\ aTR^4, \ldots \tag{4.87}$$

Summing these terms to infinity gives, at a fringe peak, the amplitude on the right as a. This is the amplitude of the incident beam and so at a fringe maximum the emergent intensity on the right equals the incident intensity (see also Equation 4.95).

The dispersion of an etalon may easily be obtained by differentiating Equation 4.86:

$$\frac{d\lambda}{d\theta} = -\frac{2t\mu}{m}\sin\theta \tag{4.88}$$

Since the material between the reflecting surfaces is usually air and the etalon is used at small angles of inclination, we have

$$\mu \approx 1 \tag{4.89}$$

$$\sin\theta \approx \theta \tag{4.90}$$

and from Equation 4.86,

$$\frac{2t}{m} \approx \lambda \tag{4.91}$$

so that

$$\frac{d\lambda}{d\theta} \approx \lambda\theta \tag{4.92}$$

Thus, the reciprocal linear dispersion for a typical system, with $\theta = 0.1°$, $f_2 = 1$ m and used in the visible, is 0.001 nm mm^{-1}, which is a factor of 100 or so larger than that achievable with more common dispersing elements.

The resolution of an etalon is rather more of a problem to estimate. Our usual measure – the Rayleigh criterion – is inapplicable since the minimum intensities between the fringe maxima do not reach zero, except for a reflectivity of 100%. However, if we consider the image of two equally bright point sources viewed through a telescope at its Rayleigh limit (Figure 2.17) then the central intensity is 81% of that of either of the peak intensities. We may therefore replace the Rayleigh criterion by the more general requirement that the central intensity of the envelope of the images of two equal sources falls to 81% of the peak intensities. Consider, therefore, an etalon illuminated by a monochromatic slit source perpendicular to the optical axis (Figure 4.16). The image will be a strip, also perpendicular to the optical axis and the intensity will vary along the strip accordingly as the emerging rays are in or out of phase with each other. The intensity variation is given by

$$I(\theta) = \frac{T^2 I_\lambda}{(1-R)^2 + 4R \sin^2\left(\dfrac{2\pi t \mu \cos\theta}{\lambda}\right)} \tag{4.93}$$

where I_λ is the incident intensity at wavelength λ. The image structure will resemble that shown in Figure 4.18. If the source is now replaced with a bichromatic one, then the image structure will be of the type shown in Figure 4.19. Consider just one of these fringes, its angular distance, θ_{max}, from the optical axis is, from Equation 4.86,

$$\theta_{max} = \cos^{-1}\left(\frac{m\lambda}{2t\mu}\right) \tag{4.94}$$

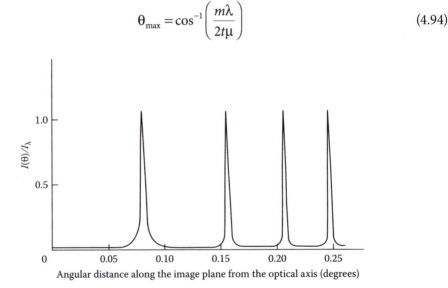

FIGURE 4.18 Image structure in a Fabry–Perot spectroscope viewing a monochromatic slit source, with $T = 0.1$, $R = 0.9$, $t = 0.1$ m, $\mu = 1$ and $\lambda = 550$ nm.

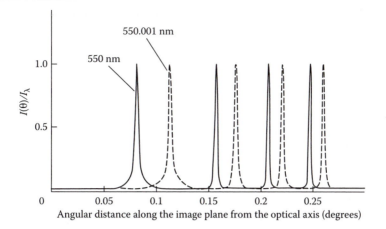

FIGURE 4.19 Image structure in a Fabry–Perot spectroscope viewing a bichromatic slit source, with $T = 0.1$, $R = 0.9$, $t = 0.1$ m, $\mu = 1$ and $\lambda = 550$ nm (full curve) and $\lambda = 550.001$ nm (broken curve).

and so the peak intensity from Equation 4.93 is

$$I(\theta_{max}) = \frac{T^2 I_\lambda}{(1-R)^2} \tag{4.95}$$

$$= I_\lambda \text{ (when there is no absorption)} \tag{4.96}$$

Let the angular half width of a fringe at half intensity be $\Delta\theta$, then a separation of twice the half-half width of the fringes gives a central intensity of 83% of either of the peak intensities, so that if α is the resolution by the extended Rayleigh criterion, we may write

$$\alpha \approx 2\Delta\theta = \frac{\lambda(1-R)}{2\pi\mu t\sqrt{R}\ \theta_{max}\cos\theta_{max}} \tag{4.97}$$

Hence from Equation 4.92 we obtain the spectral resolution

$$W_\lambda = \alpha\frac{d\lambda}{d\theta} \tag{4.98}$$

$$= \frac{\lambda^2(1-R)}{2\pi\mu t\sqrt{R}\cos\theta_{max}} \tag{4.99}$$

and so the resolution of the system (previously given the symbol R) is

$$\frac{\lambda}{\Delta\lambda} = \frac{2\pi\mu t\sqrt{R}\cos\theta_{max}}{\lambda(1-R)} \tag{4.100}$$

or, since θ_{max} is small and μ is usually close to unity,

$$\frac{\lambda}{\Delta\lambda} \approx \frac{2\pi t\sqrt{R}}{\lambda(1-R)} \tag{4.101}$$

Thus, for typical values of $t = 0.1$ m, $R = 0.9$ and for visible wavelengths we have

$$\frac{\lambda}{\Delta\lambda} \approx 10^7 \tag{4.102}$$

which is almost two orders of magnitude higher than typical values for prisms and gratings. It is comparable with the resolution for a large echelle grating whilst physically the device is much less bulky. An alternative measure of the resolution that may be encountered is the finesse. This is the reciprocal of the half-width of a fringe measured in units of the separation of the fringes from two adjacent orders. It is given by

$$\text{Finesse} = \frac{\pi\sqrt{R}}{1-R} = \frac{\lambda}{2t}\times\text{resolution} \tag{4.103}$$

For a value of R of 0.9, the finesse is therefore about 30.

The free spectral range of an etalon is small since it is operating at very high spectral orders. From Equation 4.94 we have

$$\Sigma = \lambda_1 - \lambda_2 = \frac{\lambda_2}{m} \tag{4.104}$$

where λ_1 and λ_2 are superimposed wavelengths from adjacent orders (cf. Equation 4.23). Thus, the device must be used with a cross disperser as already mentioned and/or the free spectral range increased. The latter may be achieved by combining two or more different etalons; then only the maxima that coincide will be transmitted through the whole system and the intermediate maxima will be suppressed.

Practical etalons are made from two plates of glass or quartz whose surfaces are flat to 1 or 2 per cent of their operating wavelength. They are held accurately parallel to each other by low thermal expansion spacers, with a spacing in the region of 10 to 200 mm. The inner faces are mirrors, which are usually produced by a metallic or dielectric coating. The outer faces are inclined by a very small angle to the inner faces so that the plates are the basal segments of very low angle prisms. Any multiple reflections other than the desired ones are then well displaced from the required image. The limit to the resolution of the instrument is generally imposed by departures of the two reflecting surfaces from absolute flatness. This limits the main use of the instrument to the visible and infrared regions. The absorption in metallic coatings also limits the shortwave use, so that 200 nm represents the shortest practicable wavelength even for laboratory usage. Etalons are commonly used as scanning instruments. By changing the air pressure by a few per cent, the refractive index of the material between the plates is changed and so the wavelength of a fringe at a given place within the image is altered (Equation 4.93). The astronomical applications of Fabry–Perot spectroscopes are comparatively few for direct observations. However, the instruments are used extensively in determining oscillator strengths and transition probabilities upon which much of the more conventional astronomical spectroscopy is based.

Another important application of etalons and one that does have many direct applications for astronomy is in the production of narrowband filters. These are usually known as interference filters and are etalons in which the separation of the two reflecting surfaces is very small. For materials with refractive indices near 1.5 and for near-normal incidence, we see from Equation 4.86 that if t is 167 nm then the maxima will occur at wavelengths of 500, 250, 167 nm and so on, accordingly as m is 1, 2, 3 whilst from Equation 4.97, the widths of the transmitted regions will be 8.4, 2.1, 0.9 nm for 90% reflectivity of the surfaces. Thus, a filter centred upon 500 nm with a bandwidth of 8.4 nm can be made by combining such an etalon with a simple dye filter to eliminate the shorter wavelength transmission regions (or in this example just by relying on the absorption within the glass substrates). Other wavelengths and bandwidths can easily be chosen by changing t, μ and R. Such a filter would be constructed by evaporating a partially reflective layer onto a sheet of glass. A second layer of an appropriate dielectric material such as magnesium fluoride or cryolite is then evaporated on top of this to the desired thickness, followed by a second partially reflecting layer. A second sheet of glass is then added for protection. The reflecting layers may be silver or aluminium, or they may be formed from a double layer of two materials with very different refractive indices in order to improve the overall filter transmission. In the far infrared, pairs of inductive meshes can be used in a similar way for infrared filters. The band-passes of interference filters can be made squarer by using several superimposed Fabry–Perot layers.

Recently, tunable filters have been developed (see also the Lyot birefringent filter, Section 5.3), that are especially suited to observing the emission lines of gaseous nebulae. The reflecting surfaces are mounted on stacks of piezoelectric crystals so that their separations can be altered. The Maryland-Magellan Tunable Filter (MMTF) has been operating since 2006 on the 6.5-metre Magellan-Baade telescope and covers the wavelength range from 500 to 920 nm with a bandwidth that can be varied between 0.5 and 2.5 nm. The GTC's Optical System for Imaging and low Resolution Integrated Spectroscopy (OSIRIS) has been operating for several years now and covers the 365-nm to 1-μm region with bandwidths between 2 and 0.9 nm. Most other large optical telescopes have either recently added tunable filters to their range of ancillary equipment or are in the process of doing so.

4.1.4.2 Michelson Interferometer

This Michelson interferometer should not be confused with the Michelson stellar interferometer that was discussed in Section 2.5. The instrument discussed here is similar to the device used by Michelson and Morley to try and detect the Earth's motion through the aether. Its optical principles are shown in Figure 4.20. The light from the source is split into two beams by the beam splitter and then recombined as shown. For a particular position of the movable mirror and with a monochromatic source, there will be a path difference, ΔP, between the two beams at their focus. The intensity at the focus is then

$$I_{\Delta P} = I_m \left[1 + \cos\left(\frac{2\pi\Delta P}{\lambda} \right) \right]$$ (4.105)

where I_m is a maximum intensity. If the mirror is moved, then the path difference will change and the final intensity will pass through a series of maxima and minima (Figure 4.21). If the source is bichromatic, then two such variations will be superimposed with slightly differing periods and the final output will then have a beat frequency (Figure 4.22). The difference between the two outputs (Figures 4.20 and 4.22) gives the essential principle of the Michelson interferometer when it is used as a spectroscope. Neither output in any way resembles an ordinary spectrum, yet it would be simple to recognise the first as due to a monochromatic source and the second as due to a bichromatic source. Furthermore, the spacing of the fringes could be related to the original wavelength(s) through Equation 4.105. More generally of course, sources emit a broad band of wavelengths and the final output will vary in a complex manner. To find the spectrum of an unknown source from such an output therefore requires a rather different approach than this simple visual inspection.

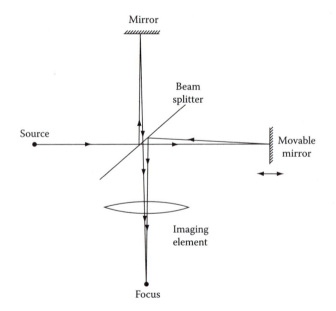

FIGURE 4.20   Optical pathways in a Michelson interferometer.

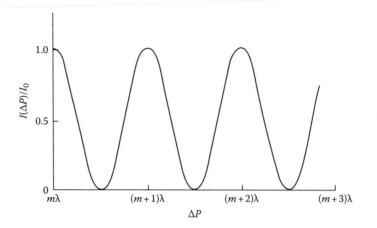

FIGURE 4.21 Variation of fringe intensity with mirror position in a Michelson interferometer.

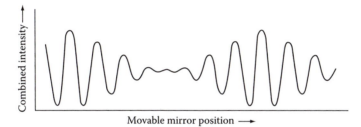

FIGURE 4.22 Output of a Michelson interferometer observing a bichromatic source.

Let us consider therefore a Michelson interferometer in which the path difference is ΔP observing a source whose intensity at wavelength λ is $I\lambda$. The intensity in the final image due to the light of a particular wavelength, $I'_{\Delta P}(\lambda)$ is then

$$I'_{\Delta P}(\lambda) = KI(\lambda)\left[1 + \cos\left(\frac{2\pi\Delta P}{\lambda}\right)\right]$$

(4.106)

where K is a constant that takes account of the losses at the various reflections, transmissions. Thus, the total intensity in the image for a given path difference is just

$$I'_{\Delta P} = \int_0^\infty I'_{\Delta P}(\lambda)d\lambda$$

(4.107)

$$= \int_0^\infty KI(\lambda)d\lambda + \int_0^\infty KI(\lambda)\cos\left(\frac{2\pi\Delta P}{\lambda}\right)d\lambda$$

(4.108)

Now the first term on the right-hand side of Equation 4.108 is independent of the path difference and is simply the mean intensity of the image. We may therefore disregard it and concentrate instead on the deviations from this average level, $I(\Delta P)$. Thus

$$I(\Delta P) = K \int_0^\infty I(\lambda) \cos\left(\frac{2\pi\Delta P}{\lambda}\right) d\lambda \qquad (4.109)$$

or in frequency terms

$$I(\Delta P) = K^I \int_0^\infty I(\nu) \cos\left(\frac{2\pi\Delta P\nu}{c}\right) d\nu \qquad (4.110)$$

Now the Fourier transform, $F(u)$, of a function, $f(t)$, (see also Section 2.1) is defined by

$$\mathbf{F}(f(t)) = F(u) = \int_{-\infty}^\infty f(t)e^{-2\pi i u t}\, dt \qquad (4.111)$$

$$= \int_{-\infty}^\infty f(t)\cos(2\pi u t)dt - i\int_{-\infty}^\infty f(t)\sin(2\pi u t)dt \qquad (4.112)$$

Thus, we see that the output of the Michelson interferometer is akin to the real part of the Fourier transform of the spectral intensity function of the source. Furthermore, by defining

$$I(-\nu) = I(\nu), \qquad (4.113)$$

we have

$$I(\Delta P) = \frac{1}{2}K^I \int_{-\infty}^\infty I(\nu)\cos\left(\frac{2\pi\Delta P\nu}{c}\right)d\nu \qquad (4.114)$$

$$= K^{II} \operatorname{Re}\left\{\int_{-\infty}^\infty I(\nu)\exp\left[-i\left(\frac{2\pi\Delta P}{c}\right)\nu\right]d\nu\right\} \qquad (4.115)$$

where K^{II} is an amalgam of all the constants. Now, by inverting the transformation

$$\mathbf{F}^{-1}(F(u)) = f(t) = \int_{-\infty}^\infty F(u)e^{2\pi i u t}\, du \qquad (4.116)$$

and taking the real part of the inversion, we may obtain the function that we require – the spectral energy distribution, or as it is more commonly known, the spectrum. Thus

$$I(\nu) = K^{III} \operatorname{Re}\left\{ \int_{-\infty}^{\infty} I\left(\frac{2\pi\Delta P}{c}\right) \exp\left[i\left(\frac{2\pi\Delta P}{c}\right)\nu \right] d\left(\frac{2\pi\Delta P}{c}\right) \right\} \tag{4.117}$$

or

$$I(\nu) = K^{IV} \int_{-\infty}^{\infty} I\left(\frac{2\pi\Delta P}{c}\right) \cos\left(\frac{2\pi\Delta P\nu}{c}\right) d(\Delta P) \tag{4.118}$$

where K^{III} and K^{IV} are again amalgamated constants. Finally, by defining

$$I\left(\frac{-2\pi\Delta P}{c}\right) = I\left(\frac{2\pi\Delta P}{c}\right) \tag{4.119}$$

we have

$$I(\nu) = 2K^{IV} \int_{0}^{\infty} I\left(\frac{2\pi\Delta P}{c}\right) \cos\left(\frac{2\pi\Delta P\nu}{c}\right) d(\Delta P) \tag{4.120}$$

and so the spectrum is obtainable from the observed output of the interferometer as the movable mirror scans through various path differences. We may now see why a Michelson interferometer when used as a scanning spectroscope is often called a Fourier transform spectroscope. The inversion of the Fourier transform is carried out on computers using the fast Fourier transform algorithm.

In practice, of course, it is not possible to scan over path differences from zero to infinity and also measurements are usually made at discrete intervals rather than continuously, requiring the use of the discrete Fourier transform equations (see Section 2.1). These limitations are reflected in a reduction in the resolving power of the instrument. To obtain an expression for the resolving power, we may consider the Michelson interferometer as equivalent to a two-aperture interferometer (Figure 4.1) since its image is the result of two interfering beams of light. We may therefore write Equation 4.20 for the resolution of two wavelengths by the Rayleigh criterion as

$$W_\lambda = \frac{\lambda^2}{2\Delta P} \tag{4.121}$$

However, if the movable mirror in the Michelson interferometer moves a distance x, then ΔP ranges from 0 to $2x$ and we must take the average value of ΔP rather than the extreme value for substitution into Equation 4.121. Thus, we obtain the spectral resolution of a Michelson interferometer as

$$W_\lambda = \frac{\lambda^2}{2x} \qquad (4.122)$$

so that the system's resolution is

$$\frac{\lambda}{W_\lambda} = \frac{2x}{\lambda} \qquad (4.123)$$

Since x can be as much as two metres, we obtain a resolution of up to 4×10^6 for such an instrument used in the visible region.

The sampling intervals must be sufficiently frequent to preserve the resolution, but not more frequent than this, or time and effort will be wasted. If the final spectrum extends from λ_1 to λ_2, then the number of useful intervals, n, into which it may be divided, is given by

$$n = \frac{\lambda_1 - \lambda_2}{W_\lambda} \qquad (4.124)$$

so that if λ_1 and λ_2 are not too different then

$$n \approx \frac{8x(\lambda_1 - \lambda_2)}{(\lambda_1 + \lambda_2)^2} \qquad (4.125)$$

However, the inverse Fourier transform gives both $I(\nu)$ and $I(-\nu)$, so that the total number of separate intervals in the final inverse transform is $2n$. Hence, we must have at least $2n$ samples in the original transformation and therefore the spectroscope's output must be sampled $2n$ times. Thus, the interval between successive positions of the movable mirrors, Δx, at which the image intensity is measured, is given by

$$\Delta x = \frac{(\lambda_1 + \lambda_2)^2}{16(\lambda_1 - \lambda_2)} \qquad (4.126)$$

A spectrum between 500 and 550 nm therefore requires step lengths of 1 μm, whilst between 2000 and 2050 nm a spectrum would require step lengths of 20 μm. This relaxation in the physical constraints required on the accuracy of the movable mirror for longer wavelength spectra, combined with the availability of other methods of obtaining visible spectra, has led to the major applications of Fourier transform spectroscopy to date being in the infrared and FIR. Even there though, the increasing size of infrared arrays is now leading to the use of more conventional diffraction grating spectroscopes.

The basic PSF of the Fourier transform spectroscope is of the form $\dfrac{\sin \Delta \lambda}{\Delta \lambda}$ (Figure 4.23), where $\Delta \lambda$ is the distance from the central wavelength, λ, of a monochromatic source. This is not a particularly convenient form for the profile and the secondary maxima may be large enough to be significant especially where the spectral energy undergoes abrupt changes, such as at ionisation edges or over molecular bands. The effect of the PSF may be reduced at some cost to the theoretical resolution by a technique known as apodisation (see also Sections 1.1, 2.1, 2.5 and 5.1). The transform is weighted by some function, ω, known as the apodisation function. Many functions can be used, but the commonest is probably a triangular weighting function; that is

$$\omega(\Delta P) = 1 - \frac{\Delta P}{2x} \tag{4.127}$$

The resolving power is halved, but the PSF becomes that for a single rectangular aperture (Figure 1.37, Equation 1.12) and has much reduced secondary maxima.

A major advantage of the Michelson interferometer over the etalon when the latter is used as a scanning instrument lies in its comparative rapidity of use when the instrument is detector-noise limited. Not only is the total amount of light gathered by the Michelson interferometer higher (the Jacquinot advantage), but even for equivalent image intensities the total time required to obtain a spectrum is much reduced. This arises because all wavelengths are contributing to every reading in the Michelson interferometer, whereas a reading from the etalon gives information for just a single wavelength. The gain of the Michelson interferometer is called the multiplex or Fellget advantage and is similar to the gain of the Hadamard masking technique over simple scanning (see Section 2.4). If t is the integration time required to record a single spectral element, then the etalon requires a total observing time of nt. The Michelson interferometer, however, requires a time of only $t/\sqrt{n/2}$ to record each sample since it has contributions from n spectral elements. It must obtain $2n$ samples so the total observing time for the same spectrum is therefore $2nt/\sqrt{n/2}$ and it has an advantage over the etalon of a factor of $\sqrt{n/8}$ in observing time.

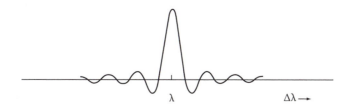

FIGURE 4.23   Basic instrumental profile (PSF) of a Fourier transform spectroscope.

Michelson interferometers have another advantage in that they require no entrance slit in order to preserve their spectral resolution. This is of considerable significance for large telescopes where the stellar image may have a physical size of a millimetre or more due to atmospheric turbulence, whilst the spectroscope's slit may be only a few tenths of a millimetre wide. Thus, either much of the light is wasted or complex image dissectors (see Section 4.2) must be used.

4.1.5 Fibre Optics

Fibre optic cables are now widely used to connect spectroscopes to telescopes, enabling the spectroscope to be mounted separately from the telescope. This reduces the problems of flexure that occur within telescope-mounted spectroscopes since the gravitational loads no longer change with the different telescope positions. It also enables the spectroscope to be kept in a temperature-controlled room and in some cases cooled and inside a vacuum chamber. Fibre optics can also be used to reformat stellar images so that all the light enters the spectroscope and to enable extended objects or multiple objects to be observed efficiently. Specific examples are discussed in the next section. Here we are concerned with the optics of fibre optic cables.

Fibre optic cables usually consist of a thin (10–500 µm) filament of glass encased in a cladding of another glass with a lower refractive index. One or more (sometimes thousands) of these strands make up the cable as a whole. Light entering the core is transmitted through the core by multiple internal reflections off the interface between the two glasses, provided that its angle of incidence exceeds the critical angle for total internal reflection. The critical angle θ_c is given by

$$\theta_c = \sin^{-1}\sqrt{(\mu_{core}^2 - \mu_{cladding}^2)} \tag{4.128}$$

where μ_{core} and $\mu_{cladding}$ are the refractive indices of the core and cladding, respectively. Fibre optic cables are usually characterised by their numerical aperture (*NA*) and this is simply equal to sin (θ_c). The minimum focal ratio, f_{min} that will be transmitted by the core is then

$$f_{min} = \frac{\sqrt{1-NA}}{2NA} \tag{4.129}$$

Commercially produced fibre optics have numerical apertures ranging from about 0.15 to 0.35, giving minimum focal ratios ranging from f3 to f0.6.

Silica glass is used as the core of the fibre for the spectral region 400 nm to 2 µm and is sufficiently transparent that cables can be tens of metres long without significant absorption. Further into the infrared, specialist glasses such as zirconium fluoride need to be used. Imperfections in the walls of the fibres and internal stresses lead to focal ratio degradation. This is a decrease in the focal ratio that can lead to light loss if the angle of incidence exceeds the critical angle. It may also cause problems in matching the output from the cable to the focal ratio of the instrument being used. Focal ratio degradation affects long focal ratio light beams worst, so short (faster) focal ratios are better for transmitting light into the fibres.

Cables comprising many individual strands will not transmit all the light that they receive because of the area occupied by the cladding. This is typically 40% of the cross-sectional area of the cable. Many astronomical applications therefore use single-strand cables with a core diameter sufficient to accept a complete stellar image. Multi-strand cables can be coherent or non-coherent. In the former, the individual strands have the same relative positions at the input end of the cable as at the output. In non-coherent cables the relationship between strand positions at the input and output ends of the cable is random. For some special applications, such as reformatting stellar images to match the shape of the spectroscope's entrance slit, the input and output faces of the fibre optic cable are of different shapes. Such cables need to be coherent in the sense that the positions of the strands at the output face are related in a simple and logical manner to those at the input face.

Optical fibres also have other potential astronomical applications that use them in rather more sophisticated ways than just simply getting light to where it is needed. We have already seen (see Section 1.1) that by varying the refractive index within the fibre, a Bragg grating can be formed which cuts out the atmospheric OH lines forming the main source of background noise in the NIR. Another possibility may result in a new type of spectroscope. The integrated photonic spectroscope uses a number of fibres of different lengths to form a phased array (see Section 1.2 for an account of a radio analogue). The fibres are fed by a two-dimensional waveguide that acts as a multiplexor and their outputs are recombined to form the spectrum in a second waveguide. Integrated photonic spectroscopes hold out the possibility of replacing the massive and cumbersome spectroscopes often used today with an instrument that is just a few centimetres in size.

EXERCISES

4.1 Calculate the heliocentric radial velocity for a star in which the Hα line (laboratory wavelength 656.2808 nm) is observed to have a wavelength of 656.1457 nm. At the time of the observation, the solar celestial longitude was 100° and the star's right ascension and declination were 13^h 30^m and +45°, respectively. The obliquity of the ecliptic is 23° 27′.

4.2 A prism spectroscope is required with a reciprocal linear dispersion of 1 nm mm^{-1} or better at a wavelength of 400 nm. The focal length of the imaging element is limited to 1 metre through the need for the spectroscope to fit onto the telescope. Calculate the minimum number of 60° crown glass prisms required to achieve this and the actually resulting reciprocal linear dispersion at 400 nm.

Note: Prisms can be arranged in a train so that the emergent beam from one forms the incident beam of the next in line. This increases the dispersion in proportion to the number of prisms, but does not affect the spectral resolution, which remains at the value appropriate for a single prism.

4.3 Show that the reciprocal linear dispersion of a Fabry–Perot etalon is given by

$$\frac{d\lambda}{dx} = \frac{\lambda l}{2 f_2^2}$$

where l is the length of the slit and the slit is symmetrical about the optical axis.

4.4 Determine the distance that the moveable mirror in a Fourier transform spectrometer needs to move in order to provide a spectral resolution of 1,000,000 between the wavelengths 800 nm and 2 μm. How many steps must the mirror make?

4.2 SPECTROSCOPES

4.2.1 Basic Design Considerations

The specification of a spectroscope usually begins from just three parameters. One is the focal ratio of the telescope upon which the spectroscope is to operate, the second is the required spectral resolution and the third is the required spectral range. Thus, in terms of the notation used in Section 4.1, we have f', W_λ and λ specified (where f' is the effective focal ratio of the telescope at the entrance aperture of the spectroscope) and we require the values of $f_1, f_2, s, R, \dfrac{d\lambda}{d\theta}, L$ and D. We may immediately write down the resolution required of the dispersion element

$$R = \frac{\lambda}{W_\lambda} \tag{4.130}$$

Now for a 60° prism from Equation 4.82 we find that at 500 nm

$$R = 6 \times 10^4\, L \text{ (crown glass)} \tag{4.131}$$

$$R = 15 \times 10^4\, L \text{ (flint glass)} \tag{4.132}$$

where L is the length of a side of a prism, in metres. So we may write

$$R \approx 10^5\, L \tag{4.133}$$

when the dispersing element is a prism. For a grating, the resolution depends upon the number of lines and the order of the spectrum (Equation 4.21). A reasonably typical astronomical grating (ignoring echelle gratings) will operate in its third order and have some 500 lines mm^{-1}. Thus, we may write the resolution of a grating as

$$R \approx 1.5 \times 10^6\, L \tag{4.134}$$

where L is the width of the ruled area of the grating, in metres. The size of the dispersing element will thus be given approximately by Equations 4.133 and 4.134. It may be calculated more accurately in second and subsequent iterations through this design process, using Equations 4.82 and 4.21.

The diameter of the exit beam from the dispersing element, assuming that it is fully illuminated, can then be obtained from

$$D = L \cos \phi \qquad (4.135)$$

where ϕ is the angular deviation of the exit beam from the perpendicular to the exit face of the dispersing element. For a prism ϕ is typically 60°, whilst for a grating used in the third order with equal angles of incidence and reflection, it is 25°. Thus, we get

$$D = 0.5\, L \text{ (prism)} \qquad (4.136)$$

$$D = 0.9\, L \text{ (grating)} \qquad (4.137)$$

The dispersion can now be determined by setting the angular resolution of the imaging element equal to the angle between two just resolved wavelengths from the dispersing element

$$\frac{\lambda}{D} = W_\lambda \frac{d\theta}{d\lambda} \qquad (4.138)$$

giving

$$\frac{d\theta}{d\lambda} = \frac{R}{D} \qquad (4.139)$$

Since the exit beam is rectangular in cross section, we must use the resolution for a rectangular aperture of the size and shape of the beam (Equation 1.13) for the resolution of the imaging element and not its actual resolution, assuming that the beam is wholly intercepted by the imaging element and that its optical quality is sufficient not to degrade the resolution below the diffraction limit.

The final parameters now follow easily. The physical separation of two just resolved wavelengths on the CCD or other imaging detector must be greater than or equal to the separation of two pixels. CCD pixels are typically around 15 to 20 μm in size, so the focal length in metres of the imaging element must be at least

$$f_2 \geq \frac{\text{Pixel size}}{W_\lambda} \frac{d\theta}{d\lambda} \qquad (4.140)$$

The diameter of the imaging element, D_2, must be sufficient to contain the whole of the exit beam. Thus, for a square cross-section exit beam

$$D_2 = \sqrt{2}\, D \tag{4.141}$$

The diameter of the collimator, D_1, must be similar to that of the imaging element in general if the dispersing element is to be fully illuminated. Thus, again

$$D_1 = \sqrt{2}\, D \tag{4.142}$$

Now in order for the collimator to be fully illuminated in its turn, its focal ratio must equal the effective focal ratio of the telescope. Hence, the focal length of the collimator, f_1, is given by

$$f_1 = \sqrt{2}\, Df' \tag{4.143}$$

Finally, from Equation 4.76 we have the slit width

$$s = \frac{f_1 \lambda}{D} \tag{4.144}$$

and a first approximation has been obtained to the design of the spectroscope.

The low light levels involved in astronomy usually require the focal ratio of the imaging elements to be small, so that it is fast in imaging terms. Satisfying this requirement usually means a compromise in some other part of the design, so that an optimally designed system is rarely achievable in practice. The slit may also need to be wider than specified in order to use a reasonable fraction of the star's light.

The limiting magnitude of a telescope–spectroscope combination is the magnitude of the faintest star for which a useful spectrum may be obtained. This is a very imprecise quantity, for it depends upon the type of spectrum and the purpose for which it is required, as well as the properties of the instrument and the detector. For example, if strong emission lines in a spectrum are the features of interest, then fainter stars may be studied than if weak absorption lines are desired. Similarly, spectra of sufficient quality to determine radial velocities may be obtained for fainter stars than if line profiles are wanted. A guide, however, to the limiting magnitude may be gained through the use of Bowen's formula

$$m = 12 + 2.5 \log_{10}\left(\frac{sD_1 T_D g q t \left(\dfrac{d\lambda}{d\theta} \right)}{f_1 f_2 \alpha H} \right) \tag{4.145}$$

where m is the faintest B magnitude that will give a usable spectrum in t seconds of exposure, T_D is the telescope objective's diameter and g is the optical efficiency of the system (i.e. the ratio of the usable light at the focus of the spectroscope to that incident upon the telescope). Typically it has a value of 0.2 (note, however, that g does not include the effect of the curtailment of the image by the slit). q is the quantum efficiency of the detector; typical values are 0.4 to 0.9 for CCDs (see Section 1.1), α is the angular size of the stellar image at the telescope's focus, typically 5×10^{-6} to 2×10^{-5} radians and H is the height of the spectrum. This formula gives quite good approximations for spectroscopes in which the star's image is larger than the slit and it is trailed along the length of the slit to broaden the spectrum. At one time this was the commonest mode of use for astronomical spectroscopes, but now many spectroscopes are fed by optical fibres that may or may not intercept the whole of the star's light. Other situations such as an untrailed image, or an image smaller than the slit, require the formula to be modified. Thus, when the slit is wide enough for the whole stellar image to pass through it, the exposure varies inversely with the square of the telescope's diameter, whilst for extended sources, it varies inversely with the square of the telescope's focal ratio (cf. Equations 1.74 and 1.77).

The slit is quite an important part of the spectroscope since in many astronomical spectroscopes it fulfils two functions. First, it acts as the entrance aperture of the spectroscope. For this purpose, its sides must be accurately parallel to each other and perpendicular to the direction of the dispersion. It is also usual for the slit width to be adjustable or slits of different widths provided so that alternative detectors may be used and/or changing observing conditions catered for. Although we have seen how to calculate the optimum slit width, it is usually better in practice to find the best slit width empirically. The slit width is optimised by taking a series of images of a sharp emission line in the comparison spectrum through slits of different widths. As the slit width decreases, the image of the line's width should also decrease at first, but should eventually become constant. The changeover point occurs when some part of the spectroscope system other than the slit starts to limit the resolution and the slit width at changeover is the required optimum value. As well as allowing the desired light through, the slit must reject unwanted radiation. The jaws of the slit are therefore usually of a knife-edge construction with the chamfering on the inside of the slit so that light is not scattered or reflected into the spectroscope from the edges of the jaws. On some instruments a secondary purpose of the slit is to assist in the guiding of the telescope on the object. When the stellar image is larger than the slit width it will overlap onto the slit jaws. By polishing the front of the jaws to an optically flat mirror finish, these overlaps can be observed via an auxiliary detector and the telescope driven and guided to keep the image bisected by the slit.

However, guiding in this manner is wasteful of the expensively collected light from the star of interest. Most modern instruments therefore use another star within the field of view to guide on. If the stellar image is then larger than the slit it can be reformatted so that all its light enters the spectroscope. There are several ways of reformatting the image. Early approaches such as that due to Bowen are still in use. The Bowen image slicer consists of a stack of overlapped mirrors (Figure 4.24) that section the image and then rearrange the sections end to end to form a linear image suitable for matching to a spectroscope slit. The

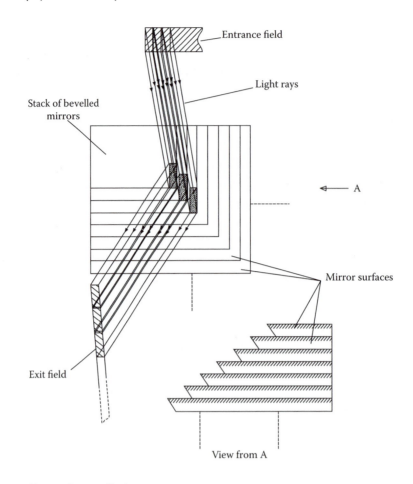

Entrance field

Light rays

Stack of bevelled mirrors

A

Mirror surfaces

Exit field

View from A

FIGURE 4.24   Bowen image slicer.

Bowen-Walraven image slicer uses multiple internal reflections. A prism with a chamfered side is used and has a thin plate attached to it. Because of the chamfered side, the plate only touches the prism along one edge. A light beam entering the plate is repeatedly internally reflected wherever the plate is not in contact with the prism, but is transmitted into the prism along the contact edge (Figure 4.25). The simplest concept is a bundle of optical fibres whose cross section matches the slit at one end and the star's image at the other. Thus, apart from the reflection and absorption losses within the fibres, all the star's light is conducted into the spectroscope. Disadvantages of fibre optics are mainly the degradation of the focal ratio of the beam due to imperfections in the walls of the fibre, so that not all the light is intercepted by the collimator, and the multilayered structure of the normal commercially available units leads to other light losses since only the central core of fibre transmits the light. Thus, specially designed fibre optic cables are usually required and these are made 'in-house' at the observatory needing them. They are usually much thicker than normal fibre optics and are formed from plastic or fused quartz. On telescopes with adaptive optics, the size of the star's image is much reduced and this allows not only the slit width to be smaller but also a more compact design to be used for the whole spectroscope.

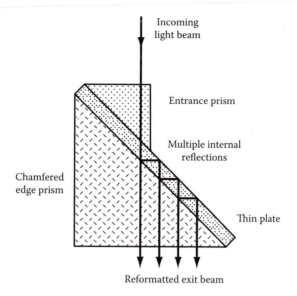

FIGURE 4.25   The Bowen-Walraven image slicer.

For extended sources it is common practice to use long slits. Provided that the image does not move with respect to the slit (and no spectrum widener is used), then each point in that portion of the image falling onto the slit has its individual spectrum recorded at the appropriate height within the final spectrum. Several slits can be used as the entrance aperture of the spectroscope provided that the spectra do not overlap. Then, all the information that is derivable from a single-slit spectrogram is available, but the whole source can be covered in a fraction of the time. The ultimate development of this procedure, known as integral field spectroscopy or 3D spectroscopy, is to obtain a spectrum for every resolved point within an extended source and this is discussed further below.

For several purposes the slit may be dispensed with and some specific designs are considered later in this section. Apart from the Fourier transform spectroscope (see Section 4.1), they fall into two main categories. In the first, the projected image size on the spectrum is smaller than some other constraint on the system's resolution. The slit and the collimator may be discarded and parallel light from the source allowed to impinge directly onto the dispersing element. In the second type of slitless spectroscope, the source produces a nebular type of spectrum (i.e. a spectrum consisting almost entirely of emission lines with little or no continuum). If the slit alone is then eliminated from the telescope–spectroscope combination, the whole of the image of the source passes into the spectroscope. The spectrum then consists of a series of monochromatic images of the source in the light of each of the emission lines. Slitless spectroscopes are difficult to calibrate so that radial velocities can be found from their spectra, but they may be very much more optically efficient than a slit spectroscope. In the latter, perhaps 1% to 10% of the incident light is eventually used in the image, but some types of slitless spectroscope can use as much as 75% of the light. Furthermore, some designs, such as the objective prism, can image as many as 10^5 stellar spectra in one exposure.

A system that is closely related to the objective prism places the disperser shortly before the focal point of the telescope. Although the light is no longer in a parallel beam, the additional aberrations that are produced may be tolerable if the focal ratio is long. Using a zero-deviation grism in combination with correcting optics enables a relatively wide field to be covered, without needing the large sizes required for objective prisms. With suitable blazing for the grating part of the grism the zero order images provide wavelength reference points for the spectra. Grisms used on the wide field imager of ESO's 2.2-metre telescope enable spectra to be obtained in this way over an unvignetted area 19 minutes of arc across with a spectral resolution of about 150 from the blue to the NIR. Similarly, the Visible Multi-Object Spectrograph (VIMOS) for the VLT uses a range of VPH transmission gratings in combination with two prisms to produce a zero-deviation disperser covering a $7' \times 8'$ field of view and various parts of the 370-nm to 1-μm spectral region at spectral resolutions ranging from 180 to 2500.

Spectroscopes, as we have seen, contain many optical elements that may be separated by large distances and arranged at large angles to each other. In order for the spectroscope to perform as expected, the relative positions of these various components must be correct and stable to within very tight limits. The two major problems in achieving such stability arise through flexure and thermal expansion. Flexure primarily affects the smaller spectroscopes that are attached to telescopes at Cassegrain foci and so move around with the telescope. Their changing attitudes as the telescope moves causes the stresses within them to alter, so that if in correct adjustment for one telescope position, they may be out of adjustment in other positions. The light beam from the telescope is often folded so that it is perpendicular to the telescope's optical axis. The spectroscope is then laid out in a plane that is parallel to the back of the main mirror. Hence, the spectroscope components can be rigidly mounted onto a stout metal plate which in turn is bolted flat onto the back of the telescope. Such a design can be made very rigid and the flexure reduced to acceptable levels. In some modern spectroscopes, active supports are used to compensate for flexure along the lines of those used for telescope mirrors (see Section 1.1).

Temperature changes affect all spectroscopes, but the relatively short light paths in the small instruments that are attached directly to telescopes mean that generally the effects are unimportant.

The large fixed spectroscopes that operate at Coudé and Nasmyth foci, or which have the light brought to them through fibre optic cables, are obviously unaffected by changing flexure and usually there is little difficulty other than that of cost in making them as rigid as desired. They experience much greater problems, however, from thermal expansion. The size of the spectrographs may be very large (see Exercise 4.4) with optical path lengths measured in tens of metres. Thus, the temperature control must be correspondingly strict. A major problem is that the thermal inertia of the system may be so large that it may be impossible to stabilise the spectroscope at ambient temperature before the night has ended. Thus, many such spectroscopes are housed in temperature-controlled sealed rooms and/or have some or all of their components cooled and enclosed in a vacuum chamber.

Any spectroscope except the Michelson interferometer can be used as a monochromator. That is, a device to observe the object in a very restricted range of wavelengths. Most scanning spectroscopes are in effect monochromators whose waveband may be varied. The most important use of the devices in astronomy, however, is in the spectrohelioscope. This builds up a picture of the Sun in the light of a single wavelength and this is usually chosen to be coincident with a strong absorption line. Further details are given in Section 5.3. A related instrument for visual use on small telescopes is called a prominence spectroscope. This has the spectroscope offset from the telescope's optical axis so that the (quite wide) entrance slit covers the solar limb. A small direct-vision prism or transmission grating then produces a spectrum and a second slit isolates an image in Hα light, allowing prominences and other solar features to be discerned.

Spectroscopy is undertaken throughout the entire spectrum. In the infrared and UV regions, techniques, designs are almost identical to those for visual work except that different materials may need to be used. Some indication of these has already been given in Section 4.1. Generally, diffraction gratings and reflection optics are preferred since there is then no worry over absorption within the optical components.

The technique of Fourier transform spectroscopy, as previously mentioned, has so far had its main applications in the infrared region. However, the Spectromètre Imageur de l'Observatoire du Mont-Mégantic (SpIOMM) on the 1.6-metre telescope at the Observatoire de Mont Mégantic in Quebec is an imaging Fourier transform spectrometer operating from 350 to 850 nm. It obtains spectra of every object within a 12-arc-minute field of view at resolutions up to 25,000. Similarly, the 1.5-metre solar telescope, Gregor* on Tenerife, uses a Fourier transform spectrometer with a potential spectral resolution up to 250,000 over the 530- to 660-nm region.

A recent development is Externally Dispersed Interferometry (EDI) which has the potential to measure radial velocities to high accuracies at low cost and can also be retrospectively added to existing spectrographs. It consists simply of sending the incoming radiation though a Michelson interferometer with fixed, but slightly different, path lengths for the two light beams before the recombined radiation enters the 'ordinary' spectrograph.

If we imagine the EDI interferometer observing a monochromatic point source then the output will be a series of point images along the image plane wherever there is constructive interference (see the dotted line curve in Figure 4.1). Because of the path difference between the light beams within the interferometer, the fringes near the optical axis will have orders larger than zero. Now imagine that the source is bichromatic. The new wavelength will also produce fringes and the zero orders of both sets of fringes will coincide, but higher orders will be displaced from each other in a line along the image plane. Adding a third wavelength to the source will add a third set of displaced fringes still along the same line along the image plane. Making the source a white light emitter will make the fringes merge continuously into each other along the line in the image plane and will also cause many different orders of fringes to overlay each other. However, if that line of radiation is now directed through the slit of the 'normal' spectrograph, the overlapping orders will be separated out (the spectrograph simply acting as a cross disperser).

---

* Not an acronym, the telescope is of Gregorian design.

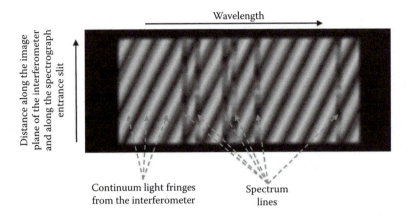

FIGURE 4.26 A schematic image from an EDI.

The appearance of the image resulting from the combined effects of the interferometer and the spectrograph is shown in Figure 4.26. The continuum portions of the spectrum result in slanting fringes and the spectrum lines cut across these fringes. Because the slant of the fringes is shallow, the darkest portions of two lines are separated in the vertical direction by more than the horizontal separation of the lines. In effect, any change in the wavelength of the line due to its Doppler shift is amplified in the vertical movement of the dark portions of the lines and hence is more easily and more accurately measureable.

At very short UV and at X-ray wavelengths, glancing optics (see Section 1.3) and diffraction gratings can be used to produce spectra using the same designs as for visual spectroscopes; however, the appearance and layout will look very different because of the off-axis optical elements, even though the optical principles are the same. Also, many of the detectors at short wavelengths have some intrinsic spectral resolution. Radio spectroscopes have been described in Section 1.2.

4.2.2 Prism-Based Spectroscopes

As noted earlier, spectroscopes using prisms as the sole dispersing element will now rarely be encountered except for some MIR instruments such as that planned for the JWST. Only a brief guide to such spectroscopes will therefore be given here – partly for historical interest, but also because many diffraction-grating-based instruments have similar designs. The basic layout of a prism-based spectroscope is shown in Figure 4.12. Many instruments have been constructed to this design with only slight modifications, the most important of which is the use of several prisms. If several identical prisms are used with the light passing through each along minimum deviation paths, then the total dispersion is that of one of the prisms multiplied by the number of prisms. The resolution is unchanged and remains that for a single prism. Thus, such an arrangement is of use when the resolution of the system is limited by some element of the spectroscope other than the prism. A rather more compact system than that shown in Figure 4.12 can be made by replacing the 60° prism by one with a 30° apex angle that is aluminised on one surface (Figure 4.27). The light therefore passes twice through the prism making its effect the equivalent of a single 60° prism.

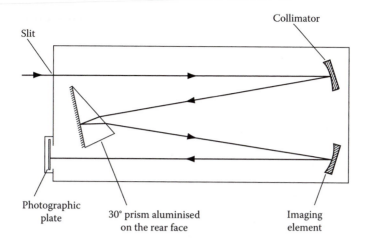

FIGURE 4.27 Compact design for the basic prism spectroscope.

However, the minimum deviation path is no longer possible, so that some astigmatism is introduced into the image, but by careful design this can be kept lower than the resolution of the system as a whole.

Another similar arrangement, widely used for long-focus spectroscopes in laboratory and solar work, is called the Littrow spectroscope, or autocollimating spectroscope. A single lens or mirror acts as both the collimator and imaging element (Figure 4.28) thus saving on both the cost and size of the system.

The deviation of the optical axis caused by the prism can be a disadvantage for some purposes. Direct-vision spectroscopes overcome this problem and have zero deviation for some selected wavelength. There are several designs but most consist of combinations of prisms made in two different types of glass with the deviations arranged so that they cancel out whilst some remnant of the dispersion remains. This is the inverse of the achromatic lens discussed in Section 1.1 and therefore usually uses crown and flint glasses for its

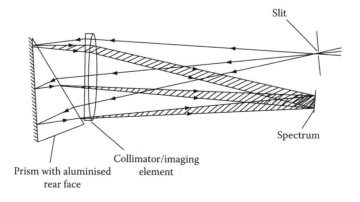

FIGURE 4.28 Light paths in a Littrow spectroscope.

prisms. The condition for zero deviation, assuming that the light passes through the prisms at minimum deviation is

$$\sin^{-1}\left[\mu_1 \sin\left(\frac{\alpha_1}{2}\right)\right] - \frac{\alpha_1}{2} = \sin^{-1}\left[\mu_2 \sin\left(\frac{\alpha_2}{2}\right)\right] - \frac{\alpha_2}{2} \qquad (4.146)$$

where α_1 is the apex angle of prism number 1, α_2 is the apex angle of prism number 2, μ_1 is the refractive index of prism number 1, and μ_2 is the refractive index of prism number 2. In practical designs the prisms are cemented together so that the light does not pass through all of them at minimum deviation. Nonetheless, Equation 4.146 still gives the conditions for direct vision to a good degree of approximation. Direct vision spectroscopes can also be based upon grisms, where the deviations of the prism and grating counteract each other.

Applications of most direct-vision spectroscopes are non-astronomical since they are best suited to visual work. They are though, being increasingly used within instruments that obtain direct images as well as spectra – such as many integral field spectroscopes. Spectra are obtained when the direct-vision spectroscope is placed into the optical path and direct images obtained when it is removed, without the need to adjust the instrument in other respects between the two operating modes.

Another application of the prism that is also now largely of historical interest, having been supplanted by integral field spectroscopy (see below), is the simplest spectroscope of all – the objective prism. This is just a thin prism that is large enough to cover completely the telescope's objective and it is positioned immediately before the telescope's entrance aperture. The starlight is already parallel so that a collimator is unnecessary, whilst the scintillation disc of the star replaces the slit. The telescope acts as the imaging element (Figure 4.29). This has the enormous advantage that a spectrum is obtained for every star in the normal field of view. Thus, if the telescope is a Schmidt camera, up to 10^5 spectra may be obtainable in a single exposure. The system has three main disadvantages. First, the dispersion is low and second, the observed star field is at an angle to the telescope axis (although direct vision objective prisms can be made from two prisms of different glasses oriented in opposite directions to each other). Finally, there is no reference point for wavelength measurements (though a number of ingenious adaptations of the device have been tried to overcome this latter difficulty).

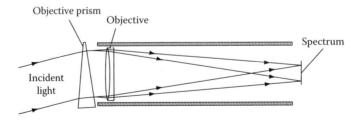

FIGURE 4.29   An objective prism spectroscope.

4.2.3 Grating Spectroscopes

Most of the gratings used in astronomical spectroscopes are of the reflection type. This is because the light can be concentrated into the desired order by blazing quite easily, whereas for transmission gratings, blazing is much more difficult and costly. Transmission gratings are, however, often used in grisms and these are finding increasing use in integral field spectroscopes.

Plane gratings are most commonly used in astronomical spectroscopes and are almost invariably incorporated into one or other of two designs discussed in the previous section, with the grating replacing the prism. These are the compact basic spectroscope (Figure 4.27), sometimes called a Czerny-Turner system when it is based upon a grating and the Littrow spectroscope (Figure 4.28) which is called an Ebert spectroscope when based upon grating and reflection optics.

Most of the designs of spectroscopes that use curved gratings are based upon the Rowland circle (Figure 4.6). The Paschen-Runge mounting in fact is identical to that shown in Figure 4.6. It is a common design for laboratory spectroscopes since wide spectral ranges can be accommodated, but its size and awkward shape make it less useful for astronomical purposes. A more compact design based upon the Rowland circle is called the Eagle spectroscope (Figure 4.30). However, the vertical displacement of the slit and the spectrum (see side view in Figure 4.30) introduces some astigmatism. The Wadsworth design abandons the Rowland circle but still produces a stigmatic image through its use of a collimator (Figure 4.31). The focal surface, however, becomes paraboloidal and some spherical

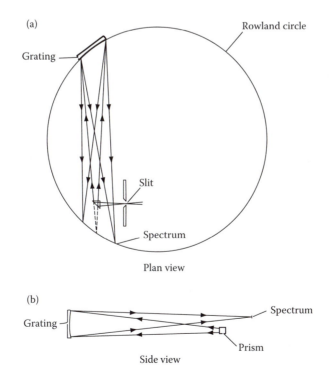

FIGURE 4.30 Optical arrangement of an Eagle spectroscope shown from above (a) and the side (b).

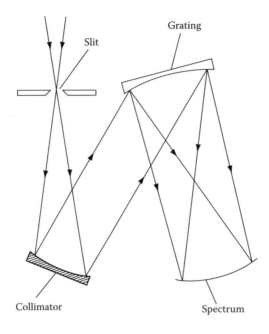

FIGURE 4.31   The Wadsworth spectroscope mounting.

aberration and coma are introduced. Furthermore, the dispersion for a given grating, if it is mounted into a Wadsworth system, is only half what it could be if the same grating were mounted into an Eagle system, since the spectrum is produced at the prime focus of the grating and not at its radius of curvature.

With designs such as the Wadsworth and its variants, the imaging element is likely to be a Schmidt camera system (see Section 1.1) in order to obtain high-quality images with a fast system. Some recent spectroscopes though have used dioptric cameras to avoid the light loss involved with the detector holder in a Schmidt system. Exotic optical materials such as calcium fluoride often need to be used in these designs to achieve the required imaging speed, flat field and elimination of other aberrations. Gratings can also be used as discussed earlier in various specialised applications such as nebular and prominence spectroscopes.

Spectroscopes are usually optimised for one spectral region. If spectra are needed that extend over a wider range than is covered by the spectroscope, then it may need adjusting to operate in another region, or a different instrument entirely may be required. To overcome this problem to some extent, several spectroscopes have two or three channels optimised for different wavelength ranges. The incoming light from the telescope is split into the channels by dichroic mirrors, so that (with some designs) the spectra can be obtained simultaneously. In other designs, the spectra are obtained in quick succession with little down time needed to adjust the spectroscope. ESO's Ultraviolet and Visual Echelle Spectroscope (UVES) for example has a blue channel covering 300 to 500 nm and a red channel covering 420 to 1100 nm.

The design of a spectroscope is generally limited by the available size and quality of the grating, and these factors are of course governed by cost. The cost of a grating in turn is dependent upon its method of production. The best gratings are originals, which are produced in the following manner. A glass or other low expansion substrate is over-coated with a thin layer of aluminium. The grooves are then scored into the surface of the aluminium by lightly drawing a diamond across it. The diamond's point is precisely machined and shaped so that the required blaze is imparted to the rulings. An extremely high-quality machine is required for controlling the diamond's movement, since not only must the grooves be straight and parallel, but also their spacings must be uniform if Rowland and Lyman ghosts are to be avoided. The position of the diamond is there-fore controlled by a precision screw thread and is monitored interferometrically. We have seen that the resolution of a grating is dependent upon the number of grooves, whilst the dispersion is a function of the groove spacing (see Section 4.1). Thus, ideally a grating should be as large as possible and the grooves should be as close together as possible (at least until their separation is less than their operating wavelength). Unfortunately, both of these parameters are limited by the wear on the diamond point. Thus, it is possible to have large coarse gratings and small fine gratings, but not large fine gratings. The upper limits on size are about 0.5 metres square and on groove spacing, about 1500 lines mm^{-1}. A typical grating for an astronomical spectroscope might be 0.1 metres across and have 500 lines mm^{-1}.

Echelle gratings are used for many recently built spectroscopes. The rectangular format (Figure 4.32) of the group of spectral segments after the cross disperser matches well to the shape of large CCD arrays, so that the latter may be used efficiently. The UVES for ESO's VLT for example operates in the blue and the red and NIR regions with two 0.2 × 0.8-metre echelle gratings. The gratings have 41 and 31 lines mm^{-1} and spectral resolutions of 80,000 and 115,000, respectively (Figure 4.32). Also, for the VLT, the X-Shooter instru-ment has three independent spectrographs, each of which uses an echelle grating with a cross disperser. The three components cover the spectral regions 300 to 559.5 nm, 559.5 to 1024 nm and 1024 to 2489 nm and receive 'their' portion of the incoming radiation via dichroic mirrors. The maximum spectral resolution is 18,200. Similarly, the Magellan Inamori Kyocera Echelle (MIKE) instrument for the Magellan telescope which started science operations in 2003 uses two echelle gratings with prism cross dispersers and a dichroic mirror to cover the spectral regions 335 to 500 nm and 490 to 950 nm at resolu-tions up to 83,000.

On a quite different scale there is the Basic Echelle Spectrograph (BACHES) instrument that is designed for use on small (≥0.2 metres) telescopes and thus potentially available to amateur astronomers. This instrument can obtain spectra over the 390- to 750-nm region for stars brighter than 5^{m} at a spectral resolution of 19,000 with a 15-minute exposure on a 0.35-metre telescope (Figure 4.33). It uses a 79 lines mm^{-1} echelle grating with a diffrac-tion grating as the cross disperser and is designed to monitor spectrum variables such as Be stars.

Often a replica grating will be adequate, especially for lower-resolution instruments. Since many replicas can be produced from a single original, their cost is a small fraction of that of an original grating. Some loss of quality occurs during replication, but this is acceptable for many purposes. A replica improves on an original in one way, however, and

FIGURE 4.32   (a) The UVES spectrograph shown on one of the Nasmyth platforms of the Kueyen VLT telescope and with its cover removed. (Reproduced by kind permission of ESO. Faces have been obscured in line with the ESO's image use policy.) (b) An echelle spectrum of supernova SN1987A in the Large Magellanic Cloud obtained with UVES on ESO's VLT. The individual segments of the spectrum can clearly be seen as almost horizontal bands. The whole coverage is from 479 nm (bottom) to 682 nm (top). Each segment covers about 10 nm with about a 2.5-nm overlap on the left-hand side with the segment above and about a 2.5-nm overlap on the right-hand side with the segment below. The supernova's spectrum is the narrow bright line running along the centre of each segment and several strong emission lines can be seen, such as Hα near the centre of the fourth segment from the top. The supernova's spectrum is a superimposed faint solar absorption spectrum arising from scattered light from the full Moon. Also, numerous emission lines from the Earth's atmosphere can be seen which are narrower than the supernova emission lines and have the same height as the solar spectrum. (Reproduced by kind permission of ESO.)

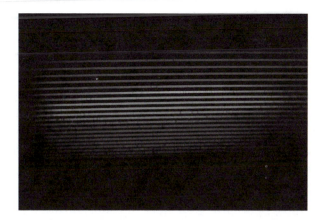

FIGURE 4.33 **(See color insert.)** A solar spectrum obtained using the BACHES echelle spectrograph. The Hα line is near the centre of the second segment from the top and the sodium D lines near the centre of the sixth segment from the top. (Reproduced by kind permission of Burwitz/Club of Aficionados in Optical Spectroscopy [CAOS]. For more information on astronomical spectroscopy using small telescopes and by amateur astronomers see http://spectroscopy.wordpress.com/.)

that is in its reflection efficiency. This is better than that of the original because the most highly burnished portions of the grooves are at the bottoms of the grooves on the original, but are transferred to the tops of the grooves on the replicas. The original has of course to be the inverse of the finally desired grating. Covering the original with a thin coat of liquid plastic or epoxy resin, which is stripped off after it has set, produces the replicas. The replica is then mounted onto a substrate, appropriately curved if necessary and then aluminised.

More recently, large high-quality gratings have been produced holographically. An intense monochromatic laser beam is collimated and used to illuminate a photoresist-covered surface. The reflections from the back of the blank interfere with the incoming radiation and the photoresist is exposed along the nodal planes of the interference field. Etching away the surface then leaves a grating that can be aluminised and used directly, or it can have replicas formed from it as above. The wavelength of the illuminating radiation and its inclination can be altered to give almost any required groove spacing and blaze angle.

A related technique to the last produces volume phase holographic gratings (VPHGs; see also Section 4.1). A layer of gelatine around 10 μm thick and containing a small proportion of ammonia or potassium dichromate is coated onto a glass substrate. It is then illuminated with an interference pattern from two laser beams. Instead of being etched, however, the layer is then treated in water and alcohol baths so that the refractive index within the layer varies according to the exposure it has received. The layer thus forms a grating with the lines of the grating produced by regions of differing refractive indices. Since the gelatine is hygroscopic, it must be protected after production of the grating by a cover sheet. VPHGs can have up to 95% efficiency in their blaze region and currently can be produced up to 300 mm in diameter and with from 100 to 6000 lines per millimetre. They can be used both as transmission and reflection and they seem likely to find increasing use within astronomical spectroscopes. It is also possible to produce two different gratings within a single element. Such gratings can be tilted with respect to each other so that the spectra are separated. Different wavelength regions can then be observed using a single spectroscope.

Most spectroscopic observations require exposures ranging from tens of seconds to hours or more. The read-out times from their detectors are therefore negligible in comparison. However, for some applications, such as observing rapidly varying or exploding stars, a series of short exposures may be needed and then the read-out and processing times can become significant. Recently, therefore, instruments optimised for high-speed spectroscopy have started to be developed. ULTRASPEC, for example, a visitor instrument on ESO's 3.6-metre telescope, uses a frame-transfer EMCCD (Section 1.1.4 and Appendix D) that enables hundreds of spectra per second to be obtained. Clearly, such short exposures can only be used for objects with very bright apparent magnitudes; however, as the 20- to 40-metre class telescopes that are currently being planned come on stream, high-speed spectroscopy is likely to be extended to fainter objects and become more widely used.

4.2.4 Integral Field Spectroscopy

Where the spectra of several individual parts of an extended object such as a gaseous nebula or galaxy are needed, then, as already discussed, a long slit may be used so that a linear segment of the object is covered by a single exposure. Where spectra of every resolution element of an object are required then repeated adjacent exposures using a long-slit spectroscope are one possible, but time-consuming, approach. Several techniques that come under the heading of integral field spectroscopy or 3D spectroscopy have thus been developed recently to obtain spectra for every resolved point within an extended source more efficiently.

More simply, several long slits can be used as the entrance aperture of the spectroscope provided that the spectra do not overlap. Then, all the information that is derivable from a single-slit spectrogram is available, but the source can be covered in a fraction of the time. Another approach is to use a scanning narrowband filter and obtain numerous images at slightly differing wavelengths. Examples of this are the OSIRIS and MMTF tunable filters discussed in Section 4.1. Alternatively, imaging detectors, such as STJs (see Section 1.1), that are also intrinsically sensitive to wavelength can be used to obtain the whole data set in a single exposure. A colour photograph or image is, of course, essentially a low-resolution 3D spectrogram. A possible extension to existing colour imaging techniques which may have a high enough spectral resolution to be useful, is through the use of dye-doped polymers. These are thin films of polymers such as polyvinyl butyral, containing a dye such as chlorin that has a very narrow absorption band. Changing conditions within the substrate cause the wavelength of the absorption band to scan through a few nanometers and potentially provide high-efficiency direct imaging combined with spectral resolutions of perhaps 10^5 or more.

Most integral field spectroscopy, however, relies on three other approaches. The first is to use an image slicer, although this will need to cover a larger area than that of the stellar image slicers discussed previously. A stack of long plane mirrors whose widths match that of the slit and that are slightly twisted with respect to each other around their short axes is used. The stack is placed at the focal pane of the telescope and the segments of the sliced image then reimaged into a line along the length of the slit. Spectrometer for Infrared Faint Field Imaging (SPIFFI), for example, which forms the heart of the Spectrograph for Integral Field Observations in the Near Infrared (SINFONI) adaptive optics integral field spectroscope on the VLT, uses two sets of 32 mirrors to split fields of view of up to 8″ × 8″ into narrow rectangular arrays matching the shape of spectroscope's entrance aperture.

The second approach is to use a large fibre optic bundle with a square or circular distribution of the fibres at the input and a linear arrangement at the output. The arrangement of the fibres at the output must correspond in a simple and logical fashion to their arrangement at the input so that the position within the final spectrum of the individual spectrum from a point in the image can be found. As with the stellar image slicers, the cladding of the fibres would mean that significant amounts of light would be lost if the fibre optics were to be placed directly at the telescope's focus. Instead, an array of small lenses is used to feed the image into the fibres. The array of square or hexagonal lenses is placed some distance behind the telescope focus and each lens images the telescope pupil onto the end of a fibre. Providing that these subimages are contained within the transmitting portions of the fibres, no light is lost due to the cladding. Segmented Pupil Image Reformatting Array Lens (SPIRAL) on the AAT for example uses an array of 512 square lenses to provide spectroscopy over an 11″ × 22″ area of the sky feeding the AAOmega spectroscope via 18 metres of fibre optic cable, whilst the recently upgraded Cyclops2, for the University College London Echelle Spectrograph (UCLES) also on the AAT, uses a lenslet array to feed 16 optical fibres covering a 5 × 5 arc-second area of the sky. The fibres reformat the area to the equivalent of a 0.6-arc-second slit.

The third approach also uses an array of small lenses but dispenses with the spectroscope slit. The lenses produce a grid of images that is fed directly into the spectroscope and results in a grid of spectra. By orienting the lens array at a small angle to the dispersion, the spectra can be arranged to lie side by side and not to overlap. Optically Adaptive System for Imaging Spectroscopy (OASIS) that was originally used on the 3.6-metre Canada-France-Hawaii Telescope (CFHT) and is now installed at the Nasmyth focus of the 4.2-metre William Herschel Telescope (WHT) and uses an 1100-hexagonal array of lenses in this fashion, whilst Supernova Integral Field Spectrograph (SNIFS) on the University of Hawaii's 2.2-metre telescope covers a 6″ × 6″ area of the sky with an array of 225 lenslets.

A recent development is the photonic lantern. The radiation is input into a multimode optical fibre. The large size* of multimode fibres enables the whole of the stellar image to be accepted. The various modes are then separated from each other within a flared section of the fibre and each is then fed into a single-mode optical fibre. The single-mode fibres are much smaller in diameter than the multimode fibre and feed the radiation directly into the spectrograph. A prototype photonic spectrograph has recently been successfully tested

---

* Multimode optical fibres can be up to 200 μm in diameter; single-mode fibres are around 10 μm across. The physical size of the image of a star from a large telescope with good adaptive optics atmospheric correction is around 100 μm.

using the 3.9-metre AAT and a NIR spectrum of π^1 Gru obtained at a spectral resolution of 2500.

The ideal output from an integral field spectroscope is a spectral cube (hence, the alternative name of 3D spectroscopy). The spectral cube has two sides that are the x and y positional coordinates of a conventional 2D image with the third side being wavelength or frequency. Thus, each pixel in the 2D image of the object or area of sky has its spectrum recorded – alternatively, we may regard the cube as a series of images at different wavelengths each separated by the spectral resolution of the system. Spectral cubes can in principle be obtained for any wavelength, but currently are only available in the visible, NIR and the microwave regions (where, for example, HARP on the JCMT observes a 16-pixel image over the 325 to 375 GHz region) (see Section 1.2).

For the future, a second generation instrument for the VLT, Multi-Unit Spectroscopic Explorer (MUSE), is currently under construction and is expected to cover a $1' \times 1'$ area of the sky at a spatial resolution of 0.2″ and a spectral resolution between 2000 and 4000 over the 465- to 930-nm region. The instrument is planned to be used with an adaptive optics image sharpener using four guide stars and a reflective image slicer that divides the total field of view into 24 subfields, each of which will then be imaged by a 2k × 4k detector. The resulting spectral cube will thus contain some 360 million elements.

4.2.5 Multi-Object Spectroscopy

Some versions of integral field spectroscopes, especially the earlier designs, use multiple slits to admit several parts of the image of the extended object into the spectroscope simultaneously. Such an instrument can easily be extended to obtain spectra of several different objects simultaneously by arranging for the slits to be positioned over the images of those objects within the telescope's image plane. Practical devices that used this approach employ a mask with narrow linear apertures cut at the appropriate points to act as the entrance slits. However, a new mask is needed for every field to be observed, or even for the same field if different objects are selected, and cutting the masks is generally a precision job requiring several hours work for a skilled technician. This approach has therefore rather fallen out of favour, although the Gemini telescopes' Multi-Object Spectroscopes (GMOSs) continue to use masks. Their masks are generated automatically using a laser cutter, with the slit positions determined from a direct image of the required area of the sky obtained by GMOS operating in an imaging mode.

The LBT's LBT NIR Utility with Camera and Integral Field Unit for Extragalactic Research (LUCIFER) can also operate with masks to provide multi-object spectroscopy over the 900-nm to 2.5-μm region as well as being available in other operating modes. Similarly, one mode of operation of the Keck telescopes' Deep Imaging Multi-Object Spectrograph (DEIMOS) instrument uses up to 130 slits cut into a mask for spectroscopy over the 410- to 1100-nm region covering an 80-square arc-minute area of the sky.

Most multi-object spectroscopy, however, is now undertaken using fibre optics to transfer the light from the telescope to the spectroscope. The individual fibre optic strands are made large enough to be able to contain the whole of the seeing disk of a stellar image or in some cases the nucleus of a distant galaxy or even the entire galaxy. Each strand then has one of its ends positioned in the image plane of the telescope so that it intercepts one of the required images. The other ends are aligned along the length of the spectroscope slit, so that hundreds of spectra may be obtained in a single exposure. Initially, the fibre optics were positioned by being plugged into holes drilled through a metal plate. But this has the same drawbacks as cutting masks for integral field spectroscopes and has been superseded. There are now two main methods of positioning the fibres, both of which are computer-controlled and allow 400 or 500 fibres to be repositioned in a matter of minutes.

The first approach is to attach the input ends of the fibre optic strands to small magnetic buttons that cling to a steel back plate. The fibres are reconfigured one at a time by a robot arm that picks up a button and moves it to its new position as required. A positional accuracy for the buttons of 10 to 20 μm is needed. This approach is used with AAOmega which uses 800 fibres attached to magnetic buttons. The OzPoz* system for ESO's VLT originally could position up to 560 fibres in the same way. It is now used with the Fibre Large Array Multi-Element Spectrograph (FLAMMES) instrument. FLAMES has several modes of operation enabling it to observe between 8 and 130 areas of the sky simultaneously with spectral resolutions up to 47,000. In some modes, fibre optic bundles replace the single fibres each covering a 2″ × 3″ area of the sky. Thus, each fibre optic bundle represents a mini integral field feeding one of two spectroscopes and so enabling the whole of the visible spectrum to be observed. The system uses two backing plates so that one can be in use whilst the other is being reset. Swapping the plates takes about 5 minutes, thus minimising the dead time between observations.

The second approach mounts the fibres on the ends of computer-controlled arms that can be moved radially and/or from side to side. The arms may be moved by small motors, or as in the Subaru telescope's Echidna system, by an electromagnetic system. The latter has 400 arms (or spines to fit in with the device's name), each of which can be positioned within a 7-mm circle to within ±10 μm. Typically, about 90% of 400 target objects can be reached in any one configuration and the light beams are fed to a NIR spectroscope. Repositioning the spines takes only 10 minutes. The SuMIRe (Subaru Measurement of Images and Redshifts) instrument is currently being developed and is expected to have 2400 fibres positioned by piezoelectric motors. The fibres will feed four separate spectrographs, each of which will have three channels and cover the spectral region 380 nm to 1.3 μm.

Two ambitious instruments for the future are Big Baryon Acoustic Oscillation Spectroscopic Survey (BigBOSS) on the 4-metre Mayall telescope and 4MOST. BigBOSS is intended to extend the SDSS's BOSS project which is attempting to investigate the properties of dark energy. BigBOSS will have 5000 fibres covering a 3° × 3° field of view and obtain spectra at a resolution of 5000 of over 20 million galaxies and 2 million quasars. 4MOST is currently undergoing a conceptual design study. Intended for use on ESO's 4-metre VISTA telescope, or other telescope of similar size, it may have up to 3000 fibres and cover a 5°

---

* The device is based upon the positioner developed for the 2dF on the AAT – hence Oz for Australia and Poz for positioner.

field of view. Its mission would be to obtain spectra of some 20 million stars at a spectral resolution of 5000.

Schmidt cameras can also be used with advantage for multi-object spectroscopy because their wide fields of view provide more objects for observation and their short focal ratios are well matched to transmission through optical fibres. Thus, for example, the 6-degree field (6dF) project developed from FLAIR (Fibre-Linked Array-Image Reformatter) on the UK Schmidt camera which could obtain up to 150 spectra simultaneously over a 40-square-degree field of view with automatic positioning of the fibre optics using the magnetic button system. FLAIR has now been replaced by the 6dF multi-object spectroscope. This still uses 150 fibres positioned using magnetic buttons but the turnaround time has been shortened so that 10 times the output from FLAIR is possible. This system is currently engaged in the Radial Velocity Experiment (RAVE) that has so far measured the radial velocities of about half a million stars.

Once the light is travelling through the fibre optic cables, it can be led anywhere. There is thus no requirement for the spectroscope to be mounted on the telescope, although this remains the case in a few systems. More frequently, the fibre optics take the light to a fixed spectroscope that can be several tens of metres away. This has the advantage that the spectroscope is not subject to changing gravitational loads and so does not flex and it can be in a temperature-controlled room. Subaru's Echidna feeds a spectroscope 50 metres away from the telescope for example. Fibre optic links to fixed spectroscopes can also be used for single-object instruments. ESO's High Accuracy Radial velocity Planet Searcher (HARPS) spectroscope on the 3.6-metre telescope is intended for extra-solar planet finding and so needs to determine stellar velocities to within ±1 m/s. Thus, a very stable instrument is required and HARPS is not only mounted 38 metres away from the telescope but is enclosed in a temperature-controlled vacuum chamber.

4.2.6 Techniques of Spectroscopy

There are several problems and techniques which are peculiar to astronomical spectroscopy and that are essential knowledge for the intending astrophysicist.

One of the greatest problems and one which has been mentioned several times already is that the image of the star may be broadened by atmospheric turbulence until its size is several times the slit width. Only a small percentage of the already pitifully small amount of light from a star therefore enters the spectroscope. The design of the spectroscope and in particular the use of a large focal length of the collimator in comparison to that of the imaging element can enable wider slits to be used. Even then, the slit is generally still too small to accept the whole stellar image and the size of the spectroscope may have to become very large if reasonable resolutions and dispersions are to be obtained. The alternative approach to the use of a large collimator focal length lies in the use of an image slicer and/or adaptive optics as discussed earlier.

Another problem in astronomical spectroscopy concerns the width of the spectrum. If the stellar image is held motionless at one point of the slit during the exposure, then the final spectrum may only be a few microns high. Not only is such a narrow spectrum very difficult to measure, but also individual spectral features are recorded by only a few pixels

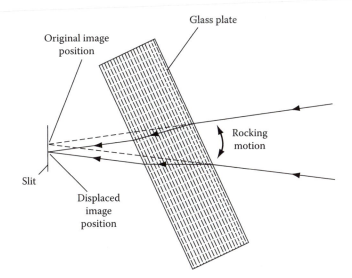

FIGURE 4.34   Image displacement by a plane-parallel glass plate.

and so the noise level is high. Both of these difficulties can be overcome, though at the expense of longer exposure times, by widening the spectrum artificially. There are several ways of doing this, such as introducing a cylindrical lens to make the images astigmatic, or by trailing the telescope during the exposure so that the image moves along the slit. For single-object spectroscopes the spectrum can be widened by using a rocker block.* This is just a thick piece of glass with polished plane-parallel sides through which the light passes. The block is oscillated about an axis parallel to the spectrum and the displacement of the light beam (Figure 4.34) moves the beam up and down, so broadening the spectrum. In multi-object spectroscopes using fibre optic links the fibre size is generally sufficient to provide a useable spectrum. Astronomical spectra are typically widened to about 0.1 to 1 mm. Clearly, however, the spectra in integral field spectroscopes cannot be widened without destroying the imaging information.

Spectrographs operating in the NIR (1–5 μm) are of conventional designs, but need to be cooled to reduce the background noise level. Typically, instruments such as ESO's ISAAC and Cryogenic Infrared Echelle Spectrograph (CRIRES) which operate from 1 to 5 μm, have their main components cooled to 70 K, whilst the detectors are held at 25 K to 4 K or less and the whole instrument is enclosed in a vacuum chamber. Long Slit Intermediate Resolution Infrared Spectrograph (LIRIS) on the WHT observes over the 0.9- to 2.4-μm region with its detectors cooled to 65 K and the rest of the instrument cooled by liquid nitrogen. Multi-Object Spectrometer for Infrared Exploration (MOSFIRE) on the Keck I telescope had first light in 2012. It uses a mask of 46 slits to cover objects over a 6.1′ × 6.1′ area in the NIR with its detector cooled to 77 K.

---

* Also called a dekker or decker.

At MIR wavelengths, the entire spectroscope may need to be cooled to less than 4 K. VISIR, which started operations in 2004, for example, has most of its structure and optics cooled to 33 K, the parts near the detectors to 15 K and the detectors to 7 K.

Ultraviolet spectroscopy is not possible from ground-based telescopes (see also Section 1.3), except in the most limited sense from the eye's cutoff wavelength of about 380 nm[*] to the point where the Earth's atmosphere becomes opaque (300–340 nm depending upon altitude, state of the ozone layer and other atmospheric conditions) – that is, the UVA and UVB regions as they are popularly known. All except the longest wavelength UV spectroscopy therefore needs spectroscopes flown on board balloons, rockets or spacecraft to lift them above most or all of the Earth's atmosphere. Two prime examples of UV spectroscopy spacecraft – both now decommissioned – were IUE and Far Ultraviolet Spectroscopic Explorer (FUSE). IUE operated from 1978 to 1996 observing spectra from 100 to 300 nm, whilst FUSE was launched in 1999 and operated until 2007 observing the 90- to 120-nm spectral region. The Cosmic Origins Spectrograph (COS) was installed on board the HST in 2009. Its design is based upon that for FUSE and it uses use windowless MCPs as its detectors and covers the 115 to 320-nm region at a spectral resolution of up to 24,000. The Russian-led World Space Observatory – Ultraviolet (WSO-UV) is planned for a 2016 launch. It is to be based upon a 1.7-metre telescope and observe over the region from 100 to 320 nm. It will have three UV spectrographs enabling high resolution ($R = 55,000$) spectroscopy of point sources, long-slit, low-resolution spectroscopy of extended sources and a slitless spectroscope for studying gaseous nebulae.

Atmospheric dispersion is another difficulty that needs to be considered. Refraction in the Earth's atmosphere changes the observed position of a star from its true position (Equation 5.13). But the refractive index varies with wavelength. For example, at standard temperature and pressure, we have Cauchy's formula (cf. the Hartmann formula, Equation 4.42) for the refractive index of the atmosphere:

$$\mu = 1.000287566 + \frac{1.3412 \times 10^{-18}}{\lambda^2} + \frac{3.777 \times 10^{-32}}{\lambda^4} \tag{4.147}$$

so that the angle of refraction changes from one wavelength to another and the star's image is drawn out into a very short vertical spectrum. In a normal basic telescope system (i.e. without the extra mirrors required, for example, by the Coudé system), the long wavelength end of the spectrum will be uppermost. To avoid spurious results, particularly when undertaking spectrophotometry, the atmospheric dispersion of the image must be arranged to lie along the slit, otherwise certain parts of the spectrum may be preferentially selected from the image. Alternatively, an atmospheric dispersion corrector may be used. This is a low but variable dispersion direct vision spectroscope that is placed before the entrance slit of the main spectroscope and whose dispersion is equal and opposite to that of the atmosphere. The VLT's X-Shooter, for example, has pairs of atmospheric dispersion corrector prisms in its UV and

---

[*] Persons having had cataract operations may be able to see further in to the UV than this, since the eye's lens absorbs shortwave radiation strongly.

visible component spectrographs. The problem, however, is usually only significant at large zenith distances, so that it is normal practice to limit spectroscopic observations to zenith angles of less than 45°. The increasing atmospheric absorption and the tendency for telescope tracking to deteriorate at large zenith angles also contributes to the wisdom of this practice.

In many cases it will be necessary to try and remove the degradation introduced into the observed spectrum because the spectroscope is not perfect. The many techniques and their ramifications for this process of deconvolution are discussed in Section 2.1.

In order to determine radial velocities, it is necessary to be able to measure the actual wavelengths of the lines in the spectrum, and to compare these with their laboratory wavelengths. The difference, $\Delta\lambda$, then provides the radial velocity, v, via the Doppler shift formula

$$v = \frac{v\lambda}{\lambda}c = \frac{\lambda_{Observed} - \lambda_{Laboratory}}{\lambda_{Laboratory}}c \qquad (4.148)$$

where c is the velocity of light. The observed wavelengths of spectrum lines are most usually determined by comparison with the positions of emission lines in an artificial spectrum. This comparison spectrum is normally that of an iron or copper arc or comes from a pressure gas emission lamp such as sodium or neon or combinations such as thorium and argon (or uranium and neon for the NIR). The light from the comparison source is fed into the spectroscope and appears as one or two spectra on one or both sides of the main spectrum (Figure 4.35). The emission lines in the comparison spectra are at their rest wavelengths and are known very precisely. The observed wavelengths of the stellar (or other object's) spectrum lines are found by comparison with the positions of the artificial emission lines. For the very high-precision radial velocities needed to detect exoplanets (see below), an absorption cell may be used to produce the comparison lines.

However, the emission lines produced by real atoms have drawbacks when used to produce comparison spectra. This problem is becoming particularly apparent with the requirement to be able to measure radial velocities to a precision of ±1 m s^{-1} or even ±0.1 m s^{-1} and for this precision to be stable over several years, which is needed for the detection of terrestrial-sized exoplanets. Thus, the lines from real atoms are not distributed uniformly and there may be long gaps between usable lines, the intensities of the lines vary widely and

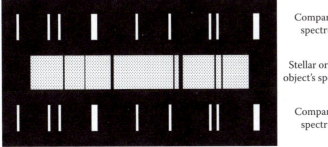

FIGURE 4.35 The wavelength comparison spectrum.

also many lines are blends whose exact median wavelength will depend upon the relative strengths of the two or more individual lines contributing to the blend – and these can vary with the physical conditions within the emission lamp. Finally, the wavelengths of the lines are not known *a priori*, but have to be measured individually in the laboratory.

At the time of writing therefore the production of a 'comb' of close, uniformly intense and regularly spaced emission features using femtosecond lasers is being investigated (see also the next subsection). The Laser Frequency Comb (LFC) is based upon the repetitive emissions from a mode-locked laser. Mode-locked lasers emit pulses of radiation separated by the round-trip time of the laser cavity. Stability is ensured by synchronising the repetitions with an atomic clock (perhaps from the GPS satellites). The Fourier transform of the output of such a laser is the frequency comb and its emission features are spaced at frequencies of $1/T$ and their spectral widths are $1/D$, where T is the laser repetition interval and D is the laser pulse duration.

For rapid determinations of radial velocity the cross-correlation spectroscope, originally devised by Roger Griffin, can be used, although this device will not be encountered much now. This instrument places a mask over the image of the spectrum and reimages the radiation passing through the mask onto a point source detector. The mask is a negative version (either a photograph or artificially generated) of the spectrum. The mask is moved across the spectrum and when the two coincide exactly, there is a very sharp drop in the output from the detector. The position of the mask at the correlation point can then be used to determine the radial velocity with a precision, in the best cases, of a few metres per second.

4.2.7 Exoplanets

Over half of the 900 or so exoplanets that have currently been discovered have been found through the radial velocity variations of their host stars as the exoplanet and star orbit their common centre of mass. The first conventional exoplanet to be found was 51 Peg b by Michel Mayor and Didier Queloz in 1995. The host star, 51 Peg, changed its radial velocity by 120 m s^{-1} every 4.2 days. They used the Elodie spectrograph for the 1.93-metre telescope at the Observatoire de Haute Provence to obtain the spectra. This spectroscope is of a relatively conventional optical design, but is housed separately from the telescope in a temperature-controlled room and fed by fibre optic cable. It was able to measure velocities to an accuracy of about 10 m s^{-1}.

The host star's velocity changes for 51 Peg b, however, are unusually large. More typical is the recently discovered exoplanet, α Cen B b, whose host star changes its velocity by just ±0.5 m s^{-1}. Spectrographs searching for exoplanets must thus be able to measure radial velocities to a precision of ±1.0 m s^{-1} to ±0.1 m s^{-1} or so. Most such instruments now in use are therefore purpose designed and in particular have high levels of stability and special arrangements for the comparison spectra.

HARPS for example, although using a thorium-argon emission lamp to produce the comparison spectrum, is housed in a vacuum chamber and has its temperature controlled to ±0.01 K. The star's light and that of the comparison spectrum are fed to the spectroscope via identical fibre optic cables. It can measure radial velocities to ±1 m s^{-1} and has discovered some 140 exoplanets to date.

An alternative means of producing the comparison spectrum is via an absorption cell. These have the advantage that the light paths for the star and comparison spectra are identical. An absorption cell is just a container for a suitable gas that has optically flat windows and through which the light from the star is passed. The gas in the cell then absorbs at its characteristic wavelengths and these lines are superimposed upon the stellar spectrum. Most absorption cells use molecular iodine at a pressure of about 10 millibars and a temperature around 50°C. The majority of iodine's spectrum lines are in the green. A cell length around 100 mm suffices to produce measureable absorption lines. Confirmation of the discovery of 51 Peg B b was made by Geoff Marcy and Paul Butler using an iodine absorption cell spectrograph on the Hamilton spectrograph of the 3.05-metre Shane telescope (Figures 4.36 and 1.69). Many large telescopes now have spectrographs equipped with absorption cells as optional facilities, such as the HIRES echelle spectrograph on the Keck I telescope which achieves ± 1 m s^{-1} accuracy and has enabled the discovery of over 200 exoplanets.

As mentioned above, a set of closely and regularly spaced emission lines can be produced by a mode-locked laser (Figure 4.37). The laser frequency comb has yet to be used as the primary wavelength standard for an astronomical spectrograph but has been tried out in the last few years at a number of observatories. In 2010, for example, stellar spectra were obtained using an LFC calibrator with the Pathfinder spectrograph on the Hobby-Eberly telescope (Figure 4.37). The LFC lines covered the region from 1.54 to 1.63 μm at a

FIGURE 4.36 The iodine absorption cell used with the Shane telescope to confirm the discovery of 51 Peg B b. The light beam shown is a digital simulation. (Reproduced by kind permission of L. Hatch. © 2006 Laurie Hatch. See www.lauriehatch.com for many more of her beautiful astro-images.)

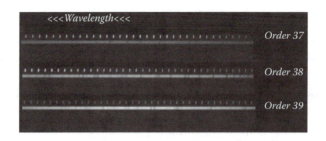

FIGURE 4.37 The spectrum of HD 168723 together with a laser comb comparison spectrum. The image shows three orders from the stellar spectrum running from 1.54 to 1.63 µm with the laser comb spectrum above the stellar spectrum in each case. (University of Colorado/NIST H band Laser Comb Spectra obtained with the Penn State Pathfinder Spectrograph on the Hobby Eberly Telescope. Reproduced by kind permission of NIST/Tech Beat.)

frequency spacing of the emission lines of 25 GHz (wavelength separations ranging from 0.18 to 0.23 nm). The precision of the results was around ±10 m s^{-1}. In the laboratory, the stability and absolute accuracy of the LFC suggested that eventually measurements of radial velocity to ±0.06 m s^{-1} may be achievable. A duplicate of the HARPS instrument has recently started operations on the 3.6-metre Telescopio Nazionale Galileo (TNG) on La Palma in the Canary islands. Known as HARPS-North (or HARPS-N), it has recently been experimentally upgraded with an LFC to provide the comparison spectrum. The LFC has emission lines at 16 GHz (0.016 nm in the green) intervals over the visible spectrum. The measurement accuracy is expected to be around ±0.25 m s^{-1} with a potential perhaps of ±0.025 m s^{-1}. An LFC has been installed on the Vacuum Tower Telescope (a 0.7-metre solar telescope on Tenerife) and is currently undergoing testing with a view to it being used regularly in the near future.

Obtaining the spectrum of an exoplanet itself is a technique that is still in its infancy, but has been accomplished. In 2010, Markus Janson and others obtained an infrared spectrum of HR8799 c using the NaCo instrument on the VLT. Direct spectroscopy of exoplanets is likely to become much more common as more and better planetary coronagraphs are produced (see Section 5.3). ESO's Spectro-Polarimetric High-contrast Exoplanet Research (SPHERE) instrument for the VLT, for example, which is currently being constructed, will use extreme adaptive optics, a planetary coronagraph, correction of instrumental aberrations and differential calibration during the data processing to undertake direct imaging, spectroscopy and polarimetry of exoplanets. Coronagraphic High Angular Resolution Imaging Spectrograph (CHARIS), a design concept for the 8.2-metre Subaru telescope with a possible completion date of 2015, is planned to be able to obtain NIR spectra of self-luminous Jupiter-mass exoplanets with a spectral resolution up to 65. Exoplanet atmospheres may be studied during their transits, when absorption within the atmosphere may add planetary absorption lines to the star's spectrum.

4.2.8 Future Developments

The major foreseeable developments in spectroscopy seem likely to lie in the direction of improving the efficiency of existing systems rather than in any radically new systems or methods. The lack of efficiency of a typical spectroscope arises mainly from the loss of light within the overall system. To improve on this requires gratings of greater efficiency, reduced scattering and surface reflection. from the optical components and so on. Since these factors are already quite good and it is mostly the total number of components in the telescope–spectroscope combination that reduces the efficiency, improvements are thus likely to be slow and gradual.

A possible alternative to the LFC could be based upon the etalon (see Section 4.1). If an etalon is illuminated by a white light point source then only those wavelengths given by Equation 4.86 will emerge from it. By operating with high orders (large values of m) numerous sharp emission lines will be produced whose wavelengths can be accurately and easily calculated. The practical realisation of this idea lies in the fibre optic Fabry–Perot (FFP) filter. This comprises a fibre optic wave guide with multilayer mirrors at each end. However, the FFP is more difficult to keep stable than the LFC and so the latter may continue to be the preferred development.

Adaptive optics is already used on most very large telescopes to reduce the seeing disk size of stellar images and so allow more compact spectroscopes to be used. This approach is likely to spread to smaller instruments in the near future. Direct energy detectors such as STJs are likely to be used more extensively, although at present they have relatively low spectral resolutions and require extremely low operating temperatures. The use of integral field and multi-object spectroscopes is likely to become more common, with wider fields of view and more objects being studied for individual instruments. The extension of high resolution spectroscopy to longer infrared wavelengths is likely to be developed, even though this may involve cooling the fibre optic connections and large parts of the telescope as well as most of the spectroscope. Plus, of course, the continuing increase in the power and speed of computers will make real-time processing of the data from spectroscopes much more commonplace.

EXERCISES

4.5 Design a spectroscope (i.e. calculate its parameters) for use at a Coudé focus where the focal ratio is f25. A resolution of 10^{-3} nm is required over the spectral range 500 to 750 nm. A grating with 500 lines mm^{-1} is available to be used in the third order and the final recording of the spectrum is to be by a CCD with 25-μm pixels.

4.6 Calculate the limiting B magnitude for the system designed in Problem 4.4 when it is used on a 4-metre telescope. The final spectrum is widened to 0.2 mm and the longest practicable exposure is 8 hours.

4.7 Calculate the apex angle of the dense flint prism required to form a direct-vision spectroscope in combination with two 40° crown glass prisms, if the undeviated wavelength is to be 550 nm. (See Section 1.1 for some of the data.)

4.8 If the Sun and Earth system (ignore all the other planets) were to be observed by an alien exoplanet hunter using host star radial velocity variations as his/her/its discovery method, how accurate would his/her/its solar radial velocity determinations need to be in order to discover the Earth?

CHAPTER 5

Other Techniques

5.1 ASTROMETRY

5.1.1 Introduction

Astrometry is probably the most ancient branch of astronomy, dating back to at least several centuries BC and possibly to a couple of millennia BC. Indeed, until the late eighteenth century, astrometry was the whole of astronomy. Yet although it is such an ancient sector of astronomy, it is still alive and well today and employing the most modern techniques, plus some that William Herschel would recognise. Astrometry is the science of measuring the positions in the sky of galaxies, stars, planets, comets, asteroids and recently, spacecraft. From these positional measurements come determinations of distance via parallax, motions in space via proper motion, orbits and hence sizes and masses within binary systems and a reference framework that is used by the whole of the rest of astronomy and astrophysics as well as by space scientists to direct and navigate their spacecraft. Astrometry also leads to the production of catalogues of the positions, and sometimes the nature, of objects that are then used for other astronomical purposes.

From the invention of the telescope until the 1970s, absolute positional accuracies of about 0.1″ (= 100 milliarc seconds) were the best that astrometry could deliver. That has now improved to better than 1 mas and space missions planned for the next couple of decades should improve that by a factor of ×100 at least. The positional accuracies of a few microarc seconds (μas) potentially allow proper motions of galaxies to be determined, though the diffuse nature of galaxy images may render this difficult. However, such accuracies will enable the direct measurement of stellar distances throughout the whole of the Milky Way galaxy and out to the Magellanic clouds and the Andromeda galaxy (M31).

Astrometry may be absolute (sometimes called fundamental) when the position of a star is determined without knowing the positions of other stars, or relative when the star's position is found with respect to the positions of its neighbours. Relative astrometry may be used to give the absolute positions of objects in the sky, provided that some of the reference stars have their absolute positions known. It may also be used for determinations of parallax, proper motion, binary star orbital motion, without needing to convert to absolute positions. An important modern application of relative astrometry is to enable the optical fibres of multi-object spectroscopes (see Section 4.2) to be positioned correctly in the focal plane of the telescope in order to intercept the light from the objects of interest. The absolute reference frame is called the International Celestial Reference System (ICRS) and is now defined using 212 extragalactic compact radio sources (known as the defining sources) with their positions determined by radio interferometry (see below) to an accuracy of about ±0.5 mas. Until 1998, the ICRS was based upon optical astrometric measurements and when space-based interferometric systems (see below) produce their results, the definition may well revert to being based upon optical measurements. The practical realisation of the ICRS for optical work is to be found in the Hipparcos catalogue (see below).

5.1.2 Background
5.1.2.1 Coordinate Systems
The measurement of a star's position in the sky must be with respect to some coordinate system. There are several such systems in use by astronomers, but the commonest is that of right ascension and declination. This system, along with most of the others, is based upon the concept of the celestial sphere (i.e. a hypothetical sphere, centred upon the Earth and enclosing all objects observed by astronomers). The space position of an object is related to a position on the celestial sphere by a radial projection from the centre of the Earth. Henceforth, in this section, we talk about the position of an object as its position on the celestial sphere and we ignore the differing radial distances that may be involved. We also extend the polar axis, the equatorial and orbital planes of the Earth until these too meet the celestial sphere (Figure 5.1). These intersections are called the celestial North Pole, the celestial equator and so on. Usually there is no ambiguity if the 'celestial' qualification is omitted, so that they are normally referred to as the North Pole, the equator and so on. The intersections (or nodes) of the ecliptic and the equator are the vernal and autumnal equinoxes. The former is additionally known as the first point of Aries from its position in the sky some 2000 years ago. The ecliptic is also the apparent path of the Sun across the sky during a year and the vernal equinox is defined as the node at which the Sun passes from the southern to the northern hemisphere. This passage occurs within a day of March 21 each year. The position of a star or other object is thus given with respect to these reference points and planes.

The declination of an object, δ, is its angular distance north or south of the equator. The right ascension, α, is its angular distance around from the meridian (or great circle) that passes through the vernal equinox and the poles, measured in the same direction as the solar motion (Figure 5.2). By convention, declination is measured from $-90°$ to $+90°$ in degrees, arc minutes and arc seconds and is positive to the north of the equator and negative to the south. Right ascension is measured from $0°$ to $360°$, in units of hours, minutes and seconds of time where

$$1 \text{ hour} = 15° \tag{5.1}$$

$$1 \text{ minute} = 15' \tag{5.2}$$

$$1 \text{ second} = 15'' \tag{5.3}$$

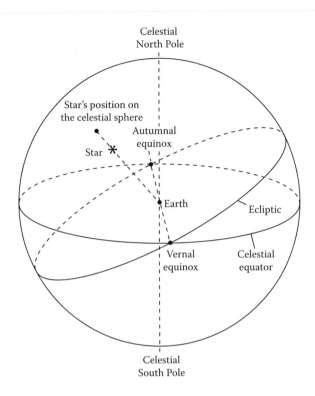

FIGURE 5.1 The celestial sphere.

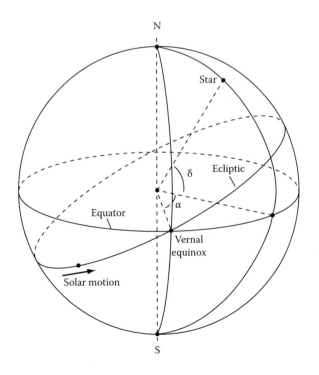

FIGURE 5.2 Right ascension and declination.

The direction of the Earth's axis moves in space with a period of about 25,750 years – a phenomenon known as precession. Hence, the celestial equator and poles also move. The positions of the stars therefore slowly change with time.[*] Catalogues of stars thus customarily give the date, or epoch, for which the stellar positions that they list are valid. To obtain the position at some other date, the effects of precession must be added to the catalogue positions

$$\delta_T = \delta_E + (\theta \sin \varepsilon \cos \alpha_E) \, T \tag{5.4}$$

$$\alpha_T = \alpha_E + [\theta \, (\cos \varepsilon + \sin \varepsilon \sin \alpha_E \tan \delta_E)] \, T \tag{5.5}$$

where α_T and δ_T are the right ascension and declination of the object at an interval T years after the epoch E, α_E, δ_E are the coordinates at the epoch and θ is the precession constant

$$\theta = 50.40'' \text{ year}^{-1} = 3.36 \text{ s year}^{-1} \tag{5.6}$$

ε is the angle between the equator and the ecliptic, more commonly known as the obliquity of the ecliptic

$$\varepsilon = 23° \, 27' \, 8'' \tag{5.7}$$

Frequently used epochs are the beginnings of the years 1900, 1950, 2000 etc. with 1975 and 2025 also being encountered. Other effects upon the position, such as nutation, proper motion etc. may also need to be taken into account in determining an up-to-date position for an object.

Two alternative coordinate systems that are in use are first celestial latitude (β) and longitude (λ), which are, respectively, the angular distances up or down from the ecliptic and around the ecliptic from the vernal equinox and second galactic latitude (b) and longitude (l), which are, respectively, the angular distances above or below the galactic plane and around the plane of the galaxy measured from the direction to the centre of the galaxy.[†]

5.1.2.2 Position Angle and Separation

The separation of a visual double star is just the angular distance between its components. The position angle is the angle from the north, measured in the sense, north → east → south → west → north, from 0° to 360° (Figure 5.3) of the fainter star with respect to the brighter star. The separation and position angle are related to the coordinates of the star by

---

[*] Note that this is an *apparent* positional change for the object only arising from the *actual* change in the coordinate system – a *true* change in the position of an object in the sky with respect to the other stars is called its proper motion.

[†] The system is based upon a position for the galactic centre of RA_{2000} 17h 45m 36s, Dec_{2000} −28° 56′ 18″. It is now known that this is in error by about 4′; however, the incorrect position continues to be used. Prior to about 1960, the intersection of the equator and the galactic plane was used as the zero point and this is about 30° away from the galactic centre. Coordinates based upon this old system are sometimes indicated by a superscript 'I' and those using the current system by a superscript 'II' (i.e. b^I or b^{II} and l^I or l^{II}).

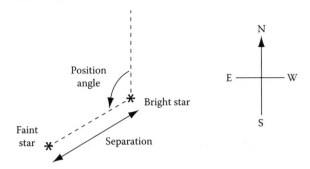

FIGURE 5.3   Position angle and separation of a visual double star (as seen in the sky directly).

$$separation = \left\{ \left[(\alpha_F - \alpha_B)\cos\delta_B \right]^2 + (\delta_F - \delta_B)^2 \right\}^{1/2} \tag{5.8}$$

$$position\ angle = \tan^{-1}\left(\frac{\delta_F - \delta_B}{(\alpha_F - \alpha_B)\cos\delta} \right) \tag{5.9}$$

where α_B and δ_B are the right ascension and declination of the brighter star, and α_F and δ_F are the right ascension and declination of the fainter star, with the right ascensions of both stars being converted to normal angular measures.

5.1.3 Transit Telescopes

Transit telescopes (Figure 5.4), which are also sometimes called meridian circles, historically provided the basic absolute measurements of stellar positions for the calibration of other astrometric methods, though they have now largely been superseded by the results from Hipparcos (see below) and radio interferometers. They developed from the mural quadrants that were used before the invention of the telescope. These were just sighting bars, pivoted at one end, restricted to moving in just one plane and with a divided arc to measure the altitude at the other end; nonetheless, they provided surprisingly good measurements when in capable hands. The one used by Tycho Brahe (1546–1601) for example was mounted on a north–south aligned wall and had a 2-metre radius. He was able to measure stellar positions using it to a precision of about ±30″.

The principle of the transit telescope is simple, but great care was required in practice if it were to produce reliable results. The instrument was almost always a refractor, because of the greater stability of the optics and it was pivoted only about a horizontal east–west axis. The telescope was thereby constrained to look just at points on the prime meridian. A star was observed as it crossed the centre of the field of view (i.e. when it transited the prime meridian) and the precise time of the passage was noted. The declination of the star is obtainable from the altitude of the telescope, while the right ascension is given by the local sidereal time at the observatory at the instant of transit,

FIGURE 5.4 A modern transit telescope – the Carlsberg Meridian Telescope on La Palma. (Reproduced by kind permission of D. W. Evans.)

$$\delta = A + \varphi - 90° \qquad\qquad (5.10)$$

$$\alpha = \text{LST} \qquad\qquad (5.11)$$

where A is the altitude of the telescope, φ is the latitude of the observatory and LST is the local sidereal time at the instant of transit. In order to achieve an accuracy of a tenth of a second of arc in the absolute position of the star, a very large number of precautions and corrections are needed. A few of the more important ones are temperature control; use of multiple or driven cross wires in the micrometer eyepiece; reversal of the telescope on its bearings; corrections for flexure, nonparallel objective and eyepiece focal planes, rotational axis not precisely horizontal, rotational axis not precisely east–west, errors in the setting circles, incorrect position of the micrometer cross wire, personal setting errors and so on. Then of course all the normal corrections for refraction, aberration etc. have also to be added.

Modern versions of the instrument, such as the Carlsberg Meridian Telescope (CMT) (Figure 5.4), use CCDs for detecting transits. The CMT is a refractor with a 17.8-cm diameter objective and a 2.66-metre focal length sited on La Palma in the Canary Islands. It is operated remotely via the internet and measures between 100,000 and 200,000 star positions in a single night. The CCD detector on the CMT uses charge transfer to move the accumulating electrons from pixel to pixel at the same rate as the images drift over the detector (also known as TDI), drift scanning and image tracking (see Section 1.1, liquid mirror telescopes, and Section 2.7). This enables the instrument to detect stars down to 17^m and also to track several stars simultaneously. The main programme of the CMT is to link the positions of the bright stars measured by Hipparcos to those of fainter stars.

5.1.4 Photographic Zenith Tube and the Impersonal Astrolabe

Two more modern instruments that until recently have performed the same function as the transit telescope are the photographic zenith tube (PZT) and the astrolabe. Both of these use a bath of mercury in order to determine the zenith position very precisely. The PZT obtained photographs of stars that transit close to the zenith and provided an accurate determination of time and the latitude of an observatory, but it was restricted in its observations of stars to those that culminated within a few tens of minutes of arc of the zenith. The PZT has been superseded by very long baseline radio interferometers (see Section 2.5) and measurements from spacecraft which are able to provide positional measurements for far more stars and with significantly higher accuracies.

The astrolabe observes near an altitude of 60° and so can give precise positions for a wider range of objects than the PZT. In its most developed form (Figure 5.5) it is known as the Danjon or impersonal astrolabe since its measurements are independent of focusing errors. Two separate beams of light are fed into the objective by the 60° prism, one of the beams having been reflected from a bath of mercury. A Wollaston prism (see Section 5.2) is used to produce two focused beams parallel to the optic axis with the other two nonparallel emergent beams being blocked off. Two images of the star are then visible in

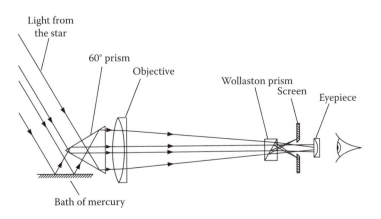

FIGURE 5.5   Optical paths in the impersonal astrolabe.

the eyepiece. The Wollaston prism is moved along the optical axis to compensate for the sidereal motion of the star. The measurement of the star's position is accomplished by moving the 60° prism along the optical axis until the two images merge into one. The position of the prism at a given time for this coincidence to occur can then be converted into the position of the star in the sky. The astrolabe has also largely fallen out of use, but one or two examples are still in operation such as the Mark III astrolabe at the Beijing National Observatory and the CCD-based Danjon astrolabe of the Observatorio Nacional, Rio de Janeiro, used for measuring the solar angular diameter.

5.1.5 Micrometers

The earliest device used on telescopes for measuring double stars and the angular sizes of extended objects was the bifilar micrometer. This comprises a fixed cross wire at the focus of an eyepiece, with a third thread parallel to one of the fixed ones. The third thread is displaced slightly from the others, although still within the eyepiece's field of sharp focus and it may be moved perpendicularly to its length by a precision screw thread. The screw has a calibrated scale so that the position of the third thread can be determined to within 1 μm. The whole assembly is mounted on a rotating turret whose angular position may also be measured. The bifilar micrometer and other variants such as double image micrometer will hardly ever be encountered now outside museums and so are not considered further here.

5.1.6 Astrographs and Other Telescopes

Almost any telescope can be used to obtain images from which the positions of objects in the sky can be determined, but some telescopes are optimised for the work. The astrograph has a wide (a degree or more) field of view so that many faint stars' positions can be determined from those of a few brighter reference stars. Astrographs designed around the beginning of the twentieth century were refractors with highly corrected multi-element objectives. Their apertures were in the region of 10 to 25 cm and focal lengths between 2 and 4 metres. Measurements using these instruments formed the basis of the ICRS. More recently, Schmidt cameras and conventional reflectors with correcting lenses have also been used.

Smaller fields of view are adequate for parallax determination. Long focus refractors are favoured for this work because of their greater stability when compared with reflectors, arising from their closed tubes, because their optics may be left undisturbed for many years and because flexure is generally less deleterious in them. However, a few specially designed reflectors, especially the US Naval Observatory's 1.55-metre Strand reflector, are used successfully for astrometry. Against the advantages of the refractor, their disadvantages must be set. These are primarily their residual chromatic aberration and coma, temperature effects and slow changes in the optics such as the rotation of the elements of the objective in their mounting cells. Refractors in use for astrometry usually have focal ratios of f15 to f20 and range in diameter from about 0.3 to 1.0 metres. Their objectives are achromatic (see Section 1.1) and filters are usually used so that the bandwidth is limited to a few tens of nanometres either side of their optimally corrected wavelength. Coma may need to be limited by stopping down the aperture, by introducing correcting lenses near the focal plane, or by compensating the measurements later, in the analysis stages, if the magnitudes and spectral types of the stars are well known. Observations are usually obtained within a few degrees of the meridian in order to minimise the effects of atmospheric refraction and dispersion and to reduce changes in flexure. The position of the focus may change with temperature and needs to be checked frequently throughout the night. Other corrections for such things as plate tilt collimation errors, astigmatism, distortion, emulsion movement etc. need to be known and their effects allowed for when reducing the raw data.

5.1.7 Interferometers

As we have seen in Section 2.5, interferometers have the capability of providing much higher resolutions than single dish-type instruments since their baselines can be extended easily and cheaply. High resolution can be translated into high positional accuracy providing that the mechanical construction of the interferometer is adequate. Interferometers also provide absolute positions for objects. The disadvantage of interferometers for astrometry is that only one object can be observed at a time (although the VLT's PRIMA and Galaxy observe a reference star as well as the object of interest simultaneously) and each observation requires several hours. Both radio and optical interferometers are used for the purpose.

Optical interferometers such as the US Naval Observatory's NPOI are essentially Michelson stellar interferometers (see Section 2.5) incorporating a delay line into one or both light beams. The delay line is simply an arrangement of several mirrors that reflects the light around an additional route. By moving one or more of the mirrors in the delay line, the delay can be varied. The observational procedure is to point both telescopes (or the flat mirrors feeding the telescopes) at the same star and adjust the delay until the fringe visibility (see Section 2.5) is maximum. The physical length of the delay line then gives the path difference between the light beams to the two telescopes. Combining this with the baseline length gives the star's altitude. Several such observations over a night will then enable the star's position in the sky to be calculated.

Currently, absolute positions may be obtained in this manner to accuracies of a few mas, with the relative positions of close stars being determined to perhaps an order of magnitude better precision. Relative measurements for widely separated objects may be obtained from the variation of the delay over several hours. The delay varies sinusoidally over a 24-hour period and the time interval between the maxima for two stars equals the difference in their right ascensions. As with conventional interferometers the use of more than two telescopes enables several delays to be measured simultaneously, speeding up the process and enabling instrumental effects to be corrected. The optical interferometer on Mount Wilson recently completed for CHARA (Center for High Angular Resolution Astronomy) for example uses six 1-metre telescopes in a two-dimensional array and has a maximum baseline of 350 metres. Gravity, a second generation instrument for the VLTI, is scheduled for first light in 2014 (see Section 2.5). It will be able to combine the light from four of the VLT instruments and observe both the object of interest and a close reference star. Operating at 2.2 μm, it is expected to provide astrometrical measurements to

10 microarc second precision. Gravity will build on the capabilities of PRIMA, a similar instrument based upon two VLT telescopes that has been operating since 2008.

Radio interferometers are operated in a similar manner for astrometry and VLBI systems (see Section 2.5) provide the highest accuracies at the time of writing. For some of the radio sources this accuracy may reach 100 μas. The measurements by VLBI of some 212 compact radio sources have thus been used to define the ICRS since January 1, 1998. Recently, the VLBA has been able to measure the parallax of a star within the Orion nebula to an uncertainty of ±0.15 mas, giving the distance to the nebula as 390 ± 23 pc, or 90 pc closer to us than had been previously thought.

A recent ingenious adaptation of VLBI has enabled details of the pulsar PSR 0834+06 to be studied with less than microarc second resolution. The approach uses the speckles introduced into the radio image by interstellar material (cf. atmospheric speckles, Figure 1.67), giving in effect an interferometer with a 5-AU baseline. For this pulsar, whose distance is 640 pc, the interstellar material is mostly around 415 pc away from us. For observations at 300 MHz (1 m), the measurement precision is about 0.3 microarc seconds. With the interstellar scattering material about 220 pc from the pulsar, this corresponds to a distance of around 10,000 km at the pulsar.

5.1.8 Space-Based Systems

By operating instruments in space and so removing the effects of the atmosphere and gravitational loading, absolute astrometry is expected to reach accuracies of a few microarc seconds in the next decade or so. Space astrometry missions, actual and planned, divide into two types: scanning (or survey) and point and stare. Scanning means that exposures are short and so only the brighter stars can be observed and with relatively low accuracy, the accuracy also depends upon the star's brightness. However, large numbers of star positions may be measured quickly. Point and stare missions, as the name suggests, look at individual stars (or a few stars very close together) for long periods of time. Such missions can observe relatively few star positions, but do so with high accuracy and to very faint magnitudes. Hipparcos and Gaia are examples of scanning systems, while the Hubble space telescope is point and stare.

The first astrometric spacecraft, called High Precision Parallax Collecting Satellite (Hipparcos), was launched by ESA in 1990. The Hipparcos telescope was fed by two flat mirrors at an angle of 29° to each other enabling two areas of the sky 58° apart to be observed simultaneously. It determined the relative phase difference between two stars viewed via the two mirrors in order to obtain their angular separation. A precision grid at the focal plane modulated the light from the stars and the modulated light was then detected from behind the grid to determine the transit times (see the Multichannel Astrometric Photometer [MAP] instrument later in this section). The satellite rotated once every 2 hours and measured all the stars down to 9.0^m about 120 times, observing each star for about 4 seconds on each occasion. Each measurement was from a slightly different angle and the final positional catalogue was obtained by the processing of some 10^{12} bits of such information. The data from Hipparcos are to be found in the Hipparcos, Tycho

FIGURE 5.6   An artist's impression of the Gaia spacecraft in orbit. (Reproduced by kind permission of ESA.)

and Tycho-2* catalogues. These may be accessed at http://archive.ast.cam.ac.uk/hipp/. The Hipparcos catalogue contains the positions, distances, proper motions and magnitudes of some 118,000 stars to an accuracy of ±0.7 mas. The Tycho catalogue contains the positions and magnitudes for about 1,000,000 stars and has positional accuracies ranging from ±7 mas to ±25 mas. The Tycho-2 catalogue utilises ground-based data as well as that from Hipparcos and has positional accuracies ranging from ±10 to ±70 mas, depending upon the star's brightness. It contains positions, magnitudes and proper motions for some two and a half million stars.

The Hubble space telescope can use the WFC3 (and previously the WFPC2) and Fine Guidance Sensors (FGS) instruments for astrometry. The planetary camera provides measurements to an accuracy of ±1 mas for objects down to magnitude 26^m, but only has a 160″ field of view. The telescope has three fine guidance sensors but only needs two for guidance purposes. The third is therefore available for astrometry. The FGS are interferometers based upon Köster prisms (see Section 2.5) and provide positional accuracies of ±5 mas.

Several other spacecraft missions have been proposed for astrometry, but most have been cancelled. However, ESA's Gaia mission (Figure 5.6) is still on stream with a currently scheduled launch date of October 2013. It will be positioned at the Sun–Earth inner Lagrange point some 1.5×10^6 km from the Earth, to avoid eclipses and occultations and so provide a stable thermal environment. It will operate in a similar fashion to Hipparcos but using two separate telescopes set at a fixed angle and with elongated apertures 1.7 metres × 0.7 m. The long axis of the aperture in each case will be aligned along

---

* A separate instrument, the star mapper, observed many more stars, but for only 0.25 seconds and its data produced the Tycho and Tycho-2 catalogues.

the scanning direction in order to provide the highest resolution. The read-out will be by CCDs using time-delayed integration. The aim is to determine positions for a billion stars to ±10 μas accuracy at visual magnitude 15^m. Gaia will also determine radial velocities for stars down to 17^m using a separate spectroscope telescope. The objective for the Gaia mission is to establish a precise 3D map of the galaxy and provide a massive data base for other investigations of the Milky Way.

In the early stages of consideration by ESA, Near Earth Astrometric Telescope (NEAT), might involve two spacecraft flying in formation about 40 metres apart and aim to reach a few times 0.1 microarc second levels of precision. If NEAT proceeds beyond the concept stage, it could be ready for launch in the early 2020s.

5.1.9 Detectors

CCDs (see Section 1.1) are now widely used as detectors for astrometry (Figure 1.17). They have the enormous advantage that the stellar or other images can be identified with individual pixels whose positions are automatically defined by the structure of the detector. The measurement of the relative positions of objects within the image is therefore vastly simplified. Software to fit a suitable point spread function (see Section 2.1) to the image enables the position of its centroid to be defined to a fraction of the size of the individual pixels.

The disadvantage of electronic images is their small scale. Most CCD chips are only a few centimetres in size (though mosaics are larger) compared with tens of centimetres for astrographic photographic plates. They therefore cover only a small region of the sky and it can become difficult to find suitable reference stars for the measurements. However, the greater dynamic range of CCDs compared with photographic emulsion and the use of anti-blooming techniques (see Section 1.1) enables the positions of many more stars to be usefully measured. With transit telescopes TDI may be used to reach fainter magnitudes. TDI is also proposed for the Gaia spacecraft detection system. Otherwise the operation of a CCD for positional work is conventional (see Section 1.1). In other respects, the processing and reduction of electronic images for astrometric purposes is the same as that for a photographic image.

Grid modulation is used by MAP at the Allegheny observatory. Light from up to 12 stars is fed into separate detectors by fibre optics as a grating with four lines per millimetre is passed across the field. Timing the disappearances of the stars behind the bars of the grating provides their relative positions to an accuracy of 3 mas. A similar technique was used by the Hipparcos spacecraft; although in this case it was the stars that moved across a fixed grid.

Photography is no longer used for astrometry. However, much material in the archives is in the form of photographs and this is still essential to long-term programmes measuring visual binary star orbits or proper motion etc. Thus, the reduction of data from photographs will continue to be needed for astrometry for some considerable time to come.

CCD imaging can be useful to observe double stars when the separation is greater than about 3 seconds of arc. Many exposures are made with a slight shift of the telescope between each exposure. Averaging the measurements can give the separation to better than a hundredth of a second of arc and the position angle to within a few minutes of arc. For double stars with large magnitude differences between their components, an objective grating can be used (see below), or the shape of the aperture can be changed so that the fainter star lies between two diffraction spikes and has an improved noise level. The latter technique is a variation of the technique of apodisation mentioned in Sections 1.1, 1.2, 2.5, 4.1 and 5.3.

5.1.10 Measurement and Reduction

Transit telescopes and interferometers give absolute positions directly as discussed above. In most other cases the position of an unknown star is obtained by comparison with reference stars that also appear on the image and whose positions are known. If absolute positions are to be determined, then the reference stars' positions must have been found via a transit instrument, interferometer or astrolabe etc. For relative positional work, such as that involved in the determination of parallax and proper motion, any very distant star may be used as a comparison star.

It is advantageous to have the star of interest and its reference stars of similar brightnesses although the much greater dynamic ranges of CCDs and other electronic detectors compared with that of photographic emulsions render this requirement less important than it was in the past. If still needed, the brightness of one or more stars may be altered through the use of variable density filters, by the use of a rotating chopper in front of the detector or by the use of an objective grating (see Section 4.2). The latter device is arranged so that pairs of images, which in fact are the first-order or higher-order spectra of the star, appear on either side of it. By adjusting the spacing of the grating, the brightness of these secondary images for the brighter stars may be arranged to be comparable with the brightness of the primary images of the fainter stars. The position of the brighter star is then taken as the average of the two positions of its secondary images. The apparent magnitude, m, of such a secondary image resulting from the nth order spectrum is given by

$$m = m_0 + 5 \log_{10} (N \eta \operatorname{cosec} \eta) \tag{5.12}$$

where m_0 is the apparent magnitude of the star, η is given by

$$\eta = \pi N d/D \tag{5.13}$$

where D is the separation of the grating bars, d is the width of the gap between the grating bars and N is the total number of slits across the objective.

Once an image of a star field has been obtained, the positions of the stars' images must be measured to a high degree of accuracy and the measurements converted into right ascension and declination or separation and position angle. For most astrometric photographic archive material, this process has already been undertaken. CCDs give positions directly once the physical structure of the device has been calibrated. Usually the position will be found by fitting a suitable point spread function (see Section 2.1) for the instrument used to obtain the overall image to the stellar images when they spread across several pixels. The stars' positions may then be determined to sub-pixel accuracy.

Howsoever the raw data may have been obtained, the process of converting the measurements into the required information is known as reduction. For accurate astrometry it is a lengthy process. A number of corrections have already been mentioned and, in addition to these, the distortion caused by the projection of the curved sky onto the flat photographic plate or CCD (known as the tangential plane), or in the case of Schmidt cameras its projection onto the curved focal plane must be corrected. For the simpler case of the astrograph, the projection is shown in Figure 5.7 and the relationship between the coordinates on the flat image and in the curved sky is given by

$$\rho = \tan^{-1}\left(\frac{y}{F}\right) \tag{5.14}$$

$$\tau = \tan^{-1}\left(\frac{x}{F}\right) \tag{5.15}$$

where x and y are the coordinates on the image, τ and ρ are the equivalent coordinates on the image of the celestial sphere, and F is the focal length of the telescope.

One instrument, which, while it is not strictly of itself a part of astrometry, is very frequently used to identify the objects that are to be measured, is called the blink comparator. In this machine the worker looks alternately at two aligned images of the same star field obtained some time apart from each other, the frequency of the interchange being about 2 Hz. Stars, asteroids or comets etc. which have moved in the interval between the images then call attention to themselves by appearing to jump backwards and forwards while the

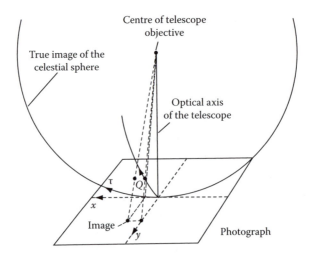

FIGURE 5.7 Projection onto a photographic plate or flat CCD.

remainder of the stars are stationary. The device can also pick out variable stars for further study by photometry (see Chapter 3). Although these remain stationary, they appear to blink on and off (hence the name of the instrument). Early blink comparators were basically binocular microscopes that had a mechanical arrangement for swapping the view of two photographic plates. Nowadays, the images are usually viewed on a computer screen and software aligns the images and provides the alternating views. Software can also be used to identify the moving or changing objects directly. But this is still one area where the human eye–brain functions as efficiently as the computer.

5.1.11 Sky Surveys and Catalogues

The end result of most astrometry is a catalogue of positions and other properties of (usually) a large number of objects in the sky. The Hipparcos, Tycho and Tycho-2 catalogues are just the latest examples of the process (see also Section 5.5). Other recent astrometric catalogues include the second issue of the U.S. Naval Observatory's Twin Astrographic Catalog (TAC 2.0) based upon photographic plates obtained with the twin astrograph and containing over 700,000 stellar positions to between ±50 mas and ±120 mas accuracy, the USNO's A2.0 catalogue containing 526 million entries, the USNO-B1.0 catalogue containing data on a billion stars with 200-mas positional accuracy and USNO CCD Astrograph Catalog (UCAC) that has the positions of some 40 million stars in the 10^m to 14^m range to ±20-mas accuracy. Older astrometric catalogues include the Fundamental Katalog series that culminated in FK5 in 1998 containing 1500 stars with positional accuracies of better than ±100 mas, the Astrographic Catalogue (AC) series, the Astronomische Gesellschaft Katalog (AGK) series, plus arguably Argelander's, Flamsteed's, Brahe's and even the original Hipparchus' catalogues, since these gave state-of-the-art stellar positions for their day. By 2020, the LSST should be swamping all these surveys and catalogues by obtaining over 1000 images per night with each image covering a 9.6-square-degree area of the sky down to a limiting magnitude of $+24.5^m$ or better. Its astrometric accuracy should be around ±10 to ±50 mas and its photometric accuracy around $±0.02^m$. The survey is planned to last for 10 years so that every area of the available sky will be imaged over two thousand times.

There have been many other sky surveys and catalogues and more are being produced all the time that are non-astrometric. That is to say their purpose is other than that of providing accurate positions and the position if determined at all for the catalogue is well below the current levels of astrometric accuracy. Indeed, many such catalogues just use the already-known astrometric positions. Non-astrometric catalogues are also produced for all regions of the spectrum, not just from optical and radio sources. There are tens of thousands of non-astrometric catalogues produced for almost as many different reasons and ranging in content from a few tens to half a billion objects. Examples of such catalogues include the Hubble space telescope second Guide Star Catalogue (GSC2) containing 500 million objects, the Two Micron All Sky Survey (2MASS) and Deep Near Infrared Survey of the Southern Sky (DENIS) in the infrared and the Sloan digital sky survey with its million red shifts of galaxies and quasars.

5.1.12 Exoplanets

Before any exoplanets had been definitively discovered, most astronomers expected that astrometry would be the method which did eventually find them. The detection would be through the cyclic change in the position of the host star in the sky as it and its exoplanet orbited around their common centre of mass – the same way in which Freidrich Bessel discovered the white dwarf, Sirius B, in 1844. In fact, at the time of writing – over two decades since the first exoplanets were found – astrometry has still to discover a single exoplanet. Astrometry has, however, detected the host star's changing position in 2002 in the case of the already known exoplanet Gliese 876 b using data from the HST. Exoplanet discoveries through astrometric measurements are thus not impossible and they may start to be made once the Gaia spacecraft begins to return data in the near future.

EXERCISES

5.1 If the axis of a transit telescope is accurately horizontal, but is displaced north or south of the east–west line by E seconds of arc, then show that for stars on the equator

$$\Delta\alpha^2 + \Delta\delta^2 \approx E^2 \sin^2 \varphi$$

where $\Delta\alpha$ and $\Delta\delta$ are the errors in the determinations of right ascension and declination in seconds of arc and φ is the latitude of the observatory.

5.2 Calculate the widths of the slits required in an objective grating for use on a 1-metre telescope if the third-order image of Sirius A is to be of the same brightness as the primary image of Sirius B. The total number of slits in the grating is 99. The apparent magnitude of Sirius A is −1.5, while that of Sirius B is +8.5.

5.3 If the alien astrophysicist involved in Exercise 4.8 were to try to detect the Earth via the movement of the Sun around the Sun–Earth centre of mass astrometrically, how accurately would he/she/it need to measure the Sun's position in his/her/its sky? Assume the alien astrophysicist is observing from a distance of 10 pc.

5.2 POLARIMETRY

5.2.1 Background

Although the discovery of polarised light from astronomical sources dates back to the beginning of the nineteenth century, when Dominique Arago detected its presence in moonlight, the extensive development of its study is a relatively recent phenomenon. This is largely due to the technical difficulties that are involved and initially, at least, to the lack of any expectation of polarised light from stars by astronomers. Many phenomena, however, can contribute to the polarisation of radiation and so, conversely, its observation can potentially provide information upon an equally wide range of basic causes.

5.2.1.1 Stokes' Parameters

Polarisation of radiation is simply the non-random angular distribution of the electric vectors of the photons in a beam of radiation. Customarily two types are distinguished – linear and circular polarisations. In the former, the electric vectors are all parallel and their direction is constant, while in the latter, the angle of the electric vector rotates with time at the frequency of the radiation. These are not really different types of phenomena, however, and all types of radiation may be considered to be different aspects of partially elliptically polarised radiation. This has two components, one of which is unpolarised, the other being elliptically polarised. Elliptically polarised light is similar to circularly polarised light in that the electric vector rotates at the frequency of the radiation, but in addition the magnitude varies at twice that frequency, so that plotted on a polar diagram the electric vector would trace out an ellipse (Figure 5.8). The properties of partially elliptically polarised light are completely described by four parameters that are called the Stokes' parameters. These fix the intensity of the unpolarised light, the degree of ellipticity, the direction of the major axis of the ellipse and the sense (left- or right-handed rotation) of the elliptically polarised light. If the radiation is imagined to be propagating along the z-axis of a three-dimensional

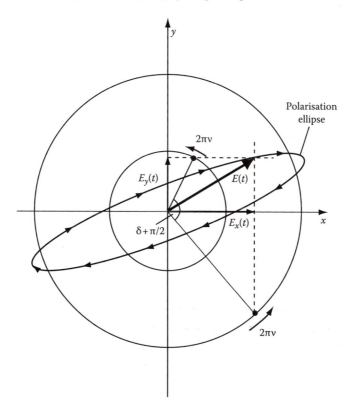

FIGURE 5.8 The x and y components of the elliptically-polarised component of partially elliptically-polarised light.

rectangular coordinate system, then the elliptically polarised component at a point along the z-axis may have its electric vector resolved into components along the x- and y-axes (Figure 5.8), these being given by

$$E_x(t) = e_1 \cos (2 \pi \nu t) \tag{5.16}$$

$$E_y(t) = e_2 \cos (2 \pi \nu t + \delta) \tag{5.17}$$

where ν is the frequency of the radiation, δ is the phase difference between the x and y components and e_1 and e_2 are the amplitudes of the x and y components. It is then a tedious but straightforward matter to show that

$$a = \left(\frac{\left(e_1^2 + e_2^2\right)}{1 + \tan^2\left[\frac{1}{2}\sin^{-1}\left\{\left[\frac{2e_1e_2}{\left(e_1^2 + e_2^2\right)}\right]\sin\delta\right\}\right]} \right)^{1/2} \tag{5.18}$$

$$b = a\tan\left[\frac{1}{2}\sin^{-1}\left\{\left[\frac{2e_1e_2}{\left(e_1^2 + e_2^2\right)}\right]\sin\delta\right\}\right] \tag{5.19}$$

$$a^2 + b^2 = e_1^2 + e_2^2 \tag{5.20}$$

$$\psi = \frac{1}{2}\tan^{-1}\left\{\left[\frac{2e_1e_2}{\left(e_1^2 - e_2^2\right)}\right]\cos\delta\right\} \tag{5.21}$$

where a and b are the semi-major and semi-minor axes of the polarisation ellipse and ψ is the angle between the x-axis and the major axis of the polarisation ellipse. The Stokes' parameters are then defined by

$$Q = e_1^2 - e_2^2 = \frac{a^2 - b^2}{a^2 + b^2}\cos(2\psi)I_p \tag{5.22}$$

$$U = 2e_1e_2\cos\delta = \frac{a^2 - b^2}{a^2 + b^2}\sin(2\psi)I_p \tag{5.23}$$

$$V = 2e_1e_2\sin\delta = \frac{2ab}{a^2 + b^2}I_p \tag{5.24}$$

where I_p is the intensity of the polarised component of the light. From Equations 5.22, 5.23 and 5.24 we have

$$I_p = (Q^2 + U^2 + V^2)^{1/2} \tag{5.25}$$

The fourth Stokes' parameter, I, is just the total intensity of the partially polarised light

$$I = I_u + I_p \tag{5.26}$$

where I_u is the intensity of the unpolarised component of the radiation. (*Note*: The notation and definitions of the Stokes'parameters can vary. While that given here is probably the most common usage, a check should always be carried out in individual cases to ensure that a different usage is not being employed.)

The degree of polarisation, π, of the radiation is given by

$$\pi = \frac{(Q^2 + U^2 + V^2)^{1/2}}{I} = \frac{I_p}{I} \tag{5.27}$$

while the degree of linear polarisation, π_L and the degree of ellipticity, π_e, are

$$\pi_L = \frac{(Q^2 + U^2)^{1/2}}{I} \tag{5.28}$$

$$\pi_e = \frac{V}{I} \tag{5.29}$$

When $V = 0$ (i.e. the phase difference, δ, is 0 or π radians) we have linearly polarised radiation. The degree of polarisation is then equal to the degree of linear polarisation and this is the quantity that is commonly determined experimentally

$$\pi = \pi_L = \frac{I_{max} - I_{min}}{I_{max} + I_{min}} \tag{5.30}$$

where I_{max} and I_{min} are the maximum and minimum intensities that are observed through a polariser as it is rotated. The value of π_e is positive for right-handed and negative for left-handed radiation.

When several incoherent beams of radiation are mixed, their various Stokes' parameters combine individually by simple addition. A given monochromatic partially elliptically polarised beam of radiation may therefore have been formed in many different ways and these are indistinguishable by the intensity and polarisation measurements alone. It is often customary therefore to regard partially elliptically polarised light as formed from two separate components, one of which is unpolarised and the other completely elliptically polarised. The Stokes' parameters of these components are then

	I	*Q*	*U*	*V*
Unpolarised component	I_u	0	0	0
Elliptically polarised component	I_p	Q	U	V

and the normalised Stokes' parameters for more specific mixtures are

	Stokes' Parameters			
Type of Radiation	*I/I*	*Q/I*	*U/I*	*V/I*
Right-hand circularly polarised (clockwise)	1	0	0	1
Left-hand circularly polarised (anticlockwise)	1	0	0	−1
Linearly polarised at an angle ψ to the x-axis	1	cos 2 ψ	sin 2 ψ	0

The Stokes' parameters are related to more familiar astronomical quantities by

$$\theta = \frac{1}{2}\tan^{-1}\left(\frac{U}{Q}\right) = \psi \tag{5.31}$$

$$e = \left\{1 - \tan^2\left[\frac{1}{2}\sin^{-1}\left(\frac{V}{(Q^2 + U^2 + V^2)^{1/2}}\right)\right]\right\}^{1/2} \tag{5.32}$$

where θ is the position angle (see Section 5.1) of the semi-major axis of the polarisation ellipse (when the x-axis is aligned north–south), e is the eccentricity of the polarisation ellipse and is 1 for linearly polarised radiation and 0 for circularly polarised radiation.

5.2.2 Optical Components for Polarimetry

Polarimeters usually contain a number of components that are optically active in the sense that they alter the state of polarisation of the radiation. They may be grouped under three headings: polarisers, converters and depolarisers. The first produces linearly polarised light, the second converts elliptically polarised light into linearly polarised light, or vice versa, while the last eliminates polarisation. Most of the devices rely upon birefringence for their effects, so we must initially discuss some of its properties before looking at the devices themselves.

5.2.2.1 Birefringence

The difference between a birefringent material and a more normal optical one may best be understood in terms of the behaviour of the Huygens' wavelets. In a normal material the refracted ray can be constructed from the incident ray by taking the envelope of the Huygens' wavelets, which spread out from the incident surface with a uniform velocity. In a birefringent material, however, the velocities of the wavelets depend upon the polarisation of the radiation and for at least one component the velocity will also depend on the orientation of the ray with respect to the structure of the material.

In some materials it is possible to find a direction for linearly polarised radiation for which the wavelets expand spherically from the point of incidence as in 'normal' optical materials. This ray is then termed the ordinary ray and its behaviour may be described by normal geometrical optics. The ray that is polarised orthogonally to the ordinary ray is then termed the extraordinary ray and wavelets from its point of incidence spread out elliptically (Figure 5.9). The velocity of the extraordinary ray thus varies with direction. We may construct the two refracted rays in a birefringent material by taking the envelopes of their wavelets as before (Figure 5.10). The direction along which the velocities of the ordinary and extraordinary rays are equal is called the optic axis of the material. When the velocity of the extraordinary ray is in general larger than that of the ordinary ray (as illustrated in Figures 5.9 and 5.10), then the birefringence is negative. It is positive when the situation is reversed. The degree of the birefringence may be obtained from the principal extraordinary refractive index, μ_E. This is the refractive index corresponding to the maximum velocity of the extraordinary ray for negative materials and the minimum velocity for positive materials. It will be obtained from rays travelling perpendicularly to the optic axis of the material. The degree of birefringence, which is often denoted by the symbol J, is then simply the difference between the principal extraordinary refractive index and the refractive index for the ordinary ray, μ_O

$$J = \mu_E - \mu_O \tag{5.33}$$

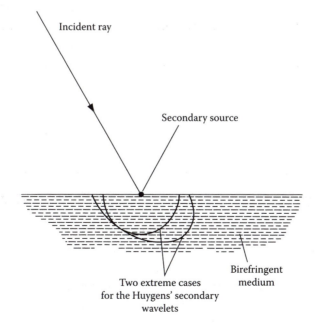

FIGURE 5.9   Huygens' secondary wavelets in a birefringent medium.

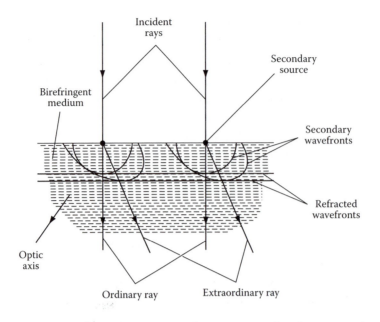

FIGURE 5.10   Formation of ordinary and extraordinary rays in a birefringent medium.

Most crystals exhibit natural birefringence and it can be introduced into many more and into amorphous substances such as glass by the presence of strain in the material. One of the most commonly encountered birefringent materials is calcite. The cleavage fragments form rhombohedrons and the optic axis then joins opposite blunt corners if all the edges are of equal length (Figure 5.11). The refractive index of the ordinary ray is 1.658, while the principal extraordinary refractive index is 1.486, giving calcite the very high degree of birefringence of −0.172.

Crystals such as calcite are uniaxial and have only one optic axis. Uniaxial crystals belong to the tetragonal or hexagonal crystallographic systems. Crystals that belong to the cubic system are not usually birefringent, while crystals in the remaining systems – orthorhombic, monoclinic, or triclinic – generally produce biaxial crystals. In the latter case, normal geometrical optics break down completely and all rays are extraordinary.

Some crystals such as quartz that are birefringent ($J = +0.009$), are in addition optically active. This is the property whereby the plane of polarisation of a beam of radiation is rotated as it passes through the material. Many substances other than crystals, including most solutions of organic chemicals, can be optically active. Looking down a beam of light, against the motion of the photons, a substance is called dextro-rotatory or right-handed if the rotation of the plane of vibration is clockwise. The other case is called laevo-rotatory or left-handed. Unfortunately and confusingly, the opposite convention is also in occasional use.

The description of the behaviour of an optically active substance in terms of the Huygens' wavelets is rather more difficult than was the case for birefringence. Incident light is split into two components as previously, but the velocities of both components vary with angle and in no direction do they become equal. The optic axis therefore has to be taken as the

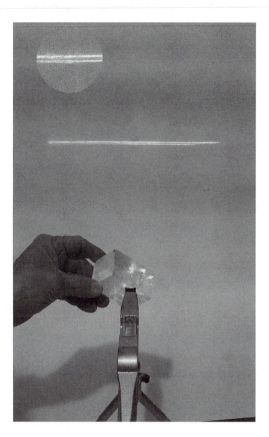

FIGURE 5.11 The beam from the laser of a surveying level being shone through a calcite cleavage rhomb. The normally single line is clearly seen split by the calcite's birefringence into the ordinary and extraordinary rays producing two lines (see magnified inset). (Reproduced by kind permission of C. E. Danes. © C. E. Danes 2013.)

direction along which the difference in velocities is minimised. Additionally, the nature of the components changes with angle as well. Along the optic axis they are circularly polarised in opposite senses, while perpendicular to the optic axis they are orthogonally linearly polarised. Between these two extremes the two components are elliptically polarised to varying degrees and in opposite senses (Figure 5.12).

5.2.2.2 Polarisers*

Polarisers are devices that only allow the passage of light that is linearly polarised in some specified direction. There are several varieties that are based upon birefringence, of which the Nicol prism is the best known. This consists of two calcite prisms cemented together by Canada balsam. Since the refractive index of Canada balsam is 1.55, it is possible for the extraordinary ray to be transmitted, while the ordinary ray is totally internally reflected (Figure 5.13). The Nicol polariser has the drawbacks of displacing the light beam and of

---

* Also often known as analysers.

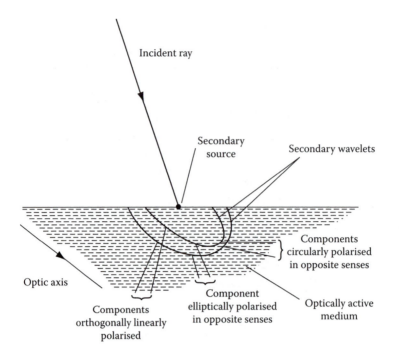

FIGURE 5.12   Huygens' secondary wavelets in an optically active medium.

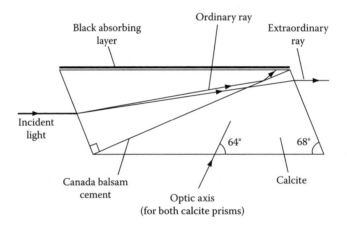

FIGURE 5.13   The Nicol prism.

introducing some elliptical polarisation into the emergent beam. Its inclined faces also introduce additional light losses by reflection. Various other designs of polarisers have therefore been developed, some of which in fact produce mutually orthogonally polarised beams with an angular separation. Examples of several such designs are shown in Figure 5.14. If both the ordinary ray and the extraordinary ray are required *and* their path lengths must be identical (perhaps so that the focal position of both is in the same plane, or because mutual interference is needed at some point) then Savart plates may be used. These are

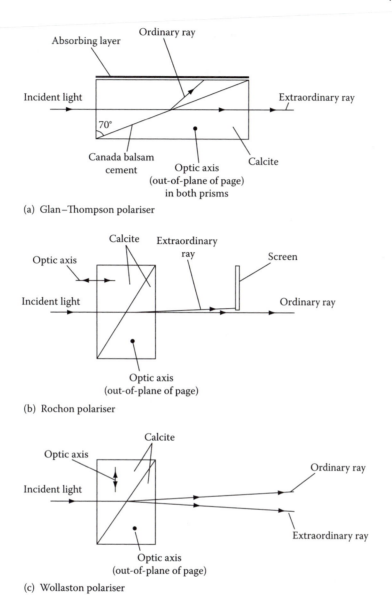

FIGURE 5.14   Examples of birefringence polarisers - (a) Glan-Thompson, (b) Rochon and (c) Wollaston.

made from two equally thick, parallel-sided plates of calcite or quartz that are cemented together. The plates are cut so that their optical axes are at 45° to their surfaces and the orientation of one plate is rotated by 90° with respect to the other. The first plate splits an incoming light beam into the ordinary and extraordinary rays as usual. Within the second plate, the ordinary ray emerging from the first plate becomes an extraordinary ray and the extraordinary ray emerging from the first plate becomes an ordinary ray. When both beams finally emerge they have therefore travelled equal optical distances within the calcite or quartz. Magnesium fluoride and quartz can also be used to form polarisers for the visible region, while lithium niobate and sapphire are used in the infrared.

The ubiquitous polarising sunglasses are based upon another type of polariser. They employ dichroic crystals that have nearly 100% absorption for one plane of polarisation and less than 100% for the other. Generally, the dichroism varies with wavelength so that these polarisers are not achromatic. Usually, however, they are sufficiently uniform in their spectral behaviour to be usable over quite wide wavebands. The polarisers are produced commercially in sheet form and contain many small aligned crystals rather than one large one. The two commonly used compounds are polyvinyl alcohol impregnated with iodine and polyvinyl alcohol catalysed to polyvinylene by hydrogen chloride. The alignment is achieved by stretching the film. The use of microscopic crystals and the existence of the large commercial market mean that dichroic polarisers are far cheaper than birefringent polarisers and so they may be used even when their performance is poorer than that of the birefringent polarisers.

Polarisation by reflection can be used to produce a polariser. A glass plate inclined at the Brewster angle will reflect a totally polarised beam. However, only a small percentage (about 7.5% for crown glass) of the incident energy is reflected. Reflection from the second surface will reinforce the first reflection, however, and then several plates may be stacked together to provide further reflections (Figure 5.15). The transmitted beam is only partially polarised. However, as the number of plates is increased, the total intensity of the reflected beams (ignoring absorption) will approach half of the incident intensity (Figure 5.16). Hence, the transmitted beam will approach complete polarisation. In practice therefore, the reflection polariser is used in transmission, since the problem of recombining the multiple reflected beams is thereby avoided and the beam suffers no angular deviation.

Outside the optical, NIR and near-UV regions, the polarisers tend to be somewhat different in nature. In the radio region, a linear dipole is naturally only sensitive to radiation

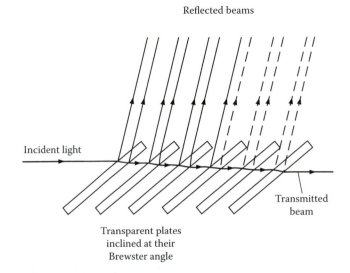

Reflected beams

Incident light

Transmitted beam

Transparent plates
inclined at their
Brewster angle

FIGURE 5.15 A reflection polariser.

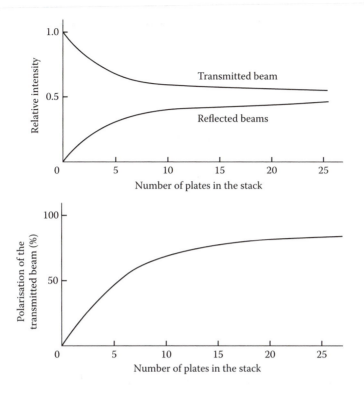

FIGURE 5.16 Properties of reflection polarisers, assuming negligible absorption and a refractive index of 1.5.

polarised along its length. In the microwave region and the medium to FIR, wire grid polarisers are suitable. These, as their name suggests, are grids of electrically conducting wires. Their spacing is about five times their thickness and they transmit the component of the radiation that is polarised along the length of the wires. They work efficiently for wavelengths longer than their spacing. In the microwave region polarisation can also be studied using TES detectors coupled with microstrip antennas. In the x-ray region, Bragg reflection is polarised (see Section 1.3) and so a rotating Bragg spectrometer can also act as a linear polarisation detector.

The behaviour of a polariser may be described mathematically by its effect upon the Stokes' parameters of the radiation. This is most easily accomplished by writing the parameters as a column vector. A matrix multiplying the vector on the left may then represent the effect of the polariser and also of the other optical components that we discuss later in this section. The technique is sometimes given the name of the Mueller calculus. There is also an alternative and to some extent complementary formulation, termed the Jones calculus. In the Mueller calculus the effect of passing the beam through several optical components is simply found by successive matrix multiplications of the Stokes' vector. The first optical component's matrix is closest to the original Stokes' vector and the matrices of subsequent components are positioned successively further to the left as the beam passes through the optical system. For a perfect polariser whose transmission axis is at an angle, θ, to the reference direction, we have

$$
\begin{bmatrix} I' \\ Q' \\ U' \\ V' \end{bmatrix} = \begin{bmatrix} \dfrac{1}{2} & \dfrac{1}{2}\cos 2\theta & \dfrac{1}{2}\sin 2\theta & 0 \\ \dfrac{1}{2}\cos 2\theta & \dfrac{1}{2}\cos^2 2\theta & \dfrac{1}{2}\cos 2\theta \sin 2\theta & 0 \\ \dfrac{1}{2}\sin 2\theta & \dfrac{1}{2}\cos 2\theta \sin 2\theta & \dfrac{1}{2}\sin^2 2\theta & 0 \\ 0 & 0 & 0 & 0 \end{bmatrix} \begin{bmatrix} I \\ Q \\ U \\ V \end{bmatrix}
\tag{5.34}
$$

where the primed Stokes' parameters are for the beam after passage through the polariser and the unprimed ones are for it before such passage.

5.2.2.3 Converters

Converters are devices that alter the type of polarisation and/or its orientation. They are also known as retarders, modulators, wave plates and phase plates. We have seen earlier (Equations 5.16 and 5.17 and Figure 5.8) that elliptically polarised light may be resolved into two orthogonal linear components with a phase difference. Altering that phase difference will alter the degree of ellipticity (Equations 5.18 and 5.19) and inclination (Equation 5.21) of the ellipse. We have also seen that the velocities of mutually orthogonal linearly polarised beams of radiation will in general differ from each other when the beams pass through a birefringent material. From inspection of Figure 5.13 it will be seen that if the optic axis is rotated until it is perpendicular to the incident radiation, then the ordinary and extraordinary rays will travel in the same direction. Thus, they will pass together through a layer of a birefringent material that is oriented in this way and will recombine upon emergence, but with an altered phase delay, due to their differing velocities. The phase delay, δ', is given to a first-order approximation by

$$
\delta' = \frac{2\pi d}{\lambda} J
\tag{5.35}
$$

where d is the thickness of the material and J is the birefringence of the material. Now let us define the x-axis of Figure 5.8 to be the polarisation direction of the extraordinary ray. We then have from Equation 5.21 the intrinsic phase difference, δ, between the components of the incident radiation

$$
\delta = \cos^{-1}\left\{ \frac{\left(e_1^2 - e_2^2\right)}{2e_1 e_2} \tan 2\psi \right\}
\tag{5.36}
$$

The ellipse for the emergent radiation then has a minor axis given by

$$b' = a' \tan \left[\frac{1}{2} \sin^{-1} \left\{ \left[\frac{2e_1 e_2}{\left(e_1^2 + e_2^2 \right)} \right] \sin(\delta + \delta') \right\} \right] \qquad (5.37)$$

where the primed quantities are for the emergent beam. So

$$b' = 0 \text{ for } \delta + \delta' = 0 \qquad (5.38)$$

$$b' = a' \text{ for } \delta + \delta' = \sin^{-1} \left[\frac{\left(e_1^2 + e_2^2 \right)}{2e_1 e_2} \right] \qquad (5.39)$$

and also

$$\psi = \frac{1}{2} \tan^{-1} \left\{ \left[\frac{2e_1 e_2}{\left(e_1^2 - e_2^2 \right)} \right] \cos(\delta + \delta') \right\} \qquad (5.40)$$

Thus

$$\psi' = -\psi \quad \text{for } \delta' = 180° \qquad (5.41)$$

and

$$a' = a \text{ and } b' = b \qquad (5.42)$$

Thus, we see that, in general, elliptically polarised radiation may have its degree of ellipticity altered and its inclination changed by passage through a converter. In particular cases it may be converted into linearly polarised or circularly polarised radiation, or its orientation may be reflected about the fast axis of the converter.

In real devices, the value of δ' is chosen to be 90° or 180° and the resulting converters are called quarter-wave plates or half-wave plates, respectively, since one beam is delayed with respect to the other by a quarter or a half of a wavelength. The quarter-wave plate is then used to convert elliptically or circularly polarised light into linearly polarised light or vice versa, while the half-wave plate is used to rotate the plane of linearly polarised light.

Many substances can be used to make converters, but mica is probably the commonest because of the ease with which it may be split along its cleavage planes. Plates of the right thickness (about 40 µm) are therefore simple to obtain. Quartz cut parallel to its optic axis can also be used, while for UV work, magnesium fluoride is suitable. Amorphous substances may be stretched or compressed to introduce stress birefringence. It is then possible to change the phase delay in the material by changing the amount of stress. Extremely high acceptance angles and chopping rates can be achieved with low power consumption if a small acoustic transducer at one of its natural frequencies drives the material. This is then commonly called a photoelastic modulator. An electric field can also induce birefringence. Along the direction of the field, the phenomenon is called the Pockels effect; while perpendicular to the field it is called the Kerr effect. Suitable materials abound and may be used to produce quarter- or half-wave plates. Generally, these devices are used when rapid switching of the state of birefringence is needed, as for example in Babcock's solar magnetometer (see Section 5.3). In glass the effects take up to a minute to appear and disappear, but in other substances the delay can be far shorter. The Kerr effect in nitrobenzene, for example, permits switching at up to 1 MHz and the Pockels effect can be used similarly with ammonium dihydrogen phosphate or potassium dihydrogen phosphate.

All the above converters will normally be chromatic and usable over only a restricted wavelength range. Converters that are more nearly achromatic can be produced which are based upon the phase changes that occur in total internal reflection. The phase difference between components with electric vectors parallel and perpendicular to the plane of incidence is shown in Figure 5.17. For two of the angles of incidence (approximately 45° and 60° in the example shown) the phase delay is 135°. Two such reflections therefore produce a total delay of 270°, or, as one may also view the situation, an advance of the second component with respect to the first by 90°. The minimum value of the phase difference shown in Figure 5.17 is equal to 135° when the refractive index is 1.497. There is then only one suitable angle of incidence and this is 51° 47′. For optimum results, the optical design should approach as close to this ideal as possible. A quarter-wave retarder that is nearly

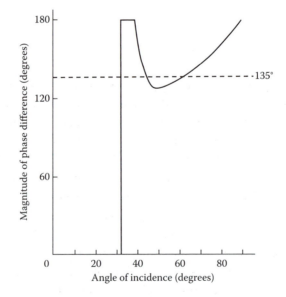

FIGURE 5.17 Differential phase delay upon internal reflection in a medium with a refractive index of 1.6.

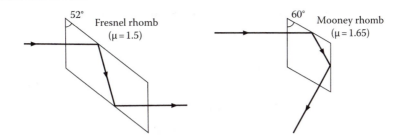

FIGURE 5.18   Quarter-wave retarders using total internal reflection.

achromatic can thus be formed using two total internal reflections at appropriate angles. The precise angles usually have to be determined by trial and error since additional phase changes are produced when the beam interacts with the entrance and exit faces of the component. Two practical designs – the Fresnel rhomb and the Mooney rhomb – are illustrated in Figure 5.18. Half-wave retarders can be formed by using two such quarter-wave retarders in succession.

Combining three retarders can make pseudo-quarter- and pseudo-half-wave plates that are usable over several hundred nanometres in the visual. The Pancharatnam design employs two retarders with the same delay and orientation, together with a third sandwiched between the first two. The inner retarder is rotated with respect to the other two and possibly has a differing delay. A composite half-wave plate with actual delays in the range 180° ± 2° over wavelengths from 300 to 1200 nm can, for example, be formed from three 180° retarders with the centre one oriented at about 60° to the two outer ones. Quartz and magnesium fluoride are commonly used materials for the production of such superachromatic wave plates.

The Mueller matrices for converters are

$$
\begin{bmatrix}
1 & 0 & 0 & 0 \\
0 & \cos^2 2\psi & \cos 2\psi \sin 2\psi & -\sin 2\psi \\
0 & \cos 2\psi \sin 2\psi & \sin^2 2\psi & \cos 2\psi \\
0 & \sin 2\psi & -\cos 2\psi & 0
\end{bmatrix}
\tag{5.43}
$$

for a quarter-wave plate and

$$
\begin{bmatrix}
1 & 0 & 0 & 0 \\
0 & \cos^2 2\psi - \sin^2 2\psi & 2\cos 2\psi \sin 2\psi & 0 \\
0 & 2\cos 2\psi \sin 2\psi & \sin^2 2\psi - \cos^2 2\psi & 0 \\
0 & 0 & 0 & -1
\end{bmatrix}
\tag{5.44}
$$

for a half-wave plate, while in the general case we have

$$
\begin{bmatrix}
1 & 0 & 0 & 0 \\
0 & \cos^2 2\psi + \sin^2 2\psi \cos\delta & (1-\cos\delta)\cos 2\psi \sin 2\psi & -\sin 2\psi \sin\delta \\
0 & (1-\cos\delta)\cos 2\psi \sin 2\psi & \sin^2 2\psi + \cos^2 2\psi \cos\delta & \cos 2\psi \sin\delta \\
0 & \sin 2\psi \sin\delta & -\cos 2\psi \sin\delta & \cos\delta
\end{bmatrix}
\tag{5.45}
$$

5.2.2.4 Depolarisers

The ideal depolariser would accept any form of polarised radiation and produce unpolarised radiation. No such device exists, but pseudo-depolarisers can be made. These convert the polarised radiation into radiation that is unpolarised when it is averaged over wavelength, time or area.

A monochromatic depolariser can be formed from a rotating quarter-wave plate that is in line with a half-wave plate rotating at twice its rate. The emerging beam at any given instant will have some form of elliptical polarisation, but this will change rapidly with time and the output will average to zero polarisation over several rotations of the plates.

The Lyot depolariser averages over wavelength. It consists of two retarders with phase differences very much greater than 360°. The second plate has twice the thickness of the first and its optic axis is rotated by 45° with respect to that of the first. The emergent beam will be polarised at any given wavelength, but the polarisation will vary very rapidly with wavelength. In the optical region, averaging over a waveband a few tens of nanometres wide is then sufficient to reduce the net polarisation to 1% of its initial value.

If a retarder is left with a rough surface and immersed in a liquid whose refractive index is the average refractive index of the retarder, then a beam of light will be undisturbed by the roughness of the surface because the hollows will be filled in by the liquid. However, the retarder will vary in its effect on the scale of its roughness. Thus, the polarisation of the emerging beam will vary on the same scale and a suitable choice for the parameters of the system can lead to an average polarisation of zero over the whole beam.

The Mueller matrix of an ideal depolariser is simply

$$
\begin{bmatrix}
1 & 0 & 0 & 0 \\
0 & 0 & 0 & 0 \\
0 & 0 & 0 & 0 \\
0 & 0 & 0 & 0
\end{bmatrix}
\tag{5.46}
$$

5.2.3 Polarimeters

A polarimeter is an instrument that measures the state of polarisation, or some aspect of the state of polarisation, of a beam of radiation. Ideally, the values of all four Stokes' parameters should be determinable, together with their variations with time, space and wavelength. In practice, this is rarely possible, at least for astronomical sources, when only the degree of linear polarisation and its direction are found most of the time. Astronomical optical polarimeters now normally use CCDs or avalanche photodiodes as their detectors but photographic plates or photomultipliers have been used in the past.

For a few astronomical sources, the degree of polarisation can be several tens of per cent (up to 60% for the Crab Nebula, for example). Mostly, however, the degree of polarisation is less than 1% and for those observers hoping to detect exoplanets through polarimetric measurements, the instruments need to detect levels of polarisation of 0.001% to 0.0001%. Polarimeters therefore vary in their designs depending upon the purpose for which they are intended. At the simplest, a sheet of Polaroid film can be inserted before the CCD detector on a small telescope and images obtained with the Polaroid rotated through 45° or 90° between successive exposures. This will certainly show the Crab Nebula's polarisation. At a slightly more sophisticated level, a higher quality polariser may be placed before the entrance aperture to a photometer or spectrograph (see Chapters 3 and 4) and estimates made of the source's degree of linear polarisation and its direction, and in the latter case, of its variation along the spectrum as well. However, most polarimeters and especially those aimed at achieving the highest levels of sensitivity have to be designed for the purpose.

5.2.3.1 Photoelectric Polarimeters

Most photoelectric polarimeters bear a distinct resemblance to photoelectric photometers (see Section 3.2) and indeed, many polarimeters can also function as photometers. The major differences, apart from the components that are needed to measure the polarisation, arise from the necessity of reducing or eliminating the instrumentally induced polarisation. The instrumental polarisation originates primarily in inclined planar reflections or in the detector. Thus, Newtonian and Coudé telescopes cannot be used for polarimetry because of their inclined subsidiary mirrors. Inclined mirrors must also be avoided within the polarimeter. Cassegrain and other similar designs should not in principle introduce any net polarisation because of their symmetry about the optical axis. However, problems may arise even with these telescopes if mechanical flexure or poor adjustments move the mirrors and/or the polarimeter from the line of symmetry.

Many detectors, especially photomultipliers (see Section 1.1), are sensitive to polarisation and furthermore this sensitivity may vary over the surface of the detector. If the detector is not shielded from the Earth's magnetic field, the sensitivity may vary with the position of the telescope as well. In addition, although an ideal Cassegrain system will produce no net polarisation, this sensitivity to polarisation arises because rays on opposite sides of the optical axis are polarised in opposite but equal ways. The detector non-uniformity may then result in incomplete cancellation of the telescope polarisation. Thus, a Fabry lens and a depolariser immediately before the detector are almost always essential components of a polarimeter.

Alternatively, the instrumental polarisation can be reversed using a Bowen compensator. This comprises two retarders with delays of about λ/8 that may be rotated with respect to the light beam and to each other until their polarisation effects are equal and opposite to those of the instrument. Small amounts of instrumental polarisation can be corrected by the insertion of a tilted glass plate into the light beam to introduce opposing polarisation. In practice, an unpolarised source is observed and the plate tilted and rotated until the observed polarisation is minimised or eliminated.

There are many detailed designs of polarimeter, but we may group them into single- and double-channel devices, with the former further subdivided into discrete and continuous systems. The single-channel discrete polarimeter is the basis of most of the other designs and its main components are shown in Figure 5.19, although the order is not critical for some of them. Since stellar polarisations are usually of the order of 1% or less, it is also usually necessary to incorporate a chopper and use a phase-sensitive detector (also known as a lock-in amplifier) plus integration in order to obtain sufficiently accurate measurements. Switching between the sky and background is also vital since the background radiation is very likely to be polarised, especially if the Moon is in the sky. The polariser is rotated either in steps or continuously, but at a rate which is slow compared with the chopping and

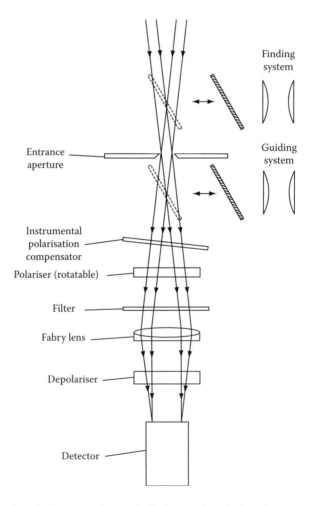

FIGURE 5.19 The basic polarimeter: schematic design and optical paths.

integration times. The output for a linearly polarised source is then modulated with the position of the polariser.

The continuous single-beam systems use a rapidly rotating polariser. This gives an oscillating output for a linearly polarised source and this may be fed directly into a computer or into a phase-sensitive detector that is driven at the same frequency. Alternatively, a quarter-wave plate and a photoelastic modulator can be used. The quarter-wave plate produces circularly polarised light and this is then converted back into linearly polarised light by the photoelastic modulator but with its plane of polarisation rotated by 90° between alternate peaks of its cycle. A polariser that is stationary will then introduce an intensity variation with time that may be detected and analysed as before. The output of the phase-sensitive detector in systems such as these is directly proportional to the degree of polarisation, with unpolarised sources giving zero output. They may also be used as photometers to determine the total intensity of the source, by inserting an additional fixed polariser before the entrance aperture, to give in effect a source with 100% polarisation and half the intensity of the actual source (provided that the intrinsic polarisation is small). An early polarimeter system that is related to these is due to Dolfus. It uses a quarter-wave plate to convert the linearly polarised component of the radiation into circularly polarised light. A rotating disc that contains alternate sectors made from zero- and half-wave plates then chops the beam. The circular polarisation alternates in its direction but remains circularly polarised. Another quarter-wave plate converts the circularly polarised light back to linearly polarised light, but the plane of its polarisation switches through 90° at the chopping frequency. Finally, a polariser is used to eliminate one of the linearly polarised components, so that the detector produces a square wave output.

The continuous single-beam polarimeter determines both linear polarisation components quasi-simultaneously, but it still requires a separate observation to acquire the total intensity of the source. In double-beam devices, the total intensity is found with each observation, but several observations are required to determine the polarisation. These instruments are essentially similar to that shown in Figure 5.19, but a double-beam polariser such as a Wollaston prism or a Savart plate is used in place of the single-beam polariser shown there. Two detectors then observe both beams simultaneously. The polariser is again rotated either in steps or continuously but slowly in order to measure the polarisation. Alternatively, a half-wave plate can be rotated above a fixed polariser. The effects of scintillation and other atmospheric changes are much reduced by the use of dual beam instruments.

The Turku Photopolarimeter (Turpol), for example, an instrument for the 2.5-metre Nordic optical telescope on La Palma, has an optical design almost identical to that shown in Figure 5.19. It uses half-wave or quarter-wave plates and a calcite block to separate the ordinary and extraordinary rays. A 25-Hz chopper alternately allows one or the other beam through to the detectors (photomultipliers) and there are dichroic mirrors to allow simultaneous U, B, V, R and I band measurements to be obtained. The wave plates are rotated in 22.5° steps. In 2011, Turpol was used to confirm earlier measurements of polarised light from the exoplanet HD 189733 b as around 0.01% ± 0.001%. Currently under construction, and intended to undertake a polarimetric survey of the southern sky down to a V magnitude of 15^m or so, is the polarimeter for the 0.84-metre robotic telescope at NOAO's Cerro Tololo Inter-American Observatory (CTIO) site. The polarimeter will use a Savart plate and a step-wise rotating half-wave plate and ±0.002% polarimetric precision is hoped for.

The current state-of-the-art is represented by PlanetPol, a polarimeter designed by James Hough at the University of Hertfordshire to detect extra solar planets. It uses a 20-kHz photoelastic modulator and a three-wedge Wollaston prism to detect the ordinary and extraordinary rays simultaneously using avalanche photodiodes and has a separate sky channel. The whole instrument rotates through 45° to measure the Q and U Stokes' parameters. Its sensitivity is 0.0001% or better, and on an 8-metre telescope, it should be able to detect the polarisation induced into the light from a star by a Jupiter-sized planet for stars brighter than 6.5^m.

5.2.3.2 CCD and Other Array Detector Polarimeters

Again, many designs are possible, but the dual beam polariser using a rotating wave plate and a Wollaston prism is in widespread use. For extended sources, the polarised images are detected directly by a CCD or infrared detector array. To avoid overlap, a mask is used at the focal plane of the telescope comprising linear obscuring bars and gaps of equal width (sometimes known as a comb dekker). The image from one of the beams then occupies the space occupied by the 'image' of the mask. Clearly, two sets of observations, shifted by the width of an obscuring bar, are then needed to show the complete extended object.

ZIMPOL uses a half-wave plate followed by a photoelastic modulator and a polarising beam splitter. The detector is an interline transfer CCD. The modulator operates at kilohertz frequencies and alternately allows through the two orthogonal linearly polarised components of the original beam. The interline transfer CCD is phased with the modulator so that each polarised component accumulates in separate rows of pixels. To improve sensitivity the second beam from the polarising beam splitter can also be collected by a second interline transfer CCD. A polarimeter of the ZIMPOL design is currently being constructed for the Characterizing Extrasolar Planets by Opto-infrared Polarimetry and Spectroscopy (CHEOPS) project for the VLT where polarisation accuracies of ±0.001% may be achieved.

In the UV little work has been undertaken to date. The solar maximum mission satellite, however, did include an UV spectrometer into which a magnesium fluoride retarder could be inserted, so that some measurements of the polarisation in strong spectrum lines could be made.

A few polarimeter designs have been produced, particularly for infrared work, in which a dichroic mirror is used. These are mirrors that reflect at one wavelength and are transparent at another. They are not related to the other form of dichroism that was discussed earlier in this section. A thinly plated gold mirror reflects infrared radiation but transmits visible radiation for example. Continuous guiding on the object being studied is therefore possible, providing that it is emitting visible radiation, through the use of such a mirror in the polarimeter. Unfortunately, the reflection introduces instrumental polarisation, but this can be reduced by incorporating a second 90° reflection whose plane is orthogonal to that of the first into the system.

5.2.4 Spectropolarimetry

Spectropolarimetry provides information on the variation of polarisation with wavelength and can be realised by several methods. Almost any spectroscope (see Section 4.2) suited for long slit spectroscopy can be adapted to spectropolarimetry. A quarter- or half-wave plate and a block of calcite are placed before the slit. The calcite is oriented so that the ordinary and extraordinary images lie along the length of the slit. Successive exposures are then obtained with the wave plate rotated between each exposure. For linear spectropolarimetry four positions separated by 22.5° are needed for the half-wave plate, while for circular spectropolarimetry just two orthogonal positions are required for the quarter-wave plate. The X-Shooter spectrograph on the VLT, for example, uses a rotating modulator and a Savart plate in the light beam before the instrument's entrance slit to enable it to operate as a spectropolarimeter. Solar spectropolarimetry is considered in Section 5.3.

5.2.5 Data Reduction and Analysis

The output of a polarimeter is usually in the form of a series of intensity measurements for varying angles of the polariser. These must be corrected for instrumental polarisation if this has not already been removed or has only been incompletely removed by the use of an inclined plate. The atmospheric contribution to the polarisation must then be removed by comparison of the observations of the star and its background. Other photometric corrections, such as for atmospheric extinction (see Section 3.2) must also be made, especially if the observations at differing orientations of the polariser are made at significantly different times so that the position of the source in the sky has changed.

The normalised Stokes' parameters are then obtained from

$$\frac{Q}{I} = \frac{I(\theta) - I(90)}{I(\theta) + I(90)} \tag{5.47}$$

and

$$\frac{U}{I} = \frac{I(45) - I(135)}{I(45) + I(135)} \tag{5.48}$$

where $I(\theta)$ is the intensity at a position angle θ for the polariser. Since for most astronomical sources the elliptically polarised component of the radiation is 1% or less of the linearly polarised component, the degree of linear polarisation, π_L, can then be found via Equation 5.28 or more simply from Equation 5.30 if sufficient observations exist to determine I_{max} and I_{min} with adequate precision. The position angle of the polarisation may be found from Equation 5.31, or again directly from the observations if these are sufficiently closely spaced. The fourth Stokes' parameter, V, is rarely determined in astronomical polarimetry.

However, the addition of a quarter-wave plate to the optical train of the polarimeter is sufficient to enable it to be found if required (see Section 5.4). The reason for the lack of interest is, as already remarked, the very low levels of circular polarisation that are usually found. However, recently, the ESO's 3.6-metre telescope has had interchangeable linear and circular polarisation detectors installed (ESO's Faint Object Spectrograph and Camera [EFOSC2]), the latter being similar to the former except for the addition of a quarter-wave plate. The first Stokes' parameter, I, is simply the intensity of the source and so if it is not measured directly as a part of the experimental procedure, can simply be obtained from

$$I = I(\theta) + I(\theta + 90°) \tag{5.49}$$

A flowchart to illustrate the steps in an observation and analysis sequence to determine all four Stokes' parameters is shown in Figure 5.20.

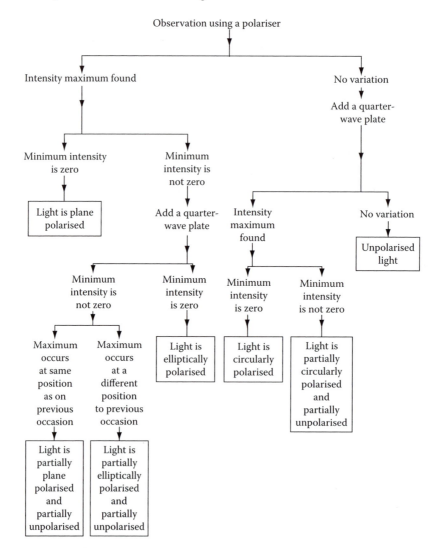

FIGURE 5.20 Flowchart of polarimetric analysis.

EXERCISES

5.4 Obtain Equation 5.21 from Equations 5.16 and 5.17.

5.5 Show, using the Mueller calculus, that the effect of two ideal half-wave plates upon a beam of radiation of any form of polarisation is zero.

5.3 SOLAR STUDIES

WARNING

Studying the Sun can be dangerous. Permanent eye damage or even blindness can easily occur through looking directly at the Sun, even with the unaided eye. Smoked glass, exposed film, CDs, space blankets, aluminised helium balloons, crossed polaroid filters, sunglasses etc. are not safe; ONLY filters made and sold specifically for viewing the Sun should ever be used.

Herschel wedges (see discussion of solar telescopes) are no longer recommended except if used on very small telescopes; also care must be taken to ensure that the rejected solar radiation does not cause burns or start fires.

Eyepiece projection is safe from the point of view of personal injury, but can damage the eyepiece and/or the telescope, especially with the shorter focal ratio instruments.

This warning should be carefully noted by all intending solar observers and passed onto inexperienced persons with whom the observer may be working. In particular, the dangers of solar observing should be made quite clear to any groups of laypeople touring the observatory, because the temptation to try looking at the Sun for themselves is very great, especially when there is an eclipse.

5.3.1 Introduction

The Sun is a reasonably typical star and as such it can be studied by many of the techniques discussed elsewhere in this book. In one major way, however, the Sun is quite different from all other stars and that is that it may easily be studied using an angular resolution smaller than its diameter. This, combined with the vastly greater amount of energy that is available compared with that from any other star, has led to the development of numerous specialised instruments and observing techniques. Many of these have already been mentioned or discussed in some detail in other sections. The main such techniques include solar cosmic-ray and neutron detection (see Section 1.4), neutrino detection (see Section 1.5), radio observations (see Sections 1.2 and 2.5), x-ray and γ-ray observations (see Section 1.3), adaptive optics (see Section 2.5), radar observations (see Section 2.8), magnetometry (see Section 5.4) and spectroscopy and prominence spectroscopes (see Section 4.2). The other remaining important specialist instrumentation and techniques for solar work are covered in this section.

The warning at the start of this section may be quantified by the European Community directive, EN 167:

> For safe viewing of the Sun with the unaided eye, filters should have a minimum size of 35 mm × 115 mm so that both eyes are covered and have maximum transmissions of

0.003% – ultraviolet (280–380 nm)
0.0032% – visible (380–780 nm)
0.027% – near infrared (780–1400 nm)

Although these maxima are safe with regard to damage to the eye, the Sun may still be uncomfortably bright and a transmittance of 0.0003% may found to be better. Since the surface brightness of the Sun is not increased by the use of a telescope (Equation 1.74 and related discussion), the same limits are safe for telescopic viewing, although the larger angular size of the image may increase the total energy entering the eye to the point of discomfort, so filters should generally be denser than this directive suggests. A transmission of 0.003% corresponds to an optical density for the filter of 4.5. Full aperture filters (see below) sold for solar observing generally have optical densities of 5 to 6.

5.3.2 Solar Telescopes

Almost any telescope may be used to image the Sun. While specialised telescopes are often built for professional work (see later in this section), small conventional telescopes are frequently used by amateurs or for teaching or training purposes. Since it is dangerous to look at the Sun through an unadapted telescope (see the warning above), special methods must be used. By far the best approach is to use a full aperture filter sold by a reputable supplier specifically for solar observing (see below), but eyepiece projection may also be possible. In this the image is projected onto a sheet of white card held behind the eyepiece. This method may be preferred when demonstrating to a group of people, since all can see the image at the same time and they are looking away from the Sun itself in order to see its image. However, telescopes such as Schmidt-Cassegrains and Maksutovs have short focal length primary mirrors and these can concentrate the light internally sufficiently to damage the instrument. The use of a telescope for solar projection may well invalidate the manufacturer's guarantee and the manufacturer should always be consulted on this in advance of using the telescope for solar work. Even for other telescope designs, the heat passing through the eyepiece may be sufficient to damage it, especially with the more expensive multi-component types. The Herschel wedge, in the past a common telescope accessory, is no longer recommended. Filters placed at the eyepiece end of the telescope, even if of adequate optical density, should never be used. They can become very hot and shatter and the unfiltered image within the telescope can cause damage to the instrument.

The Herschel wedge or solar diagonal is shown in Figure 5.21. It is a thin prism with unsilvered faces. The first face is inclined at 45° to the optical axis and thus reflects about 5% of the solar radiation into an eyepiece in the normal star diagonal position. The second

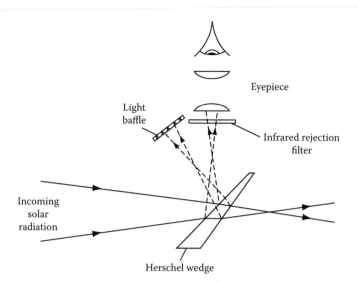

FIGURE 5.21 Optical paths in a Herschel wedge.

face also reflects about 5% of the radiation, but its inclination to the optical axis is not 45° and so it is easy to arrange for this radiation to be intercepted by a baffle before it reaches the eyepiece. The remaining radiation passes through the prism and can be absorbed in a heat trap, or more commonly (and dangerously) just allowed to emerge as an intense beam of radiation at the rear end of the telescope. In order to reduce the solar intensity to less than the recommended maximum, the wedge must be used on a very small telescope at a high magnification – a 50-mm telescope would need a minimum magnification of ×300 to be safe for example. The device must also incorporate a separate infrared filter. Since modern 'small' telescopes range from 100 to 300 mm or more in aperture, they are *much* too large to use with a Herschel wedge. Even if a larger telescope were to be stopped down to 50 mm, the Herschel wedge still has the disadvantage of producing a real image inside the telescope that may damage the telescope and/or invalidate its guarantee. The recommendation now therefore is that Herschel wedges should *not* be used for solar observing. If you have a Herschel wedge already, then it would be best to get the front surface of the wedge aluminised and then use it as a star diagonal.

The principle of eyepiece projection is shown in Figure 5.22 and it is the same as the eyepiece projection method of imaging (see Section 2.2). The eyepiece is moved outwards from its normal position for visual use and a real image of the Sun is obtained that may easily be projected onto a sheet of white paper or cardboard etc. for viewing purposes. From Equation 2.26, we have the effective focal length (EFL) of the telescope

$$EFL = \frac{Fd}{e} \qquad (5.50)$$

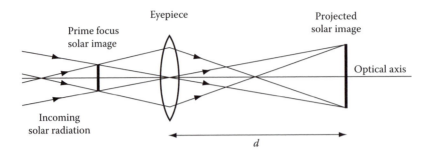

FIGURE 5.22   Projection of the solar image.

where F is the focal length of the objective, e is the focal length of the eyepiece and d is the projection distance (see Figure 5.22). The final image size, D, for the Sun is then simply (for D, F, d and e all in the same linear units)

$$D = \frac{0.0093\,Fd}{e} \tag{5.51}$$

There are two drawbacks to this method. The first is that a real image of the Sun is formed at the prime focus of the objective. If this is larger than the eyepiece acceptance area, or if this image of the Sun is intercepted by a part of the telescope's structure during acquisition, then structural damage may occur through the intense heat in the beam. This is especially a problem for modern short focal ratio Schmidt-Cassegrain and Maksutov telescopes. The second drawback is that the eyepiece lenses will absorb a small fraction of the solar energy and this may stress the glass to the point at which it may fracture. Even if this does not occur, the heating will distort the lenses and reduce the quality of the final image. The technique is thus normally limited to telescopes with apertures of 75 mm or less. Larger instruments will need to be stopped down to about this area if they are to be so used. As noted above, the use of a commercially produced telescope for solar observation by eyepiece projection may invalidate its guarantee, even if the telescope is not damaged and the manufacturer should always be consulted before using the telescope in this fashion.

Full aperture filters are produced commercially for use on small telescopes. They cover the whole aperture of the telescope as their name suggests and may be made sufficiently opaque for use on telescopes up to 0.5 metres in diameter. They are much the preferred approach to converting a smallish telescope to solar observation. One of the earliest types to be produced and which is still available today consists of a very thin Mylar film that has been coated with a reflective layer of aluminium, stainless steel etc. The film is so thin that very little degradation of the image occurs; however, this also makes the film liable to damage and its useful life when in regular use can be quite short. Such filters are produced with single or double layers of reflective coating. The double layer is to be preferred since any pinholes in one layer are likely to be covered by the second layer and vice versa. Reflective coatings on optically flat glass substrates have become available recently, though they are usually somewhat more expensive than the coated plastic filters. Finally, the filter may simply be a thin sheet of absorbing plastic (do not be tempted to use any old piece of black plastic that may be lying around because it may not be sufficiently opaque in the UV or infrared – always purchase a purpose-made filter from a reliable supplier). An additional note of warning: do not forget to put a full aperture filter on the guide/finder telescope (or blank it off with an opaque screen). Without such a filter the guide/finder telescope must *not* be used to align the telescope. Instead, the telescope's shadow should be circularised – this is usually sufficient to bring the solar image into the field of view of the (filtered) main telescope. Alternatively, if a small portable telescope's mounting is correctly aligned and has accurate setting circles, or for larger permanently mounted instruments, the Sun may be found in the telescope by setting onto its current position (listed in the *Astronomical Almanac* each year).

Larger instruments built specifically for solar observing come in several varieties. They are designed to try and overcome the major problem of the solar observer, which is the turbulence within and outside the instrument, while taking advantage of the observational opportunity afforded by the plentiful supply of radiation. The most intractable of the problems is that of external turbulence. This arises from the differential heating of the observatory, its surroundings and the atmosphere which leads to the generation of convection currents. The effect reduces with height above the ground and so one solution is to use a tower telescope. This involves a long focal length telescope that is fixed vertically with its objective supported on a tower at a height of several tens of metres. The solar radiation is fed into it using a coelostat (see Section 1.1) or heliostat. The latter uses a single-plane mirror rather than the double mirror of the coelostat (cf. siderostat, Section 1.1), but has the disadvantage that the field of view rotates as the instrument tracks the Sun. The 150-foot (46 m) tower telescope on Mount Wilson for example was constructed by George Ellery Hale in 1911 with a 0.3-metre lens. It has been observing the Sun every clear day since 1912 and continues to do so to this day.

The other main approach to the turbulence problem (the two approaches are not necessarily mutually exclusive) is to try and reduce the extent of the turbulence. The planting of a low growing tree or bush covering the ground is one method of doing this; another is to surround the observatory by water. The Big Bear Solar Observatory is thus built on an artificial island in Big Bear Lake, California and houses several solar telescopes including the 1.6-metre New Solar Telescope (see below). Painting the telescope and observatory with titanium-dioxide-based white paint will help to reduce heating since this reflects the solar incoming radiation, but allows the long-wave infrared radiation from the building to be radiated away. A gentle breeze also helps to clear away local turbulence and some tower telescopes such as the 0.45-metre Dutch Open Solar Telescope on La Palma are completely exposed to the elements to facilitate the effect of the wind.

Within the telescope, the solar energy can create convection currents, especially from heating the objective. Sometimes these can be minimised by careful design of the instrument, or the telescope may be sealed and filled with helium whose refractive index is only 1.000036 compared with 1.000293 for air and whose high conductivity helps to keep the components cool and minimise convection. But the ultimate solution to them is to evacuate the optical system to a pressure of a few millibars, so that all the light paths after the objective, or a window in front of the objective, are effectively in a vacuum. Most modern solar telescopes are evacuated such as the 0.7-metre German Vacuum Tower Telescope (VTT) and the 0.9-metre French Télescope Héliographique pour l'Etude du Magnétisme et des Instabilités Solaires (THEMIS), both on Tenerife, while the 0.76-metre Dunn solar telescope in New Mexico was the first vacuum telescope to be constructed (in 1969) and is still in active use. The largest vacuum solar telescope is currently the 1-metre Gregorian instrument at the Fuxian solar observatory in Southwest China.

Given the problems that solar telescopes have with excessive energy, it might seem that at least they would not suffer from too little light for their observations. However, the use of very narrowband filters and short exposures means that some types of solar observations *are* limited by too few photons. Solar telescopes therefore are starting to follow the trend of nighttime instruments towards larger and larger sizes. The recently completed New Solar telescope at the Big Bear Observatory thus has a 1.6-metre aperture. It is of an off-axis Gregorian design on an equatorial mounting. It is too large to be a vacuum telescope so the mirrors have thermal control systems that keep them to within 1 degree of the ambient air temperature and air knives* ensure streamlined air flows over the mirrors. Gregor has also recently been commissioned. It is of a 1.5-metre Gregorian design and operates without any enclosure, even in quite strong winds, in order to minimise turbulence. However, the venerable McMath-Pierce solar telescope, constructed in 1961 with a 1.61-metre primary mirror remains, at the time of writing, the world's largest solar telescope. It uses a 2.1-metre heliostat mounted at the top of a 30-metre tower to feed its main mirror which is at the bottom of an 80-metre long slanting tunnel lying partly underground.

The McMath-Pierce telescope, however, will not hold its world record for much longer. Planning permission has recently been granted for the Advanced Technology Solar Telescope (ATST) to start being built on Hawaii. The ATST is planned to have a 4.24-metre primary mirror (although the clear aperture will be 4 metres) and to be of an off-axis Gregorian design. It is expected to be able to resolve 30-km (50 milliarc second) features on the Sun using an adaptive optics system to overcome turbulence. First light for the ATST is currently planned for 2017. Other possible projects that are still in the very early stages of planning include the 4-metre European solar telescope, a 2-metre solar telescope in India and a 5- to 8-metre solar telescope in China.

Optically solar telescopes are fairly conventional (see Section 1.1) except that very long focal lengths can be used to give large image scales since there is usually plenty of available light. It is generally undesirable to fold the light paths since the extra reflections will introduce additional scattering and distortion into the image. Thus, the instruments are often very cumbersome and many are fixed in position and use a coelostat to acquire the solar radiation.

Conventional large optical telescopes can be used for solar infrared observations if they are fitted with a suitable mask. The mask covers the whole of the aperture of the telescope

---

* Essentially, a pipe containing air at a high pressure that has a line of small holes along one side producing a series of air jets.

and protects the instrument from the visible solar radiation. It also contains one or more apertures filled with infrared transmission filters. The resulting combination provides high-resolution infrared images without running the normal risk of thermal damage when a large telescope is pointed towards the Sun.

Smaller instruments may conveniently be mounted onto a solar spar. This is an equatorial mounting that is driven so that it tracks the Sun and which has a general-purpose mounting bracket in the place normally reserved for a telescope. Special equipment may then be attached as desired. Often several separate instruments will be mounted on the spar together and may be in use simultaneously.

Numerous spacecraft, including many manned missions, have carried instruments for solar observing. Amongst the spacecraft whose primary mission was aimed at the Sun, we may pick out as examples the Solar Maximum Mission (SMM), Ulysses, Yokhoh and Solar and Heliospheric Observatory (SOHO). SMM operated from 1980 to 1989 and observed the Sun from the UV through to gamma rays, as well as monitoring the total solar luminosity. Ulysses (1992–2009) had an orbit that took it nearly over the solar poles enabling high solar latitude regions to be observed clearly for the first time. As well as monitoring solar x-rays, Ulysses also studied the solar wind. Yokhoh (1991–2002) was designed to detect high-energy radiation from flares and the corona, while SOHO (1995 to date) monitors solar oscillations (see below) and the solar wind. The two Solar Terrestrial Relations Observatory (STEREO) (2006–to date) spacecraft are in orbits just inside and just outside that of the Earth around the Sun. They therefore gain and lose on the position of the Earth in its orbit by 22° per year (60,000,000 km per year). Their spatial separation of tens of millions of kilometres enables them to obtain 3D images of the Sun and in particular of coronal mass ejections. Hinode (sunrise in Japanese) was also launched in 2006 carrying a 0.5-metre optical telescope – the largest solar telescope in space. Additionally, Hinode has an x-ray telescope and an EUV spectroscope for studying the solar corona.

For the future, the Interface Region Imaging Spectrograph (IRIS) is a NASA solar spacecraft launched in June 2013 and which achieved first light three weeks later. It will study the motion of solar material using a UV telescope and an imaging spectroscope. ESA's planned Solar Orbiter will approach closer to the Sun than any other spacecraft has ever done. With a launch expected for 2017, it will use gravity assists from the Earth and Venus and reach its final orbit in 2020. At its closest approach to the Sun it will be just 42,000,000 km above the photosphere (4,000,000 km closer than Mercury at perihelion). More importantly, its angular orbital velocity at perihelion will match the angular rotational velocity of the Sun,* so that it will hover above the same point on the Sun, permitting that region to be observed for days at a time. Solar Orbiter is expected to carry a high-resolution imager and spectrograph, coronagraphs and a magnetometer and also to observe the particles in the solar wind directly. Solar Orbiter, however, may quickly be surpassed by a possible NASA mission called Solar Probe Plus which, hiding behind a carbon composite heat shield, could approach to within 4,000,000 km of the Sun. A possible launch date could be 2018. NASA's Solar Sentinels project envisages six spacecraft studying the Sun and working together

---

* Like a geostationary satellite above the Earth.

with other solar spacecraft including the Solar Orbiter. Four of the sentinels would be stationed inside Venus' orbit, the fifth would be on the far side of the Sun and the sixth close to the Earth.

However, spacecraft are not the only option. The Sunrise (not to be confused with Hinode described earlier) telescope is a balloon-borne 1-metre Gregorian instrument flown for 6 days above northern Canada in 2009 that imaged the Sun at five wavelengths ranging from 214 to 397 nm, while the EUV-imaging Hi-C discussed in Section 1.3 was launched on a 5-minute suborbital flight by a sounding rocket in 2011.

5.3.3 Spectrohelioscope

A spectrohelioscope is a monochromator that is adapted to provide an image of the whole or of a substantial part of the Sun. It operates as a normal spectroscope, but with a second slit at the focal plane of the spectrum that is aligned with the position of the desired wavelength in the spectrum (Figure 5.23). The spectroscope's entrance slit is then oscillated so that it scans across the solar image. As the entrance slit moves, so also will the position of the spectrum. Hence, the second slit must move in sympathy with the first if it is to continue to isolate the required wavelength. As an alternative to moving the slits, rotating prisms can be used to scan the beam over the grating and the image plane. If the frequency of the oscillation is higher than 15 to 20 Hz, then the monochromatic image may be viewed directly by the eye. Alternatively, an image may be taken and a spectroheliogram produced.

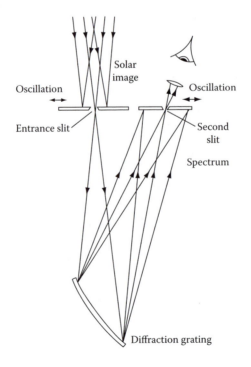

FIGURE 5.23 Principle of the spectrohelioscope.

Usually the wavelength that is isolated in a spectrohelioscope is chosen to be that of a strong absorption line, so that particular chromospheric features such as prominences, plages, mottling etc. are highlighted. In the visual region, birefringent filters (see below) can be used in place of spectrohelioscopes, but for UV imaging and scanning, satellite-based spectrohelioscopes are used exclusively.

5.3.4 Narrowband Filters

The spectrohelioscope produces an image of the Sun over a very narrow range of wavelengths. However, a similar result can be obtained by the use of a very narrowband filter in the optical path of a normal telescope. Since the bandwidth of the filter must lie in the region of 0.01 to 0.5 nm if the desired solar features are to be visible, normal dye filters or interference filters (see Section 4.1) are not suitable. Instead a filter based upon a birefringent material (see Section 5.2) has been developed by Bernard Lyot. It has various names – quartz monochromator, Lyot monochromator, or birefringent filter being amongst the commonest.

The filter's operational principle is based upon a slab of quartz or other birefringent material that has been cut parallel to its optic axis. As we saw in Section 5.2, the extraordinary ray will then travel more slowly than the ordinary ray and so the two rays will emerge from the slab with a phase difference (Figure 5.24). The rays then pass through a sheet of Polaroid film whose axis is midway between the directions of polarisation of the ordinary and extraordinary rays. Only the components of each ray which lie along the Polaroid's

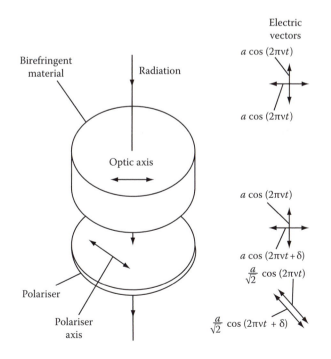

FIGURE 5.24   Basic unit of a birefringent monochromator.

axis are transmitted by it. Thus, the rays emerging from the Polaroid film have parallel polarisation directions and a constant phase difference and so they can mutually interfere. If the original electric vectors of the radiation, E_o and E_e, along the directions of the ordinary and extraordinary rays' polarisations, respectively, are given at some point by

$$E_o = E_e = a \cos(2\pi\nu t) \tag{5.52}$$

then after passage through the quartz, we will have at some point

$$E_o = a \cos(2\pi\nu t) \tag{5.53}$$

and

$$E_e = a \cos(2\pi\nu t + \delta) \tag{5.54}$$

where δ is the phase difference between the two rays. After passage through the polaroid, we then have

$$E_{45,o} = \frac{a}{\sqrt{2}} \cos(2\pi\nu t) \tag{5.55}$$

$$E_{45,e} = \frac{a}{\sqrt{2}} \cos(2\pi\nu t + \delta) \tag{5.56}$$

where $E_{45,x}$ is the component of the electric vector along the Polaroid's axis for the xth component (ordinary or extraordinary). Thus, the total electric vector of the emerging radiation, E_{45} will be given by

$$E_{45} = \frac{a}{\sqrt{2}} \left[\cos(2\pi\nu t) + \cos(2\pi\nu t + \delta) \right] \tag{5.57}$$

and so the emergent intensity of the radiation, I_{45}, is

$$I_{45} = 2 a^2 \text{ for } \delta = 2n \pi \tag{5.58}$$

$$I_{45} = 0 \text{ for } \delta = (2n + 1)\pi \tag{5.59}$$

where n is an integer. Now

$$\delta = \frac{2\pi c \Delta t}{\lambda} \tag{5.60}$$

where Δt is the time delay introduced between the ordinary and extraordinary rays by the material. If v_o and v_e are the velocities of the ordinary and extraordinary rays in the material then

$$\Delta t = \frac{T}{v_o} - \frac{T}{v_e} \tag{5.61}$$

$$= T \left(\frac{v_e - v_o}{v_e v_o} \right) \tag{5.62}$$

$$= \frac{TJ}{c} \tag{5.63}$$

where J is the birefringence of the material (see Section 5.2) and T is the thickness of the material. Thus

$$\delta = \frac{\left[2\pi c \left(\dfrac{TJ}{c} \right) \right]}{\lambda} \tag{5.64}$$

$$= \frac{2\pi TJ}{\lambda} \tag{5.65}$$

The emergent ray therefore reaches a maximum intensity at wavelengths, λ_{max}, given by

$$\lambda_{max} = \frac{TJ}{n} \tag{5.66}$$

and is zero at wavelengths, λ_{min}, given by

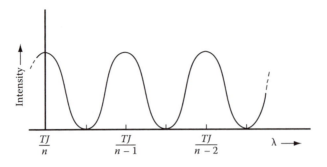

FIGURE 5.25 Spectrum of the emerging radiation from a basic unit of a birefringent monochromator.

$$\lambda_{min} = \frac{2TJ}{(2n+1)} \tag{5.67}$$

(see Figure 5.25).

Now if we require the eventual filter to have a whole bandwidth of $\Delta\lambda$ centred on λ_c, then the parameters of the basic unit must be such that one of the maxima in Figure 5.25 coincides with λ_c and the width of one of the fringes is $\Delta\lambda$; that is

$$\lambda_c = \frac{TJ}{n_c} \tag{5.68}$$

$$\Delta\lambda = TJ \left[\frac{1}{n_c - 1/2} - \frac{1}{n_c + 1/2} \right] \tag{5.69}$$

Since selection of the material fixes J and n_c is deviously related to T, the only truly free parameter of the system is T, and thus for a given filter we have

$$T = \frac{\lambda_c}{2J} \left[\frac{\lambda_c}{\Delta\lambda} + \left(\frac{\lambda_c^2}{\Delta\lambda^2} + 1 \right)^{1/2} \right] \tag{5.70}$$

and

$$n_c = \frac{TJ}{\lambda_c} \tag{5.71}$$

Normally, however

$$\lambda_c \gg \Delta\lambda \tag{5.72}$$

and quartz is the birefringent material, for which in the visible

$$J = +0.0092 \tag{5.73}$$

so that

$$T \approx \frac{109\lambda_c^2}{\Delta\lambda} \tag{5.74}$$

Thus, for a quartz filter to isolate the Hα line at 656.2 nm with a bandwidth of 0.1 nm, the thickness of the quartz plate should be about 470 mm.

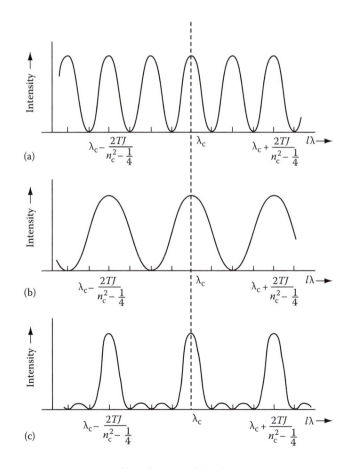

FIGURE 5.26 Transmission curves of birefringent filter basic units. (a) Basic unit of thickness T, (b) basic unit of thickness $T/2$, and (c) combination of the units in (a) and (b).

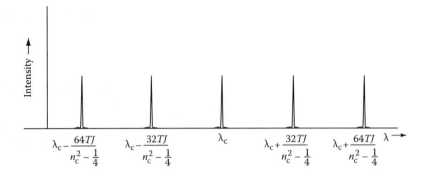

FIGURE 5.27 Transmission curve of a birefringent filter comprising six basic units with a maximum thickness T.

Now from just one such basic unit, the emergent spectrum will contain many closely spaced fringes as shown in Figure 5.25. But if we now combine it with a second basic unit oriented at 45° to the first and whose central frequency is still λ_c, but whose bandwidth is $2\Delta\lambda$, then the final output will be suppressed at alternate maxima (Figure 5.26). From Equation 5.74 we see that the thickness required for the second unit, for it to behave in this way, is just half the thickness of the first unit. Further basic units may be added, with thicknesses 1/4, 1/8, 1/16 etc. that of the first, whose transmissions continue to be centred upon λ_c, but whose bandwidths are 4, 8, 16 etc. times that of the original unit. These continue to suppress additional unwanted maxima. With six such units, the final spectrum has only a few narrow maxima that are separated from λ by multiples of $32\Delta\lambda$ (Figure 5.27). At this stage the last remaining unwanted maxima are sufficiently well separated from λ_c for them to be eliminated by conventional dye or interference filters, so that just the desired transmission curve remains.

One further refinement to the filter is required and that is to ensure that the initial intensities of the ordinary and extraordinary rays are equal, since this was assumed in obtaining Equation 5.52. This is easily accomplished, however, by placing an additional sheet

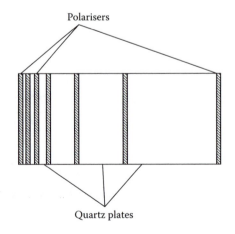

FIGURE 5.28 Six-element quartz birefringent filter.

of Polaroid before the first unit that has its transmission axis at 45° to the optical axis of that first unit. The complete unit is shown in Figure 5.28. Neglecting absorption and other losses, the peak transmitted intensity is half the incident intensity because of the use of the first Polaroid. The whole filter, however, uses such a depth of quartz that its actual transmission is a few per cent. The properties of quartz are temperature-dependent and so the whole unit must be enclosed and its temperature controlled to a fraction of a degree when it is in use. The temperature dependence, however, allows the filter's central wavelength to be tuned slightly by changing the operating temperature. Much larger changes in wavelength may be encompassed by splitting each retarder into two opposed wedges. Their combined thicknesses can then be varied by displacing one set of wedges with respect to the other in a direction perpendicular to the axis of the filter. The wavelength may also be decreased by slightly tilting the filter to the direction of the incoming radiation, so increasing the effective thicknesses of the quartz plates. For the same reason, the beam of radiation must be collimated for its passage through the filter or the bandwidth will be increased.

A closely related filter is due to Solč.* It uses only two polarisers placed before and after the retarders. All the retarders are of the same thickness and they have their optic axes alternately oriented with respect to the axis of the first polariser in clockwise and anticlockwise directions at some specified angle. The output consists of a series of transmission fringes whose spacing in wavelength terms increases with wavelength. Two such units of differing thicknesses can then be combined so that only at the desired wavelength do the transmission peaks coincide.

Another device for observing the Sun over a very narrow wavelength range is the Magneto-Optical Filter (MOF). This, however, can only operate at the wavelengths of strong absorption lines due to gases. It comprises two polarisers on either side of a gas cell. The polarisers are oriented orthogonally to each other if they are linear or are left- and right-handed if circular in nature. In either case, no light is transmitted through the system. A magnetic field is then applied to the gas cell, inducing the Zeeman effect. The Zeeman components of the lines produced by the gas are then linearly and/or circularly polarised (see Section 5.4) and so permit the partial transmission of light at their wavelengths through the whole system. The gases currently used in MOFs are vapours of sodium or potassium.

Relatively inexpensive Hα filters can be made as solid Fabry–Perot etalons (see Section 4.1). A thin fused-silica spacer between two optically flat dielectric mirrors is used together with a blocking filter. Peak transmissions of several tens of per cent and bandwidths of better than a tenth of a nanometre can be achieved in this way.

5.3.5 Coronagraph

The coronograph enables observations of the corona to be made at times other than during solar eclipses (see stellar coronagraphs, Section 2.7). It does this by producing an artificial eclipse. The principle is very simple: an occulting disk at the prime focus of a telescope obscures the photospheric image while allowing that of the corona to pass by. The practice, however, is considerably more complex, since scattered and/or diffracted light etc. in the

---

* Pronounced 'sholts'.

instrument and the atmosphere can still be several orders of magnitude brighter than the corona. Extreme precautions have therefore to be taken in the design and operation of the instrument in order to minimise this extraneous light.

The most critical of these precautions lies in the structure of the objective. A single simple lens objective is used in order to minimise the number of surfaces involved and it is formed from a glass blank that is as free from bubbles, and other imperfections as possible. The surfaces of the lens are polished with extreme care in order to eliminate all scratches and other surface markings. In use, they are kept dust-free by being tightly sealed when not in operation and by the addition of a very long tube lined with grease in front of the objective to act as a dust trap.

The occulting disc is a polished metal cone or an inclined mirror so that the photospheric radiation may be safely reflected away to a separate light and heat trap. Diffraction from the edges of the objective is eliminated by imaging the objective with a Fabry lens after the occulting disk and by using a Lyot* stop that is slightly smaller than the image of the objective to remove the edge effects. Alternatively, the objective can be apodised (see Section 2.5); its transparency decreases from its centre to its edge in a Gaussian fashion. This leads to full suppression of the diffraction halo although with some loss in resolution. Recently, the use of aspherical mirrors for this same purpose has been investigated (see PIAA, Section 2.7). A second occulting disk (the Lyot spot) before the final objective removes the effects of multiple reflections within the first objective. The final image of the corona is produced by this second objective and this is placed after the diffraction stop. The full system is shown in Figure 5.29.

A Lyot coronagraph with an additional occulting disk placed before the objective so that the whole instrument is shielded from direct photospheric light is called the Newkirk design. It has been used for many of the solar coronagraphs flown on spacecraft including the solar maximum mission (1980–1989). A Lyot-type coronagraph based upon mirrors and two Newkirk coronagraphs form the Large Angle Spectroscopic Coronagraph (LASCO) instrument on board the still-operating SOHO spacecraft, while the STEREO spacecraft carries classical Lyot coronagraphs as well as externally occulted coronagraphs.

For ground-based instruments, the atmospheric scattering can only be reduced by a suitable choice of observing site. Thus, the early coronagraphs are to be found at high altitude observatories.

The use of a simple lens results in a chromatic prime focus image. A filter must therefore normally be added to the system. This is desirable in any case since the coronal spectrum is largely composed of emission lines superimposed upon a diluted solar photospheric spectrum. Selection of a narrowband filter that is centred upon a strong coronal emission line therefore considerably improves the contrast of the final image. White light or wideband imaging of the corona is only possible using Earth-based instruments on rare occasions and it can only be attempted under absolutely optimum observing conditions. Spacecraft-borne instruments may be so used more routinely.

---

* After Bernard Lyot, who obtained the first photograph of the solar corona outside a total eclipse with his newly constructed coronagraph in 1931.

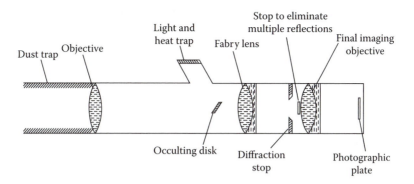

FIGURE 5.29   Schematic optical arrangement of a coronagraph.

Improvements to the basic coronagraph may be justified for balloon or spacecraft-borne instruments, since the sky background is then less than the scattered light in a more normal device and they have taken two different forms. A reflecting objective can be used. This is formed from an uncoated off-axis parabola. Most of the light passes through the objective and is absorbed. Bubbles and striations in the glass are of much less importance since they cause scattering primarily in the forward direction. The mirror is uncoated since metallic films are sufficiently irregular to a cause a considerable amount of scattering. In other respects, the coronagraph then follows the layout of Figure 5.29.

The second approach to the improvement of coronagraphs is quite different and consists simply of producing the artificial eclipse outside the instrument rather than at its prime focus (see the Newkirk design). An occulting disk is placed well in front of the first objective of an otherwise fairly conventional telescope. The disk must be large enough to ensure that the first objective lies entirely within its umbral shadow. The inner parts of the corona will therefore be badly affected by vignetting. However, this is of little importance since these are the brightest portions of the corona and it may even be advantageous since it will reduce the dynamic range that must be covered by the detector. A simple disk produces an image with a bright central spot due to diffraction, but this can be eliminated by using a disk whose edge is formed into a zigzag pattern of sharp teeth, or by the use of multiple occulting disks. By such means the instrumentally scattered light can be reduced to 10^{-4} of that of a basic coronagraph.

The final image in a coronagraph may be imaged directly, but more commonly is fed to a spectroscope, photometer or other ancillary instrument. From the Earth the corona may normally only be detected out to about one solar radius, but satellite-based coronagraphs have been used successfully out to six solar radii or more.

Devices similar to the coronagraph are also sometimes carried on spacecraft so that they may observe the atmospheres of planets, whilst shielding the detector from the radiation from the planet's surface. The contrast is generally far smaller than in the solar case so that such planetary coronagraphs can be far simpler in design.

5.3.6 Pyrheliometer/Radiometer

The pyrheliometer/radiometer is an instrument intended to measure the total flux of solar radiation at all wavelengths. In practice, current devices measure the energy from the microwave to the soft x-ray region. Modern pyrheliometers are active cavity radiometers. The radiation is absorbed within a conical cavity. The cavity is held in contact with a heat sink and maintained at a temperature about 1 K higher than the heat sink by a small heater. The difference between the power used by the heater when the cavity is exposed to the Sun and that when a shutter closes it off provides the measure of the solar energy. ESA's PROBA-2 spacecraft, launched in 2009, for example, carries an ultraviolet to x-ray radiometer monitoring the Sun over four wavebands, while Centre National d'Études Spatiales' (CNES) Picard spacecraft launched in 2010 carries a radiometer with a bolometric detector.

5.3.7 Solar Oscillations

Whole-body vibrations of the Sun reveal much about its inner structure. They are small-scale effects and their study requires very precise measurements of the velocities of parts of the solar surface. The resonance scattering spectrometer originated by the solar group at the University of Birmingham is capable of detecting motions of a few tens of millimetres per second. It operates by passing the 770-nm potassium line from the Sun through a container of heated potassium vapour. A magnetic field is directed through the vapour and the solar radiation is circularly polarised. Depending upon the direction of the circular polarisation either the light from the long-wave wing (I_L) or the shortwave wing (I_S) of the line is resonantly scattered. By switching the direction of the circular polarisation rapidly, the two intensities may be measured nearly simultaneously. The line of sight velocity is then given by

$$v = k \frac{I_S - I_L}{I_S + I_L} \tag{5.75}$$

where k is a constant with a value around 3 km s^{-1}.

Birmingham Solar Oscillations Network (BISON) has six automated observatories around the world so that almost continuous observations of the Sun can be maintained (live feeds from BISON can be seen at http://bison.ph.bham.ac.uk/live/index.php). It measures the average radial velocity over the whole solar surface using resonance scattering spectrometers. The Global Oscillation Network Group (GONG) project likewise has six observatories and measures radial velocities on angular scales down to 8″ using Michelson interferometers to monitor the position of the nickel 676.8-nm spectrum line. The SOHO spacecraft carries three helioseismology instruments – Global Oscillations at Low Frequencies (GOLF), Michelson Doppler Interferometer (MDI) and Variability of Solar Irradiance and Gravity Oscillations (VIRGO). GOLF is a resonance scattering device based upon sodium D line that detects motions to better than 1 mm s^{-1} over the whole solar surface. MDI is a Michelson Doppler imager that measures magnetic fields as well as velocities, while VIRGO measures the solar constant and detects oscillations via variations in the Sun's brightness.

5.3.8 Other Solar Observing Methods

Slitless spectroscopes (see Section 4.2) are of considerable importance in observing regions of the Sun whose spectra consist primarily of emission lines. They are simply spectroscopes in which the entrance aperture is large enough to accept a significant proportion of the solar image. The resulting spectrum is therefore a series of monochromatic images of this part of the Sun in the light of each of the emission lines. They have the advantage of greatly improved speed over a slit spectroscope combined with the obvious ability to obtain spectra simultaneously from different parts of the Sun. Their most well known application is during solar eclipses, when the 'flash spectrum' of the chromosphere may be obtained in the few seconds immediately after the start of totality or just before its end. More recently, they have found widespread use as spacecraft-borne instrumentation for observing solar flares in the UV part of the spectrum.

One specialised satellite-borne instrument based upon slitless spectroscopes is the Ly α camera. The Lyman α line is a strong emission line. If the image of the whole disk of the Sun from an UV telescope enters a slitless spectroscope, then the resulting Ly α image may be isolated from the rest of the spectrum by a simple diaphragm, with very little contamination from other wavelengths. A second identical slitless spectroscope whose entrance aperture is this diaphragm and whose dispersion is perpendicular to that of the first spectroscope will then provide a stigmatic spectroheliogram at a wavelength of 121.6 nm.

A solar 'instrument' that has recently become available is the virtual solar observatory (see Section 5.5). This is a software system linking various solar data archives and which provides tools for searching and analysing the data. It may be accessed by anyone who is interested at http://vso.nascom.nasa.gov/cgi-bin/search.

In the radio region, solar observations tend to be undertaken by fairly conventional equipment although there are now a few dedicated solar radio telescopes. The Siberian Solar Radio Telescope, for example, comprises two hundred and fifty six 2.5-metre dishes laid out in the shape of a cross (cf. Mills Cross, Section 1.2) and monitors solar activity at 5.7 GHz (53 mm) with a 15″ resolution. There is also a proposal, which is currently at the concept stage, for an array of many antennas forming a radio aperture synthesis system either on the surface of the Moon or carried on board a number of micro spacecraft for observing the inner parts of the solar wind.

One exception to the usual conventionality of solar radio instruments, however, is the use of multiplexed receivers to provide a quasi-instantaneous radio spectrum of the Sun. This is of particular value for the study of solar radio bursts, since their emitted bandwidths may be quite narrow and their wavelengths drift rapidly with time. These radio spectroscopes and the acousto-optical radio spectroscope were discussed more fully in Section 1.2. The data from them is usually presented as a frequency/time plot, from which the characteristic behaviour patterns of different types of solar bursts may easily be recognised.

This account of highly specialised instrumentation and techniques for solar observing could be extended almost indefinitely and might encompass all the equipment designed for parallax observations, solar radius determinations, oblateness determinations, eclipse work, tree rings and C^{14} determination and so on. However, at least in the author's opinion, these are becoming too specialised for inclusion in a general book like this and so the reader is referred to more specialised texts and scientific journals for further information about them.

EXERCISE

5.6 Calculate the maximum and minimum thicknesses of the elements required for an H α birefringent filter based upon calcite, if its whole bandwidth is to be 0.05 nm and it is to be used in conjunction with an interference filter whose whole bandwidth is 3 nm. The birefringence of calcite is −0.172.

5.4 MAGNETOMETRY

5.4.1 Background

The measurement of astronomical magnetic fields is accomplished in two quite separate ways. The first is direct measurement by means of an apparatus carried by spacecraft, while the second is indirect and is based upon the Zeeman effect of a magnetic field upon spectrum lines (or more strictly upon the inverse Zeeman effect, since it is usually applied to absorption lines). A third approach suggested recently is to observe the x-rays arising from the solar wind interactions with the Earth's magnetosheath, and thereby infer the shape of the terrestrial magnetic fields involved, but this has yet to be tried in practice.

5.4.1.1 Zeeman Effect

The Zeeman effect describes the change in the structure of the emission lines in a spectrum when the emitting object is in a magnetic field. The simplest change arises for singlet lines; that is, lines arising from transitions between levels with a multiplicity of one, or a total spin quantum number, M_s, of zero for each level. For these lines the effect is called the normal Zeeman effect. The line splits into two or three components depending on whether the line of sight is along, or perpendicular to, the magnetic field lines. An appreciation of the basis of the normal Zeeman effect may be obtained from a classical approach to the problem. If we imagine an electron in an orbit around an atom, then its motion may be resolved into three simple harmonic motions along the three coordinate axes. Each of

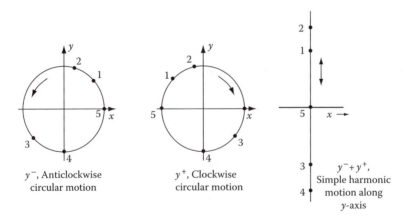

FIGURE 5.30 Resolution of simple harmonic motion along the y-axis into two equal but opposite circular motions.

these in turn we may imagine to be the sum of two equal but opposite circular motions (Figure 5.30). If we now imagine a magnetic field applied along the *z*-axis, then these various motions may be modified by it. First, the simple harmonic motion along the *z*-axis will be unchanged since it lies along the direction of the magnetic field. The simple harmonic motions along the *x*- and *y*-axes, however, are cutting across the magnetic field and so will be altered. We may best see how their motion changes by considering the effect upon their circular components. When the magnetic field is applied, the radii of the circular motions remain unchanged, but their frequencies alter. If ν is the original frequency of the circular motion and *H* is the magnetic field strength, then the new frequencies of the two resolved components ν^+ and ν^- are

$$\nu^+ = \nu + \Delta\nu \tag{5.76}$$

$$\nu^- = \nu - \Delta\nu \tag{5.77}$$

where

$$\Delta\nu = \frac{eH}{4\pi m_e c} \tag{5.78}$$

$$= 1.40 \times 10^{10}\, H \quad HzT^{-1} \tag{5.79}$$

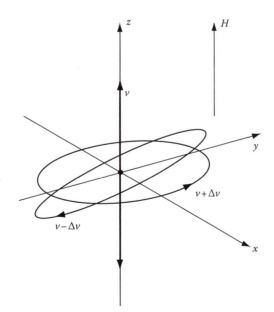

FIGURE 5.31 Components of electron orbital motion in the presence of a magnetic field.

Thus, we may combine the two higher frequency components arising from the x and y simple harmonic motions to give a single elliptical motion in the xy plane at a frequency of $\nu + \Delta\nu$. Similarly, we may combine the lower frequency components to give another elliptical motion in the xy plane at a frequency of $\nu - \Delta\nu$. Thus, the electron's motion may be resolved, when it is in the presence of a magnetic field along the z-axis, into two elliptical motions in the xy plane, plus simple harmonic motion along the z-axis (Figure 5.31), the frequencies being $\nu + \Delta\nu$, $\nu - \Delta\nu$ and ν, respectively. Now if we imagine looking at such a system, then only those components that have some motion across the line of sight will be able to emit light towards the observer, since light propagates in a direction perpendicular to its electric vector. Hence, looking along the z-axis (i.e. along the magnetic field lines) only emission from the two elliptical components of the electron's motion will be visible. Since the final spectrum contains contributions from many atoms, these will average out to two circularly polarised emissions, shifted by $\Delta\nu$ from the normal frequency (Figure 5.32), one with clockwise polarisation and the other with anticlockwise polarisation. Lines of sight perpendicular to the magnetic field direction that is within the xy plane will in general view the two elliptical motions as two collinear simple harmonic motions, while the z-axis motion will remain as simple harmonic motion orthogonal to the first two motions. Again, the spectrum is the average of many atoms and so will therefore comprise three linearly polarised lines. The first of these is at the normal frequency of the line and is polarised parallel to the field direction. It arises from the z-axis motion. The other two lines are polarised at right angles to the first and are shifted in frequency by $\Delta\nu$ (Figure 5.32) from the normal position of the line. When observing the source along the line of the magnetic field, the two spectrum lines have equal intensities, while when observing perpendicular to the magnetic field, the central line has twice the intensity of either of the other components. Thus, imagining the magnetic field progressively reducing then as the components

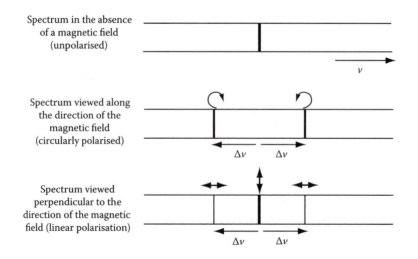

FIGURE 5.32 The normal Zeeman effect.

remix an unpolarised line results, as one would expect. This pattern of behaviour for a spectrum line originating in a magnetic field is termed the normal Zeeman effect.

In astronomy, absorption lines, rather than emission lines, are the main area of interest and the inverse Zeeman effect describes their behaviour. This, however, is precisely the same as the Zeeman effect except that emission processes are replaced by their inverse absorption processes. The above analysis may therefore be equally well applied to describe the behaviour of absorption lines. The one major difference from the emission line case is that the observed radiation remaining at the wavelength of one of the lines is preferetially polarised in the opposite sense to that of the Zeeman component, since the polarised Zeeman component is being *subtracted* from unpolarised radiation.

If the spectrum line does not originate from a transition between singlet levels (i.e. $M_s \neq 0$), then the effect of a magnetic field is more complex. The resulting behaviour is known as the anomalous Zeeman effect, but this is something of a misnomer since it is anomalous only in the sense that it does not have a classical explanation. Quantum mechanics describes the effect completely.

The orientation of an atom in a magnetic field is quantised in the following manner. The angular momentum of the atom is given by

$$\left[J(J+1)\right]\frac{h}{2\pi} \tag{5.80}$$

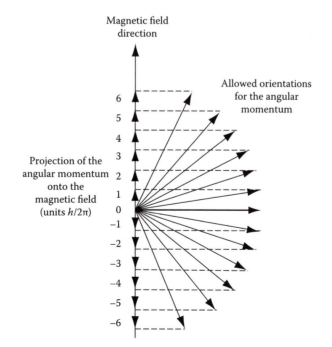

FIGURE 5.33 Space quantisation for an atom with $J = 6$.

where J is the inner quantum number and its space quantisation is such that the projection of the angular momentum onto the magnetic field direction must be an integer multiple of $h/2\pi$ when J is an integer, or a half-integer multiple of $h/2\pi$ when J is a half-integer. Thus, there are always $(2J + 1)$ possible quantised states (Figure 5.33). Each state may be described by a total magnetic quantum number, M, which for a given level can take all integer values from $-J$ to $+J$ when J is an integer, or all half-integer values over the same range when J is a half-integer. In the absence of a magnetic field, electrons in these states all have the same energy (i.e. the states are degenerate) and the set of states forms a level. Once a magnetic field is present, however, electrons in different states have different energies, with the energy change from the normal energy of the level, ΔE, being given by

$$\Delta E = \frac{eh}{4\pi m_e c} MgH \tag{5.81}$$

where g is the Landé factor, given by

$$g = 1 + \frac{J(J+1) + M_s(M_s+1) + L(L+1)}{2J(J+1)} \tag{5.82}$$

where L is the total azimuthal quantum number. Thus, the change in the frequency, Δv, for a transition to or from the state is

$$\Delta v = \frac{e}{4\pi m_e c} MgH \tag{5.83}$$

$$= 1.40 \times 10^{10} \, MgH \quad HzT^{-1} \tag{5.84}$$

Now for transitions between such states we have the selection rule that M can change by 0 or ±1 only. Thus, we may understand the normal Zeeman effect in the quantum mechanical case simply from the allowed transitions (Figure 5.34), plus the fact that the splitting of each level is by the same amount, since for the singlet levels

$$M_s = 0 \tag{5.85}$$

so that

$$J = L \tag{5.86}$$

and so

$$g = 1 \tag{5.87}$$

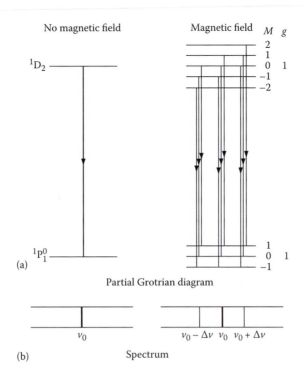

FIGURE 5.34 Quantum mechanical explanation of the normal Zeeman effect (a) the Grotrian diagram, and (b) the lines in the spectrum.

Each of the normal Zeeman components for the example shown in Figure 5.34 is therefore triply degenerate and only three lines result from the nine possible transitions. When M_s is not zero, g will in general be different for the two levels and the degeneracy will cease. All transitions will then produce separate lines and hence we get the anomalous Zeeman effect (Figure 5.35). Only one of many possible different patterns is shown in the figure; the details of any individual pattern will depend upon the individual properties of the levels involved.

As the magnetic field strength increases, the pattern changes almost back to that of the normal Zeeman effect. This change is known as the Paschen-Back effect. It arises as the magnetic field becomes strong enough to decouple L and M_s from each other. They then no longer couple together to form J, which then couples with the magnetic field, as just described, but couple separately and independently to the magnetic field. The pattern of spectrum lines is then that of the normal Zeeman effect, but with each component of the pattern formed from a narrow doublet, triplet etc. accordingly as the original transition was between doublet, triplet etc. levels. Field strengths of around 0.5 T or more are usually necessary for the complete development of the Paschen-Back effect.

At very strong magnetic field strengths ($>10^3$ T), the quadratic Zeeman effect will predominate. This displaces the spectrum lines to higher frequencies by an amount $\Delta\nu$, given by

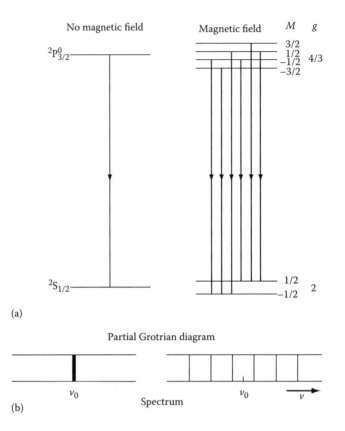

(a)

(b)

FIGURE 5.35 Quantum mechanical explanation of the anomalous Zeeman effect (a) the Grotian diagram, and (b) the lines in the spectrum.

$$\Delta \nu = \frac{\varepsilon_o h^3}{8\pi^2 m_e^3 c^2 e^2 \mu_o} n^4 (1 + M^2) H^2 \tag{5.88}$$

$$= 1.489 \times 10^4 n^4 (1 + M^2) H^2 \quad Hz \tag{5.89}$$

where n is the principal quantum number.

5.4.2 Magnetometers

Amongst the direct measuring devices, the commonest type is the flux gate magnetometer illustrated in Figure 5.36. Two magnetically soft cores have windings coiled around them as shown. Winding A is driven by an alternating current. When there is no external magnetic field, winding B has equal and opposite currents induced in its two coils by the alternating magnetic fields of the cores – and so there is no output. The presence of an external magnetic field introduces an imbalance into the currents, giving rise to a net alternating current output. This may then easily be detected and calibrated in terms of the external

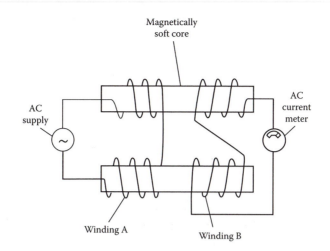

FIGURE 5.36   Flux gate magnetometer.

field strength. Spacecraft usually carry three such magnetometers oriented orthogonally to each other so that all the components of the external magnetic field may be measured. The four spacecraft of NASA's Magnetospheric Multiscale (MMS) mission for example will each be equipped with flux gate magnetometers capable of measuring field strengths to ±5 pT when they are launched in 2014.

Another type of magnetometer that is often used on spacecraft is based upon atoms in a gas oscillating at their Larmor frequency. It is often used for calibrating flux gate magnetometers in orbit. The vector helium magnetometer operates by detecting the effect of a magnetic field upon the population of a metastable state of helium. A cell filled with helium is illuminated by 1.08 μm radiation that pumps the electrons into the metastable state. The efficiency of the optical pump is affected by the magnetic field and the population of that state is monitored by observing the absorption of the radiation. A variable artificially generated magnetic field is swept through the cell until the external magnetic field is nullified. The strength and direction of the artificial field are then equal and opposite to the external field. Any magnetometer on board a spacecraft usually has to be deployed at the end of a long boom after launch in order to distance it from the magnetic effects of the spacecraft itself.

An alternative technique altogether, is to look at the electron flux. Since electron paths are modified by the magnetic field, their distribution can be interpreted in terms of the local magnetic field strength. This enables the magnetic field to be detected with moderate accuracy over a large volume, in comparison with the direct methods that give very high accuracies, but only in the immediate vicinity of the spacecraft.

Most of the successful indirect work in magnetometry based upon the (inverse) Zeeman effect has been undertaken by Harold and Horace Babcock* or uses instruments based upon their designs. For stars, the magnetic field strength along the line of sight may be determined via the longitudinal Zeeman effect. The procedure is made very much more

---

* 1882 to 1968 and 1912 to 2003, respectively.

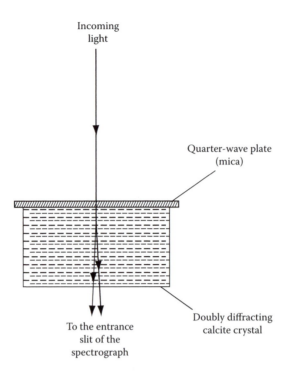

Incoming
light

Quarter-wave plate
(mica)

Doubly diffracting
calcite crystal

To the entrance
slit of the
spectrograph

FIGURE 5.37   Babcock's differential analyser.

difficult than in the laboratory by the widths of the spectrum lines. For a magnetic field of 1 T, the change in the frequency of the line is 1.4×10^{10} Hz (Equation 5.79), which for lines near 500 nm corresponds to a separation for the components in wavelength terms of only 0.02 nm. This is smaller than the normal line width for most stars. Thus, even strong magnetic fields do not cause the spectrum lines to split into separate components, but merely to become somewhat broader than normal. The technique is saved by the opposite senses of circular polarisation of the components which enables them to be separated from each other. Babcock's differential analyser therefore consists of a quarter-wave plate (see Section 5.2) followed by a doubly refracting calcite crystal. This is placed immediately before the entrance aperture of a high-dispersion spectroscope (Figure 5.37). The quarter-wave plate converts the two circularly polarised components into two mutually perpendicular linearly polarised components. The calcite is suitably oriented so that one of these components is the ordinary ray and the other the extraordinary ray. The components are therefore separated by the calcite into two beams. These are both accepted by the spectroscope slit and so two spectra are formed in close juxtaposition, with the lines of each slightly shifted with respect to the other due to the Zeeman splitting. This shift may then be measured and translated back to give the longitudinal magnetic field intensity. A lower limit on the magnetic field intensity of about 0.02 T is detectable by this method, while the strongest stellar fields found have strengths of a few teslas. By convention the magnetic field is positive when it is directed towards the observer.

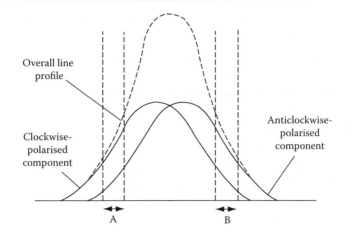

FIGURE 5.38   Longitudinal Zeeman components.

The relatively new technique of Zeeman-Doppler imaging potentially enables maps of the distribution of magnetic fields over the surfaces of stars to be constructed, though often with considerable ambiguities left in the results. The basis of the technique is best envisaged by imagining a single magnetic region on the surface of a star. The Zeeman effect will induce polarisation into the lines emitted or absorbed from this region, as already discussed. However, when the rotation of the star brings the region first into view, it will be approaching us. The features from the region will therefore be blue-shifted (i.e. Doppler shifted) compared with their 'normal' wavelengths. As the rotation continues to move the region across the disk of the star, the spectral features will change their wavelengths, going through their 'normal' positions to being red-sifted just before they disappear around the opposite limb of the star. If there are several magnetic regions, then each will usually have a different Doppler shift at any given moment and the changing Doppler shifts can be interpreted in terms of the relative positions of the magnetic regions on the star's surface (i.e. a map). The reduction of the data from this type of observations is a complex procedure and MEM is usually needed to constrain the range of possible answers. Spectropolarimeters such as SemelPol* on the AAT, Echelle SpectroPolarimetric Device for the Observation of Stars (ESPaDOnS) on the CFHT and Narval on the Pic du Midi's 2-metre Lyot telescope have been used in this manner recently, but the technique is still in its infancy.

On the Sun much greater sensitivity is possible and fields as weak as 10^{-5} T can be studied. George Ellery Hale first detected magnetic fields in sunspots in 1908, but most of the modern methods and apparatus are again due to the Babcocks. Their method relies upon the slight shift in the line position between the two circularly polarised components, which causes an exaggerated difference in their relative intensities in the wings of the lines (Figure 5.38). The solar light is passed through a differential analyser as before. This time, however, the quarter-wave plate is composed of ammonium dihydrogen phosphate and this requires an electric potential of some 9 kV across it in order to make it sufficiently

---

* Not an anagram – named after its designer, Meir Semel.

birefringent to work. Furthermore, only one of the beams emerging from the differential analyser is fed to the spectroscope. Originally a pair of photomultipliers that accepted the spectral regions labelled A and B in Figure 5.38 detected the line, now CCDs or other array detectors are used. Thus, the wing intensities of, say, the clockwise-polarised component are detected when a positive voltage is applied to the quarter-wave plate, while those of the anticlockwise component are detected when a negative voltage is applied. The voltage is switched rapidly between the two states and the outputs are detected in phase with the switching. Since all noise except that in phase with the switching is automatically eliminated and the latter may be reduced by integration and/or the use of a phase-sensitive detector (see Section 3.2), the technique is very sensitive. The entrance aperture to the apparatus can be scanned across the solar disk so that a magnetogram can be built up with a typical resolution of a few seconds of arc and of a few microtesla. Interline transfer CCD detectors are used which have every alternate row of pixels covered over. The accumulating charges in the pixels are switched between an exposed row and a covered row in phase with the switching between the polarised components. Thus, at the end of the exposure, alternate rows of pixels contain the intensities of the clockwise and anticlockwise components. Similar instruments have been devised for stellar observations, but so far are not as sensitive as the previously mentioned system.

Several other types of magnetometer have been devised for solar work. For example, at the higher field strengths, Leighton's method provides an interesting pictorial representation of the magnetic patterns. Two spectroheliograms are obtained in opposite wings of a suitable absorption line and through a differential analyser that is arranged so that it only allows the passage into the system of the stronger circularly polarised component in each case. A composite image is then made of both spectroheliograms, with one as a negative and the other as a positive. Areas with magnetic fields less than about 2 mT are seen as grey, while areas with stronger magnetic fields show up light or dark according to

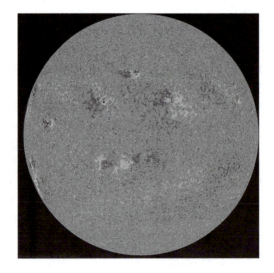

FIGURE 5.39 A line-of-sight solar magnetogram obtained by the HMI instrument on the SDO spacecraft on January 1, 2013. (Reproduced by kind permission of NASA.)

their polarity. More modern devices use video recording but the basic technique remains unaltered. Vector-imaging magnetographs image the Sun over a very narrow wavelength range and then rapidly step that image through a magnetically sensitive spectrum line. This enables both the direction and strength of the magnetic field to be determined over the selected region. Usually this region is just a small part of the whole solar disk because of the enormous amounts of data that such instruments can generate. The Helioseismic and Magnetic Imager (HMI) on board NASA's SDO, however, does obtain vector magnetograms of the whole solar disk (Figure 5.39).

In the radio region, the Zeeman effect affects lines such as the 1.42-GHz emission from hydrogen (usually better known as the twenty-one centimetre line). Their Zeeman splitting is found in a similar manner to Babcock's solar method – by observing in the steepest part of the wings. Interstellar magnetic fields of about 10^{-9} T are detectable in this way.

The very strongest fields, up to 10^8 T, are expected to exist in white dwarfs and neutron stars. However, there is usually little spectral structure in such objects from which the fields might be detected. The quadratic Zeeman effect, however, may then be sufficient for circular polarisation of the continuum to be distinguished at points where it is varying rapidly in intensity. Field strengths of upwards of 1000 T are found in this way.

A few devices have been constructed to detect electric fields via the Stark effect upon the Paschen hydrogen lines. They are basically similar to magnetometers and for solar studies can reach sensitivities of 500 V m^{-1}.

5.4.3 Data Reduction and Analysis

Once a flux gate magnetometer has been calibrated, there is little involved in the analysis of its results except to convert them to field strength and direction. However, it may be necessary to correct the readings for influence from the spacecraft and for the effects of the solar wind and the solar or interplanetary magnetic fields. Measurements with accuracies as good as 10^{-11} T are possible when these corrections are reliably known.

With the first of the Babcocks' methods, in which two spectra are produced side by side, the spectra are simply measured as if for radial velocity. The longitudinal magnetic field strength can then be obtained via Equation 5.84. The photoelectric method's data are rather more complex to analyse. No general method can be given since it will depend in detail upon the experimental apparatus and method and upon the properties of the individual line (or continuum) being observed.

With strong solar magnetic fields, such as those found in sunspots, information may be obtainable in addition to the longitudinal field strength. The line in such a case may be split into two or three components and when these are viewed through a circular analyser (i.e. an analyser which passes only one circularly polarised component) the relative intensities of the three lines of the normal Zeeman effect are given by Seare's equations

$$I_v = \frac{1}{4}(1 \pm \cos\theta)^2 I \tag{5.90}$$

$$I_c = \frac{1}{2}(\sin^2\theta)I \qquad\qquad (5.91)$$

$$I_r = \frac{1}{4}(1 \pm \cos\theta)^2 I \qquad\qquad (5.92)$$

where I_v, I_c and I_r are the intensities of the high-frequency, central and low-frequency components of the triplet, I is the total intensity of all the components and θ is the angle of the magnetic field's axis to the line of sight. Thus, the direction of the magnetic field as well as its total strength may be found.

5.5 COMPUTERS AND THE INTERNET

5.5.1 Introduction

Information technology (IT), computers, the Internet and their applications are now essential to just about every aspect of astrophysics and indeed to most of the rest of science and to life in general. Many of the uses of computers, such as controlling telescopes, processing data etc. have already been mentioned within other sections of this book. Internet links are now being used quite extensively to transfer data between the individual instruments making up interferometric arrays and in some cases for controlling remote/robotic instruments. There are several programmes aimed at enabling schools to use telescopes (sometimes of quite large sizes) via the Internet such as the two 2-metre Faulkes robotic optical telescopes and the NRAO's 20-metre radio telescope at Green Bank. There are also bibliographic data bases such as arXiv (http://arxiv.org/) and Astrophysical Data System (ADS) (http://adsabs.harvard.edu/index.html) that are essential facilities for any astrophysicist needing to keep up to date in his/her specialism. For anyone wishing to keep informed on a wide range of astronomical topics then Astrobites (http://astrobites.com/) sends out a brief explanation and up-to-date account of selected subjects on a daily basis.

There are a great many other astronomical resources to be found on the Internet and far too little space here to mention even a fraction of them. Furthermore, most come and go or change in other ways rapidly, so the reader will have to maintain his/her own list of favourites. However, there are two additional promising IT applications that deserve a section to themselves. These are the availability of digital sky surveys and the related development of 'virtual' observatories.

A major concern of astronomy, dating back to Hipparchus and earlier, has always been the cataloguing or listing of objects in the sky together with their positions and properties. The quantum changes that have occurred in the last few years are that many of these catalogues are now on open access through the internet and personal computers affordable by individuals are now powerful enough to enable them to process the vast amounts of data present in the catalogues. The possibility of conducting real research is therefore open to anyone with the interest and some spare time; no longer just to professional scientists. Of course there are still problems, the main one being data links. Some of the surveys contain a terabyte (10^{12} bytes) of data and using a 10 Mb s^{-1} broadband connection,* this would take about 10 days to download. Data archives such as SIMBAD and MAST (see below) contain 10 TB at the time of writing and are being added to at a rate of 5 or more TB per year. The cumulative information content within astronomy is now hundreds of terabytes and one proposal alone (the Large Synoptic Survey Telescope) could be generating 5 TB of data *per day* within a few years. Fortunately, for many purposes, it is not necessary to download a whole survey in order to work with it; a small subset can be sufficient.

5.5.2 Digital Sky Surveys

We have already seen (see Section 5.1) that the Hipparcos, Tycho and other astrometric catalogues are available via the Internet. Other large surveys presented on the internet include

- The Hubble space telescope GSC2, containing 500 million objects

- 2MASS, containing 300 million entries

- The SDSS, 100 million objects

- The Palomar Digital Sky Survey (DPOSS), 2000 million objects

- The second issue of the U.S. Naval Observatory's Twin Astrographic Catalog (TAC 2.0) based upon photographic plates obtained with the twin astrograph and containing over 700,000 stellar positions, to between ±50- and ±120-mas accuracy

- The USNO's A2.0 catalogue, containing 526 million entries

- The USNO-B1.0 catalogue, containing data on a billion stars with 200-mas positional accuracy

- The UCAC (USNO CCD Astrograph Catalog) with the positions of some 40 million stars in the 10^m to 14^m range, to ±20 mas accuracy

We may expect that many of the gigantic surveys, catalogues and databases that will result from projects soon to commence such as the JWST, Gaia, SKA and the LSST will also quickly become available via the Internet.

However, not all digital sky surveys are the prerogative of the professional astronomer. The Amateur Sky Survey (TASS) (http://stupendous.rit.edu/tass/tass.shtml) uses small cameras and CCD detectors to record some 200 Mbytes of data per night while searching for comets and variable stars.

A list of the surveys and catalogues available online that currently contains 11,000 entries is given by VizieR at the CDS (see below).

---

* Bits and bytes can easily cause confusion. In information technology a bit is a single binary digit (i.e. a zero or a one). A byte is a collection of bits. Nowadays, a byte is almost always made up from 8 bits, but there have been variations on this in the past. A byte therefore represents a binary number between 0 and 255. A bit is symbolised by the lowercase b, while a byte is symbolised by the uppercase B. The volume of a data set is usually given in bytes, while the capacity of a broadband line is usually given in bits. Thus, a 10-Mb s^{-1} broadband line actually transmits 1.25 MB s^{-1}.

Anyone with access to the Internet and a computer can join in the work for some surveys. These citizen science projects work in several different ways but generally make use of spare time on individual personal computers. Some data processing software is downloaded onto the computer via the Internet from the host institution's computer together with some data from a large survey type project. When the personal computer is on standby, it processes the batch of data that it has been sent and then sends the processed data back to the host computer. A new batch of data will then be sent to the personal computer for processing at the next opportunity. Sometimes the work for the citizen science project requires little or no input from the owner of the personal computer – he/she is, in effect, simply donating processing power to the project. Other types of project require much more human input so that the volunteer involved is a part of the project science team and just happens to communicate with the rest of the team via the Internet. In either case, the distributed network enables projects to have access to far more resources than could be afforded through dedicated facilities.

Amongst the earliest citizen science projects is SETI@home (http://setiathome.berkeley. edu/). It was started in 1999 and currently has over 3,300,000 host computers. It is searching for radio signals from extraterrestrial intelligences though without any success to date. Other large projects include Einstein@home (http://boinc.berkeley.edu/wiki/Einstein@ Home) with over 3,600,000 host computers and which, since 2005, searches data from the LIGO and GEO600 gravitational wave detectors (see Section 1.6) for evidence of gravitational waves (again so far without success) and MilkyWay@home (http://milkyway.cs.rpi. edu/milkyway/) with 300,000 host computers analysing data from the SDSS in order to map the 3D structure of the Milky Way galaxy.

Projects requiring more input from the volunteer include the Galaxy Zoo (http://www. galaxyzoo.org/) which is classifying galaxies and Planet Hunters (http://www.planethunters. org/) which is searching data from the Kepler spacecraft for exoplanets.

There are now hundreds of citizen science projects covering most of the sciences and lists of them can be found at the time of writing at Berkeley Open Infrastructure for Network Computing (BOINC) websites (http://boinc.berkeley.edu/and http://www.allprojectstats. com/), at the Zooniverse website (https://www.zooniverse.org/?lang=en) and as 'List of Distributed Computing Projects' on Wikipedia, although these sites are not all-inclusive.

Many surveys have been gathered together to be available through single sites such as Centre de données astronomiques de Strasbourg (CDS)* that operates Set of Identifications, Measurements and Bibliography for Astronomical Data (SIMBAD) (http://simbad. u-strasbg.fr/Simbad), National Space Science Data Center (NSSDC) (http://nssdc.gsfc.nasa. gov) and MAST (http://archive.stsci.edu). Anyone who is interested can easily look up the details on an individual object within these catalogues via the Internet. However, if more than a few tens of objects are needed, then this approach becomes cumbersome and time consuming, and larger projects therefore need more powerful search and processing software. Some of the sites provide this on an individual basis, but with little consistency in

---

* The CDS also archives bibliographical information, enabling searches for publications relating to objects of interest to be conducted.

what is available and how to use it. Recently, therefore, Virtual Observatories have started to come on-stream providing much greater software and other support, although for some of them their use is limited to accredited scientists.

5.5.3 Virtual Observatories

Virtual observatories are interfaces whereby huge data sets such as the digital sky surveys and others, such as collections of spectra from the IUE and the HST spacecraft, results from other spacecraft such as ROSAT, Infrared Space Observatory (ISO), Chandra etc. interferometric data from radio and optical telescopes and so on, can be handled and mined for new information. The virtual observatory is a set of data archives, software tools, hardware and staff that enables data in archives to be found and processed in many different ways. Amongst the type of functions that a virtual observatory can provide there are

- Standardising the formats of differing data sets
- Finding different observations of a given object or a set of objects and sorting through them
- Comparing and combining the information about objects available in various catalogues and other databases
- Comparing and combining observations taken at different times
- Combining archive data with new data obtained from telescopes
- Correlating data from different sources
- Classifying objects
- Performing statistical analyses
- Measuring images
- Image processing

While the types of scientific study that are possible include

- Multiwavelength studies
- Large statistical studies
- Searches for changing, moving, rare, unusual or new objects and types of objects
- Combining data in new ways leading to unexpected discoveries

Virtual observatories started becoming available a few years ago and are now proliferating rapidly. Much of the information on virtual observatories is collated by the International Virtual Observatory Alliance (IOVA) and that may be accessed at http://www.ivoa.net/.

IVOA attempts to develop universal standards for virtual observatories so that they can work together and lists 20 programmes at the time of writing (although the United Kingdom's Astrogrid is no longer being funded). Most of the programmes are national but the Euro-VO (http://www.euro-vo.org/) is pan-European. The United States' programme is called the Virtual Astronomical Observatory (VAO) and is to be found at http://www.usvao.org/. Details and contacts for the others may be found on the IVOA website above.

For anyone with an interest in the Earth's Moon, NASA's Lunar Mapping and Modeling website at http://lunarscience.nasa.gov/articles/lunar-mapping-and-modeling-project/is well worth looking at. It is a simple virtual observatory containing data from many lunar missions together with tools to produce 3D visualisations, to overlay images, to determine Sun angles etc.

Finally, the most widely used virtual observatory of all is Google Sky. This forms a part of Google Earth which may be accessed at http://earth.google.com/. Google Sky provides a complete map of the sky containing some 100 million stars and 200 million galaxies. The more prominent nebulae and galaxies can be selected for higher resolution images and for notes about the natures of the objects. Best of all – it is available to anyone and it is *free*!

Appendix A: Julian Date

The Julian date is the number of days elapsed since noon on 24 November 4714 BC (on the Gregorian calendar) or since noon on 1 January 4713 BC (on the Julian calendar). The modified Julian date is a variation of the Julian date that starts at midnight on 17 November 1858. The modified Julian date is thus the Julian date minus 2,400,000.5 days.

Date (January 1 at Noon Gregorian Reckoning)	Julian Day Number	Date (January 1 at Noon Gregorian Reckoning)	Julian Day Number
2050	2,469,808.0	1600	2,305,448.0
2025	2,460,677.0	1200	2,159,351.0
2000	2,451,545.0	800	2,013,254.0
1975	2,442,414.0	400	1,867,157.0
1950	2,433,283.0	0	1,721,060.0
1925	2,424,152.0	400 BC	1,574,963.0
1900	2,415,021.0	800 BC	1,428,866.0
1875	2,405,890.0	1200 BC	1,282,769.0
1850	2,396,759.0	1600 BC	1,136,672.0
1825	2,387,628.0	2000 BC	990,575.0
1800	2,378,497.0	2400 BC	844,478.0
1775	2,369,366.0	2800 BC	698,381.0
1750	2,360,234.0	3200 BC	552,284.0
1725	2,351,103.0	3600 BC	406,187.0
1700	2,341,972.0	4000 BC	260,090.0
1675	2,332,841.0	4400 BC	113,993.0
1650	2,323,710.0	4714 BC (24 Nov)	0.0
1625	2,314,579.0		

For days subsequent to 1 January, the following number of days should be added to the Julian day number.

Date (Noon)	Number of Days (Non-Leap Years)	Number of Days (Leap Years)
1 Feb	31	31
1 Mar	59	60
1 Apr	90	91
1 May	120	121
1 June	151	152
1 July	181	182
1 Aug	212	213
1 Sept	243	244
1 Oct	273	274
1 Nov	304	305
1 Dec	334	335
1 Jan	365	366

Appendix B: Answers to the Exercises

1.1. 69.21″ and 21.4″ (do not forget the Purkinje effect)

1.2. 1.834 and −3.387 m

1.3. 0.31 mm

1.4. 94 mm, 850× (for the exit pupil to be smaller than the eye's pupil)

1.5. About 75° (see Equation 1.173)

1.6. No (E-ELT Rayleigh resolution at 1.5 μm = 2.9 mas, angular separation of α Cen B and α Cen B b = 9.7 mas)

1.7. Mean = 804.25 pc, σ = 151.33 pc, S = 53.50 pc, Distance = 800 ± 50 pc

1.8. 57, 83, 99, 107 and 108 mm. Minimum altitude ≈ 60°

1.10. 84 m along the length of its arms

1.12. 0.002 m³

1.14. 250/s

1.16. 1.6×10^{-23}

1.17. About 44^{37}_{18} *Ar* atoms

1.19. The calculated data are tabulated below:

Planet	Mass (M_p/M_{Sun})	Period (Days)	L_G (W)
Mercury	1.6×10^{-7}	88	62
Venus	2.4×10^{-6}	225	600
Earth	3.0×10^{-6}	365	180
Mars	3.2×10^{-7}	687	0.25
Jupiter	9.6×10^{-4}	4330	5300
Saturn	2.9×10^{-4}	10,760	20
Uranus	4.3×10^{-5}	30,700	0.015
Neptune	5.3×10^{-5}	60,200	0.002
Pluto	2.5×10^{-6}	90,700	10^{-6}

2.4. 28 mW, 72 km

2.5.1. (a) 35 mm, (b) 52 mm

2.5.2. 3.1×10^{16} m (or 1 pc)

2.5.3. 5×10^{17} m (or 16 pc)

3.1. +25.9, −20.6

3.2. 3050 pc

3.3. $(U - B) = -0.84$, $(B - V) = -0.16$

$Q = -0.72$

$(B - V)_o = -0.24$, $E_{B-V} = 0.08$

$E_{U-B} = 0.06$, $(U - B)_o = -0.90$

Spectral type B3 V

Temperature 20,500 K

Distance (average) 230 pc

$U_o = 2.82$, $B_o = 3.72$

$V_o = 3.96$, $M_V = -2.8$

4.1. −80.75 km/s

4.2. One prism, 0.79 nm mm^{-1}

4.4. Distance = 700 mm, Number of steps = 1,715,686

4.5. $f' = 25$, $W_\lambda = 10^{-12}$ m, $\lambda = 625$ nm

$R = 6.25 \times 10^5$, $L = 0.42$ m, $D = 0.38$ m

$d\theta/d\lambda = 1.66 \times 10^6$ rad m^{-1}

$d\lambda/d\theta = 6.08 \times 10^{-7}$ m rad^{-1}

$f_1 = 13.4$ m, $f_2 = 12.2$ m

$D_1 = 0.54$ m, $D_2 = 0.54$ m

$S = 20$ μm

Problems to be solved if possible in the next iteration through the exercise: grating is too large (normal upper limit on L is 0.1 m). Slit width is too small. The overall size of the instrument will lead to major thermal control problems. The projected slit width on the CCD is slightly smaller than the CCD's pixel size.

4.6. +5.3 (taking $q = 0.002$ and $\alpha = 5 \times 10^{-6}$)

4.7. 64.96°

4.8. The Sun's orbital velocity around the Sun–Earth centre of mass is about 90 mm s^{-1}; thus, so a velocity resolution of ±10 to ±30 mm s^{-1} should be good enough

5.2. 0.26 mm

5.3. The Sun's maximum angular shift as seen by the alien would be 6 mas so a measurement accuracy of around ±1 mas should suffice

5.6. 50.6 mm, 0.79 mm

Appendix C: Acronyms

Early editions of this book tried to avoid the use of acronyms, believing that except to those people working in the particular field, they obscure understanding without actually saving much time or effort. Unfortunately, few other astronomers seem to follow this belief, instead appearing to delight in inventing ever more tortuous acronyms to name their instruments, techniques etc. Second- and third-level acronyms (i.e. an acronym that contains an acronym that contains yet another acronym) are now being encountered. The use of acronyms in astronomy is nowadays so prevalent and the distortions so often introduced to the actual names that they were reluctantly included in the fourth edition and have now been extended to the fifth and sixth editions. The acronyms are defined when first encountered or within their main section, but are also defined below, so that their meaning can be retrieved after the original definition can no longer be found within the text.

A list of thousands of mostly astronomical acronyms can be found at the time of writing at http://www.maa.mhn.de/FAQ/acronyms.html, http://cfa-www.harvard.edu/~gpetitpas/Links/Astroacro.html, http://faqs.org/faqs/space/acronyms/ and many other websites – alternatively, a search for 'astronomy acronyms' or for the acronym itself will usually be successful. There are also some 200,000 general acronyms listed at http://www.acronymfinder.com/ and numerous lists of acronyms used within special projects such as the HST, AAO, ESO etc. that may easily be found by an Internet search.

Below is the list of acronyms to be encountered within this book.

2dF	2-degree field (AAT)
2MASS	2 Micron All Sky Survey
6dF	6-degree field (UKST)
AAO	Australian Astronomical Observatory
AAT	Anglo-Australian Telescope
ACIS	Advanced CCD Imaging Spectrometer
ADC	Analogue-to-digital converter
ADI	Angular differential imaging
ADS	Astrophysical Data System
ADU	Analogue-to-digital unit
AERA	Auger Engineering Radio Array
AGK	Astronomische Gesellschaft Katalog
AIA	Atmospheric Imaging Assembly
AIPS	Astronomical Image Processing System
AMADEUS	ANTARES Modules for the Acoustic Detection Under the Sea

A-MKD	APEX MKID
AMS-02	Alpha Magnetic Spectrometer
ANITA	Antarctic Impulsive Transient Antenna
ANTARES	Astronomy with a Neutrino Telescope and Abyssal Environmental Research
AOS	Acousto-Optical Radio Spectrometer
APD	Avalanche photodiode
APEX	Atacama Pathfinder Experiment
APS	Active Pixel Sensor
AquEYE	Asagio Quantum Eye
ARA	Askaryan Radio Array
ARCONS	Array Camera for Optical to Near Infrared Spectrophotometry
ARIANNA	Antarctic Ross Ice-Shelf Antenna Neutrino Array
ASA	American Standards Association
ASKAP	Australian SKA Pathfinder
ASTROD-GW	Astronomical space test of relativity using optical devices
ATST	Advanced Technology Solar Telescope
AURIGA	Antenna Ultracriogenica Risonante per l'Indagine Gravitazionale Astronomica
BAT	Burst Alert Telescope
BGO	Bismuth germinate
bHROS	Bench-mounted High Resolution Optical Spectrometer
BIB	Blocked impurity band device
BigBOSS	Big Baryon Acoustic Oscillations Spectroscopic Survey
BISON	Birmingham Solar Oscillations Network
BLAST	Balloon-borne Large Aperture Submillimeter Telescope
BOINC	Berkeley Open Infrastructure for Network Computing
BOSS	Baryon Acoustic Oscillations Spectroscopic Survey
BWFN	Beam width at first nulls
BWHP	Beam width at half-power points
CANGAROO	Collaboration of Australia and Nippon for a Gamma Ray Observatory in the Outback
CAOS	Club of Aficionados in Optical Spectroscopy
CASA	Common Astronomy Software Applications
CCAT	Cornel Caltech Atacama Telescope
CCD	Charge-coupled device
CDS	Centre de Données astronomiques de Strasbourg
CELT	California Extremely Large Telescope (now known as the TMT)
CFHT	Canada-France-Hawaii Telescope
CGRO	Compton Gamma-Ray Observatory
CHARA	Centre for High Angular Resolution Astronomy
CHARIS	Coronagraphic High Angular Resolution Imaging Spectrograph
CHEOPS	Characterizing Extrasolar Planets by Opto-infrared Polarimetry and Spectroscopy
CIE	Commission Internationale de l'Éclairage
CMB	Cosmic microwave background radiation
CMOS	Complementary metal-oxide-semiconductor
CMOS-APS	Complementary metal-oxide-semiconductor–active pixel sensor

CMT	Carlsberg Meridian Telescope
CNES	Centre National d'Études Spatiales
COBE	Cosmic Background Explorer Satellite
CoGeNT	Coherent Germanium Neutrino Technology
CONICA	Near-Infrared Imager and Spectrograph
COS	Cosmic Origins Spectrograph
CPCCD	Column parallel CCD
CRAF	Committee on Radio Astronomy Frequencies'
CRESST II	Cryogenic Rare Event Search with Superconducting Thermometers
CRIRES	Cryogenic Infrared Echelle Spectrograph
CTA	Čerenkov Telescope Array
CTIO	Cerro Tololo Inter-American Observatory
CW	Continuous wave (radar)
CZT	Cadmium-zinc-tellurium detectors
DAMA	Dark matter
DECIGO	Deci-Hertz Interferometer Gravitational Wave Observatory
DEIMOS	Deep Imaging Multi-Object Spectrograph
DENIS	Deep Near Infrared Survey
DIN	Deutsche Industrie Norm
DMP	Dark matter particle
DPOSS	Digital Palomar Observatory Sky Survey
e-APD	Electron avalanche photodiode
EAS	Extensive air shower
EBCCD	Electron bombarded charge-coupled device
EBEX	E and B Experiment
EDI	Externally Dispersed Interferometry
E-ELT	European Extremely Large Telescope
EFOSC2	ESO's Faint Object Spectrograph and Camera
EI	Exposure index number
EISCAT	European Incoherent Scatter
ELF	Extremely low frequency
EMCCD	Electron multiplying charge-coupled device
ESA	European Space Agency
ESO	European Southern Observatory
ESTEC	European Space Research and Technology Centre
EURECA	European Underground Rare Event Calorimeter Array
EUV	Extreme ultraviolet (also XUV)
EVE	Extreme Ultraviolet Variability Experiment
EXCEDE	Exoplanetary Circumstellar Environments and Disk Explorer
FAST	Five-hundred metre Aperture Spherical Telescope
FCRAO	Five College Radio Astronomy Observatory
FFP	Fibre optic Fabry–Perot filter
FGS	Fine guidance sensors
FIR	Far infrared
FITS	Flexible Image Transport System
FK	Fundamental Katalog
FLAIR	Fibre-Linked Array-Image Reformatter

FLAMES	Fibre Large Array Multi Object Spectrograph
FUSE	Far Ultraviolet Spectroscopic Explorer
FUV	Far ultraviolet
GALLEX	Gallium Experiment
GBM	Gamma-ray Burst Monitor
GeMS	Gemini Multi-Conjugate Adaptive Optics System
GEMS	Gravity and Extreme Magnetism Small explorer
GIF	Graphic Interchange Format
GLACIER	Grand Liquid Argon Charge Imaging Experiment
GMOS	Gemini Multi-Object Spectroscopes
GMT	Giant Magellan Telescope
GNO	Gallium Neutrino Observatory
GOLF	Global Oscillations at Low Frequencies
GONG	Global Oscillation Network Group
GSC	(Hubble) Guide Star Catalogue
GRB	Gamma-ray burst
GREAT	German Receiver for Astronomy at Terahertz Frequencies
GTC	Gran Telescopio Canarias
GVD	Gigaton Volume Detector
GZK	Greisen Zatsepin Kuzmin cut-off
HARP	Heterodyne Array Receiver Programme
HARPS	High Accuracy Radial Velocity Planet Searcher
HAWC	High Altitude Water Čerenkov
HEB	Hot electron bolometer
HEDTEX	Hobby-Eberly Telescope Dark Energy Experiment
HEFT	High Energy Focussing Telescope
HEMT	High electron mobility transistor
HERO	High-Energy Replicated Optics
HESS	High Energy Stereoscopic System
HFET	Heterostructure field effect transistor
Hi-C	High Resolution Coronal Imager
HiCIAO	High Contrast Instrument for the Subaru Next generation Adaptive Optics
HIFI	Heterodyne Instrument for the Far Infrared
HIPO	High Speed Imaging camera for Occultations
Hipparcos	High-Precision Parallax Collecting Satellite
HMI	Helioseismic and Magnetic Imager
HST	Hubble Space Telescope
IBC	Impurity Band Conduction device
ICARUS	Imaging Cosmic and Rare Underground Signals
ICCD	Intensified charge-coupled device
ICRS	International Celestial Reference System
IF	Intermediate frequency
ILMT	International Liquid Mirror Telescope
IMB	Irvine-Michigan-Brookhaven
INO	Indian Neutrino Observatory
INTEGRAL	International Gamma Ray Astrophysics Laboratory

IPCS	Image Photon Counting System
IR	Infrared
IRAF	Image Reduction and Analysis Facility
IRAM	Institut de Radio Astronomie Millimétrique
IRIS	Interface Region Imaging Spectrograph
ISAAC	Infrared Spectrometer and Array Camera
ISI	Infrared Spatial Interferometer
ISO	(i)–Infrared Space Observatory
	(ii)–International Standards Organization
IT	Information technology
IUE	International Ultraviolet Explorer
JCG	Johnson-Cousins-Glass photometric system
JCMT	James Clerk Maxwell Telescope
JPCam	Javalambre-PAU Camera
KID	Kinetic inductance detector
JAXA	Japan Aerospace Exploration Agency
JEM-EUSO	Japanese Experiment Module – Extreme Universe Space Observatory
JPEG	Joint Photographic Experts Group
JWST	James Webb Space Telescope
KAGRA	Kamioka Gravitational Wave Telescope
Kamiokande	Kamioka Neutrino Detector
KAO	Kuiper Airborne Observatory
KAP	Potassium (K) acid phthalate
KARST	Kilometer-square Area Radio Synthesis Telescope
KAT-7	Karoo Array Telescope Seven
KELT	Kilodegree Extremely Little Telescope
KM3NeT	Cubic Kilometre Neutrino Telescope
LABOCA	Large APEX Bolometer Camera
LAGUNA	Large Apparatus studying Grand Unification and neutrino Astrophysics
LAMA	Large Aperture Mirror Array
LAMOST	Large Sky Area Multi-Object Fibre Spectroscopic Telescope
LASCO	Large Angle Spectroscopic Coronagraph
LAT	Large Area Telescope
LAXPC	Large Area X-ray Proportional Counter
LBT	Large Binocular Telescope
LCGT	Large scale Cryogenic Gravity wave Telescope
LENA	Low Energy Neutrino Astronomy
LFC	Laser Frequency Comb
LHC	Large Hadron Collider
LIGO	Laser Interferometer Gravitational-wave Observatory
LIRIS	Long Slit Intermediate Resolution Infrared Spectrograph
LISA	Laser Interferometer Space Antenna
LOFAR	Low Frequency Array
LOFT	Large Observatory for X-ray Timing
LSPE	Large Scale Polarisation Explorer

LSST	Large Synoptic Survey Telescope
LUCIFER	LBT NIR Utility with Camera and Integral Field Unit for Extragalactic Research
LWA	Long Wavelength Array
LWIR	Long Wavelength Infrared
LZT	Large Zenith Telescope
MACAO	Multi-Application Curvature Adaptive Optics
MACHO	Massive Astrophysical Compact Halo Objects
MAGIC	Major Atmospheric Gamma Imaging Čerenkov telescope
MAMA	Multi-anode micro-channel array
MAP	Multichannel Astrometric Photometer
MARSIS	Mars Advanced Radar for Subsurface and Ionosphere Sounding
mas	Milliarc second
MAST	Multimission Archive for the Space Telescope
MATISSE	Multi Aperture mid-Infrared SpectroScopic Experiment
MCAO	Multi-Conjugate Adaptive Optics
MCP	Microchannel plate
MDI	Michelson Doppler Interferometer
MEGS	Multiple EUV Grating Spectrograph
MEM	Maximum entropy method
MEMPHYS	Megaton Mass Physics
MERLIN	Multi-Element Radio Linked Interferometer
Micromegas	Micro-mesh gaseous structure
MIDAS	Microwave Detection of Air Showers
MINOS	Main Injector Neutrino Oscillation Search
MIR	Mid-infrared
MIT	Massachusetts Institute of Technology
MKID	Microwave kinetic inductance detector
MMIC	Monolithic microwave integrated circuits
MMS	Magnetospheric Multiscale mission
MOAO	Multi-Object Adaptive Optics
MOF	Magneto-Optical Filter
MOS	Metal oxide-silicon transistor
MOSFIRE	Multi-Object Spectrometer for Infrared Exploration
MTF	Modulation transfer function
MUSE	Multi Unit Spectroscopic Explorer
MUSIC	Multiband Submillimeter Inductance Camera
MUV	Middle ultraviolet
MWIR	Middle wavelength infrared
NA	Numerical aperture
NaCo	Nasmyth Adaptive Optics System (NAOS) and Coudé Near Infrared Camera (CONICA)
NANOGrav	North American Nanohertz Observatory for Gravitational Waves
NAOS-CONICA	Nasmyth Adaptive Optics System-Near-Infrared Imager and Spectrograph
NASA	National Aeronautics and Space Administration
NEAT	Near Earth Astrometric Telescope

NFIRAOS	Narrow Field Infrared Adaptive Optics System
NGAO	New Generation Adaptive Optics
NGO	New Gravitational Wave Observatory
NIKA	Néel IRAM KID Arrays
NIR	Near-infrared
NIRISS	Near-InfraRed Imager and Slitless Spectrograph
NNLS	Non-negative least square
NOAO	National Optical Astronomy Observatories
NPOI	Navy Prototype Optical Interferometer
NRAO	National Radio Astronomy Observatory
NSSDC	National Space Science Data Centre
NTD	Neutron transmutation doping
NuSTAR	Nuclear Spectroscopic Telescope Array
NUV	Near ultraviolet
OASIS	Optically Adaptive System for Imaging Spectroscopy
OGLE	Optical Gravitational Lensing Experiment
OSIRIS	Optical System for Imaging and low Resolution Integrated Spectroscopy
OTCCD	Orthogonal Transfer CCD
PAIRITEL	Peters Automated Infrared Imaging Telescope
PACS	Photodetector Array Camera and Spectrometer
PAMELA	Payload for Anti-Matter Exploration and Light-nuclei Astrophysics
Pan-STARRS	Panoramic Survey Telescope and Rapid Response System
PDI	Polarimetric differential imaging
PIAA	Phase-Induced Amplitude Apodization
PIONIER	Precision Integrated-Optics Near-infrared Imaging ExpeRiment
POETS	Portable Occultation, Eclipse and Transit System
PoGOLite	Polarized Gamma Ray Observer
PRIMA	Phase-Referenced Imaging and Microarcsecond Astrometry
PROBA-2	Project for Onboard Autonomy
PROMPT	Panchromatic Robotic Optical Monitoring and Polarimetry Telescopes
PSF	Point spread function
PSST	Planet Search Survey Telescope
PZT	Photographic zenith tube
QUIET	Q/U Imaging Experiment
QDIPS	Quantum dot infrared photodetectors
QWIPS	Quantum well infrared photodetectors
RAVE	Radial Velocity Experiment
RHESSI	Reuven Ramaty High Energy Solar Spectroscopic Imager
RICH	Ring Imaging Čerenkov
RL	Richardson Lucy algorithm
ROSAT	Röntgen Satellite
RXTE	Rossi X-ray Timing Explorer
SAGE	Soviet-American Gallium Experiment
SALT	South African Large Telescope

SAR	Synthetic aperture radar
SASIR	Synoptic All-Sky Infrared Imaging Survey
SCUBA	Submillimetre Common User Bolometer Array
SDD	Silicon drift detector
SDI	Spectral differential imaging
SDO	Solar Dynamics Observatory
SDSS	Sloan Digital Sky Survey
SEQUOIA	Second Quabbin Optical Imaging Array
SETI	Search for Extraterrestrial Intelligence
SIMBAD	Set of Identifications, Measurements and Bibliography for Astronomical Data
SINFONI	Spectrograph for Integral Field Observations in the Near Infrared
SIS	Superconductor-insulator-superconductor device
SKA	Square Kilometre Array
SLR	Single-lens reflex (camera)
SMA	Submillimeter Array
SMM	Solar Maximum Mission
SNIFS	Supernova Integral Field Spectrograph
SNO	Sudbury Neutrino Observatory
SNSPD	Superconducting Nanowire Single-Photon Detector
SNU	Solar neutrino unit
SOFIA	Stratospheric Observatory for Infrared Astronomy
SOHO	Solar and Heliospheric Observatory
SolO	Solar Orbiter
SPATS	South Pole Acoustic Test Setup
SPICE	Spectral Imaging of the Coronal Environment
SPAD	Single photon avalanche photodiode
SPHERE	Spectro-Polarimetric High-contrast Exoplanet Research
SPIFFI	Spectrometer for Infrared Faint Field Imaging
SpIOMM	Spectromètre Imageur de l'Observatoire du Mont-Mégantic
SPIRAL	Segmented Pupil Image Reformatting Array Lens
SPIRE	Spectral and Photometric Imaging Receiver
SSD	Silicon strip detector
SSM	Scanning Sky Monitor
STARE	Stellar Astrophysics and Research on Exoplanets
STJ	Superconducting tunnel junction
SUSI	Sydney University Stellar Interferometer
SuperTIGER	Super Trans-Iron Galactic Element Recorder
SuperWASP	Super Wide Angle Search for Planets
SWAP	Sun Watcher with Active Pixels and Image Processing
SWIR	Short wavelength infrared
TAC	Twin Astrographic Catalog
TASS	The Amateur Sky Survey
TDI	Time-delayed integration
TES	Transition edge sensor
THEMIS	Télescope Héliographique pour l'Etude du Magnétisme et des Instabilités Solaires

TIFF	Tagged Image File Format
TMA	Three-Mirror Anastigmat
TMT	Thirty Metre Telescope (previously known as the California Extremely Large Telescope) (CELT)
TNG	Telescopio Nazionale Galileo
TRACE	Transition Region and Corona Explorer
TrES	Trans-Atlantic Exoplanet Survey
Turpol	Turku Photopolarimeter
UCAC	USNO CCD Astrograph Catalog
UCLES	University College London Echelle Spectrograph
UKIRT	United Kingdom Infrared Telescope
UKST	United Kingdom Schmidt Telescope
ULE	Ultra-low expansion fused silica
USNO	U.S. Naval Observatory
UV	Ultraviolet
UV-A	Ultraviolet band A
UV-B	Ultraviolet band B
UV-C	Ultraviolet band C
UVES	Ultraviolet/Visual Echelle Spectroscope
VAO	Virtual Astronomical Observatory
VIRGO	Variability of Solar Irradiance and Gravity Oscillations
VERITAS	Very Energetic Radiation Imaging Telescope Array System
VIMOS	Visible Multi-Object Spectrograph
VISIR	VLT Mid-Infrared Imager and Spectrometer
VISTA	Visible and Infrared Survey Telescope for Astronomy
VLA	Very Large Array
VLBA	Very Long Baseline Array
VLT	Very Large Telescope
VLTI	Very Large Telescope Interferometer
VLWIR	Very long wavelength infrared
VPHG	Volume phase holographic grating
VTT	Vacuum Tower Telescope
VUV	Vacuum ultraviolet
WEBT	Whole Earth Blazar Telescope
WFC	Wide Field Camera
WFC3	Wide Field Camera 3
WFPC2	Wide Field Planetary Camera 2
WHT	William Herschel Telescope
WIMP	Weakly interacting massive particle
WISE	Wide-field Infrared Survey Explorer
WMAP	Wilkinson Microwave Anisotropy Probe
XEUS	X-ray Evolving Universe Spectrometer
XUV	Extreme ultraviolet (also EUV)
ZIMPOL	Zurich Polarimeter

Appendix D: CCD Glossary

Here, gathered together, are brief definitions/descriptions of the various varieties of CCDs.

Back-illuminated CCD: A CCD that is illuminated from the back so that the light does not have to pass through the device's electrode structure. The thickness of the silicon substrate has to be thinned to about 10 to 20 μm.

Buried-channel CCD: A CCD that has a positively charged layer of n-type silicon between the substrate and the insulating layer forcing the charge packets deeper into the substrate.

CCD: Charge coupled device – the basic device.

CCD mosaic: A grid of multiple closely packed individual CCD arrays used to enable a larger area of the sky to be imaged with a single exposure than that covered by a single CCD array.

CMOS-APD: Complementary metal-oxide-semiconductor–active pixel sensor. A variant on the CCD, much used in cell-phone and web cameras, which has found application as an astronomical UV and x-ray detector. It has the same detection mechanism as the CCD, but the pixels are read out individually.

CPCCD: Column parallel CCDs have independent outputs for each column of pixels thus allowing read-out times as short as 50 μs.

Deep-depletion CCD: A variant of the basic CCD with improved x-ray or NIR sensitivity. These devices have a thick silicon substrate which has a high resistivity and use a bias voltage.

EBCCD: Electron bombarded CCD. These have a negatively charged photocathode placed before the CCD. The photoelectron is accelerated by the voltage difference between the photocathode and the CCD and hits the CCD at high energy, producing many electron-hole pairs in the CCD for each incident photon.

EMCCD: Electron multiplying CCD – a synonym for an EBCCD or an L3CCD.

Frame transfer CCD: A CCD that has a storage area into which its electron charges can be dumped before they are read out individually thereby speeding up the read-out process.

ICCD: Intensified CCD – a synonym for an EBCCD or a CCD coupled to an image intensifier.

Interline transfer CCD: These devices have an opaque column adjacent to each detecting column. The charge can be rapidly transferred into the opaque columns and read out from there more slowly while the next exposure is obtained using the detecting columns. Mostly used for video cameras and the like.

L3CCD: Low-light-level CCD or LLLCCD. A device that amplifies the basic output of the CCD by passing it through up to 600 stages, each of which an amplification of a few per cent, within its extended output register. It is best suited to applications requiring the counting of individual photons.

LLLCCD: A synonym for L3CCD.

Low-light-level CCD: A synonym for L3CCD.

OTCCD: Orthogonal transfer CCDs are designed to facilitate TDI by enabling the charge packets to be transferred in up to eight directions (up/down, left/right and at 45° between these directions).

Peristaltic CCD: A variant on the buried-channel CCD wherein additional electrodes enhance the speed of the electron transfer enabling the devices to operate at 100 MHz or more.

p-channel CCD: A synonym for a deep-depletion CCD.

pn-CCD: A synonym for a deep-depletion CCD.

Rear-illuminated CCD: A synonym for a back-illuminated CCD.

Skipper CCD: A CCD in which the lowest intensity parts of the image are repeatedly read out (in order to reduce noise) while the high-intensity parts are read out just once and then skipped for the remaining cycles.

TDI: Time-delayed integration – a technique used for fixed of semi-fixed telescopes whereby the motion of objects in the sky is followed by transferring the charges in the CCD detector at the same speed as the image drifts across the focal plane.

Three-edge-buttable CCD: A CCD constructed so that all the contacts are brought to one edge. The other three edges may therefore be placed against (butted up to) other, similar, CCDs to form a mosaic that has only very small gaps between its elements.

Three-phase CCD: A CCD requiring three separate power supplies to move the electrons through the device.

Two-phase CCD: A CCD requiring just one clock and two power supplies but needing pairs of electrodes, one of which is at the surface and one slightly buried below the surface, in order to determine the direction of movement of the electrons through the device.

Virtual phase CCD: This device requires just one set of electrodes. Additional wells with a fixed potential are produced by p and n implants directly into the silicon substrate.

Appendix E: Bibliography

Some selected journals, books and articles that provide further reading for various aspects of this book, or from which further information may be sought, are listed below.

RESEARCH JOURNALS

Astronomical Journal
Astronomy and Astrophysics
Astrophysical Journal
Icarus
Monthly Notices of the Royal Astronomical Society
Nature
Publications of the Astronomical Society of the Pacific
Science
Solar Physics

POPULAR JOURNALS

Astronomy
Astronomy Now
Ciel et Espace
New Scientist
Scientific American
Sky at Night magazine
Sky and Telescope

OTHER BOOKS BY C. R. KITCHIN

Early Emission Line Stars 1982 (Adam Hilger) ISBN 0141-1128;8.
Stars, Nebulae and the Interstellar Medium 1987 (IOP Publishing) ISBN 0-85274-580-X.
Journeys to the Ends of the Universe 1990 (Adam Hilger) ISBN 0-7503-0037-X.
Optical Astronomical Spectroscopy 1995 (Institute of Physics Publishing) ISBN 0-7503-0345-X.
Photo-Guide to the Constellations 1997 (Springer) ISBN 3-540-76203-5.
Seeing Stars (with R. Forrest) 1997 (Springer) ISBN 3-540-76030-X.
Solar Observing Techniques 2001 (Springer) ISBN 1-85233-035-X.
Illustrated Dictionary of Practical Astronomy 2002 (Springer) ISBN 1-85233-559-8.
Galaxies in Turmoil 2007 (Springer) ISBN 1-84628-670-0.
Exoplanets: Finding, Exploring and Understanding Alien Worlds 2012 (Springer) ISBN 978-1-4614-0643-3.
Telescope and Techniques (3rd Edition) 2013 (Springer) ISBN 978-1-4614-4890-7.

EPHEMERIDES

Astronomical Almanac (published for each year), H.M.S.O./U.S. Government Printing Office.
Handbook of the British Astronomical Association (published for each year), British Astronomical Association.
Yearbook of Astronomy (published each year) Macmillan.

STAR AND OTHER CATALOGUES, ATLASES, SKY GUIDES

Cox, J. *Philip's Pocket Star Atlas*, 2012 (Phillip's) ISBN 13-978-1849072397.
Dunlop, S., Tirion, W. and Ruki, A. *Collins' Atlas of the Night Sky*, 2005 (Harper-Collins) ISBN 13-978-0007172238.
Inglis, M. *Field Guide to the Deep Sky Objects*, 2011 (Springer) ISBN 13-978-146142656.
Mullaney, J. *The Cambridge Double Star Atlas*, 2009 (Cambridge University Press) ISBN 13-978-0521493437.
Ridpath, I. (Ed.), *Norton's Star Atlas 20th Edition*, 2003 (Addison Wesley) ISBN 13-978-0131451643.
Tirion, W. *The Cambridge Star Atlas*, 2011 (Cambridge University Press) ISBN 13-978-0521173636.

REFERENCE BOOKS

Allen, C. W. *Allen's Astrophysical Quantities*, 2001 (Springer) ISBN 13-978-0387987460.
Mitton, J. *Cambridge Illustrated Dictionary of Astronomy*, 2007 (Cambridge University Press) ISBN 13-978-0521823647.
Murdin, P. (Ed.). *Encyclopaedia of Astronomy and Astrophysics*, 2001 (Nature and IoP Publishing) ISBN 13-978-0333750889.
Ridpath, I. *Oxford Dictionary of Astronomy*, 2012 (Oxford University Press) ISBN 13-978-0199609055.

INTRODUCTORY ASTRONOMY BOOKS

Carroll, B. W. and Ostlie, D. A. *An Introduction to Modern Astrophysics*, 2006 (Pearson) ISBN 13-978-0321442840.
Clark, S. *The Sun Kings*, 2009 (Princeton University Press) ISBN 13-978-0691141268.
Freedman, R. A., Geller, R. and Kaufmann, W. J. III. *Universe*, 2010 (WH Freeman) ISBN 13-978-1429231534.
Green, S. F. and Jones, M. H. *An Introduction to the Sun and Stars*, 2004 (Cambridge University Press) ISBN 13-978-0521546225.
Inglis, M. *Astrophysics is Easy! An Introduction for the Amateur Astronomer*, 2007 (Springer) ISBN 13-978-1852338909.
Morison, I. *Introduction to Astronomy and Cosmology*, 2008 (Wiley-Blackwell) ISBN 13-978-0470033340.
Nicolson, I. *Dark Side of the Universe*, 2007 (Canopus Publishing) ISBN 13-978-0954984632.
Rothery, D. A., McBride, N. and Gilmour, I. *An Introduction to the Solar System*, 2011 (Cambridge University Press) ISBN 13-978-1107600928.

INTRODUCTORY PRACTICAL ASTRONOMY BOOKS

Arditti, D. *Setting Up a Small Observatory; From Concept to Construction*, 2008 (Springer) ISBN 13-978-0387345215.
Buick, T. *How to Photograph the Moon and Planets with Your Digital Camera*, 2011 (Springer) ISBN 13-978-1441958273.
Burke, B. F., Graham-Smith, F. *An Introduction to Radio Astronomy*, 2009 (Cambridge University Press) ISBN 13-978-0521878081.
Clark, R. L. *Amateur Telescope Making in the Internet Age: Finding Parts, Getting Help and More*, 2011 (Springer) ISBN 13-978-1441964144.

Cooke, A. *Visual Astronomy Under Dark Skies: A New Approach to Observing Deep Space*, 2008 (Springer) ISBN 13-978-1848008090.

Harrison, K. M. *Grating Spectroscopes and How to Use Them*, 2012 (Springer) ISBN 13-978-1461413967.

Joardar, S. *Radio Astronomy: An Introduction*, 2013 (Mercury Learning and Information) ISBN 13-978-1936420353.

North, G. *Observing Variable Stars, Novae and Supernovae*, 2004 (Cambridge University Press) ISBN 13-978-0521820479.

Robinson, K. *Spectroscopy, the Key to the Stars: Reading the Lines in Stellar Spectra*, 2007 (Springer).

Roth, G. D. *Handbook of Practical Astronomy*, 2009 (Springer) ISBN 13-978-3540763772.

Seip, S. Digital *Astrophotography: A Guide to Capturing the Cosmos*, 2007 (Rocky Nook) ISBN 13-978-1933952161.

Steinicke, W. and Jakiel, R. *Galaxies and How to Observe Them*, 2007 (Springer) ISBN 13-978-1852337520.

Taylor, P. O. *Observing the Sun*, 2008 (Cambridge University Press) ISBN 13978-0521056366.

Tonkin, S. F. *Practical Amateur Spectroscopy*, 2002 (Springer) ISBN 13-978-1852334895.

ADVANCED PRACTICAL ASTRONOMY BOOKS (OR BOOKS WITH A SIGNIFICANT SECTION ON ADVANCED PRACTICAL ASTRONOMY)

Creighton, J. D. E. and Anderson, W. G. *Gravitational-Wave Physics and Astronomy: An Introduction to Theory, Experiment and Data Analysis*, 2011 (Wiley), ISBN 13-978-3527408863.

Rieke, G. H. *Detection of Light: From the Ultraviolet to the Submillimeter*, 2002 (Cambridge University Press) ISBN 13-978-0521017107.

DATA REDUCTION AND ANALYSIS

Bodenheimer, P. *et al. Numerical Methods in Astrophysics: An Introduction*, 2006 (Taylor & Francis) ISBN 13-978-0750308830.

Privett, G. *Creating and Enhancing Digital Astro Images: A Guide for Practical Astronomers*, 2007 (Springer) ISBN 13-978-1846285806.

Wall, J. and Jenkins, C. *Practical Statistics for Astronomers*, 2012 (Cambridge University Press) ISBN 13: 978-0521732499.

Warner, B. A *Practical Guide to Lightcurve Photometry and Analysis*, 2008 (Springer) ISBN 13-978-0387510125.

Index

Page numbers <u>underlined</u> and in **bold** indicate the start of several pages on the entry.